Lecture Notes in Artificial Intell

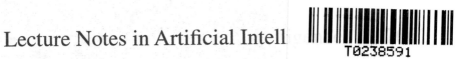

T0238591

Subseries of Lecture Notes in Computer Science

LNAI Series Editors

Randy Goebel
University of Alberta, Edmonton, Canada
Yuzuru Tanaka
Hokkaido University, Sapporo, Japan
Wolfgang Wahlster
DFKI and Saarland University, Saarbrücken, Germany

LNAI Founding Series Editor

Joerg Siekmann
DFKI and Saarland University, Saarbrücken, Germany

Li Chen   Alexander Felfernig   Jiming Liu
Zbigniew W. Raś (Eds.)

Foundations
of Intelligent Systems

20th International Symposium, ISMIS 2012
Macau, China, December 4-7, 2012
Proceedings

 Springer

Series Editors

Randy Goebel, University of Alberta, Edmonton, Canada
Jörg Siekmann, University of Saarland, Saarbrücken, Germany
Wolfgang Wahlster, DFKI and University of Saarland, Saarbrücken, Germany

Volume Editors

Li Chen
Hong Kong Baptist University
224 Waterloo Road, Kowloon, Hong Kong
E-mail: lichen@comp.hkbu.edu.hk

Alexander Felfernig
Graz University of Technology
Institute for Software Technology
Inffeldgasse 16b, 8010 Graz, Austria
E-mail: alexander.felfernig@ist.tugraz.at

Jiming Liu
Hong Kong Baptist University
224 Waterloo Road, Kowloon, Hong Kong
E-mail: jiming@comp.hkbu.edu.hk

Zbigniew W. Raś
University of North Carolina, Charlotte, NC 28223, USA and
Warsaw University of Technology, Nowowiejska 15/19, 00-665 Warsaw, Poland
E-mail: ras@uncc.edu

ISSN 0302-9743   e-ISSN 1611-3349
ISBN 978-3-642-34623-1   e-ISBN 978-3-642-34624-8
DOI 10.1007/978-3-642-34624-8
Springer Heidelberg Dordrecht London New York

Library of Congress Control Number: 2012950992

CR Subject Classification (1998): H.2.7-8, I.2.6-8, I.2.4, H.3.3-5, H.4.1-3, H.5.3-5

LNCS Sublibrary: SL 7 – Artificial Intelligence

Typesetting: Camera-ready by author, data conversion by Scientific Publishing Services, Chennai, India

Printed on acid-free paper

Springer is part of Springer Science+Business Media (www.springer.com)

Preface

This volume contains the papers selected for presentation at the 20th International Symposium on Methodologies for Intelligent Systems (ISMIS 2012), held in Macau, December 4–7, 2012. The symposium was organized by the Department of Computer Science at Hong Kong Baptist University, and held in conjunction with the Web Intelligence 2012 Congress (WIC 2012) that included four other intelligent informatics-related conferences: Active Media Technology 2012 (AMT 2012), Brain Informatics 2012 (BI 2012), IEEE/WIC/ACM Web Intelligence 2012 (WI 2012), and IEEE/WIC/ACM Intelligent Agent Technology 2012 (IAT 2012). In 2012, ISMIS was also a special event of the Alan Turing Year (Centenary of Alan Turing's birth).

ISMIS is a conference series that started in 1986. It is an established and prestigious conference for exchanging the latest research results in building intelligent systems. The scope of ISMIS is intended to represent a wide range of topics on applying artificial intelligence techniques to areas as diverse as decision support, automated deduction, reasoning, knowledge-based systems, machine learning, computer vision, robotics, planning, databases, information retrieval, and so on. ISMIS provides a forum and a means for exchanging information for those interested purely in theory, those interested primarily in implementation, and those interested in specific research and industrial applications.

The following major areas were selected in the regular track of ISMIS 2012: knowledge discovery and data mining, intelligent information systems, text mining and language processing, knowledge representation and integration, music information retrieval, and recommender systems. Moreover, five special sessions were organized: Technology Intelligence and Applications, Product Configuration, Human Factors in Information Retrieval, Social Recommender Systems, and Warehousing and OLAPing Complex, Spatial and Spatio-Temporal Data. Out of 88 submissions, 51 contributed papers were accepted for publication, among which 41 were accepted as regular papers and 10 were accepted as short papers. Every paper was reviewed by at least two reviewers in the international Program Committee. The ISMIS 2012 conference was additionally accompanied by two pre-conference workshops: the International Workshop on Privacy-Aware Intelligent Systems and the International Workshop on Domain Knowledge in Knowledge Discovery. More notably, we had two distinguished researchers to give the keynotes. One was the Turing keynote speaker, Prof. Edward Feigenbaum, (Stanford University, USA), who was the 1994 Turing Award Winner and the other was Prof. Alfred Kobsa (University of California, Irvine, USA).

We wish to express our special thanks to the major committee members of ISMIS 2012: William Kwok Wai Cheung (Workshop Chair), Howard Leung (Publicity Chair), Hao Wang (Sponsorship Chair), Alvin Chin (Industrial Track Chair), and Kelvin Chi Kuen Wong (Organizing Chair). We also thank the organizers of special sessions, namely, Hanmin Jung (Special Session on Technology Intelligence and Applications); Andreas Falkner and Deepak Dhungana (Special Session on Product Configuration); Anne Boyer and Sylvain Castagnos (Special Session on Human Factors in Information Retrieval); Linyuan Lü and Zi-Ke Zhang (Special Session on Social Recommender Systems); Alfredo Cuzzocrea (Special Session on Warehousing and OLAPing Complex, Spatial and Spatio-Temporal Data). We are grateful to Ryan Leong Hou for the creation and maintenance of the conference website and the local organization work, and Juzhen Dong for the maintenance of the paper reviewing system. Our sincere thanks go to the WIC 2012 Chairs: Wei Zhao, Philip Chen, Yuan Yan Tang, Ning Zhong, and Jiming Liu, and the ISMIS 2012 Steering Committee members: Aijun An, Andrzej Skowron, Dominik Ślęzak, Donato Malerba, Floriana Esposito, Giovanni Semeraro, Henryk Rybinski, Jaime Carbonell, Jan Rauch, Lorenza Saitta, Maria Zemankova, Marzena Kryszkiewicz, Nick Cercone, Petr Berka, Stan Matwin, Tapio Elomaa, and Zbigniew W. Raś. Moreover, our thanks are due to Alfred Hofmann of Springer for his continuous support and to Anna Kramer for her work on the proceedings.

September 2012

Li Chen
Alexander Felfernig
Jiming Liu
Zbigniew W. Raś

Organization

ISMIS 2012 was organized by the Department of Computer Science, Hong Kong Baptist University, Hong Kong.

Executive Committee

General Chair
Zbigniew W. Raś — University of North Carolina, Charlotte, USA; and Warsaw University of Technology, Poland

Conference Chair
Jiming Liu — Hong Kong Baptist University, Hong Kong

Program Co-chairs
Li Chen — Hong Kong Baptist University, Hong Kong
Alexander Felfernig — Graz University of Technology, Austria

Workshop Chair
William Kwok Wai Cheung — Hong Kong Baptist University, Hong Kong

Publicity Chair
Howard Leung — City University of Hong Kong, Hong Kong

Sponsorship Chair
Hao Wang — Lenovo Corporate Research, China

Industrial Track Chair
Alvin Chin — Nokia Research Center, China

Organizing Chair
Kelvin Chi Kuen Wong — Hong Kong Baptist University, Hong Kong

Special Session Chairs

Hanmin Jung	Korea Institute of Science and Technology Information, Korea
Andreas Falkner	Siemens Corporate Research and Technologies, Vienna, Austria
Deepak Dhungana	Siemens Corporate Research and Technologies, Vienna, Austria
Anne Boyer	LORIA, Lorraine University, France
Sylvain Castagnos	LORIA, Lorraine University, France
Linyuan Lü	University of Fribourg, Switzerland
Zi-Ke Zhang	Hangzhou Normal University, China
Alfredo Cuzzocrea	ICAR-CNR and University of Calabria, Italy

Steering Committee

Aijun An	York University, Canada
Andrzej Skowron	University of Warsaw, Poland
Dominik Ślęzak	Infobright Inc., Canada and University of Warsaw, Poland
Donato Malerba	University of Bari, Italy
Floriana Esposito	University of Bari, Italy
Giovanni Semeraro	University of Bari, Italy
Henryk Rybinski	Warsaw University of Technology, Poland
Jaime Carbonell	Carnegie Mellon University, USA
Jan Rauch	University of Economics, Prague, Czech Republic
Lorenza Saitta	University of Piemonte Orientale, Italy
Maria Zemankova	NSF, USA
Marzena Kryszkiewicz	Warsaw University of Technology, Poland
Nick Cercone	York University, Canada
Petr Berka	University of Economics, Prague, Czech Republic
Stan Matwin	University of Ottawa, Canada
Tapio Elomaa	Tampere University of Technology, Finland
Zbigniew W. Raś	UNC-Charlotte, USA and Warsaw University of Technology, Poland

Program Committee

Luigia Carlucci Aiello	Sapienza University of Rome, Italy
Reda Alhajj	University of Calgary, Canada
Aijun An	York University, Canada
Troels Andreasen	Roskilde University, Denmark
Salima Benbernou	University of Paris V, France

Reviewers of Special Sessions

Technology Intelligence and Applications

Brahmananda Sapkota	University of Twente, The Netherlands
Chengzhi Zhang	Nanjing University of Science and Technology, China
Haklae Kim	Samsung Electronics Inc., Korea
In-Su Kang	Computer Science and Engineering, Kyungsung University, Korea
Kazunari Sugiyama	National University of Singapore, Singapore
Michaela Geierhos	Munich University, Germany
Qing Li	Southwestern University of Finance and Economics (SWUFE), China
Sa-Kwang Song	Korea Institute of Science and Technology Information, Korea
Seung-Hoon Na	National University of Singapore, Singapore
Seungwoo Lee	Korea Institute of Science and Technology Information, Korea
Yeong Su Lee	Munich University, Germany
Zhangbing Zhou	Institut TELECOM & Management SudParis, France
Zhixiong Zhang	National Science Library, China
Zhu Lijun	ISTIC, China

Product Configuration

Michel Aldanondo	Toulouse University - Mines Albi, France
David Benavides	University of Seville, Spain
Gerhard Friedrich	University of Klagenfurt, Austria
Georg Gottlob	University of Oxford, UK
Lothar Hotz	University of Hamburg, HITeC e.V., Germany
Lars Hvam	Technical University of Denmark, Denmark
Alexander Nöhrer	Johannes Kepler University Linz, Austria
Fabrizio Salvador	IE Business School, Madrid, Spain
Carsten Sinz	KIT Karlsruhe, Institute of Theoretical Informatics (ITI), Germany
Markus Stumptner	University of South Australia, Australia
Juha Tiihonen	Aalto University, School of Science, Finland
Mitchell Tseng	Hong Kong University of Science and Technology, Hong Kong

Human Factors in Information Retrieval

Anne Boyer	LORIA, Lorraine University, Nancy, France
Shlomo Berkovsky	CSIRO, TasICT Centre, Hobart, Australia
Robin Burke	DePaul University, Chicago, USA

Sylvain Castagnos   LORIA, Lorraine University, Nancy, France
Li Chen   Hong Kong Baptist University, Hong Kong
Nathalie Denos   LIG, Grenoble, France
Dietmar Jannach   University of Dortmund, Germany
Alfred Kobsa   University of California, Irvine, USA
Judith Masthoff   University of Aberdeen, Scotland, UK
Cecile Paris   CSIRO ICT Centre, Australia
Liana Razmerita   CBS, Copenhagen, Denmark
Francesco Ricci   University of Bozen-Bolzano, Italy
Michael Thelwall   University of Wolverhampton, UK
Patrick Gallinari   LIP6 lab in Paris, France
Catherine Berrut   LIG, Grenoble, France

Social Recommender Systems

Ciro Cattuto   ISI Foundation, Turin, Italy
Chih-Chun Chen   Centre d'Analyse et de Mathématique Sociales,
                             CNRS, France
Duan-Bing Chen   Web Sciences Center, UESTC, China
Xue-Qi Cheng   Chinese Academy of Sciences, China
Giulio Cimini   University of Fribourg, Switzerland
Tobias Galla   University of Manchester, UK
Liang Gou   Pennsylvania State University, USA
Roger Guimera   Northwestern University, USA/Universitat
                             Rovira i Virgili, Spain
Petter Holme   Umeå University, Sweden
Baoxin Li   Arizona State University, USA
Chuang Liu   Hangzhou Normal University, China
Jian-Guo Liu   Oxford University, UK
Matúš Medo   University of Fribourg, Switzerland
Ming-Sheng Shang   Web Sciences Center, UESTC, China
Meng Su   Peking University, China
Bin Wu   Beijing University of Posts and
                             Telecommunications, China
Ke Xu   Beijing University of Aeronautics and
                             Astronautics, China
Li Yu   Renmin University of China, China
Chi Ho Yueng   Aston University, Birmingham, UK
Yi-Cheng Zhang   University of Fribourg, Switzerland
Tao Zhou   Web Sciences Center, UESTC, China

Warehousing and OLAPing Complex, Spatial and Spatio-Temporal Data

Alberto Abello	Polytechnical University of Catalonia, Spain
Mohammed Al-Kateb	University of Vermont, USA
Fabrizio Angiulli	University of Calabria, Italy
Ladjel Bellatreche	ENSMA, France
Alfredo Cuzzocrea	ICAR-CNR and University of Calabria, Italy
Curtis Dyreson	Utah State University, USA
Filippo Furfaro	University of Calabria, Italy
Anne Laurent	LIRMM, France
Jens Lechtenborger	University of Münster, Germany
Jason li	Drexel University, USA
Pat Martin	Queen's University, Canada
Mirek Riedewald	Northeastern University, USA
Alkis Simitsis	HP Labs, USA
David Taniar	Monash University, Australia
Panos Vassiliadis	University of Ioannina, Greece
Wei Wang	University of New South Wales, Australia
Karine Zeitouni	University of Versailles Saint-Quentin-en-Yvelines, France
Bin Zhou	University of Maryland, USA
Esteban Zimanyi	Université Libre de Bruxelles, Belgium

Table of Contents

Knowledge Discovery and Data Mining

An Anti-tampering Algorithm Based on an Artificial Intelligence
Approach 1
  Andrea Moio, Attilio Giordana, and Dino Mendola

Domain Knowledge and Data Mining with Association Rules –
A Logical Point of View 11
  Jan Rauch

The Problem of Finding the Sparsest Bayesian Network for an Input
Data Set is NP-Hard 21
  Paweł Betliński and Dominik Ślęzak

Mining Top-K Non-redundant Association Rules 31
  Philippe Fournier-Viger and Vincent S. Tseng

ABML Knowledge Refinement Loop: A Case Study 41
  *Matej Guid, Martin Možina, Vida Groznik, Dejan Georgiev,
  Aleksander Sadikov, Zvezdan Pirtošek, and Ivan Bratko*

Min-Max Itemset Trees for Dense and Categorical Datasets 51
  Jennifer Lavergne, Ryan Benton, and Vijay V. Raghavan

TRARM-RelSup: Targeted Rare Association Rule Mining Using
Itemset Trees and the Relative Support Measure 61
  Jennifer Lavergne, Ryan Benton, and Vijay V. Raghavan

Incremental Rules Induction Method Based on Three Rule Layers 71
  Shusaku Tsumoto and Shoji Hirano

RPKM: The Rough Possibilistic K-Modes 81
  Asma Ammar, Zied Elouedi, and Pawan Lingras

Exploration and Exploitation Operators for Genetic Graph Clustering
Algorithm... 87
  Jan Kohout and Roman Neruda

Local Probabilistic Approximations for Incomplete Data 93
  *Patrick G. Clark, Jerzy W. Grzymala-Busse, and
  Martin Kuehnhausen*

On the Relations between Retention Replacement, Additive
Perturbation, and Randomisations for Nominal Attributes in Privacy
Preserving Data Mining.. 99
    Piotr Andruszkiewicz

Intelligent Information Systems

Decision Support for Extensive Form Negotiation Games.............. 105
    Sujata Ghosh, Thiri Haymar Kyaw, and Rineke Verbrugge

Intelligent Alarm Filter Using Knowledge-Based Alert Verification in
Network Intrusion Detection....................................... 115
    Yuxin Meng, Wenjuan Li, and Lam-for Kwok

Clustering-Based Media Analysis for Understanding Human Emotional
Reactions in an Extreme Event 125
    Chao Gao and Jiming Liu

An Event-Driven Architecture for Spatio-temporal Surveillance of
Business Activities 136
    Gabriel Pestana, Joachim Metter, Sebastian Heuchler, and
    Pedro Reis

Text Mining and Language Processing

Multiple Model Text Normalization for the Polish Language........... 143
    Łukasz Brocki, Krzysztof Marasek, and Danijel Koržinek

Discovering Semantic Relations Using Prepositional Phrases........... 149
    Janardhana Punuru and Jianhua Chen

DEBORA: Dependency-Based Method for Extracting Entity-
Relationship Triples from Open-Domain Texts in Polish.............. 155
    Alina Wróblewska and Marcin Sydow

Knowledge Representation and Integration

Lexical Ontology Layer – A Bridge between Text and Concepts 162
    Grzegorz Protaziuk, Anna Wróblewska, Robert Bembenik,
    Henryk Rybiński, and Teresa Podsiadły-Marczykowska

Using Web Mining for Discovering Spatial Patterns and Hot Spots for
Spatial Generalization 172
    Jan Burdziej and Piotr Gawrysiak

AGNES: A Novel Algorithm for Visualising Diversified Graphical
Entity Summarisations on Knowledge Graphs 182
    Grzegorz Sobczak, Mariusz Pikuła, and Marcin Sydow

Large Scale Skill Matching through Knowledge Compilation 192
    Eufemia Tinelli, Simona Colucci, Silvia Giannini,
    Eugenio Di Sciascio, and Francesco M. Donini

Evaluating Role Based Authorization Programs 202
    Chun Ruan

Music Information Retrieval

A Comparison of Random Forests and Ferns on Recognition of
Instruments in Jazz Recordings 208
    Alicja A. Wieczorkowska and Miron B. Kursa

Advanced Searching in Spaces of Music Information 218
    Mariusz Rybnik, Wladyslaw Homenda, and Tomasz Sitarek

Mood Tracking of Musical Compositions 228
    Jacek Grekow

Recommender Systems

Foundations of Recommender System for STN Localization during
DBS Surgery in Parkinson's Patients 234
    Konrad Ciecierski, Zbigniew W. Raś, and Andrzej W. Przybyszewski

From Music to Emotions and Tinnitus Treatment, Initial Study 244
    Deepali Kohli, Zbigniew W. Raś, Pamela L. Thompson,
    Pawel J. Jastreboff, and Alicja A. Wieczorkowska

Adapting to Natural Rating Acquisition with Combined Active
Learning Strategies... 254
    Mehdi Elahi, Francesco Ricci, and Neil Rubens

Special Sessions

Technology Intelligence and Applications

Intelligence in Interoperability with AIDA........................ 264
    Hugo Peixoto, Manuel Santos, António Abelha, and José Machado

An Intelligent Patient Monitoring System 274
    Rui Rodrigues, Pedro Gonçalves, Miguel Miranda, Carlos Portela,
    Manuel Santos, José Neves, António Abelha, and José Machado

Corpus Construction for Extracting Disease-Gene Relations 284
    Hong-Woo Chun, Sa-Kwang Song, Sung-Pil Choi, and Hanmin Jung

On Supporting Weapon System Information Analysis with Ontology
Model and Text Mining.. 293
    Jung-Whoan Choi, Seungwoo Lee, Dongmin Seo, Sa-Kwang Song,
    Hanmin Jung, Sang Hwan Lee, and Pyung Kim

Product Configuration

Some Experimental Results Relevant to the Optimization of
Configuration and Planning Problems 301
    Paul Pitiot, Michel Aldanondo, Elise Vareilles, Linda Zhang, and
    Thierry Coudert

Resolving Anomalies in Configuration Knowledge Bases 311
    Alexander Felfernig, Florian Reinfrank, and Gerald Ninaus

Configuration Repair via Flow Networks 321
    Ingo Feinerer, Gerhard Niederbrucker, Gernot Salzer, and
    Tanja Sisel

Analyzing the Accuracy of Calculations When Scoping Product
Configuration Projects... 331
    Martin Bonev and Lars Hvam

A Two-Step Hierarchical Product Configurator Design Methodology 343
    Yue Wang and Mitchell M. Tseng

Human Factors in Information Retrieval

Anonymous Preference Elicitation for Requirements Prioritization 349
    Gerald Ninaus, Alexander Felfernig, and Florian Reinfrank

Clustering Users to Explain Recommender Systems' Performance
Fluctuation 357
    Charif Haydar, Azim Roussanaly, and Anne Boyer

Social Recommender Systems

Network-Based Inference Algorithm on Hadoop 367
    Zhen Tang, Qingxian Wang, and Shimin Cai

A Hybrid Decision Approach to Detect Profile Injection Attacks in
Collaborative Recommender Systems 377
    Sheng Huang, Mingsheng Shang, and Shimin Cai

Recommendation of Leaders in Online Social Systems 387
    Hao Liu, Fei Yu, An Zeng, and Linyuan Lü

Application of Recommendation System: An Empirical Study of the
Mobile Reading Platform 397
    Chun-Xiao Jia, Chuang Liu, Run-Ran Liu, and Peng Wang

Measuring Quality, Reputation and Trust in Online Communities 405
    Hao Liao, Giulio Cimini, and Matúš Medo

Warehousing and OLAPing Complex, Spatial and Spatio-temporal Data

Providing Timely Results with an Elastic Parallel DW 415
    João Pedro Costa, Pedro Martins, José Cecilio, and Pedro Furtado

Discovering Dynamic Classification Hierarchies in OLAP Dimensions ... 425
    *Nafees Ur Rehman, Svetlana Mansmann, Andreas Weiler, and
Marc H. Scholl*

BIAccelerator – A Template-Based Approach for Rapid ETL
Development 435
    Reinhard Stumptner, Bernhard Freudenthaler, and Markus Krenn

BPMN Patterns for ETL Conceptual Modelling and Validation 445
    Bruno Oliveira and Orlando Belo

Efficiently Compressing OLAP Data Cubes via R-Tree Based Recursive
Partitions 455
    Alfredo Cuzzocrea and Carson K. Leung

Author Index 467

An Anti-tampering Algorithm
Based on an Artificial Intelligence Approach

Andrea Moio, Attilio Giordana, and Dino Mendola

Penta Dynamic Solutions srl,
Viale T. Michel, 11
15121 - Alessandria, Italy

Abstract. Home automation poses requirements, which are typically
solved by AI methods. The paper focuses on the problem of protecting
video-surveillance systems against tampering actions, and proposes a
new algorithm. This is based on a model of the environment observed by
the camera, which must be protected. The model is automatically learned
by observing the video stream generated by the camera. The method is
now implemented in a commercial system are the results reported from
seven experimental sites shows an excellent performance outperforming
state of the art algorithms described in the literature.

1 Introduction

Ambient intelligence and, more specifically, home intelligence is a new challenging area, which offers the possibility of applying Artificial Intelligence techniques
[1].

In fact, the goal of home intelligence is to make a house aware of what is
appending inside, in order to improve the quality of life of inhabitants by offering
new facilities, preventing possible accidents, and detecting hazards. Nevertheless,
providing awareness entails providing the house with the capability of learning
the model of the different agents inside it, and of detecting and interpreting
any possible deviation from the expectation. These are major tasks for Artificial
Intelligence [2,3].

Along this line, this paper discusses how Machine Learning has been integrated in a subsystem designed for preventing tampering a video-surveillance
system, which is a major component of a house automation platform. However,
it is worth noticing that the focus is not on new methodologies proposed in the
Machine Learning area, but on a real application that is now integral part of a
commercial product [4].

Preventing tampering of video-cameras is a critical and important problem,
which has been investigated by many authors, as reviewed in [5]. Nevertheless,
no completely satisfactory solutions, capable of dealing with any environmental
condition exists. For this reason, we decided to develop a new method. Differently
from other approaches to the same problem [5,6], which focus on the detection
of abrupt changes as a consequence of a tamper, the anti-tamper algorithm

L. Chen et al. (Eds.): ISMIS 2012, LNAI 7661, pp. 1–10, 2012.

we propose is based on a model of the environment the visual apparatus must observe, when it is working correctly. The model accounts for a set of items, which must be detectable under different light conditions during the day and the night. When these items are not detectable for a given period of time, a security warning is sent to the house administrator.

The model is automatically learned and updated by continuously observing the environment, without requiring any intervention by the part of the inhabitants. The learning algorithm is set in the Bayesian framework and implements a variant of the classical instance based learning approach [7]. It is implemented as a module of a home automation system called HORUS[1] and is now in beta-test in several sites of customers of Penta Dynamic Solutions. Some results of the in field testing, which show the effectiveness of the approach, are provided.

In the following, Section 2 discusses the problem the anti-tamper algorithm needs to solve, and provides a high level algorithm overview. Section 3 describes the learning algorithm. Finally, an overview of the current deployment of the system is provided in Section 4. Some conclusions are drawn in Section 5.

2   Problem Setting

There are several problems bound to video-surveillance, which are currently actively investigated. A first kind of problem is related to recognizing some item, or detecting some event in the observed scenario, such as a person or an intrusion, and so on. A second kind of task is related to the monitoring of the health status of the visual apparatus, detecting actions aimed at compromising its functioning. This task is usually referred as anti-tampering, and even if quite strongly related to the first one is usually approached with different methods. A video-camera can be tampered in many different ways. Typical examples are:

- *Defocusing.* The focus of the camera objective is set in such a way the image is blurred.
- *Misplacing.* The position or the orientation of the camera is modified so that part of the scenario is no longer observable.
- *Masking.* The objective is partially or totally covered, hiding then the observed scenario. A malicious way of masking the camera may be obtained by spraying a partially transparent liquid on the objective. This action degrades the image quality without causing a total masking, which is easier to detect.
- *Blinding.* The camera is still observing the scenario, but the visual sensor is run out range, so that the image is no more interpretable. For instance, this can be done by illuminating the objective with a laser pointer.

A tampering action may occur both when the video-surveillance is active and when it is non-active. In the first case, intruders tend to use attack methods like blinding, which allow them to tamper the camera without entering in the observed area. In the second case, attacks of the other kinds are preferred, and are carried on choosing periods in the day, in which the action is difficult to be noticed.

----

[1] Horus is a trademark of Penta Dynamic Solutions S.r.L.

2.1   Existing Approaches

The currently available anti-tampering techniques follow two alternative approaches: (i) device specific approach; (ii) device independent approach. In the first one, the anti-tamper algorithm is integrated in the software on-board to the camera, which encodes the video stream. The advantage of this approach is that many actions and changes in the video stream due to a change in the control strategy (for instance, the adaptation to the light conditions) can be immediately distinguished from tampering attempts, like defocusing. The disadvantage is that, on the one hand, the limited power of the cpu poses severe limitations, so that only simple kinds of attack can be effectively handled. On the other hand, this kind of solution bounds the user to a specific hardware. The second approach aims at finding camera independent methods for detecting tampering actions, not accounting for the specific features of the camera [5]. In many cases, the technique consists in identifying abrupt changes in the video stream, which usually correspond to the signature of a tampering action. The recent sequence of images recorded from the camera is used to construct the background image the algorithm is expected to see. When the current image changes too much from the background, a tamper-alarm is sent. The limitation of this technique is that a tampering action executed very slowly can be accepted as a gradual change in the background. Moreover, temporary changes in the video stream, of short duration, are considered tamper attempts causing false alarms. This frequently happens when the light conditions change quickly as, for instance, at dawn or at sunset.

2.2   Horus Approach

Horus approach differs from the previous ones in the sense that it is based on a static model of the environment, which must be observable when the camera is working properly. Then, an alert condition due to a temporary blinding of the camera can be dismissed if the observation becomes *normal* again in a short time. Nevertheless, the algorithm does not risk of accepting a tampering action, executed very slowly, as an evolution of the background.

A model of the world observed by a video-camera is a collection of scenarios, each one corresponding a specific light condition, which may occur depending on the hour of the day and on the weather.

The anti-tamper algorithm is set in a Bayesian framework and uses the model for computing the probability of observing the image currently recorded from the camera. This is done according to the conditional probability chain described in Figure 1.

The algorithm firstly computes the probability for each scenario being the one corresponding to the current light conditions. Then, given the probability distribution on the scenarios, the probability of observing specific items in the image recorded from the video-camera is computed.

In this framework, every scenario is characterized by two sets of features: a set of global features characterizing the light condition, and a set of local features

Fig. 1. Conditional probability dependences between the current image and the environment model

characterizing specific items, referred as *targets* in the following, which are easily detectable and are stable with respect to the position.

The system enters in an alert condition when too many *targets*, which should be detectable in the current light condition, are not detected. When an alert condition holds for a too long period of time, alarms warning are sent to an administrator. Site specific rules, accounting for both the duration of the alert and for the period of the day, decide when to send alarm warnings.

The algorithm keeps the capability of self-adapting to possible changes in the environment. This is necessary because, on the one hand, the light conditions characterizing the scenarios evolve during the seasons, and, on the other hand, an environment where humans and animals are living is continuously evolving: new objects may appear in the scenarios and other ones may disappear permanently. Then, the target set must evolve consequently. Nevertheless, this is done avoiding the risk of adapting to tampering conditions. Two adaptation mechanisms have been implemented:

1. Real time adaptation to small changes in the video-camera orientation, as it may happen because of wind or other accidental causes.
2. Long term adaptation to deep changes in the environment.

The first capability is implemented using what we will call the model *invariant*, which is a subset of selected items that appear to be detectable under any light condition. The anti-tamper algorithm first tries to detect the position of the *invariant* in the current image. Then, the image recorded from the camera is realigned by applying a linear transformation. However, only small translations and rotations of the *invariant* position, with respect to the nominal ones, are accepted. If the detected changes are too large, a tamper alarm is immediately sent.

The long term adaptation is based on the assumption that environment changes slowly, so that, in general, a model learned today, would be a good model also for tomorrow. Then, every day, a new model of the environment is constructed by the learning algorithm, while the anti-tamper algorithm uses the model constructed the previous day. If no serious tamper conditions have been detected at the end of the day, the new model is considered a valid one and becomes the new working model for the anti-tamper algorithm.

3   The Algorithm

The anti-tamper algorithm consists of two steps, as described in Figure 2. As mentioned in Section 2, the scenarios of the environment model are described by a set of global features and by a set of local features. The former ones are used to compare scenarios to the current image C, recorded from the camera, according the light conditions. The second ones characterizes specific items that must be found in the scene. In the first step, the algorithm computes a probability distribution on the scenarios of the model considering the conditioning due to the current image C. This is done using the global features characterizing the light condition. Then, at the second level the algorithm matches the current image from the camera to the *targets* characterizing the selected scenarios. This is done again in a probabilistic framework. If too many targets are not clearly detected the system enters in alert status. Alert warnings are sent to a rule-based decision module, which decides when to send an alarm. The rules of the decision module are site-dependent and account both for the duration of the alert status and for the time of the day. For instance, during the day alarms are sent when the alert condition lasts for long time, so that a permanent damage of the camera is suspected, while, during the night, alarms are sent as soon as the alert condition lasts more than a few seconds.

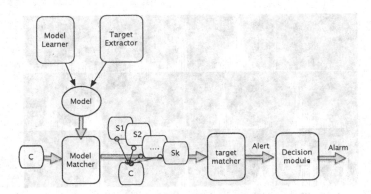

Fig. 2. Anti-tampering algorithm architecture

The anti-tamper algorithm is associated to a learning algorithm that is able to automatically construct the models and the targets in the scenarios, and to estimate the required probability distributions.

3.1 Environment Model

The model M of the environment observed by a video-camera is a set of items called *scenarios* characterizing the possible views the video-camera will observe depending on the light conditions, which change according to the time of the day and to the weather. A scenario is a pair $S = \langle H, T \rangle$, being H a feature set characterizing the light conditions in which S is considered reliable, and T is a data structure we will call *target* characterizing the image details, which must be observed considering the light conditions characterized by H. In addition to the scenarios, M contains the model *invariant* I. This is a structure containing only the targets, which are observable in all the scenarios of M.

The light conditions may change from one to another region of the image recorded by a camera, depending upon the sun position and other factors. Then, H is constructed in order to account for this phenomenon. More precisely, H is a matrix $h_{i,j}$ ($0 \le i \le m$ and $0 \le j \le n$), of size $m \times n$ being m and n two assigned integers. Each element $h_{i,j}$ corresponds to a rectangular region a_{ij} of the region observed by the camera and characterizes the light condition in a_{ij} by means of a vector of integers obtained by chaining the triplet of histograms $\langle h_{ij}{}^{R}, h_{ij}{}^{G}, h_{ij}{}^{B} \rangle$ of the three fundamental colors RGB(see Figure 3). The values of m and n are usually selected considering the site observed by the video-camera. If the image illumination is homogenous, small m and n are sufficient, otherwise larger values provide a greater accuracy but increase the computational complexity. In practice, setting $n = 8$ and $m = 6$ proved to be a reasonable compromise for the majority of the environments in which Horus has been tested.

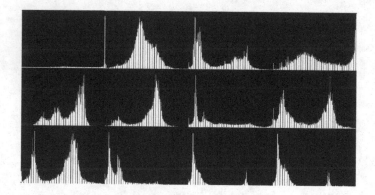

Fig. 3. Example of feature set H constructed from $3x4$ RGB histograms

The target T describes the items that must be detectable under the light conditions characterized by H. The focus is set on objects that have a well definite shape, not changing during the time, and have a permanent position. As described into detail in Section 3.3, T is obtained from a contour map of the image and is encoded as a matrix of real numbers $\tau(x, y)$ in the interval $[0, 1]$. Each $\tau(x, y)$ represents the probability of finding the corresponding pixel $p(x, y)$ included in the contour map $\mathcal{T}(C)$ extracted from the current image C from the camera. The model *invariant* **I** is the map of the pixel having probability greater than 0.5 in all scenarios.

An example of two target T_1 and T_2, extracted from two different scenarios, is reported in Figure 4. The probability values are represented as gray levels in a black and white image ($\tau = 1$ corresponds to a white pixel, while $\tau = 0$ corresponds to a black pixel).

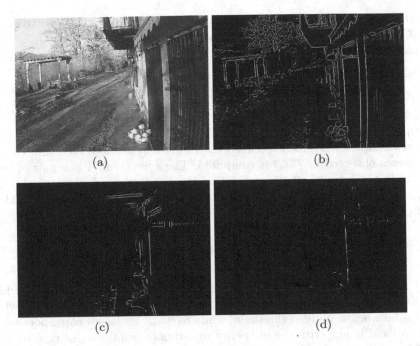

Fig. 4. Example of targets characterizing the scenarios in the environment model: (a) Image recorded from the camera; (b) T_1 target corresponding to the image (a); (c) T_2 target for an image recorded at sunset; (d) model *invariant*

3.2   Matching the Camera Image to the Model

The first step consists in computing the $m \times n$ RGB histograms corresponding to $H(C)$ and extracting the contours map on $\mathcal{T}(C)$. Then, the conditional distribution $P(S_k | H(C))$ is computed for all $S_k \in M$. For each pair $\langle H(C), H(S_k) \rangle$ the matrix d_{ij} is computed, being d_{ij} the Bhattacharyya distance [8] between two corresponding histograms $h_{ij}(C)$, $h_{ij}(S)$. The Bhattacharyya distance is computed using the formula:

$$d(\boldsymbol{h_{ij}}(C), \boldsymbol{h_{ij}}(S_k)) = \sqrt{1 - \frac{1}{\sqrt{\boldsymbol{h_{ij}}(C)\boldsymbol{h_{ij}}(S)\rho^2}} \sum_{l=1}^{\rho} \sqrt{h_{ijl}(C)h_{ijl}(S_k)}} \quad (1)$$

being ρ the number of components in the vector $\boldsymbol{h_{ij}}$ and h_{ijl} the l element of $\boldsymbol{h_{ij}}$. The global distance $D(C, S_k)$ between C and S_k is computed as the average of all values d_{ij}.

The probability of S_k is computed as:

$$P(S_k|C) = 1 - \frac{D(C, S_k)}{\sum_{r=1}^{|M|} D(C, S_r)} \quad (2)$$

i.e, 1 minus the distance of S_k from C normalized with respect the sum of the distances of all scenarios from C. The assumption underlaying (2), which is reasonable but remains somehow arbitrary, is that the probability distribution $P(S_k|C)$ is linearly dependent on the Bhattacharyya distance.

Then, given $P(S_k|C)$ the expectation

$$T(C) = \sum_{k=1}^{|M|} T(S_k)P(S_k|C) \tag{3}$$

is computed for the targets in $\mathcal{T}(C)$ in C.

Afterwards, the agreement between the target expectation $T(C)$ and the actual targets observed in $\mathcal{T}(C)$ is computed. The agreement A is a pair of reals $\langle \alpha_w, \alpha_b \rangle$ in the interval $[0,1]$ representing the proportion of white pixels, and black pixels in $\mathcal{T}(C)$ in agreement with the expectation $T(C)$, respectively. More specifically α_w, α_b are computed according to the the following formula:

$$\alpha_w = \frac{1}{M_w} \sum_{p(x,y)=b \in \mathcal{T}(C)} \tau(x,y), \qquad \alpha_b = \frac{1}{M_b} \sum_{p(x,y)=b) \in \mathcal{T}(C)} 1 - \tau(x,y) \tag{4}$$

In (4) M_w and M_b correspond to the maximum possible agreement for white and black pixels in $\mathcal{T}(C)$, respectively. A tamper condition is detected when α_w (α_b) falls below a given threshold, while α_b (α_w) increases correspondingly. The threshold is not critical and is easy to estimate from a set of positive and negative examples, recorded on site from the camera.

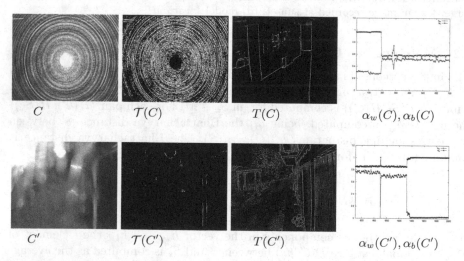

Fig. 5. Example of time evolution of the agreement measures α_w and α_b in presence of two typical tampering attempts: Images generated by a laser blinding (C) and by a selz spray (C')

An example of the temporal evolution of α_w, or α_b in presence of two typical tampering attempts (laser blinding and masking using a selz spray), is described in Figure 5.

3.3   Model Learning Algorithm

The basic information source for learning and updating the environment model M is the sequence of the last ν frames recorded from the camera, being ν the only parameter, which needs to be specified, when the system is installed. We will denote as $\mathbf{C}_t(\nu)$ the learning sequence at time t, consisting on the ν frames recorded before t.

The distance measure defined by expressions (1) is also used for learning M by direct observation of the video stream recorded from a camera.

The learning algorithm works as in the following. Let M be the set of scenarios characterizing the model at time t. If for sequence $\mathbf{C}_t(\nu)$ no scenario $S \in M$ is closer than a minimum distance $\delta = 0.25$ to any one of the frames in $\mathbf{C}_t(\nu)$, a new scenario S_t is constructed from $\mathbf{C}_t(\nu)$ and added to M. In order to avoid M reaching an excessive size, periodically the scenarios, which for long time (one week) have not been selected by the matching algorithm, are deleted.

When a new scenario is added to M, the target T is constructed by estimating the probabilities associated to the pixels on the sequence $\mathbf{C}_t(\nu)$ of frames used for learning the new scenario. Then the model *invariant* $\mathbf{I}$ is updated in order to account for the new scenario. Every pixel which occurs with probability higher then $\tau = 0.5$ in all scenarios is included in the *invariant* matrix with probability $\tau = 1.0$.

As previously mentioned the *invariant* has an important role for dynamically recalibrating the position of the image C observed from the camera.

4   In Field Testing

At the moment, HORUS anti tamper module has been deployed in field in seven sites. Six of them are outdoor environments in country houses owned by customers, who volunteered to participate to the test. The seventh one is an indoor environment located in the labs of Penta Dynamic Solutions.

Outdoor sites are particularly difficult because of the presence of moving objects, like trees, flies, and birds, and the continuously changing light conditions due to sunset, sunrise and clouds. Previous experimentation with state of the art algorithms was quite unsatisfactory. The one, which performed best among those we tested [5], is the one by Ribnick [6]. Nevertheless, the rate of false alarms was still unacceptable for the customers participating to the test. The major weakness we detected was the lack of a long term environment model for checking when the camera condition becomes normal again after an alert status.

In the current installations, after the initial tuning of the system, which takes around three days for learning the initial model and the alert threshold, and for tuning the rules of the Decision Module, the rate of false alarms is practically null. On the other hand the system has always been able to detect any sort of simulated tampering actions.

Therefore, the described algorithm is going now to be integrated in the commercial release of HORUS (Horus-3.0).

5   Conclusion

We have shown how the availability of a model of the environment greatly improves the reliability of an anti-tampering algorithm, implemented in the framework of a home automation system. The important result is that the model is learned and maintained automatically by the system without requiring any action by the users. The algorithm is now integrated in a commercial tool installed in several houses.

The learned lesson is that we can expect a dramatic increase of the role of Machine Learning methods in the future evolution of ambient intelligence and home automation. In fact this is the only way to develop smart environments, without requiring a difficult tuning phase involving both users and domain experts.

References

1. Cook, D.J., Augusto, J., Jakkula, V.: Ambient intelligence: Technologies, applications, and opportunities. Pervasive and Mobile Computing 5(4), 277–299 (2009)
2. Simpson, R., Schreckenghost, D., LoPresti, E.F., Kirsch, N.: Plans and Planning in Smart Homes. In: Augusto, J.C., Nugent, C.D. (eds.) Designing Smart Homes. LNCS (LNAI), vol. 4008, pp. 71–84. Springer, Heidelberg (2006)
3. Ramos, C.: Ambient Intelligence – A State of the Art from Artificial Intelligence Perspective. In: Neves, J., Santos, M.F., Machado, J.M. (eds.) EPIA 2007. LNCS (LNAI), vol. 4874, pp. 285–295. Springer, Heidelberg (2007)
4. Mendola, D., Monfrecola, A., Moio, A., Giordana, A.: Horus: an agent based system for home automation. Technical report TR-01-2012 (2012)
5. Saglam, A.: Adaptive camera tamper detection for video surveillance (June 2009)
6. Ribnick, E., Atev, S., Masoud, O., Papanikolopoulos, N., Voyles, R.: Real-time detection of camera tampering. In: Proceedings of the IEEE International Conference on Video and Signal Based Surveillance, AVSS 2006, p. 10. IEEE Computer Society, Washington, DC (2006)
7. Aha, D., Kibler, D., Albert, M.: Instance-based learning algorithms. Machine Learning 6, 37–66 (1991)
8. Djouadi, A., Snorrason, O., Garber, F.: The quality of training-sample estimates of the bhattacharyya coefficient. IEEE Transactions on Pattern Analysis and Machine Intelligence 12(1), 92–97 (1990)

Domain Knowledge and Data Mining with Association Rules – A Logical Point of View

Jan Rauch

Faculty of Informatics and Statistics, University of Economics, Prague*
nám W. Churchilla 4, 130 67 Prague 3, Czech Republic
rauch@vse.cz

Abstract. A formal framework for data mining with association rules is presented. All important steps of CRISP-DM are covered. Role of formalized domain knowledge is described. Logical aspects of this approach are emphasized. Possibilities of application of logic of association rules in solution of related problems are outlined. The presented approach is based on identifying particular items of domain knowledge with sets of rules which can be considered their consequences.

1 Introduction

A formal framework FOFRADAR (FOrmal FRAmework for Data mining with Association Rules) is shortly introduced in [3]. It covers all important steps described by CRISP-DM methodology, see Fig. 1. FOFRADAR is based on

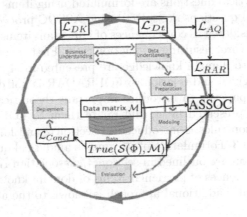

Fig. 1. FOFRADAR and CRISP-DM

* The work described here has been supported by Grant No. 201/08/0802 of the Czech Science Foundation.

enhancement of a logical calculus of association rules [2] by several languages and procedures. Most of them are introduced below.

Language $\mathcal{L}_{DK}$ – formulas of $\mathcal{L}_{DK}$ correspond to items of domain knowledge. These items are in a form understandable to domain experts. The expression $BMI \uparrow\uparrow Diast$ meaning that if patient's BMI (i.e. Body Mass Index) increases, then his/here diastolic blood pressure increases too is an example. We consider formulas of $\mathcal{L}_{DK}$ as results of business understanding.

Language $\mathcal{L}_{Dt}$ – formulas of $\mathcal{L}_{Dt}$ correspond to relevant information on analyzed data. Information that given data concerns only male patients is an example. Formulas of $\mathcal{L}_{Dt}$ can be considered as results of data understanding.

Language $\mathcal{L}_{AQ}$ – formulas of $\mathcal{L}_{AQ}$ correspond to reasonable analytical questions. Analytical questions are created using formulas of $\mathcal{L}_{DK}$ and $\mathcal{L}_{Dt}$.

Language $\mathcal{L}_{RAR}$ – formulas of this language define sets of Relevant Association Rules. **Procedure** AQ_RAR assigns to a given analytical question $\mathcal{Q}$ (i.e. a formula of the language $\mathcal{L}_{AQ}$) a formula $AQ\_RAR(\mathcal{Q})$ of $\mathcal{L}_{RAR}$ which defines a set of relevant association rules to be verified in a given data matrix $\mathcal{M}$ to solve a given analytical question $\mathcal{Q}$.

Procedure ASSOC – its input consists of a formula Φ of a language $\mathcal{L}_{RAR}$ and of an analyzed data matrix $\mathcal{M}$. Formula Φ defines a set $\mathcal{S}(\Phi)$ of association rules to be verified in $\mathcal{M}$ to solve the given analytical question. Output of the ASSOC is a set $True(\mathcal{S}(\Phi), \mathcal{M})$ of all rules $\varphi \approx \psi$ belonging to $\mathcal{S}(\Phi)$ which are true in data matrix $\mathcal{M}$. The procedure ASSOC is introduced in [1]. The procedure 4ft-Miner is used in [4] as an implementation of ASSOC.

Language $\mathcal{L}_{Concl}$ – formulas of $\mathcal{L}_{Concl}$ correspond to conclusions which can be made on the basis of the output set $True(\mathcal{S}(\Phi), \mathcal{M})$ of the ASSOC procedure.

Let us note that there are additional procedures transforming formulas of one language into formulas of another language. This can be seen as a formal background of a description of the whole data mining process

Reasonable analytical questions are formulated using items of domain knowledge. The analytical questions are solved using ASSOC procedure. Sets of rules which can be be considered as consequences of important items of domain knowledge are used to interpret resulting association rules. First successful experiments with this approach to domain knowledge are presented in [4]. However, no theoretical details are given and relation to FOFRADAR is not described in [4].

The goal of this paper is to introduce important theoretical details relevant to the experiments presented in [4] and relate them to FOFRADAR. Logical calculus of association rules is introduced in section 2 and language of domain knowledge in section 3. Formulation of reasonable analytical questions and principles of their solution are outlined in section 4. Association rules which can be considered as consequences of particular items of domain knowledge are studied in section 5. No similar additional approach is known to the author.

2   Logical Calculus of Association Rules

Association rules concern data matrices. To keep explanation simple, we consider data matrices with values – natural numbers only. There is only a finite number of possible values (i.e. categories) for each column. We assume that the number of categories of a column is t and the categories in this column are integers $1, \ldots, t$. Then all the possible values in the data matrix are described by the number of its columns and by the numbers of categories for each column. These numbers determine a type of a data matrix and also a type of a logical calculus.

A *type of a logical calculus of association rules* is a K-tuple $\mathcal{T} = \langle t_1, \ldots, t_K \rangle$ where $K \geq 2$ is a natural number and $t_i \geq 2$ are natural numbers for $i = 1, \ldots, K$. A *data matrix of the type* $\mathcal{T}$ is a $K + 1$-tuple $\mathcal{M} = \langle M, f_1, \ldots, f_K \rangle$, where M is a non-empty finite set and f_i is an unary function from M to $\{1, \ldots, t_i\}$ for $i = 1, \ldots, K$. Elements of M correspond to rows of $\mathcal{M}$ (i.e. to observed objects) and functions $f_1, \ldots, f_K$ correspond to columns of data matrix $\mathcal{M}$.

Language $\mathcal{L}_{\mathcal{T}}$ of association rules of type $\mathcal{T}$ is given by basic attributes $A_1, \ldots, A_K$ and by 4ft-quantifiers $\approx_1, \ldots, \approx_Q$. We say that A_i is of type t_i for $i = 1, \ldots, K$. If A is a basic attribute of $\mathcal{L}_{\mathcal{T}}$ and $\alpha \subset \{1, \ldots, t\}$ then $A(\alpha)$ is a basic Boolean attribute of $\mathcal{L}_{\mathcal{T}}$. Each basic Boolean attribute of $\mathcal{L}_{\mathcal{T}}$ is a Boolean attribute of $\mathcal{L}_{\mathcal{T}}$. If φ and ψ are Boolean attributes of $\mathcal{L}_{\mathcal{T}}$ then $\neg \varphi$, $\varphi \wedge \psi$, and $\varphi \vee \psi$ are Boolean attributes of $\mathcal{L}_{\mathcal{T}}$. If φ and ψ are Boolean attributes and $\approx$ is a 4ft-quantifier of $\mathcal{L}_{\mathcal{T}}$ then $\varphi \approx \psi$ is an association rule of $\mathcal{L}_{\mathcal{T}}$. Here φ is called *antecedent* and ψ is called *succedent*.

Let $\mathcal{M} = \langle M, f_1, \ldots, f_K \rangle$ be a data matrix of type $\mathcal{T}$. Then a value of attribute A_i for row $o \in M$ is $f_i(o_j)$. A value of a basic Boolean attribute $A_i(\alpha)$ for row $o \in M$ is defined as $f_{A_i(\alpha)}(o)$ where $f_{A_i(\alpha)}(o) = 1$ if $f_i(o) \in \alpha$ and $f_i A(\alpha)(o) = 0$ otherwise. Values of derived Boolean attributes $\neg \varphi$, $\varphi \wedge \psi$, and $\varphi \vee \psi$ for row $o \in M$ are defined as values of functions $f_{\neg \varphi}(o)$, $f_{\varphi \wedge \psi}(o)$, and $f_{\varphi \vee \psi}(o)$ respectively where $f_{\neg \varphi}(o) = 1 - f_{\neg \varphi}(o)$, $f_{\varphi \wedge \psi}(o) = \min\{f_\varphi(o), f_\psi(o)\}$, and $f_{\varphi \vee \psi}(o) = \max\{f_\varphi(o), f_\psi(o)\}$. For examples see Fig. 2 where type $\mathcal{T} = \langle 10, 20, \ldots, 30 \rangle$. Function f_ω is an *interpretation of Boolean attribute* ω in $\mathcal{M}$. If $o \in M$ and $f_\omega(o) = 1$ then we say that o *satisfies* ω in $\mathcal{M}$.

	A_1 A_2 $\ldots$ A_K	$A_1(1,2,3)$ $A_2(17,18,19)$			$\mathcal{M}$	ψ	$\neg \psi$
obj.	f_1 f_2 $\ldots$ f_K	$f_{A_1(1,2,3)}$ $\quad f_{A_2(17,18,19)}$					
o_1	2 17 $\ldots$ 30	1 $\qquad\qquad$ 1			φ	a	b
$\vdots$	$\vdots$ $\vdots$ $\ddots$ $\vdots$	$\vdots$ $\qquad\qquad$ $\vdots$			$\neg \varphi$	c	d
o_n	10 19 $\ldots$ 15	0 $\qquad\qquad$ 1					

Data matrix $\mathcal{M}$ and interpretations of attributes $\qquad\qquad 4ft(\varphi, \psi, \mathcal{M})$

Fig. 2. Data matrix $\mathcal{M}$ and *4ft-table* $4ft(\varphi, \psi, \mathcal{M})$ of φ and ψ in $\mathcal{M}$

Logical calculus $\mathcal{LC}_{\mathcal{T}}$ of association rules of type $\mathcal{T}$ is given by a language $\mathcal{L}_{\mathcal{T}}$ of association rules of type $\mathcal{T}$, a set $\mathrm{M}_{\mathcal{T}}$ of all data matrices $\mathcal{M} = \langle M, f_1, \ldots, f_K \rangle$ of type $\mathcal{T}$ and by associated functions $F_{\approx_1}, \ldots, F_{\approx_Q}$ of 4ft-quantifiers $\approx_1, \ldots, \approx_Q$ of $\mathcal{L}_{\mathcal{T}}$ respectively. Association function $F_\approx$ of 4ft-quantifier $\approx$ maps a set of all quadruples $\langle a, b, c, d \rangle$ of non-negative integer numbers satisfying $a + b + c + d > 0$ into the set $\{0, 1\}$.

Let φ and ψ be Boolean attributes. Then the *4ft-table* $4ft(\varphi,\psi,\mathcal{M})$ of φ and ψ in $\mathcal{M}$ is a quadruple $4ft(\varphi,\psi,\mathcal{M}) = \langle a,b,c,d \rangle$ where a is the number of rows of $\mathcal{M}$ satisfying both φ and ψ, b is the number of rows satisfying φ and not satisfying ψ etc., see Fig. 2.

A *value* $Val(\varphi \approx \psi, \mathcal{M})$ *of an association rule* $\varphi \approx \psi$ *of* $\mathcal{L}_T$ *in a data matrix* $\mathcal{M} \in \mathsf{M}_T$ is defined such that $Val(\varphi \approx \psi, \mathcal{M}) = F_\approx(a,b,c,d)$ where $\langle a,b,c,d \rangle = 4ft(\varphi,\psi,\mathcal{M})$ is the 4ft-table of φ and ψ in data matrix $\mathcal{M}$ and $F_\approx$ is the associated function of the 4ft-quantifier $\approx$. The *association rule* $\varphi \approx \psi$ *is true in data matrix* $\mathcal{M}$ if $Val(\varphi \approx \psi, \mathcal{M}) = 1$, otherwise the *association rule* $\varphi \approx \psi$ *is false in data matrix* $\mathcal{M}$. Two important 4ft-quantifiers are introduced below, about 20 additional 4ft-quantifiers are defined in [1,2,5].

4ft-quantifier $\Rightarrow_{p,B}$ of *founded implication* is defined in [1] such that $F_{\Rightarrow_{p,B}}(a,b,c,d) = 1$ if and only if $\frac{a}{a+b} \geq p \wedge a \geq B$ for $0 < p \leq 1$ and $B > 0$ [1]. Rule $\varphi \Rightarrow_{p,B} \psi$ means that at least $100p$ per cent of rows satisfying φ satisfy also ψ and that there are at least B rows of $\mathcal{M}$ satisfying both φ and ψ.

4ft-quantifier $\sim^+_{p,B}$ of *above average dependence* is defined in [5] for $0 < p$ and $B > 0$ by the condition $\frac{a}{a+b} \geq (1+p)\frac{a+c}{a+b+c+d} \wedge a \geq B$. This means that the relative frequency of the rows satisfying ψ among the rows satisfying φ is at least $100p$ per cent higher than the relative frequency of the rows satisfying ψ in the whole data matrix $\mathcal{M}$ and that there are at least B rows satisfying both φ and ψ.

There are important results concerning correct deduction rules $\frac{\varphi \approx \psi}{\varphi' \approx \psi'}$ where both $\varphi \approx \psi$ and $\varphi' \approx \psi'$ are association rules. Deduction rule $\frac{\varphi \approx \psi}{\varphi' \approx \psi'}$ is correct if it holds for each data matrix $\mathcal{M}$: *if* $\varphi \approx \psi$ *is true in* $\mathcal{M}$ *then also* $\varphi' \approx \psi'$ *is true in* $\mathcal{M}$. The rule $\frac{A(1) \Rightarrow_{p,B} B(1)}{A(1) \Rightarrow_{p,B} B(1,2)}$ is a very simple example of a correct deduction rule. There are relatively simple criteria making possible to decide if a given deduction rule $\frac{\varphi \approx \psi}{\varphi' \approx \psi'}$ is correct or not [2].

3   Language of Domain Knowledge

There are three types of domain knowledge – types of attributes, groups of attributes and mutual simple influence of attributes. *Types of attributes* of a language $\mathcal{L}_{DK}$ of domain knowledge related to a logical calculus $\mathcal{LC}_T$ of association rules of type $T = \langle t_1, \ldots, t_K \rangle$ is a K-tuple $\mathcal{NO} = \langle \tau_1, \ldots, \tau_K \rangle$ where $\tau_i \in \{N, O\}$. If $\tau_i = N$ then *the attribute A_i is nominal*, if $\tau_i = O$ then *the attribute A_i is ordinal*. If an attribute A is ordinal then we assume that it is reasonable to use ordering of its categories $1, \ldots, t$. If an attribute A is nominal, then we do not assume to deal with ordering of its categories $1, \ldots, t$.

If $\mathcal{LC}_T$ is a logical calculus of association rules of type T with basic attributes $A_1, \ldots, A_K$ then language $\mathcal{L}_{DK}$ of domain knowledge associated to $\mathcal{LC}_T$ contains a definition of L *basic groups of attributes* and definitions of *additional groups of attributes*. Basic groups of attributes are defined as sets $G_1, \ldots G_L$, where $G_i \subset \{A_1, \ldots, A_K\}$ and $L < K$, $\cup_{i=1}^L G_i = \{A_1, \ldots, A_K\}$, and $G_i \cap G_j = \emptyset$ for $i \neq j$ and $i, j = 1, \ldots, L$. Each non-empty subset $G \subset \{A_1, \ldots, A_K\}$ satisfying $G \neq G_i$ for $i = 1, \ldots L$ is an *additional group of attributes*.

Data matrix *Entry* used in [4] is an example of a data matrix. It is a part of data set STULONG concerning male patients, see http://euromise.vse.cz/challenge2004/. *Entry* has 64 columns corresponding to 64 basic attributes and 1417 rows corresponding to patients. The application described in [4] deals with three groups of attributes. Group *Measurement* consists of three ordinal attributes - *BMI* i.e. Body Mass Index, with 13 categories, *Subsc* i.e. skinfold above musculus subscapularis (in mm) with 14 categories, and *Tric* i.e. skinfold above musculus triceps (in mm) with 12 categories. Group *Difficulties* has three attributes with 2-5 categories - *Asthma*, *Chest pain* and *Lower limbs pain*. Group *Blood* has two attributes - *Diast* i.e. diastolic blood pressure with 7 categories and *Syst* i.e. systolic blood pressure with 9 categories.

Mutual simple influence among attributes is expressed by *SI-formulas*. There are several types of SI-formulas. Below, we assume that A_i, A_j, $i \neq j$ are ordinal attributes and φ is a Boolean attribute. Examples of types of SI-formulas follow:

- *ii*-formula (i.e. increases - increases) $A_i \uparrow\uparrow A_j$ meaning *if A_i increases then A_j increases too*, $BMI \uparrow\uparrow Diast$ being an example.
- *id*-formula (i.e. increases - decreases) $A_i \uparrow\downarrow A_j$ meaning *if A_i increases then A_j decreases*
- i^+b^+-formula has a form $A_i \uparrow^+ \varphi$ and its meaning is: *if A increases, then relative frequency of φ increases too*
- i^+b^--formula, i^-b^+-formula, i^-b^--formula have form $A_i \uparrow^- \varphi$, $A_i \downarrow^+ \varphi$, $A_i \downarrow^- \varphi$ respectively, their meaning is analogous to that of $A \uparrow^+ \varphi$
- bb^+-formula has a form $\varphi \rightarrow^+ \psi$ and its meaning is: *if φ is true, then relative frequency of ψ increases*
- bb^--formula has a form $\varphi \rightarrow^- \psi$ and its meaning is: *if φ is true, then relative frequency of ψ decreases*.

4   Analytical Questions and Their Solutions

Items of domain knowledge are used to formulate reasonable analytical questions. We introduce here two types of such questions.

GG-question – its general form is $[\mathcal{M} : G'_1, \ldots, G'_U \approx^? G''_1, \ldots G''_V]$ where $\mathcal{M}$ is a data matrix and $G'_1, \ldots, G'_U$ and $G''_1, \ldots G''_V$ are groups of attributes. An example is a formula $\mathcal{Q}_1 = [Entry : Measurement, Difficulties \approx^? Blood]$ of a language $\mathcal{L}_{AQ}$ meaning: *In the data matrix* Entry, *are there any interesting relations between combinations of values of attributes of groups* Measurement *and* Difficulties *on one side and combinations of values of attributes of group* Blood *on the other side?*

Negative GG_SI-question – a general form of these questions is $[\mathcal{M} : G'_1, \ldots, G'_U \approx^? G''_1, \ldots G''_V; \not\Rightarrow \Omega_1, \ldots, \Omega_K]$ where $\mathcal{M}$, $G'_1, \ldots, G'_U$ and $G''_1, \ldots G''_V$ are as above and $\Omega_1, \ldots, \Omega_K$ are SI-formulas. An example is $\mathcal{Q}_2 = [Entry : Measurement, Difficulties \approx^? Blood ; \not\Rightarrow BMI \uparrow\uparrow Diast]$ meaning: *In the data matrix* Entry, *are there any relations between combinations of values of attributes of groups* Measurement *and* Difficulties *on one side and values* of attributes of group Blood *on the other side which are not consequences of* BMI ↑↑ Diast?

The question Q_2 is solved in [4]. The procedure 4ft-Miner [5] dealing with association rules is used, it is an implementation of the procedure ASSOC introduced in [1]. We consider each valid configuration of input parameters of 4ft-Miner as a formula Φ of a special language $\mathcal{L}_{RAR}$ defining a set $\mathcal{S}(\Phi)$ of relevant association rules to be verified in a given data matrix $\mathcal{M}$. Output of the procedure 4ft-Miner is a set $True(\mathcal{S}(\Phi), \mathcal{M})$ of all association rules belonging to $\mathcal{S}(\Phi)$ which are true in $\mathcal{M}$.

Each formula Φ of the language $\mathcal{L}_{RAR}$ is a quadruple $\langle \Phi_A, \Phi_S, \approx, Par \rangle$ where Φ_A and Φ_S are formulas of $\mathcal{L}_{RAR}$ defining sets $\mathcal{S}(\Phi_A)$ and $\mathcal{S}(\Phi_S)$ of Boolean attributes respectively, $\approx$ is a 4ft-quantifier and Par is a formula of $\mathcal{L}_{RAR}$ specifying some additional conditions. The formula Φ define a set of rules $\varphi \approx \psi$ such that $\varphi \in \mathcal{S}(\Phi_A)$, $\psi \in \mathcal{S}(\Phi_S)$, φ and ψ have no common attributes and the additional conditions given by Par are satisfied.

The set $\mathcal{S}(\Phi_A)$ is given by groups of attributes $G'_1, \ldots, G'_U$ from which Boolean attributes φ are generated. We assume that $\varphi = \varphi_1 \wedge \ldots \wedge \varphi_U$ where φ_i is generated from the group G'_i, $i = 1, \ldots, U$, some of φ_i can be omitted. The parameters determine a minimal and a maximal number of φ_i in φ and a way in which φ_i is generated from the group G'_i, similarly for ψ. Let us assume that the group G'_i consists of attributes $A'_1, \ldots, A'_h$. Then $\varphi_i = \kappa_1 \wedge \ldots \wedge \kappa_h$ or $\varphi_i = \kappa_1 \vee \ldots \vee \kappa_h$, where κ_j is a basic Boolean attribute derived from A'_j, $j = 1, \ldots, h$. Some of κ_j can be omitted. The formula Φ_A determines the choice conjunctions/disjunctions for each G'_i, a minimal and a maximal number of κ_j in φ_i and a way in which the particular κ_j are created from A'_j. The set $\mathcal{S}(\Phi_S)$ is given in a similar way by groups of attributes $G''_1, \ldots, G''_U$.

A formula Φ_E defining a set of rules relevant to the solution of the question Q_2 is described in details in [4]. A set of rules $\varphi_M \wedge \varphi_D \Rightarrow_{0.9,30} \psi_B$ is defined where φ_M is a conjunction of 1 - 3 Boolean attributes derived from particular attributes of the group *Measurement*, each φ_D is a disjunction of 0 - 3 Boolean attributes derived from attributes of the group *Difficulties* and ψ_B is a conjunction of 1 - 2 Boolean attributes derived from attributes of the group *Blood*. The procedure 4ft-Miner produced a set $True(\mathcal{S}(\Phi_E), Entry)$ of all 158 rules which belong to $\mathcal{S}(\Phi_E)$ and which are true in *Entry*. Let us note that $456 * 10^6$ association rules were generated and tested. In addition, lot of relevant rules were skipped due to various optimizations implemented in the 4ft-Miner procedure.

The rule $BMI(21; 22\rangle \wedge Subsc(< 14) \Rightarrow_{0.97,33} Diast\langle 65; 75)$ has the highest confidence. This rule means that 34 patients satisfy $BMI(21; 22\rangle \wedge Subsc(< 14)$ and 33 from them satisfy also $Diast\langle 65; 75)$. Here $BMI(21; 22\rangle$ is a basic Boolean attribute meaning that the value of BMI is in interval $BMI(21; 22\rangle$ etc. To solve the question Q_2, we have to decide which association rules from $True(\mathcal{S}(\Phi_E), Entry)$ can be considered as consequences of BMI ↑↑ $Diast$.

There is a closely related task - i.e. to identify a set $Cons(\Omega, \approx)$ of association rules $\varphi \approx \psi$ which can be considered as consequences of a given SI-formula Ω. Having solved this task we can accept that all rules from the set

$True(\mathcal{S}(\varPhi_E), Entry) \cap Cons(BMI \uparrow\uparrow Diast, \Rightarrow_{0.9,30})$ can be considered as consequences of $BMI \uparrow\uparrow Diast$.

In addition, let $\varPhi = \langle \varPhi_A, \varPhi_S, \approx, Par \rangle$. Then, having a set $Cons(\Omega, \approx)$ we can test how much rules is in the set $True(\mathcal{S}(\varPhi), \mathcal{M}) \cap Cons(\Omega, \approx)$ and make a conclusion like: *There are lot of consequences of Ω in $True(\mathcal{S}(\varPhi), \mathcal{M})$ and thus it is reasonable to start a research of a hypothesis Ω.* This is important also when solving questions like $\mathcal{Q}_1$.

5   Consequences of Items of Domain Knowledge

We assume to have an SI-formula Ω of $\mathcal{L}_{DK}$. Our task is to find a set $Cons(\Omega, \approx)$ of all rules $\varphi \approx \psi$ which can be considered as consequences of Ω. The goal of this section is to present logical aspects of a solution of this task.

The set $Cons(\Omega, \approx)$ is defined in four steps:

1. A set $AC(\Omega, \approx)$ of *atomic consequences of* Ω for $\approx$ is defined as a set of very simple rules $\kappa \approx' \lambda$ which can be, according to a domain expert, considered as direct consequences of Ω, see section 5.1.
2. A set $AgC(\Omega, \approx)$ of *agreed consequences of* Ω for $\approx$ is defined. A rule $\rho \approx' \sigma$ belongs to $AgC(\Omega, \approx)$ if $\rho \approx' \sigma \notin AC(\Omega, \approx)$ and there is $\kappa \approx' \lambda \in AC(\Omega, \approx)$ such that $\rho \approx' \sigma$ does not logically follow from $\kappa \approx' \lambda$, however, according to a domain expert, it is possible to agree that $\rho \approx' \sigma$ says nothing new in addition to $\kappa \approx' \lambda$. There are some details in section 5.2.
3. A set $LgC(\Omega, \approx)$ of *logical consequences of* Ω for $\approx$ is defined. A rule $\varphi \approx' \psi$ belongs to $LgC(\Omega, \approx)$ if $\varphi \approx' \psi \notin (AC(\Omega, \approx) \cup AgC(\Omega, \approx))$ and there is $\tau \approx' \omega \in (AC(\Omega, \approx) \cup AgC(\Omega, \approx))$ such that $\varphi \approx' \psi$ logically follows from $\tau \approx' \omega$ i.e. $\frac{\tau \approx' \omega}{\varphi \approx' \psi}$ is a correct deduction rule. See section 5.3 for details.
4. We define $Cons(\Omega, \approx) = AC(\Omega, \approx) \cup AgC(\Omega, \approx) \cup LgC(\Omega, \approx)$.

We use an SI-formula $A_1 \uparrow\uparrow A_2$ as an example of Ω and a quantifier $\Rightarrow_{p_0, B_0}$ to introduce important details related to this approach in the rest of this section.

5.1   Atomic Consequences

Let A_1, A_2 be attributes with categories $1, \ldots, 12$. Let us define subsets of categories $low = \{1, 2, 3, 4\}$, $medium = \{5, 6, 7, 8\}$, $high = \{9, 10, 11, 12\}$. The set $AC(A \uparrow\uparrow B, \Rightarrow_{p_0, B_0})$ can be then defined as a union

$$AC(A \uparrow\uparrow B, \Rightarrow_{p_0, B_0}) = low \times low \ \cup \ medium \times medium \ \cup \ high \times high \ .$$

Here $low \times low$ is a set of all rules $A_1(\alpha_1) \Rightarrow_{p, B} A_2(\alpha_2)$ where both $\alpha_1 \subseteq \{1, 2, 3, 4\}$ and $\alpha_2 \subseteq \{1, 2, 3, 4\}$ and in addition $p \geq p_0$ and $B \geq B_0$, similarly for $medium \times medium$ and $high \times high$.

This approach is applied for $BMI \uparrow\uparrow Diast$ when solving the question $\mathcal{Q}_2$ in [4]. Six "rectangles of rules" are used instead of three "rectangles of rules" $low \times low$, $medium \times medium$ and $high \times high$.

Please note that $AC(A_1 \uparrow\uparrow A_2, \Rightarrow_{p_0,B_0})$ is a set of rules $A_1(\gamma) \Rightarrow_{p,B} A_2(\delta)$ where $p \geq p_0 \wedge B \geq B_0$ and γ and δ satisfy conditions given by the sets $low = \{1,2,3,4\}$, $medium = \{5,6,7,8\}$, $high = \{9,10,11,12\}$. Here 4ft-quantifier $\Rightarrow_{p_0,B_0}$ corresponds to 4ft-quantifier $\approx$ and $\Rightarrow_{p,B}$ corresponds to $\approx'$, see point 1 above.

5.2   Agreed Consequences

It χ is a Boolean attribute then it is easy to show that the association rule $A_1(\alpha_1) \wedge \chi \Rightarrow_{p,B} A_2(\alpha_2)$ does not logically follow from $A_1(\alpha_1) \Rightarrow_{p,B} A_2(\alpha_2)$. The core of the proof is the fact that if there are at least B rows in a data matrix $\mathcal{M}$ satisfying $A_1(\alpha_1) \wedge A_2(\alpha_2)$ then there need not be also at least B objects satisfying $A_1(\alpha_1) \wedge \chi \wedge A_2(\alpha_2)$. However, in *some* cases it can be reasonable from a domain expert point of view to agree that $A_1(\alpha_1) \wedge \chi \Rightarrow_{p,B} A_2(\alpha_2)$ is a consequence of $A_1(\alpha_1) \Rightarrow_{p,B} A_2(\alpha_2)$. In such a case we call the rule $A_1(\alpha_1) \wedge \chi \Rightarrow_{p,B} A_2(\alpha_2)$ an *agreed consequence* of $A_1(\alpha_1) \Rightarrow_{p,B} A_2(\alpha_2)$.

Let us assume that the rule $BMI(\langle 23;24\rangle, \langle 24;25\rangle) \Rightarrow_{0.9,30} Diast(\langle 75;85\rangle)$ shortly $BMI\langle 23;25\rangle \Rightarrow_{0.95,35} Diast\langle 75;85\rangle$ belongs to $AC(BMI \uparrow\uparrow Diast, \Rightarrow_{0.9,30})$, see [4]. The rule $BMI\langle 23;25\rangle \wedge Chest\ pain(not\ present) \Rightarrow_{0.9,30} Diast\langle 75;85\rangle$ does not logically follow from $BMI\langle 23;25\rangle \Rightarrow_{0.9,30} Diast\langle 75;85\rangle$, see above. However, it says nothing new because truthfulness of *Chest pain(not present)* has no influence on a relation of *BMI* and *Diast*. This means that we can add the rule $BMI\langle 23;25\rangle \wedge Chest\ pain(not\ present) \Rightarrow_{0.9,30} Diast\langle 75;85\rangle$ to the set $AgC(BMI \uparrow\uparrow Diast, \Rightarrow_{0.9,30})$ of agreed consequences of $BMI \uparrow\uparrow Diast$ for $\Rightarrow_{0.9,30}$. This way the set $AgC(BMI \uparrow\uparrow Diast, \Rightarrow_{0.9,30})$ can be defined in a co-operation with a domain expert.

Again, let us note that the set $AgC(BMI \uparrow\uparrow Diast, \Rightarrow_{0.9,30})$ is a set of rules $\rho \Rightarrow_{p,B} \sigma$ where $p \geq 0.9 \wedge B \geq 30$, see point 2 above. Similarly, a set $AgC(A_1 \uparrow\uparrow A_2, \Rightarrow_{p_0,B_0})$ is a set of rules $\rho \Rightarrow_{p,B} \sigma$ where $p \geq p_0 \wedge B \geq B_0$.

5.3   Logical Consequences

A set $LgC(A_1 \uparrow\uparrow A_2, \Rightarrow_{p_0,B_0})$ of logical consequences of $A_1 \uparrow\uparrow A_2$ for $\Rightarrow_{p_0,B_0}$ is defined such that a rule $\varphi \Rightarrow_{p,B} \psi$ belongs to $LgC(A_1 \uparrow\uparrow A_2, \Rightarrow_{p_0,B_0})$ if $\varphi \Rightarrow_{p,B} \psi \notin (AC(A_1 \uparrow\uparrow A_2, \Rightarrow_{p_0,B_0}) \cup AgC(A_1 \uparrow\uparrow A_2, \Rightarrow_{p_0,B_0}))$ and there is $\tau \Rightarrow_{p,B} \omega \in (AC(A_1 \uparrow\uparrow A_2, \Rightarrow_{p_0,B_0}) \cup AgC(A_1 \uparrow\uparrow A_2, \Rightarrow_{p_0,B_0}))$ such that $\varphi \Rightarrow_{p,B} \psi$ logically follows from $\tau \Rightarrow_{p,B} \omega$ i.e. $\frac{\tau \Rightarrow_{p,B} \omega}{\varphi \Rightarrow_{p,B} \psi}$ is a correct deduction rule.

Our goal is to discuss how to decide if a given rule $\varphi \Rightarrow_{p,B} \psi$ belongs to $LgC(A_1 \uparrow\uparrow A_2, \Rightarrow_{p_0,B_0})$. There are theorems proved in [2] saying that $\frac{\tau \Rightarrow_{p,B} \omega}{\varphi \Rightarrow_{p,B} \psi}$ is a correct deduction rule if and only if (T_1) or (T_2) are satisfied:

(T_1): both $\Gamma_1(\tau, \omega, \varphi, \psi)$ and $\Gamma_2(\tau, \omega, \varphi, \psi)$ are tautologies of a propositional calculus

(T_2): $\Delta(\tau, \omega)$ is a tautology of a propositional calculus where $\Gamma_1(\tau, \omega, \varphi, \psi)$, $\Gamma_2(\tau, \omega, \varphi, \psi)$ and $\Delta(\tau, \omega)$ are formulas of a propositional calculus created from Boolean attributes $\tau, \omega, \varphi, \psi$ - formulas of a logical calculus of association rules. However, a direct application of this theorem is very cumbersome due to complexity of formulas $\Gamma_1(\tau, \omega, \varphi, \psi)$, $\Gamma_2(\tau, \omega, \varphi, \psi)$ and $\Delta(\tau, \omega)$.

In the rest of this section we informally introduce an approach to simplify decision if $\frac{\tau \Rightarrow_{p,B} \omega}{\varphi \Rightarrow_{p,B} \psi}$ is a correct deduction rule or not. We deal only with rules $\tau \Rightarrow_{p,B} \omega \in AC(A_1 \uparrow\uparrow A_2, \Rightarrow_{p_0, B_0})$, a solution of this task for a case when $\tau \Rightarrow_{p,B} \omega \in AgC(A_1 \uparrow\uparrow A_2, \Rightarrow_{p_0, B_0})$ is similar. This means that $\tau \Rightarrow_{p,B} \omega$ is in a form $A_1(\gamma) \Rightarrow_{p,B} A_2(\delta)$, see the end of section 5.1.

We use the syntax of rules $\varphi \Rightarrow_{p,B} \psi$ produced by 4ft-Miner, note that this syntax is very rich. Thus it is $\varphi \in \mathcal{S}(\Phi_A)$ and $\psi \in \mathcal{S}(\Phi_S)$, see section 4. In addition we concentrate on a case when basic Boolean attribute $A_1(\mu)$ created from A_1 is a part of φ and basic Boolean attribute $A_2(\nu)$ created from A_2 is a part of ψ, the remaining cases can be treated similarly or even more simple.

This means that

- φ has one of the forms $A_1(\mu)$, $A_1(\mu) \wedge \theta_1$, $A_1(\mu) \vee \theta_1$, $(A_1(\mu) \vee \theta_1) \wedge \theta_2$
- ψ has one of the forms $A_2(\nu)$, $A_2(\nu) \wedge \pi_1$, $A_2(\nu) \vee \pi_1$, $(A_2(\nu) \vee \pi_1) \wedge \pi_2$.

where θ_1, θ_2, π_1, and π_2 are Boolean attributes. This means that the rule $\varphi \Rightarrow_{p,B} \psi$ is in one of 16 forms: $A_1(\mu) \Rightarrow_{p,B} A_2(\nu)$, $A_1(\mu) \Rightarrow_{p,B} A_2(\nu) \wedge \pi_1$, $A_1(\mu) \Rightarrow_{p,B} A_2(\nu) \vee \pi_1$, $A_1(\mu) \Rightarrow_{p,B} (A_2(\nu) \vee \pi_1) \wedge \pi_2$, $A_1(\mu) \wedge \theta_1 \Rightarrow_{p,B} A_2(\nu)$, $\ldots$, $(A_1(\mu) \vee \theta_1) \wedge \theta_2 \Rightarrow_{p,B} (A_2(\nu) \vee \pi_1) \wedge \pi_2$.

Thus, the task of deciding if $\frac{\tau \Rightarrow_{p,B} \omega}{\varphi \Rightarrow_{p,B} \psi}$ is a correct deduction rule can be split into 16 tasks of deciding if deduction rules $\frac{A_1(\gamma) \Rightarrow_{p,B} A_2(\delta)}{A_1(\mu) \Rightarrow_{p,B} A_2(\nu)}$, $\frac{A_1(\gamma) \Rightarrow_{p,B} A_2(\delta)}{A_1(\mu) \Rightarrow_{p,B} A_2(\nu) \wedge \pi_1}$, $\frac{A_1(\gamma) \Rightarrow_{p,B} A_2(\delta)}{A_1(\mu) \Rightarrow_{p,B} A_2(\nu) \vee \pi_1}$, $\ldots$, $\frac{A_1(\gamma) \Rightarrow_{p,B} A_2(\delta)}{(A_1(\mu) \vee \theta_1) \wedge \theta_2 \Rightarrow_{p,B} (A_2(\nu) \vee \pi_1) \wedge \pi_2}$ are correct or not. These tasks are relatively simple. An example is the rule $\frac{A_1(\gamma) \Rightarrow_{p,B} A_2(\delta)}{A_1(\mu) \Rightarrow_{p,B} A_2(\nu) \wedge \pi_1}$. It is easy to show that the there is a data matrix $\mathcal{M}$ in which there is no row satisfying $A_2(\nu) \wedge \pi_1$. Thus, the rule $\frac{A_1(\gamma) \Rightarrow_{p,B} A_2(\delta)}{A_1(\mu) \Rightarrow_{p,B} A_2(\nu) \wedge \pi_1}$ is not correct.

Let us note that Boolean attributes $\neg A_1(\mu)$ and $\neg A_2(\nu)$ can also be used in rules $\varphi \approx \psi$ the procedure 4ft-Miner deals with [5]. Thus, there are 4 additional possibilities for the form of φ: $\neg A_1(\mu)$, $\neg A_1(\mu) \wedge \theta_1$, $\neg A_1(\mu) \vee \theta_1$, $(\neg A_1(\mu) \vee \theta_1) \wedge \theta_2$. The same is true for ψ. This means that we have to deal with 8×8 forms of deduction rules instead of 16 forms introduced above. However, for each of corresponding 64 deduction rules is a decision about its correctness similar to that above, principles used in [2] can be applied.

6   Conclusions

We have introduced an approach to dealing with domain knowledge in data mining with association rules - couples of related general Boolean attributes. The practical experience with this approach was first time published in [4]. The presented approach is based on dealing with formalized items of domain knowledge which are not in a form of association rules. Particular items of domain knowledge are mapped to sets of association rules which can be considered as their consequences. These sets of rules are then used to make reasonable conclusions. Logical deduction in a special logical calculus formulas of which correspond to association rules is used in this approach.

In addition, FOFRADAR i.e. the formal frame for the whole process of data mining with rules is introduced. FOFRADAR is based on enhancement of the logical calculus of association rules by several additional languages and procedures.

The association rules are understood as general relations of Boolean attributes derived from columns of the analyzed data matrix. Such rules can be also understood as "granules of knowledge". There are various types of relations Boolean attributes, each correspond to one 4ft-quantifier. Logical aspects of mapping items of domain knowledge to sets of rules were discussed in details for one type of 4ft-quantifiers i.e. for founded implication.

There are two main directions of further work. The first one concerns detailed elaboration of the presented approach for additional 4ft-quantifiers. The second direction concerns further development of the FOFRADAR framework with the goal to use it to automate the whole data mining process with association rules, see also [6,7].

References

1. Hájek, P., Havránek, T.: Mechanising Hypothesis Formation - Mathematical Foundations for a General Theory. Springer, Heidelberg (1978)
2. Rauch, J.: Logic of Association Rules. Applied Intelligence 22, 9–28 (2005)
3. Rauch, J.: Consideration on a Formal Frame for Data Mining. In: Proceedings of Granular Computing 2011, pp. 562–569. IEEE Computer Society, Piscataway (2011)
4. Rauch, J., Šimůnek, M.: Applying Domain Knowledge in Association Rules Mining Process – First Experience. In: Kryszkiewicz, M., Rybinski, H., Skowron, A., Raś, Z.W. (eds.) ISMIS 2011. LNCS, vol. 6804, pp. 113–122. Springer, Heidelberg (2011)
5. Rauch, J., Šimůnek, M.: An Alternative Approach to Mining Association Rules. In: Lin, T.Y., et al. (eds.) Data Mining: Foundations, Methods, and Applications, pp. 219–238. Springer, Heidelberg (2005)
6. Šimůnek, M., Tammisto, T.: Distributed Data-Mining in the LISp-Miner Systém Using Techila Grid. In: Zavoral, F., et al. (eds.) Networked Digital Technologies, pp. 15–21. Springer, Berlin (2010)
7. Šimůnek, M., Rauch, J.: EverMiner – Towards Fully Automated KDD Process. In: Funatsu, K., Hasegava, K. (eds.) New Fundamental Technologies in Data Mining, pp. 221–240. InTech, Rijeka (2011)

The Problem of Finding the Sparsest Bayesian Network for an Input Data Set is NP-Hard*

Paweł Betliński[1] and Dominik Ślęzak[1,2]

[1] Institute of Mathematics, University of Warsaw
Banacha 2, 02-097 Warsaw, Poland
[2] Infobright Inc.
Krzywickiego 34 pok. 219, 02-078 Warsaw, Poland

Abstract. We show that the problem of finding a Bayesian network with minimum number of edges for an input data set is NP-hard. We discuss the analogies of formulation and proof of our result to other studies in the areas of Bayesian networks and knowledge discovery.

Keywords: Bayesian networks, Markov boundaries, Knowledge discovery, Optimization criteria, Minimum dominating sets, NP-hardness.

1 Introduction

Bayesian networks are a well-known methodology for conducting approximate reasoning and representing data dependencies in a graph-theoretic way. There are a number of approaches to investigate complexity of learning Bayesian network structures basing on information about the underlying probability distributions, as well as heuristic algorithms attempting to derive such structures from the empirical data. In this paper, we focus on showing that the problem of finding a Bayesian network with minimum number of edges for an input data set is NP-hard. Such a formulation of an optimization problem differs from those known so far for Bayesian networks. It seems to provide a theoretical background for the cases where the data itself is the only available source of information about a hypothetic network structure. It also seems to correspond to those algorithms known from the literature, which focus on input data sets rather than information about probability distributions that generate them.

The paper is organized as follows: In Section 2, we introduce the basics of Bayesian networks and discuss why one may be interested in possibly sparse Bayesian network structures, which – as directed acyclic graphs – contain no redundant edges. In Section 3, we outline different formulations of the problems of searching for optimal Bayesian networks that were reported in the literature so far. In Section 4, we proceed with our own version of such a problem and the proof of its NP-hardness with respect to the number of variables in the input data. In Section 5, we conclude the paper with final comments.

* This work was partially supported by the Polish National Science Centre grant 2011/01/B/ST6/03867.

L. Chen et al. (Eds.): ISMIS 2012, LNAI 7661, pp. 21–30, 2012.
© Springer-Verlag Berlin Heidelberg 2012

2   Why Sparse Bayesian Networks Are Useful?

Bayesian network [9,10] is a directed acyclic graph (DAG), where vertices represent random variables, together with parameters describing probability distribution of each of vertices subject to its parents in the graph. In practice, vertices often correspond to attributes in a data table. In such a case, the studies are usually restricted to random variables with finite sets of values.

Bayesian networks are commonly used to model approximate reasoning employing the structure of a network and the probability distribution parameters [5]. Sparser networks provide better reasoning efficiency and clarity. Even more important is simplicity of the corresponding distributions, which is related to sparsity, although it can be used as an evaluation criterion by itself.

A sparse Bayesian network is able to represent distribution in a compressed way. For n binary variables, direct representation of their distribution requires storing $2^n - 1$ numbers, while in Bayesian network in which each vertex has up to k parents – at most $n2^k$. The usage of sparse Bayesian networks becomes even more beneficial in the case of non-binary variables.

Bayesian networks can be also used as knowledge bases about independencies [16]. Sparser networks are known to represent richer knowledge. Conditional independencies can be important in many areas, not only in relation to probabilities [12]. Also, it is interesting to consider networks, which represent approximate conditional independencies and multivalued dependencies [14].

Let us recall some formal concepts together with some known theoretical properties, which give better understanding of sparse Bayesian networks, and are the basis for further considerations. Bayesian network represents joint probability distribution of its vertices – random variables $X_1, ..., X_n$, as follows:

$$Pr(X_1, ..., X_n) = Pr(X_1 \mid \pi(X_1)) \times ... \times Pr(X_n \mid \pi(X_n)), \tag{1}$$

where $\pi(X_i)$ is a subset of variables – parents of vertex X_i in the network (it might be an empty set). We say that DAG $\mathbb{G}$ with n vertices models n-dimensional probability distribution $\mathbb{P}$, if it is possible to choose such parameters describing conditional distribution of each vertex given its parents in $\mathbb{G}$, that Bayesian network created in this way represents $\mathbb{P}$ in the sense of formula (1). The set of all distributions modeled by DAG $\mathbb{G}$ will be denoted as $MOD(\mathbb{G})$.

Let $\mathbb{G} = (V, E)$ be any DAG such that $|V| \geq 2$. By trail between vertices X_{i_1} and X_{i_k} in $\mathbb{G}$ we mean any sequence of different vertices $[X_{i_1}, ..., X_{i_k}]$, where $k \geq 2$ and for all $2 \leq j \leq k$ it holds that $(X_{i_{j-1}}, X_{i_j}) \in E$ or $(X_{i_j}, X_{i_{j-1}}) \in E$. Consider a subset (it might be empty) $C \subseteq V$ of size at most $|V| - 2$ and any $X, Y \in V \setminus C$, $X \neq Y$. We say that a trail t between X and Y is blocked by C, if there holds at least one of the following conditions: there exists $Z \in C$, which is element of t with not converging arrows; or there exists Z which is element of t different than X and Y and with converging arrows, such that Z and every descendant of Z does not belong to C. Let $A \neq \emptyset$, $B \neq \emptyset$, and C be mutually exclusive subsets of V. We say that A and B are d-separated by C in $\mathbb{G}$, if for every $X \in A$ and $Y \in B$ each trail between them is blocked by C.

The set of all d-separations in the given DAG $\mathbb{G}$ – that is the set of all triples of mutually exclusive subsets of vertices $A \neq \emptyset$, $B \neq \emptyset$, and C such that A and B are d-separated by C – will be denoted as $IND(\mathbb{G})$. The set of all conditional independencies in the given joint probability distribution $\mathbb{P}$ – that is the set of all triples of mutually exclusive subsets of variables $A \neq \emptyset$, $B \neq \emptyset$, and C such that A and B are independent given C – will be denoted as $IND(\mathbb{P})$. There is an essential connection between the graph property of d-separation, and the joint probability distribution property of conditional independence:

Theorem 1. *[7] For every DAG $\mathbb{G}$: $MOD(\mathbb{G}) = \{\mathbb{P} : IND(\mathbb{P}) \supseteq IND(\mathbb{G})\}$.*

For every DAG $\mathbb{G}$, each vertex X is d-separated from its non-descendants excluding $\pi(X)$ by $\pi(X)$ (also when $\pi(X)$ is empty). These special d-separations are called Markov conditions. We denote them as $MC(\mathbb{G})$.

Markov conditions are the next reason for searching for sparse Bayesian networks. Theorem 1 together with Markov conditions gives us that in a probability distribution represented by a Bayesian network every variable X is independent from the set of its non-descendants in the network excluding $\pi(X)$, given $\pi(X)$. This means that parents in the network gather direct causes in the represented distribution. Therefore, it is desired to find such a network, in which the sets of direct causes are as small as possible – that is we want to find real causes, not just their supersets. This is exactly what a sparse network can give. The set of Markov conditions is a kind of generator of the set of all d-separations:

Theorem 2. *[7] For every DAG $\mathbb{G}$: $MOD(\mathbb{G}) = \{\mathbb{P} : IND(\mathbb{P}) \supseteq MC(\mathbb{G})\}$.*

Finally, we present one more well known and important result, which shows a great correlation between three concepts: set of d-separations, set of modeled distributions, and the sparsity. Hard implication in this theorem is from (a) or (b) to (c) – it was a hypothesis finally proved in [2]. The concept 'covered edge' used here means such edge from X to Y, that $\pi(X) = \pi(Y) \setminus \{X\}$.

Theorem 3. *[2] For DAGs $\mathbb{G}_1$ and $\mathbb{G}_2$ with the same set of vertices the following conditions are equivalent: (a) $IND(\mathbb{G}_1) \supseteq IND(\mathbb{G}_2)$, (b) $MOD(\mathbb{G}_1) \subseteq MOD(\mathbb{G}_2)$, (c) $\mathbb{G}_1$ can be obtained from $\mathbb{G}_2$ after a sequence of operations, where each operation is removing one edge or reversing currently covered edge, provided that after each step of the sequence a modified graph remains a DAG.*

So, if $IND(\mathbb{G}_1) \supseteq IND(\mathbb{G}_2)$, then $\mathbb{G}_1$ is, forgetting directions of edges, a subgraph of $\mathbb{G}_2$. Theorems 1 and 3 give us then that the sparsest possible candidate for the network representing joint probability distribution $\mathbb{P}$ is a network which corresponds to such DAG $\mathbb{G}$, that $IND(\mathbb{G}) = IND(\mathbb{P})$. Such a DAG is called a perfect map of $\mathbb{P}$. Unfortunately, for many distributions there does not exist a perfect map, so other characteristic of the sparsest graph is necessary.

3 Optimization Criteria

Sparsity of Bayesian networks may be an important criterion of their evaluation. In particular, one may consider it as the means for searching for optimal Bayesian network structures for the given information about the joint probability distributions or the corresponding available data. In this section, we recall the basic scheme of Bayesian network learning, the corresponding optimization problems studied in the literature, as well as the analogous optimization problems taken from other methodologies that seem to be related to the idea of focusing on a knowledge model simplicity.

We assume that on the entry we have some information system of m objects and n attributes, where each attribute has some finite set of values. Such systems are often assumed to be a result of generating independently n rows from some unknown joint probability distribution $\mathbb{P}$. What we would like to obtain is a Bayesian network, which (approximately) represents $\mathbb{P}$. Learning of such network is typically done in two steps [7]: learning the network's structure, that is a DAG with n vertices, and then learning parameters, e.g., by direct estimation of conditional probabilities as frequencies in the information system. In the first step, we want to learn as sparse as possible DAG modeling desired distribution. However, as discussed in the end of Section 2, the problem is what characteristic of the sparsest graph we should use. The following classical approach is related to simplicity and statistical quality of probabilistic distributions.

By a minimal graph for distribution $\mathbb{P}$ we understand such DAG $\mathbb{G}$, that $IND(\mathbb{G}) \subseteq IND(\mathbb{P})$ and there is no such DAG $\mathbb{H}$, that $IND(\mathbb{G}) \subset IND(\mathbb{H}) \subseteq IND(\mathbb{P})$. If there is a perfect map for some distribution, then in particular it is a minimal graph for this distribution. If distribution does not have a perfect map, then minimal graphs seem to be quite reasonable substitute for a sparse structure. By dimension of a DAG $\mathbb{G}$ we understand the number of parameters in the Bayesian network corresponding to $\mathbb{G}$. We call $\mathbb{G}$ parametrically optimal for $\mathbb{P}$, if $\mathbb{G}$ is a DAG of minimal dimension from all graphs modeling $\mathbb{P}$. Every parametrically optimal DAG is also minimal [7]. So, parametrical optimality is even more restricted sparseness condition.

A parametrically optimal DAG, in contrary to a perfect map (which is parametrically optimal itself), always exists. Determining graphs satisfying this condition was the inspiration, as the result of which there has been created several measures of adequacy of given network structure. This adequacy is in the context of representing in compact way distribution of attributes of an information system. Typically these measures are represented as a function $score(.,.)$, where arguments are the information system and a candidate for sparse DAG that models the corresponding joint distribution over the system's attributes.

We call $score(.,.)$ consistent, if for every finite set of nominal attributes A there exists M, such that for any $m \geq M$ the following holds: if DAG $\mathbb{G}_1$ models $\mathbb{P}_m$ and DAG $\mathbb{G}_2$ does not, or if both $\mathbb{G}_1$ and $\mathbb{G}_2$ model $\mathbb{P}_m$ but $\mathbb{G}_1$ has smaller dimension, then $score(\mathbb{D}_m, \mathbb{G}_1) > score(\mathbb{D}_m, \mathbb{G}_2)$, where $\mathbb{D}_m$ is any information system of m objects and set of attributes A, and $\mathbb{P}_m$ is distribution of attributes induced by $\mathbb{D}_m$. For a sufficiently large (in the sense of number of objects) information system $\mathbb{D}$ a consistent measure $score(\mathbb{D},.)$ reaches its maximum for a parametrically optimal DAG for distribution induced by $\mathbb{D}$. So asymptotically, when a number of objects goes to infinity, consistent measure behaves reasonably. It reaches maximum for parametrically optimal DAG for distribution, from which each row is independently generated.

One of the most examined measures of this type is the Bayesian scoring criterion, also known as the Bayesian Dirichlet metric [4]. For appropriately chosen parameters it is a consistent measure. The problem of finding a network structure maximizing this criterion for the input information system is NP-hard – even in the case, when we restrict searching only to such graphs, where every vertex has at most two parents [1]. But this was not a result obtained for an arbitrary consistent measure. Moreover, consider the case, when we restrict input information systems only to such, for which there is sufficient amount of objects to guarantee regular behavior of the measure resulting from its consistency. There was still some hope, that in such case finding optimal Bayesian network structure will be easier. Unfortunately, it has turned out later, that even when restricting input to such tables, and for an arbitrary consistent measure, the optimization problem is still NP-hard [3]. A general topic of hardness of the Bayesian network learning is nowadays probably mostly identified with this result.

The Bayesian scoring criterion introduced in [4] can be expressed as approximately proportional to a difference between the overall entropy of a network, defined as the sum of entropies of conditional distributions assigned to particular vertices subject to their parents, and the entropy of joint probability distribution. In [13], a corresponding notion of an approximate Markov boundary was studied, by using the terminology of decision reducts known from the theory of rough sets [8]. Precisely, for an input data set with some distinguished attribute d, an approximate entropy reduct was defined as an irreducible subset B of all other attributes A that provides almost the same quantity of conditional entropy $H(d|B)$ as the initial $H(d|A)$. NP-hardness of the problem of finding an optimal approximate entropy reduct B was shown for the criteria related to minimization of a number of attributes in B and minimization of a number of parameters necessary to describe the probability distribution of d subject to B. In [14], as a special case of the notion of an approximate Bayesian network, DAGs generating almost the same entropy as the entropy of joint distribution were considered. This is in direct correspondence to the above-mentioned approximate Markov boundaries and the Bayesian scoring criterion used here as the network's approximate validity constraint. It was shown that such networks can graphically represent the approximate conditional independence statements.

4   The Main Result

The main aim of this paper is to propose an alternative approach to understand, formalize and prove NP-hardness of the problem of learning Bayesian networks. Our view seems to be simpler than the classic way based on consistent measures recalled in Section 3. Let us first give some motivations.

Consistent measures are able to reflect different approaches, which approximately find optimal Bayesian network structure. However, one may often need to go through advanced statistical computations in order to understand their background. The idea of using some special measures, which asymptotically lead to parametrically optimal networks in order to express a sparseness criterion may be questionable especially for practitioners. On the other hand, the criterion that we would like to consider is very natural. It takes the following form:

MINIMAL BAYESIAN NETWORK (MBN)

Given on the entry information system, where each attribute has finite set of possible values, find DAG with minimal number of edges modeling joint probability distribution of attributes induced by this information system.

We will now prove NP-hardness of the MBN problem. The proof is by polynomial reduction of the following well-known NP-hard problem to MBN:

MINIMUM DOMINATING SET (MDS)

Given on the entry undirected graph, find minimal dominating set for this graph, that is such subset of vertices of the graph, that every vertex of this graph is either in this subset, or it is a neighbor of some element of this subset.

Assume that on the entry of MDS we have $\mathbb{H} = (V, E)$, where $V = \{v_1, ..., v_n\}$. Our reduction encodes $\mathbb{H}$ as an information system $A_{\mathbb{H}} = (U_{\mathbb{H}}, A_{\mathbb{H}})$, $U_{\mathbb{H}} = \{u_1, ..., u_{2n+1}\}$, $A_{\mathbb{H}} = \{a_1, ..., a_n, b\}$, where: for $1 \leq i, j \leq n$, $a_j(u_i) = 1$ if $i = j$ or $(v_i, v_j) \in E$, otherwise $a_j(u_i) = 0$; $a_i(u_{n+1}) = b(u_i) = 0$ for $1 \leq i \leq n$; $b(u_{n+1}) = 1$; $a_i(u_{n+1+i}) = 3$ for $1 \leq i \leq n$; all remaining fields have value 2. The joint probability distribution of attributes induced by $A_{\mathbb{H}}$ we denote as $\mathbb{P}_{\mathbb{H}}$. We will use the same notation $\{a_1, ..., a_n, b\}$ for the set of attributes of an information system, and the corresponding set of vertices of the network.

The presented derivation of MDS to MBN uses the same representation of an input graph in a tabular data form as in the case of the complexity results reported in [15] for the problem of finding the most meaningful association reducts. An association reduct is a pair (B, C) of disjoint attribute sets such that the values of attributes in C are determined by those in B, where B cannot be reduced and C cannot be extended without losing the required level of determinism. In such a case, we draw relationship between a degree of meaningfulness of (B, C) and cardinalities of B and C, attempting to find the cases with minimal amount of attributes in B and maximal amount of attributes in C. After some modifications, we could also use this kind of derivation to study NP-hardness of the problem of finding the most meaningful probabilistic conditional independencies or, e.g., multivalued dependencies [11]. A relatively simpler tabular representation of input graphs was used also within the theory of rough sets to prove NP-hardness of the problems of finding minimal decision reducts and minimal Markov boundaries interpreted as probabilistic decision reducts [13].

The remainder of this section includes the proofs of four lemmas that lead step by step to showing that the MBN problem is NP-hard.

Lemma 1. *Let $\mathbb{G}$ be an arbitrary DAG which models distribution $\mathbb{P}_{\mathbb{H}}$. Then every pair of different vertices a_i, a_j is connected by an edge in $\mathbb{G}$.*

Proof. Assume that some pair of vertices a_i, a_j is not connected by an edge in some DAG $\mathbb{G}$ modeling $\mathbb{P}_{\mathbb{H}}$. Then a_i and a_j are d-separated by some subset C of the remaining vertices in $\mathbb{G}$ (it is a well-known implication, see [7]).

According to Theorem 1, this means that in $\mathbb{P}_\mathbb{H}$ a_i and a_j are independent given C. This leads to contradiction – thanks to n last rows of $\mathbb{A}_\mathbb{H}$, every two different variables a_i and a_j are dependent given any subset of remaining variables: $a_j = 2$ with probability 1 given that $a_i = 3$, while $a_j = 3$ with nonzero probability given that $a_i = 2$. This holds given that all variables from a subset are equal to 2. $\qquad\square$

For the purposes of the next lemma, let us recall that by a Markov blanket of $T \in A$ for a vector of variables A one means a subset $M \subseteq A \setminus \{T\}$ such that T and $A \setminus (\{T\} \cup M)$ are independent given M. Also recall that Markov boundaries mentioned in the previous sections are irreducible Markov blankets.

Lemma 2. *For arbitrary subset* $\{i_1, ..., i_k\} \subseteq \{1, ..., n\}$: $\{v_{i_1}, ..., v_{i_k}\}$ *is a dominating set in* $\mathbb{H}$ $\Longleftrightarrow$ $\{a_{i_1}, ..., a_{i_k}\}$ *is a Markov blanket of* b *for distribution* $\mathbb{P}_\mathbb{H}$.

Proof. $\Longrightarrow$ We will show that if $\{v_{i_1}, ..., v_{i_k}\}$ is a dominating set in $\mathbb{H}$, then for any configuration of values of attributes from $M \overset{def}{=} \{a_{i_1}, ..., a_{i_k}\}$ that appears in $\mathbb{A}_\mathbb{H}$ the number of possible values of b is reduced to 1. Let us check all cases: When all attributes from M are equal to 2 or 3, then b must be equal to 2. When all attributes from M are equal to 0 or 1 and at least one attribute from M is equal to 1, then b must be equal to 0. When all attributes from M are equal to 0, then, thanks to the assumption that $\{v_{i_1}, ..., v_{i_k}\}$ is a dominating set in $\mathbb{H}$, the only possibility is object u_{n+1}, so b must be equal to 1.
$\Longleftarrow$ Assume that subset of vertices $\{v_{i_1}, ..., v_{i_k}\}$ is not a dominating set in $\mathbb{H}$. Then there is at least one vertex v_l in $\mathbb{H}$, which is not connected by any edge with any element of this subset. So in $\mathbb{A}_\mathbb{H}$ object u_l has value 0 on attributes from M. Another object with the same pattern on M is u_{n+1}. Therefore, there is some nonzero probability of both $b = 0$ and $b = 1$ subject to the vector of 0's on M. Now, let us consider $M \cup \{a_l\}$. When $a_l = 1$ (there is at least one row satisfying this – u_l), then b is equal to 0 with probability 1. Hence, M is not a Markov blanket of b, which is contradiction. $\qquad\square$

Lemma 3. *Let* $\{v_{i_1}, ..., v_{i_k}\}$ *be an arbitrary dominating set in* $\mathbb{H}$. *There exists DAG* $\mathbb{G}$ *modeling* $\mathbb{P}_\mathbb{H}$, *in which the set of neighbors of* b *is* $\{a_{i_1}, ..., a_{i_k}\}$.

Proof. From every vertex in the set $\{a_{i_1}, ..., a_{i_k}\}$ we draw edge to every vertex not belonging to this set. All not connected yet pairs of vertices from the set $\{a_1, ..., a_n\}$ we connect with edge in such way, that whole graph remains DAG.

Below we prove that this is desired graph $\mathbb{G}$ – we show that it models $\mathbb{P}_\mathbb{H}$. In order to do this, according to Theorem 2, it is sufficient to check that $\mathbb{P}_\mathbb{H}$ satisfies Markov conditions $MC(\mathbb{G})$. So we simply check now these conditions.

Thanks to Lemma 2 we know that $\pi(b) = \{a_{i_1}, ..., a_{i_k}\}$ is a Markov blanket of variable b, so in particular b is independent from its non-descendants excluding $\pi(b)$ given $\pi(b)$ – that is Markov condition for b is satisfied.

For variables from the set $\{a_{i_1}, ..., a_{i_k}\}$ Markov condition is satisfied in trivial way, because for each variable from this set set of its non-descendants excluding parents is empty (these variables are connected with all other in $\mathbb{G}$).

For each remaining variable a_l we have that there is only one non-descendant of a_l which is not a parent of a_l – vertex b. So, in order to check Markov condition for a_l we need to prove independence of a_l and b given $\pi(a_l)$. We obtain it as follows. We have already checked Markov condition for b, that is we know that b is independent from its non-descendants excluding $\pi(b)$ given $\pi(b)$. But because b does not have descendants, we automatically obtain that b is independent from all other than $\pi(b)$ variables given $\pi(b)$. This in particular means, that a_l and b are independent given $\pi(a_l)$ - because $\pi(b) \subseteq \pi(a_l)$. □

For information system $\mathbb{A} = (U, A)$ by $\mathbb{A}[V, B]$ for $V \subseteq U$ and $B \subseteq A$ we will denote $\mathbb{A}$ limited to the set of objects V and the set of attributes B. We say that $D \subseteq A$ dominates in this system, if for every object having all attributes from D equal to 0 all remaining attributes are also equal to 0. We say that $D \subseteq A$ dominates $a \in A$, if D dominates in $\mathbb{A}[U, D \cup \{a\}]$. We say that $D \subseteq A$ dominates $a \in A$ on a subset of objects $V \subseteq U$, if D dominates in $\mathbb{A}[V, D \cup \{a\}]$.

Lemma 4. *Let $\mathbb{G}$ be an arbitrary DAG modeling $\mathbb{P}_\mathbb{H}$. Let $\{a_{i_1}, ..., a_{i_k}\}$ be the set of neighbors of b in $\mathbb{G}$. Then subset $\{v_{i_1}, ..., v_{i_k}\}$ is a dominating set in $\mathbb{H}$.*

Proof. Graph $\mathbb{G}$ is DAG, so there exists ancestral order of its vertices, such that every variable has all its parents before itself in the order. Let this order be $a_{j_1}, ..., a_{j_t}, b, a_{j_{t+1}}, ..., a_{j_n}$. In particular, before vertex b there are placed all its parents, while after b – all its children. We will first consider the case, when vertex b has at least one parent and at least one child.

The first observation is that $\pi(b)$ dominates in $\mathbb{A}_\mathbb{H}[\{u_1, ..., u_n\}, \{a_{j_1}, ..., a_{j_t}\}]$. The reason is following. According to the Markov condition b is d-separated in $\mathbb{G}$ from variables $\{a_{j_1}, ..., a_{j_t}\}$ excluding $\pi(b)$ by $\pi(b)$, so the corresponding conditional independence holds for $\mathbb{P}_\mathbb{H}$. Assume that $\pi(b)$ does not dominate in $\mathbb{A}_\mathbb{H}[\{u_1, ..., u_n\}, \{a_{j_1}, ..., a_{j_t}\}]$. This would mean that there is such object u_p in $\mathbb{A}_\mathbb{H}$ for which all attributes corresponding to $\pi(b)$ in $\mathbb{G}$ are equal to 0 and some other attribute $a_l \in \{a_{j_1}, ..., a_{j_t}\}$ is equal to 1. Given that all parents of b are equal to 0, we have that with nonzero probability b has value 1 (see object u_{n+1}). But when we add one more condition $a_l = 1$ (there is at least one row satisfying this – u_p), then with probability 1 b has value 0. So b is not independent from variables $\{a_{j_1}, ..., a_{j_t}\}$ excluding $\pi(b)$ given $\pi(b)$ (because b is not independent from a_l given $\pi(b)$) – which is contradiction.

The second observation is that for all l satisfying $t + 1 \leq l \leq n$ we have that if a_{j_l} is not a child of b in $\mathbb{G}$, then it is dominated by its parents on the set of objects $\{u_1, ..., u_n\}$: From Markov condition for a_l and the fact that $\mathbb{G}$ models $\mathbb{P}_\mathbb{H}$ we obtain that in this distribution a_l is independent from its non-descendants excluding $\pi(a_l)$ given $\pi(a_l)$ – so in particular a_l and b are independent given $\pi(a_l)$. Thus, like in the case of the first observation, the parents of a_l dominate a_l on the set of objects $\{u_1, ..., u_n\}$.

Now we will inductively prove that for each l, $t \leq l \leq n$, the set of dominating attributes in $\mathbb{A}_\mathbb{H}[\{u_1, ..., u_n\}, \{a_{j_1}, ..., a_{j_l}\}]$ corresponds to neighbors of b in $\mathbb{G}$. For $l = n$ the fact that $\{a_{i_1}, ..., a_{i_k}\}$ dominates in $\mathbb{A}_\mathbb{H}[\{u_1, ..., u_n\}, \{a_{j_1}, ..., a_{j_n}\}] = \mathbb{A}_\mathbb{H}[\{u_1, ..., u_n\}, \{a_1, ..., a_n\}]$ means that $\{v_{i_1}, ..., v_{i_k}\}$ is a dominating set in $\mathbb{H}$.

The induction starts with the first out of the above observations. Now, let us assume that neighbors of b dominate in $\mathbb{A}_\mathbb{H}[\{u_1, ..., u_n\}, \{a_{j_1}, ..., a_{j_l}\}]$ for some l, $t \leq l < n$. If $a_{j_{l+1}}$ is a child of b, then neighbors of b dominate in $\mathbb{A}_\mathbb{H}[\{u_1, ..., u_n\}, \{a_{j_1}, ..., a_{j_{l+1}}\}]$. If $a_{j_{l+1}}$ is not a child of b, then, by following the second out of the above observations, $a_{j_{l+1}}$ is dominated by the remaining attributes in $\mathbb{A}_\mathbb{H}[\{u_1, ..., u_n\}, \{a_{j_1}, ..., a_{j_{l+1}}\}]$, which – thanks to the assumption – are dominated by neighbors of b. Therefore, neighbors of b dominate in $\mathbb{A}_\mathbb{H}[\{u_1, ..., u_n\}, \{a_{j_1}, ..., a_{j_{l+1}}\}]$.

What remains to check are special cases: When b has only parents in $\mathbb{G}$, then b can be placed at the end of ancestral order, so the first step of described above induction ends the proof. When b has only children in $\mathbb{G}$, then the only thing necessary to change is the first step of induction. We do it as follows: Vertex b has no parents, so we put it at the beginning in the ancestral order – our order is now $b, a_{j_1}, ..., a_{j_n}$. Note that b and a_{j_1} are then connected by the edge (b is the parent) because otherwise, from the Markov condition for a_{j_1}, we would have that b and a_{j_1} are independent, which is not true (in $\mathbb{P}_\mathbb{H}$ b depends on all variables). So, as for $l = 1$ in the inductive procedure, $\{a_{j_1}\}$ dominates in $\mathbb{A}_\mathbb{H}[\{u_1, ..., u_n\}, \{a_{j_1}\}]$ as the set of neighbors of b. This argumentation shows also that it is not possible for b to not have any neighbor in $\mathbb{G}$. $\qquad\square$

Now we can finally formulate and prove the following result:

Theorem 4. *The* MBN *problem is NP-hard.*

Proof. From Lemmas 3 and 4 we have, that if we take DAG of minimal number of neighbors of b from all DAGs modeling $\mathbb{P}_\mathbb{H}$, then these neighbors correspond to the minimal dominating set in $\mathbb{H}$. Lemma 1 in addition gives us guarantee that DAG of minimal number of edges from all modeling $\mathbb{P}_\mathbb{H}$ has also minimal number of neighbors of b. $\qquad\square$

5 Conclusions

We focused on showing NP-hardness of the problem of the sparsest Bayesian network with respect to the number of attributes in an input data set. We also showed in what sense the presented proof methodology remains analogous to other complexity results related to derivation of the most meaningful knowledge about data dependencies from large data sets. Although it seems to be a purely theoretical study, our result can provide a useful intuition how to construct efficient heuristic algorithms for extracting Bayesian networks from data.

One of open questions is how to approach the complexity of learning approximate Bayesian networks, which represent input data sets up to some predefined degree [14]. According to the discussion in Section 3, the search for such DAG structures seems to be a useful compromise between understanding a Bayesian network's optimality by means of scoring functions and by means of approximate consistency with the currently observed data. Some NP-hardness results are known for a relatively simpler case of searching for optimal approximate Markov boundaries or, using a different terminology, approximate decision reducts [13]. The layout of the proof in Section 4 gives some hope that a similar result could be obtained at the level of approximate Bayesian networks too.

References

1. Chickering, D.M.: Learning Bayesian Networks is NP-Complete. In: Fisher, D., Lenz, H. (eds.) Learning from Data: Artificial Intelligence and Statistics V, pp. 121–130. Springer (1996)
2. Chickering, D.M.: Optimal Structure Identification with Greedy Search. Journal of Machine Learning Research 3, 507–554 (2002)
3. Chickering, D.M., Heckerman, D., Meek, C.: Large-sample Learning of Bayesian Networks is NP-Hard. The Journal of Machine Learning Research 5, 1287–1330 (2004)
4. Cooper, F.G., Herskovits, E.: A Bayesian Method for the Induction of Probabilistic Networks from Data. Machine Learning 9, 309–347 (1992)
5. Jensen, F.V.: Bayesian Networks and Decision Graphs. Statistics for Engineering and Information Science. Springer (2001)
6. Kloesgen, W., Żytkow, J.M. (eds.): Handbook of Data Mining and Knowledge Discovery. Oxford University Press (2002)
7. Neapolitan, R.: Learning Bayesian Networks. Prentice Hall (2004)
8. Pawlak, Z., Skowron, A.: Rudiments of Rough Sets. Information Sciences 177(1), 3–27 (2007)
9. Pearl, J.: Bayesian Networks: A Model of Self-activated Memory for Evidential Reasoning. In: Proc. of the 7th Conf. of the Cognitive Science Society, pp. 329–334. University of California, Irvine (1985)
10. Pearl, J.: Probabilistic Reasoning in Intelligent Systems: Networks of Plausible Inference. Morgan Kaufmann (1988)
11. Savnik, I., Flach, P.A.: Discovery of Multivalued Dependencies from Relations. Intell. Data Anal. 4(3-4), 195–211 (2000)
12. Shenoy, P.P.: Conditional Independence in Valuation-based Systems. International Journal of Approximate Reasoning 10, 203–234 (1994)
13. Ślęzak, D.: Approximate Entropy Reducts. Fundamenta Informaticae 53(3-4), 365–390 (2002)
14. Ślęzak, D.: Degrees of Conditional (In)dependence: A Framework for Approximate Bayesian Networks and Examples Related to the Rough Set-based Feature Selection. Information Sciences 179(3), 197–209 (2009)
15. Ślęzak, D.: Rough Sets and Functional Dependencies in Data: Foundations of Association Reducts. In: Gavrilova, M.L., Tan, C.J.K., Wang, Y., Chan, K.C.C. (eds.) Transactions on Computational Science V. LNCS, vol. 5540, pp. 182–205. Springer, Heidelberg (2009)
16. Studeny, M.: Probabilistic Conditional Independence Structures. Springer (2005)

Mining Top-K Non-redundant Association Rules

Philippe Fournier-Viger[1] and Vincent S. Tseng[2]

[1] Dept. of Computer Science, University of Moncton, Canada
[2] Dept. of Computer Science and Info. Engineering, National Cheng Kung University, Taiwan
philippe.fournier-viger@umoncton.ca, tsengsm@mail.ncku.edu.tw

Abstract. Association rule mining is a fundamental data mining task. However, depending on the choice of the thresholds, current algorithms can become very slow and generate an extremely large amount of results or generate too few results, omitting valuable information. Furthermore, it is well-known that a large proportion of association rules generated are redundant. In previous works, these two problems have been addressed separately. In this paper, we address both of them at the same time by proposing an approximate algorithm named TNR for mining top-k non redundant association rules.

Keywords: association rules, top-k, non-redundant rules, algorithm.

1   Introduction

Association rule mining [1] consists of discovering associations between sets of items in transactions. It is one of the most important data mining tasks. It has been integrated in many commercial data mining software and has numerous applications [2].

The problem of *association rule mining* is stated as follows. Let $I = \{a_1, a_2, \ldots a_n\}$ be a finite set of items. A transaction database is a set of transactions $T=\{t_1, t_2 \ldots t_m\}$ where each transaction $t_j \subseteq I$ ($1 \leq j \leq m$) represents a set of items purchased by a customer at a given time. An itemset is a set of items $X \subseteq I$. The *support of an itemset* X is denoted as $sup(X)$ and is defined as the number of transactions that contain X. An association rule $X \rightarrow Y$ is a relationship between two itemsets X, Y such that $X, Y \subseteq I$ and $X \cap Y = \emptyset$. The *support of a rule* $X \rightarrow Y$ is defined as $sup(X \rightarrow Y) = sup(X \cup Y) / |T|$. The confidence of a rule $X \rightarrow Y$ is defined as $conf(X \rightarrow Y) = sup(X \cup Y) / sup(X)$. The *problem of mining association rules* [1] is to find all association rules in a database having a support no less than a user-defined threshold *minsup* and a confidence no less than a user-defined threshold *minconf*. For instance, Figure 1 shows a transaction database (left) and some association rules found for *minsup* = 0.5 and *minconf* = 0.5 (right).

Despite that much research has been done on association rule mining, an important issue that has been overlooked is how users should choose the *minsup* and *minconf* thresholds to generate a desired amount of rules [3, 4, 5, 6]. This is an important problem because in practice users have limited resources (time and storage space) for analyzing the results and thus are often only interested in discovering a certain amount of rules, and fine tuning the parameters is time-consuming. Depending on the choice of the thresholds, current algorithms can become very slow and generate an extremely large amount of results or generate none or too few results, omitting valuable

L. Chen et al. (Eds.): ISMIS 2012, LNAI 7661, pp. 31–40, 2012.

information. To address this problem, it was proposed to replace the task of *association rule mining* with the task of *top-k association rules mining*, where k is the number of association rules to be found, and is set by the user [3, 4, 5, 6]. Several top-k rule mining algorithms were proposed [3, 4, 5, 6]. However, a major problem remains with these algorithms. It is that top-k association rules often contain a large proportion of redundant rules. i.e. rules that provide information that is redundant to the user. This problem has been confirmed in our experimental study (cf. section 4). We found that up to 83 % of top-k association rules are redundant for datasets commonly used in the association rule mining literature. This means that the user has to analyze a large proportion of redundant rules.

ID	Transactions
t_1	$\{a, b, c, e, f, g\}$
t_2	$\{a, b, c, d, e, f\}$
t_3	$\{a, b, e, f\}$
t_4	$\{b, f, g\}$

$\rightarrow$

ID	Rules	Support	Confidence
r_1	$\{a\} \rightarrow \{e, f\}$	0.75	1
r_2	$\{a\} \rightarrow \{c, e, f\}$	0.5	0.6
r_3	$\{a, b\} \rightarrow \{e, f\}$	0.75	1
r_4	$\{a\} \rightarrow \{c, f\}$	0.5	0.6

Fig. 1. (a) A transaction database and (b) some association rules found

The problem of redundancy in association rule mining has been studied extensively and various definitions of redundancy have been proposed [7, 8, 9, 10]. However, it remains an open challenge to combine the idea of mining a set of non-redundant rules with the idea of top-k association rule mining, to propose an efficient algorithm to mine top-k non-redundant association rules. The benefit of such an algorithm would be to present a small set of k non-redundant rules to the user. However, devising an algorithm to mine these rules is difficult. The reason is that eliminating redundancy cannot be performed as a post-processing step after mining the top-k association rules, because it would result in less than k rules. The process of eliminating redundancy has therefore to be integrated in the mining process.

In this paper, we undertake this challenge by proposing an approximate algorithm for mining the top-k non redundant association rules that we name TNR (*Top-k Non-redundant Rules*). It is based on a recently proposed approach for generating association rules that is named "rule expansions", and adds strategies to avoid generating redundant rules. An evaluation of the algorithm with datasets commonly used in the literature shows that TNR has excellent performance and scalability.

The rest of the paper is organized as follows. Section 2 reviews related works and presents the problem definition. Section 3 presents TNR. Section 4 presents an experimental evaluation. Section 5 presents the conclusion.

2   Related Works and Problem Definition

Top-k association rule mining. Several algorithms have been proposed for top-k association rule mining [3, 4, 5]. However, most of them do not use the standard definition of an association rule. For instance, KORD [3, 4] finds rules with a single item in the consequent, whereas the algorithm of You et al. [5] mines association rules from a stream instead of a transaction database. To the best of our knowledge, only TopKRules [5] discovers top-k association rules based on the standard definition of an association rule (with multiple items, in a transaction database). TopKRules takes as parameters k and *minconf*, and it returns the k rules with the highest support that meet the *minconf* threshold. The reason why this algorithm defines the task of mining the top k rules on the support instead of the confidence is that *minsup* is much more difficult to set than *minconf* because *minsup* depends on database characteristics that are unknown to most users, whereas *minconf* represents the minimal confidence that users want in rules and is therefore generally easy to determine. TopKRules defines the problem of top-k association rule mining as follows [5]. The *problem of top-k association rule mining* is to discover a set L containing k rules in T such that for each rule $r \in L$ | $conf(r) \geq minconf$, there does not exist a rule $s \notin L$ | $conf(s) \geq minconf \wedge sup(s) > sup(r)$.

Discovering Non-redundant Association Rules. The problem of redundancy has been extensively studied in association rule mining [7, 8, 9, 10]. Researchers have proposed to mine several sets of non-redundant association rules such as the Generic Basis [7], the Informative Basis [7], the Informative and Generic Basis [8], the Minimal Generic Basis [9] and Minimum Condition Maximum Consequent Rules [10]. These rule sets can be compared based on several criteria such as their compactness, the possibility of recovering redundant rules with their properties (support and confidence), and their meaningfulness to users (see [10] for a detailed comparison). In this paper, we choose to base our work on Minimum Condition Maximum Consequent Rules (MCMR). The reason is that the most important criteria in top-k association rule mining is the meaningfulness of rules for users. MCMR meet this goal because it defines non-redundant rules as rules with a minimum antecedent and a maximum consequent [10]. In other words, MCMR are the rules that allow deriving the maximum amount of information based on the minimum amount of information. It is argued that these rules are the most meaningful for several tasks [10]. In the context of top-k association rule mining, other criteria such as the compactness and recoverability of redundant rules are not relevant because the goal of a top-k algorithm is to present a small set of k meaningful rules to the user. MCMR are defined based on the following definition of redundancy [10]. An association rule $r_a : X \rightarrow Y$ is *redundant with respect to another rule* $r_b : X_1 \rightarrow Y_1$ if and only if $conf(r_a) = conf(r_b) \wedge sup(r_a) = sup(r_b) \wedge X_1 \subseteq X \wedge Y \subseteq Y_1$. **Example.** Consider the association rules presented in the right part of Figure 1. The rule $\{a\} \rightarrow \{c, f\}$ is redundant with respect to $\{a\} \rightarrow \{c, e, f\}$. Moreover, the rule $\{a, b\} \rightarrow \{e, f\}$ is redundant with respect to $\{a\} \rightarrow \{e, f\}$.

Problem Definition. Based on the previous definition of redundancy and the definition of top-k association rule mining, we define the problem of *top-k non redundant association mining* as follows. The *problem of mining top-k non-redundant association rules* is to discover a set L containing k association rules in a transaction database. For each rule $r_a \in L$ | $conf(r_a) \geq minconf$, there does not exist a rule $r_b \notin L$ | $conf(r_b) \geq minconf \wedge sup(r_b) > sup(r_a)$, otherwise r_b is redundant with respect to r_a. Moreover, $\nexists r_c, r_d \in L$ such that r_c is redundant with respect to r_d.

3   The TNR Algorithm

To address the problem of top-k non-redundant rule mining, we propose an algorithm named TNR. It is based on the same depth-first search procedure as TopKRules. The difference between TNR and TopKRules lies in how to avoid generating redundant rules. The next subsection briefly explains the search procedure of TopKRules.

3.1   The Search Procedure

To explain the search procedure, we introduce a few definitions. A rule $X \rightarrow Y$ is of *size* $p*q$ if $|X| = p$ and $|Y| = q$. For example, the size of $\{a\} \rightarrow \{e, f\}$ is $1*2$. Moreover, we say that a rule of size $p*q$ is *larger* than a rule of size $r*s$ if $p > r$ and $q \geq s$, or if $p \geq r$ and $q > s$. An association rule r is *valid* if $sup(r) \geq minsup$ and $conf(r) \geq minconf$.

The search procedure takes as parameter a transaction database, an integer k and the *minconf* threshold. The search procedure first sets an internal *minsup* variable to 0 to ensure that all the top-k rules are found. Then, the procedure starts searching for rules. As soon as a rule is found, it is added to a list of rules L ordered by the support. The list is used to maintain the top-k rules found until now and all the rules that have the same support. Once k valid rules are found in L, the internal *minsup* variable is raised to the support of the rule with the lowest support in L. Raising the *minsup* value is used to prune the search space when searching for more rules. Thereafter, each time that a valid rule is found, the rule is inserted in L, the rules in L not respecting *minsup* anymore are removed from L, and *minsup* is raised to the support of the rule with the lowest support in L. The algorithm continues searching for more rules until no rule are found. This means that it has found the top-k rules in L. The top-k rules are the k rules with the highest support in L.

To search for rules, the search procedure first scans the database to identify single items that appear in at least *minsup* transactions. It uses these items to generate rules of size $1*1$ (containing a single item in the antecedent and a single item in the consequent). Then, each rule is recursively grown by adding items to its antecedent or consequent (a depth-first search). To determine the items that should be added to a rule, the search procedure scans the transactions containing the rule to find single items that could expand its antecedent or consequent. The two processes for expanding rules are named *left expansion* and *right expansion*. These processes are applied recursively to explore the search space of association rules. Left and right expansions are formally defined as follows. A *left expansion* is the process of adding an item i to the left side of a rule $X \rightarrow Y$ to obtain a larger rule $X \cup \{i\} \rightarrow Y$. A *right expansion* is the process of adding an item i to the right side of a rule $X \rightarrow Y$ to obtain a larger rule $X \rightarrow Y \cup \{i\}$.

The search procedure described above is correct and complete for mining top-k association rules [5]. It is very efficient because the internal *minsup* variable is raised during the search. This allows pruning large part of the search space instead of generating all association rules. This pruning is possible because the support is monotonic with respect to left and right expansions (see [5] for details).

3.2   Adapting the Search Procedure to Find Top-*k* Non-redundant Rules

We now explain how we have adapted the search procedure to design an efficient top-k non-redundant rule mining algorithm. These modifications are based on the following observation.

Property 3. During the search, if only non-redundant rules are added to L, then L will contain the top-k non-redundant rules when the search procedure terminates. **Rationale.** The search procedure is correct and complete for mining the top-k association rules [5]. If only non-redundant rules are added to L instead of both redundant and non-redundant rules, it follows that the result will be the top-k non-redundant rules instead of the top-k association rules.

Based on this observation, we have aimed at modifying the search procedure to ensure that only non-redundant rules are added to L. This means that we need to make sure that every generated rule r_a is added to L only if $sup(r_a) \geq minsup$ and r_a is not redundant with respect to another rule. To determine if r_a is redundant with respect to another rule, there are two cases to consider.

The *first case* is that r_a is redundant with respect to a rule r_b that was generated before r_a. By the definition of redundancy (cf. Definition 2), if r_a is redundant with respect to r_b, then $sup(r_b) = sup(r_a)$. Because $sup(r_a) \geq minsup$, it follows that $sup(r_b) \geq minsup$, and that $r_b \in L$. Therefore, the first case can be detected by implementing the following strategy.

Strategy 1. For each rule r_a that is generated such that $sup(r_a) \geq minsup$, if $\exists \; r_b \in L \mid sup(r_b) = sup(r_a)$ and r_a is redundant with respect to r_b, then r_a is not added to L. Otherwise, r_a is added to L.

The *second case* is that r_a is redundant with respect to a rule r_b that has not yet been generated. There are two ways that we could try to detect this case. The first way is to postpone the decision of adding r_a to L until r_b is generated. However, this would not work because it is not known beforehand if a rule r_b will make r_a redundant and r_b could appear much later after r_a. The second way is to scan transactions to determine if item(s) could be added to r_a to generate a rule r_b such tat r_a redundant is redundant with respect to r_b. However, this approach would also be inefficient because the confidence is non-monotonic with respect to left/right expansions [5]. This mean that it is possible that r_b may contain several more items than r_a. For this reason, it would be too costly to test all the possibilities of adding items to r_a to detect the second case.

Because the second case cannot be checked efficiently, our solution is to propose an approximate approach that is efficient and to prove that this approach can generate exact results if certain conditions are met. The idea of this approach is the following. Each rule r_a that satisfy the requirements of Strategy 1 is added to L without checking the second case. Then, eventually, if a rule r_b is generated such that the rule r_a is still in L and that r_a is redundant with respect to r_b, then r_a is removed from L. This idea is formalized as the following strategy.

Strategy 2. For each rule r_b that is generated such that $sup(r_b) \geq minsup$, if $\exists \; r_a \in L \mid sup(r_b) = sup(r_a)$ and r_a is redundant with respect to r_b, then r_a is removed from L.

By incorporating Strategy 2, the algorithm becomes approximate. The reason is that each rule r_a that is removed by Strategy 2 previously occupied a place in the set L. By its presence in L, the rule r_a may have forced raising the internal *minsup* variable. If that happened, then the algorithm may have missed some rules that have a support lower than r_a but are non-redundant.

Given that the algorithm is approximate, it would be desirable to modify it to be able to increase the likelihood that the result is exact. To achieve this, we propose to add a parameter that we name Δ that increase by Δ the number of rules k that is necessary to raise the internal *minsup* variable. For example, if the user sets $k = 1000$ and $\Delta = 100$, it will now be required to have $k + \Delta = 1100$ rules in L to raise the internal *minsup* variable instead of just $k{=}1000$. This means that up to 100 redundant rules can be at the same time in L and the result will still be exact. This latter observation is formalized by the following property.

Property 4. If the number of redundant rules in L is never more than Δ rules, then the algorithm result is exact and the k rules in L having the highest support will be the top-k non redundant rules. **Rationale**. As previously explained, the danger is that too many redundant rules are in L such that they would force to raise *minsup* and prune part of the search space containing top-k non-redundant rules. If there is no more than Δ redundant rules at the same time in L and that $k + \Delta$ are needed to raise *minsup*, then redundant rules cannot force to raise *minsup*.

The previous property proposes a condition under which Strategy 2 is guaranteed to generate an exact result. Based on this property, an important question is "Would it be efficient to integrate a check for this condition in the algorithm?". The answer is that it would be too costly to verify that no groups of more than Δ redundant rules are present at the same time in L. The reason is that rules are only known to be redundant when they are removed by Strategy 2. Therefore, checking Property 4 would be very expensive to perform. For this reason, we choose to utilize the following weaker version of Property 4 in our implementation.

Property 5. If the number of redundant rules removed by Strategy 2 during the execution of the algorithm is less or equal to Δ, then the final result is exact and the first k rules of L will be the top-k non redundant rules. **Rationale**. It can be easily seen that if Property 5 is met, Property 4 is also met.

Property 5 can be easily incorporated in the algorithm. To implement this functionality, we have added a counter that is incremented by 1 after every rule removal from L by Strategy 2. Then, when the algorithm terminates, the counter is compared with Δ. If the counter value is lower or equal to Δ, the user is informed that the result is guaranteed to be exact. Otherwise, the user is informed that the result may not be exact. In this case, the user has the option to rerun the algorithm with a higher Δ value. In the experimental study presented in section 4 of this paper, we will address the question of how to select Δ.

The previous paragraphs have presented the main idea of the TNR algorithm. Due to space limitation, we do not provide the pseudo-code of TNR. But the Java source code of our implementation can be downloaded freely from http://www.philippe-fournier-viger.com/spmf/.

Note that in our implementation, we have added a few optimizations that are used in TopKRules [5] and that are compatible with TNR. The first optimization is to try to generate the most promising rules first when exploring the search space of association rules. This is because if rules with high support are found earlier, the algorithm can raise its internal *minsup* variable faster to prune the search space. To perform this, an internal variable R is added to store all the rules that can be expanded to have a chance of finding more valid rules. This set is then used to determine the rules that are the most likely to produce valid rules with a high support to raise *minsup* more quickly and prune a larger part of the search space [5]. The second optimization is to use bit vectors as data structure for representing the set of transactions that contains each rule (*tidsets*) [11]. The third optimization is to implement L and R with data structures supporting efficient insertion, deletion and finding the smallest element and maximum element. In our implementation, we used a Red-black tree for L and R.

4   Experimental Evaluation

We have carried several experiments to assess the performance of TNR and to compare its performance with TopKRules under different scenarios. For these experiments, we have implemented TNR in Java. For TopKrules, we have obtained the Java implementation from their authors. All experiments were performed on a computer with a Core i5 processor running Windows 7 and 2 GB of free RAM. Experiments were carried on four real-life datasets commonly used in the association rule mining literature, namely *Chess, Connect, Mushrooms* and *Pumsb* (available at: http://fimi.ua.ac.be/data/). The datasets' characteristics are summarized in Table 1.

Experiment 1: What is the percentage of redundant rules? The goal of the first experiment was to assess the percentage of rules returned by TopKRules that are redundant to determine how important the problem of redundancy is. For this experiment, we ran TopKRules on the four datasets with *minconf* = 0.8 and k = 2000, and then examined the rules returned by the algorithm. We chose *minconf* = 0.8 and k = 2000 because these values are plausible values that a user could choose. The results for *Chess, Connect, Mushrooms* and *Pumsb* are that respectively, 13.8 %, 25.9 %, 82.6 % and 24.4 % of the rules returned by TopKRules are redundant. These results indicate that eliminating redundancy in top-k association rule mining is a major problem.

Table 1. Datasets' Characteristics

Datasets	Number of transactions	Number of distinct items	Average transaction size
Chess	3,196	75	37
Connect	67,557	129	43
Mushrooms	8,416	128	23
Pumsb	49,046	7,116	74

Experiment 2: How many rules are discarded by Strategy 1 and Strategy 2 for Δ = 0? The second experiment's goal was to determine how many rules were discarded by Strategy 1 and Strategy 2 for each dataset. To perform this study, we set *minconf* = 0.8, k = 2000 and Δ = 0. We then recorded the number of rules discarded. Results are shown in Table 2. From this experiment, we can see that the number of discarded rules is high for all datasets and that it is especially high for dense datasets (e.g. *Mushrooms*) because they contains more redundant rules compared to sparse datasets (e.g. *Pumsb*).

Experiment 3: What if we use the Δ parameter? The next experiment consisted of using the Δ parameter to see if an exact result could be guaranteed by using Property 5. For this experiment, we used k = 2000 and *minconf* = 0.8. We then set Δ to values slightly larger than the number of rules discarded by Strategy 2 in Experiment 2.

For example, for the *Pumsb* dataset, we set $\Delta = 4000$. The total runtime was 501 s and the maximum memory usage was around 1.3 GB. The number of rules eliminated by Strategy 1 and Strategy 2 was respectively 3454 and 16,066. Because these values are larger than Δ, the result could not be guaranteed to be exact. Moreover, the execution time and memory requirement significantly increased when setting $\Delta = 4000$ (with $\Delta = 0$, the runtime is 125 s and the maximum memory usage is 576 MB). Furthermore, we tried to continue raising Δ and still got similar results.

We did similar experiments with *Chess*, *Connect*, *Mushrooms* and observed the same phenomenon. Our conclusion from this experiment is that in practice using the Δ parameter does not help to guarantee an exact result. This means that although the algorithm is guaranteed to find k rules that are non-redundant, there rules are not guaranteed to be the *top-k* non-redundant rules.

Table 2. Rules discarded by each strategy for *minconf* = 0.8, *k*=2000 and $\Delta = 0$

Dataset	# rules discarded by Strategy 1	# rules discarded by Strategy 2
Chess	961	10454
Connect	2732	15275
Mushrooms	39848	38627
Pumsb	803	3629

Experiment 4: Performance comparison with TopKRules. The next experiment consisted of comparing the performance of TNR with TopKRules. The parameter *minconf* and *k* were set to 0.8 and 2000 respectively. The execution times and maximum memory usage of both algorithms are shown in Table 3. The results show that there is a significant additional cost for using TNR. The reason is that checking Strategy 1 and Strategy 2 is costly. Moreover, because a large amount of rules are discarded as shown in the second experiment, the algorithm needs to generate much more rules before it can terminate.

Table 3. Performance comparison

Datasets	Runtime (s)		Maximum Memory Usage (MB)	
	TNR	TopKRules	TNR	TopKRules
Chess	8	1.49	269	72.12
Connect	283	25.51	699	403.38
Mushrooms	105	3.46	684	255
Pumsb	125	46.39	576	535

Experiment 5: Influence of the number of transactions. Next, we ran TNR on the datasets while varying the number of transactions in each dataset. We used *k*=500, *minconf*=0.8. We varied the database size by using 70%, 85 % and 100 % of the transactions in each dataset. Results are shown in Figure 2. Globally we found that for all datasets, the execution time and memory usage increased more or less linearly for TNR. This shows that TNR has good scalability.

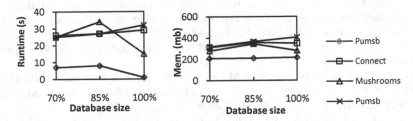

Fig. 2. Influence of the number of transactions

Discussion. Our conclusion from these experiments is that TNR is more costly than TopKRules. But it provides the benefit of eliminating redundancy.

5   Conclusion

Two important problems with classical association rule mining algorithm are that (1) it is usually difficult and time-consuming to select the parameters to generate a desired amount of rules and (2) there can a large amount of redundancy in the results. Previously, these two problems have been addressed separately. In this paper, we have addressed them together by proposing an approximate algorithm named TNR for mining the top-*k* non-redundant association rules.   The algorithm is said to be approximate because it is guaranteed to find non-redundant rules. But the rules found may not be the *top-k* non redundant rules. We have compared the performance of TNR with TopKRules and found that TNR is more costly than TopKRules. However, it provides the benefit of eliminating a great deal of redundancy. Source code of TNR and TopKRules can be downloaded at http://www.philippe-fournier-viger.com/spmf/ as part of the SPMF data mining platform.

References

[1] Agrawal, R., Imielminski, T., Swami, A.: Mining Association Rules Between Sets of Items in Large Databases. In: Proc. ACM Intern. Conf. on Management of Data, pp. 207–216. ACM Press (June 1993)

[2] Han, J., Kamber, M.: Data Mining: Concepts and Techniques, 2nd edn. Morgan Kaufmann Publ., San Francisco (2006)

[3] Fournier-Viger, P., Wu, C.-W., Tseng, V.S.: Mining Top-K Association Rules. In: Kosseim, L., Inkpen, D. (eds.) Canadian AI 2012. LNCS, vol. 7310, pp. 61–73. Springer, Heidelberg (2012)

[4] Webb, G.I., Zhang, S.: k-Optimal-Rule-Discovery Data Mining and Knowledge Discovery 10(1), 39–79 (2005)

[5] Webb, G.I.: Filtered top-k association discovery. WIREs Data Mining and Knowledge Discovery 1, 183–192 (2011)

[6] You, Y., Zhang, J., Yang, Z., Liu, G.: Mining Top-k Fault Tolerant Association Rules by Redundant Pattern Disambiguation in Data Streams. In: Proc. 2010 Intern. Conf. Intelligent Computing and Cognitive Informatics, pp. 470–473. IEEE Press (March 2010)

[7] Bastide, Y., Pasquier, N., Taouil, R., Stumme, G., Lakhal, L.: Mining Minimal Non-redundant Association Rules Using Frequent Closed Itemsets. In: Palamidessi, C., Moniz Pereira, L., Lloyd, J.W., Dahl, V., Furbach, U., Kerber, M., Lau, K.-K., Sagiv, Y., Stuckey, P.J. (eds.) CL 2000. LNCS (LNAI), vol. 1861, pp. 972–986. Springer, Heidelberg (2000)

[8] Gasmi, G., Yahia, S.B., Nguifo, E.M., Slimani, Y.: $\mathcal{IGB}$: A New Informative Generic Base of Association Rules. In: Ho, T.-B., Cheung, D., Liu, H. (eds.) PAKDD 2005. LNCS (LNAI), vol. 3518, pp. 81–90. Springer, Heidelberg (2005)

[9] Cherif, C.L., Bellegua, W., Ben Yahia, S., Guesmi, G.: VIE_MGB: A Visual Interactive Exploration of Minimal Generic Basis of Association Rules. In: Proc. of the Intern. Conf. on Concept Lattices and Application (CLA 2005), pp. 179–196 (2005)

[10] Kryszkiewicz, M.: Representative Association Rules and Minimum Condition Maximum Consequence Association Rules

[11] Lucchese, C., Orlando, S., Perego, R.: Fast and Memory Efficient Mining of Frequent Closed Itemsets. IEEE Trans. Knowl. and Data Eng. 18(1), 21–36 (2006)

ABML Knowledge Refinement Loop: A Case Study

Matej Guid[1], Martin Možina[1], Vida Groznik[1], Dejan Georgiev[2],
Aleksander Sadikov[1], Zvezdan Pirtošek[2], and Ivan Bratko[1]

[1] Faculty of Computer and Information Science, University of Ljubljana, Slovenia
[2] Department of Neurology, University Medical Centre Ljubljana, Slovenia

Abstract. Argument Based Machine Learning (ABML) was recently demonstrated to offer significant benefits for knowledge elicitation. In knowledge acquisition, ABML is used by a domain expert in the so-called ABML *knowledge refinement loop*. This draws the expert's attention to the most critical parts of the current knowledge base, and helps the expert to argue about critical concrete cases in terms of the expert's own understanding of such cases. Knowledge elicited through ABML refinement loop is therefore more consistent with expert's knowledge and thus leads to more comprehensible models in comparison with other ways of knowledge acquisition with machine learning from examples. Whereas the ABML learning method has been described elsewhere, in this paper we concentrate on detailed mechanisms of the ABML knowledge refinement loop. We illustrate these mechanisms with examples from a case study in the acquisition of neurological knowledge, and provide quantitative results that demonstrate how the model evolving through the ABML loop becomes increasingly more consistent with the expert's knowledge during the process.

1 Introduction

Machine learning has long ago been proposed as a way of addressing the problem of knowledge acquisition [1]. While it was shown that it can be successful in building knowledge bases [2], the major problem with this approach is that automatically induced models rarely conform to the way an expert wants the knowledge organized and expressed. Models that are incomprehensible have less chance to be trusted by experts and users alike [3]. In striving for better accuracy, modern trends in machine learning pay only limited attention to the comprehensibility and the intuitiveness of prediction models [4].

A common view is that a combination of a domain expert and machine learning would yield the best results [5]. Argumentation Based Machine Learning (ABML) [6] naturally fuses argumentation and machine learning. One of the advantages over traditional machine learning methods is better comprehensibility of the obtained models. Improvement in comprehensibility is especially important in the light of knowledge acquisition. Through argumentation, ABML enables the expert to articulate his or her knowledge easily and in a very natural way. Moreover, it prompts the expert to share exactly the knowledge that is most useful for the machine to learn, thus significantly saving the time of the expert.

ABML is comprised of two main parts: the modified machine learning algorithm that can handle and use arguments, and the iterative ABML loop that manages the

L. Chen et al. (Eds.): ISMIS 2012, LNAI 7661, pp. 41–50, 2012.

interaction between the expert(s) and the machine. The algorithm usually takes all the credit for successful results, however, the iterative loop is at least as important. After all, the loop is what the expert and the knowledge engineer actually use during the process of knowledge elicitation, while the inner workings of the algorithm remain hidden in the background. Therefore, in this paper, the focus is (solely) on the loop part of the ABML process.

The paper represents a continuation of the work presented in [7] and [8]. Here we provide a step-by-step presentation of the knowledge elicitation process with ABML. Through a case study of acquiring knowledge for a neurological decision support system, we analyze and clearly demonstrate the benefits of this particular style of the interaction between the experts and the machine learning algorithm. Along the way, the reader is alerted to some typical and atypical situations and is shown how to deal with them. The paper illustrates what is expected from the experts and knowledge engineers alike, demonstrates the required level of their involvement, and conveys the natural feel of the human-computer interaction. We also demonstrate quantitatively how the model obtained with the ABML knowledge elicitation process becomes more and more consistent with the expert's knowledge.

The organization of the paper is as follows. Chapter 2 describes the case study domain and experimental setup, and Chapter 3 shortly describes the ABML algorithm, focusing mainly on its modifications in view of the current application. The ABML iterative process, the main part of the paper, is in Chapter 4. We finish with an evaluation of the knowledge elicitation process and conclusions.

2 Domain Description and Experimental Setup

We are developing a neurological decision support system (DSS) to help the neurologists differentiate between three types of tremors: essential, Parkinsonian, and mixed tremor (comorbidity, see [8] for more detail). The system is intended to act as a second opinion and a teaching tool for the neurologists. Although several sets of guidelines for diagnosing both essential and Parkinsonian tremor do exist [9], none of them enjoys general consensus in the neurological community.

The data set consisted of 114 patients. These were divided into a learning set with 47 examples and a test set with 67 examples. The class distribution was: 50 patients diagnosed with essential tremor (ET), 23 patients with Parkinsonian tremor (PT), and 41 patients with a mixed-type tremor (MT). The patients were described by 45 attributes.

According to the domain experts, some of the characteristics reflected in these attributes speak in favour of a particular tremor type as follows.

Essential tremor is characterized by *postural tremor, kinetic tremor, harmonics, essential spiral drawings*, positive family *anamnesis* etc.

Parkinsonian tremor is characterized by *resting tremor, bradykinesia, rigidity, Parkinsonian spiral drawings* etc.

Mixed tremor implies presence of both essential tremor and Parkinsonian tremor.

However, according to domain experts it is difficult to combine these characteristics into sensible rules for successful diagnosis.

3   Argument Based Machine Learning (ABML)

Argument Based Machine Learning (ABML)[6] is machine learning extended with concepts from argumentation. In ABML, arguments are used as means for experts to elicit some of their knowledge through explanations of the learning examples. The experts need to focus on one specific case at the time only and provide knowledge that seems relevant for this case. We use the ABCN2 [6] method, an argument based extension of the well-known CN2 method, that learns a set of unordered probabilistic rules from examples with attached arguments, also called *argumented examples*.

The problem domain described in this paper contains a class variable with three values. According to the domain expert opinion, it was appropriate to translate our three-class problem into two binary-class problems solved by two binary classifiers. The first binary classifier distinguishes between ET and non-ET, the second between PT and non-PT. A new case is then probabilistically classified roughly as follows. The first classifier assigns probability $p(ET)$ to class ET, the second $p(PT)$ to class PT. The predicted probability of MT is then $p(MT) = 1 - p(ET) - p(PT)$. ET and MT are merged into EMT class, while PT and MT are merged into PMT class. The two binary classifiers are independent, so it may happen that $p(MT) < 0$. In such cases the three probabilities are adjusted to satisfy the formal properties of probabilities (see [8] for details).

3.1   ABML Knowledge Refinement Loop

The ABML knowledge refinement loop consists of the following steps:

Step 1: Learn a hypothesis with ABCN2 using given data.

Step 2: Find the "most critical" example and present it to the expert. If a critical example can not be found, stop the procedure.

Step 3: Expert explains the example; the explanation is encoded in arguments and attached to the learning example.

Step 4: Return to step 1.

In the sequel, we explain (1) how we select critical examples, and (2) how we obtain all necessary information for the chosen example.

Identifying Critical Examples. A critical example is an example the current hypothesis can not explain very well. As our method gives probabilistic class prediction, we first identify the most problematic example as one with highest probabilistic error. To estimate the probabilistic error we used a k-fold cross-validation repeated n times (e.g. $n = 4, k = 5$), so that each example is tested n times. The critical example is thus selected according to the following two rules.

1. If the problematic example is from class MT, it becomes the critical example.
2. Otherwise, the method will seek out which of the rules is the culprit for example's misclassification. As the problematic rule is likely to be bad, since it covers our problematic example, the critical example will become an example from PT or MT class (or ET or MT) covered by the problematic rule. Then, the expert will be asked to explain what are the reasons for the patient's diagnosis. Domain expert's explanations should result in replacing the problematic rule with a better one for the PMT (or EMT) class, which will not cover the problematic example.

Are the expert's arguments good enough or should they be improved? Here we describe in details the third step of the above algorithm:

Step 3a: Explaining a critical example. If the example is from the MT class, the expert can be asked to explain its Parkinsonian and essential signs (which happens when the problematic example is from MT) or to explain only one of the diseases. In the other two cases (ET or PT), the expert always explains only signs relevant to the example's class. The expert then articulates a set of reasons confirming the example's class value. The provided argument should contain a minimal number of reasons to avoid overspecified arguments.

Step 3b: Adding arguments to an example. The argument is given in natural language and needs to be translated into domain description language (attributes). If the argument mentions concepts currently not present in the domain, these concepts need to be included in the domain (as new attributes) before the argument can be added to the example.

Step 3c: Discovering counter examples. Counter examples are used to spot if an argument is sufficient to successfully explain the critical example or not. If not, ABCN2 will select a counter example. A counter example has the opposite class of the critical example, however it is covered by the rule induced from the given arguments.

Step 3d: Improving arguments with counter examples. The expert needs to revise his initial argument with respect to the counter example.

Step 3e: Return to step 3c if counter example found.

4   Knowledge Elicitation Process for Differentiating Tremors

In this section, we analyze the complete knowledge elicitation process for differentiating between essential and Parkinsonian tremors. We identify the main effects of each iteration on the process.

Iteration 1. Example E.65 (classified as MT) was the first critical example selected by our algorithm. The expert was asked to describe which features are in favor of ET *and* which features are in favor of PT. He explained that the presence of harmonics speaks in favor of ET, while the presence of bradykinesia speaks in favor of PT. Both features were selected as the most influential ones.

The presence of harmonics was represented by four attributes in the data set, each one with possible values of *true* and *false*. The expert explained that just one of these feature values being *true* already suffices to decide in favor of ET. Similarly, the presence of bradykinesia was indicated with two attributes, one for the left side and one for the right side, with possible values in range from 0 (not indicated) to 5 (high). The expert explained that the side does not play any particular role for differentiating between ET and PT, and that any value higher than zero already speaks in favor of PT.

The expert's explanation served the knowledge engineer to induce two new attributes: (1) HARMONICS, with possible values *true* (indicating the presence of harmonics) and *false*, and (2) BRADYKINESIA, with possible values *true* (bradykinesia is present on the left side *or* on the right side) and *false* (bradykinesia was not indicated on *either* side).

At the same time the original six attributes (indicating harmonics and bradykinesia) were excluded from the domain, since it is their *combination* (reflected in the expert's argument) that provides relevant information according to the expert.

Based on the expert's explanation, the reasons (1) "HARMONICS is *true*" and (2) "BRADYKINESIA is *true*" were added as the arguments for ET and PT, respectively, to the critical example E.65.

The method selected E.67 as a counter example for the expert's argument in favor of ET, and E.12 as the counter example for his argument in favor of PT. The expert was now asked to compare the counter example E.67 with the critical example E.65, and to explain what is the most important feature in favor of ET in E.65 that does not apply to E.67. Similarly, he was asked to explain what is the most important feature in favor of ET in E.65 that does not apply to E.12.

It turned out that both counter examples occurred as consequences of the following errors in the data set. In case of E.67, one of the original four attributes for harmonics was set to *true*, although the actual value was discovered to be *false* upon examination. Consequently, the value of the newly added attribute HARMONICS in E.67 had to be corrected from *true* to *false*. In case of E.12, upon the examination of the feature values, the expert realized that some strong arguments in favor of PT were overlooked at the time of diagnosis. After careful deliberation, the class of E.12 was modified from ET to MT by the expert.

In this case, the method actually helped to discover errors in the data set. Improving the arguments turned out to be unnecessary: the correction of the aforementioned errors in the data set resulted in no further counter examples. Thus, Iteration 1 was concluded.

Iteration 2. Upon entering into Iteration 2, E.61 (MT) was selected as a critical example. The expert was asked to describe which features are in favor of ET. He gave two features as an explanation: the presence of postural tremor and the presence of resting tremor. Similarly as in Iteration 1, these two features were each represented by two attributes. Again, the expert explained that neither the side nor the magnitude of a non-zero value play any particular role in differentiating between ET and PT.

The expert's explanation served to induce two derived attributes: (1) POSTURAL, with possible values *true* (indicating the presence of postural tremor) and *false*, and (2) RESTING, with possible values *true* (indicating the presence of resting tremor) and *false*. The original four attributes (indicating the presence of postural tremor and resting tremor) were excluded from the domain. The reason "POSTURAL is *true* and RESTING is *true*" was added as the argument for ET to the critical example E.61.

The expert's argument did not prove to be sufficient to produce a rule with pure distribution: the method selected E.32 as the counter example. The expert was asked to compare the counter example E.32 with the critical example E.61.

The expert spotted the presence of bradykinesia in E.32 as the most important difference, and thus extended his argument to "POSTURAL is *true* and RESTING is *true* and BRADYKINESIA is *false*."

A new counter example was found by the method: E.51. It turned out that this counter example occurred as a consequence of another misdiagnosis. After reviewing the feature values describing the patient's conditions, the expert modified the class of E.51 from MT to PT. No further counter examples were found.

Iteration 3. Critical example E.55 (MT) was presented to the expert. The expert gave two features in favor of ET: the presence of postural tremor and the presence of kinetic tremor. He also gave two features in favor of PT: the presence of bradykinesia and the presence of rigidity in upper extremities. Kinetic tremor and rigidity were each represented by two attributes, similarly as in some aforementioned cases.

The knowledge engineer induced two attributes: (1) KINETIC, with possible values *true* (indicating the presence of kinetic tremor) and *false*, and (2) RIGIDITY, with possible values *true* (indicating rigidity in upper extremities) and *false*. The original four attributes (indicating the presence of kinetic tremor and rigidity in upper extremities) were excluded from the domain.

Expert's explanation lead to (1) "POSTURAL is *true* and KINETIC is *true*," and (2) "BRADYKINESIA is *true* and RIGIDITY is *true*" as the arguments for ET and PT, respectively, to the critical example E.55.

The method selected E.63 as a counter example for the argument in favor of ET. The expert explained what is the most important feature in favor of ET in the critical example E.55 that does not apply to E.63. He contemplated that ET typically occurs much earlier than PT, and advocated that if tremor occurs before the age of 50 (as in E.55), it is usually ET. There were no counter examples for the argument in favor of PT.

The knowledge engineer realized that there was no suitable attribute in the domain that would express exactly what the expert had just explained. There were similar attributes AGE (indicating the age of the patient) and TREMOR.PERIOD (indicating the number of years since the tremor was diagnosed). They were used to construct a new attribute TREMOR.START, indicating the patient's age when the tremor was diagnosed.

The argument in favor of ET was extended to "POSTURAL is *true* and KINETIC is *true* and TREMOR.START < 50." No more counter examples were found.

Iteration 4. Critical example E.51 (MT) was presented to the expert. The expert explained that positive anamnesis and postural tremor are in favor of ET, while qualitative assessment (given by the neurologist at the time of the examination of a patient) is in favor of PT. The arguments to E.51 therefore became (1) "ANAMNESIS is *positive* and POSTURAL is *true*" for ET, and (2) "QUALITATIVE.ASSESSMENT is *PT*" for PT.

With the help of the counter example E.62 the argument for ET was extended to "ANAMNESIS is *positive* and POSTURAL is *true* and BRADYKINESIA is *false*." Another counter example E.21, for argument in favor of PT, turned out to be misdiagnosed. The class of E.21 was modified from ET to MT. There were no further counter examples.

Iteration 5. Critical example E.42 (MT) was presented to the expert. The qualitative assessment by the neurologist was given as sufficient argument in favor of ET. The assessment of free-hand spiral drawings were in favor of PT, as can be seen from the following explanation by the expert: "The assessment of the free-hand spiral in some of the four observations in the original data is Parkinsonian, and none of them indicative of essential tremor."

This explanation lead to a new attribute SPIRO.FREE.PT.ONLY. By analogy, another attribute, SPIRO.FREE.ET.ONLY, was introduced, while the four original attributes were excluded from the domain upon consultation with the expert. The arguments to E.42 became "QUALITATIVE.ASSESSMENT is *ET*" for ET, and "SPIRO.FREE.PT.ONLY is *true*" for PT.

No counter examples opposing the argument for ET were found, while E.33 was given as the counter example against the argument for PT. The expert mentioned template-based spiral drawings: "The assessment of the template-based spiral in some of the four observations (attributes) in the original data are essential in E.42, and none of them is Parkinsonian. This does not apply to E.33."

Similarly as in the above case, new attributes SPIRO.TEMPLATE.PT.ONLY and SPIRO.TEMPLATE.ET.ONLY were introduced, and the four original attributes were excluded from the domain. The argument attached to E.42 for PT was extended to "SPIRO.FREE.PT.ONLY is *true* and SPIRO.TEMPLATE.ET.ONLY is *false*." This time no counter examples were found by the algorithm.

Iteration 6. Critical example was E.39 (MT). The expert's argument for ET were postural tremor and the qualitative assessment in favor of ET. The presence of resting tremor was the argument for PT.

No counter examples opposing the argument in favor of ET were found. Counter example against the argument in favor of PT became E.45. When comparing this counter example with the critical example, the expert spotted an important difference: the lack of harmonics in the critical example, and their presence in the counter example. The argument in favor of PT was thus extended to "RESTING is *true* and ANY.HARMONICS is *false*." No new counter examples were found.

Iteration 7. Iteration 7 turned out to be exceptionally short. When the expert was asked to give arguments in favor of ET and PT for the critical example E.26 (MT), he realized that there were no valid arguments in favor of ET. The class was therefore changed from MT to PT.

Iteration 8. Critical example E.30 (ET) was presented to the expert. The expert was asked to describe which features are in favor of ET. He gave two features as an explanation: the presence of kinetic tremor and the lack of bradykinesia. No counter examples were found.

Iteration 9. Iteration 9 also demanded very little time from the expert. He was presented with critical example E.36 (MT). The expert gave two features in favor of ET: the presence of kinetic tremor and the presence of postural tremor. No counter examples were found to the expert's arguments.

Iteration 10. The expert observed critical example E.21 (MT). The expert gave the following explanation in favor of ET: "Qualitative assessment of both free-hand and template-based spiral in some of the observations (attributes) are essential, and none of them is Parkinsonian." The knowledge engineers thus attached the following argument to the critical example: "SPIRO.FREE.ET.ONLY is *true* and SPIRO.TEMPLATE.ET.ONLY is *true*." No counter examples and no further critical examples were found.

Review of the Model. At the end of the iterative procedure, the expert is asked to review the final model. Upon examination of the rules, the expert noticed the following rule that was in contradiction with his general knowledge about the domain.

IF TREMOR.START > 61 THEN class = EMT; [16,0]

The newly added attribute TREMOR.START occurred in this counter-intuitive rule (according to the expert) and now became the subject of careful examination.

The expert realized that the values of the attribute TREMOR.PERIOD from which TREMOR.START was calculated, may indeed reflect the number of years since the tremor was *diagnosed*, but this attribute actually does not reflect the age when the tremor actually *started*. The reason for this was pointed out by the expert: the patients with ET tend to visit the neurologist only when the tremor starts to cause them problems in everyday life, and this is usually several years *after* it actually first occurred. While it is commonly accepted that ET typically occurs much earlier than PT, the attribute TREMOR.START simply cannot reflect the time of its occurrence. The expert and the knowledge engineer decided to exclude the attribute TREMOR.START from the domain.

As a consequence, another critical example emerged. The expert was now asked to improve the argument "POSTURAL is *true* and KINETIC is *true*," given to critical example E.55 in Iteration 3, again having E.63 as a counter example. He realized that there is another important difference between E.63 and E.55: the presence of resting tremor in E.63. He extended his argument to "POSTURAL is *true* and KINETIC is *true* and RESTING is *false*." The method found no counter examples to the expert's argument.

The expert revised the newly induced rules and found them all to be acceptable. No further critical examples were found, and the knowledge elicitation process concluded.

5   Results

The number of rules after each iteration varied from 12 to 15. Table 1 shows the final model, *i.e.*, rules obtained after the end of the knowledge elicitation process. The domain expert evaluated each rule according to the following criteria.

Counter-intuitive: an illogical rule that is in contradiction with expert knowledge.
Reasonable: a rule consistent with expert knowledge, but insufficient to decide in favor of ET or PT on its basis alone.
Adequate: a rule consistent with expert knowledge, ready to be used as a strong argument in favor of ET or PT.

Figure 1 demonstrates how the model became increasingly more consistent with the expert's knowledge during the knowledge elicitation process. The expert's evaluations of the initial and final rules are significantly different (p=0.026 using Mann-Whitney-Wilcoxon non-parametric test). All rules in the final model are consistent with the domain knowledge, and five of them were marked by the expert as sufficiently meaningful to determine the type of tremor by themselves. The other nine rules (marked as *reasonable* by the expert) could alone not be improved by the method to *adequate* rules, as there were no counter examples in the data set, and therefore the method did not have any reason to further specialize these rules. A larger number of learning examples might, however, result in a greater number of *adequate* rules.

During the process of knowledge elicitation, 15 arguments were given by the expert, 14 new attributes were included into the domain, and 21 attributes were excluded from the domain. After each iteration, the obtained model was evaluated on the test data set. If all the rules that triggered were for the class EMT (PMT), then the example was classified as ET (PT). In cases where the rules for both classes triggered, the example was classified as MT.

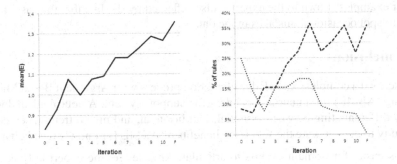

Fig. 1. The graph on the left side shows the average of expert's evaluations of the rules (0 - counter-intuitive, 1 - reasonable, 2 - adequate) obtained after each iteration of the knowledge elicitation process. The graph on the right side shows the percentage of counter-intuitive (the lower curve) and adequate (the upper curve) rules among all rules obtained after each iteration.

Table 1. The rules after the end of the knowledge elicitation process. The condition and class columns show the condition and the consequent parts of a rule. Columns + and − stand for the number of positive and negative examples covered, respectively. All rules have pure distributions – they do not cover any examples from the opposite class. Column E stands for the expert's evaluation (0 - counter-intuitive, 1 - reasonable, 2 - adequate) of a rule.

#	Condition	Class	+	−	E
1	IF QUALITATIVE.ASSESSMENT = *ET*	EMT	21	0	1
2	IF BRADYKINESIA = *false*	EMT	18	0	1
3	IF BRADYKINESIA = *true* AND RIGIDITY = *true*	PMT	17	0	2
4	IF QUALITATIVE.ASSESSMENT = *ET* AND POSTURAL = *true*	EMT	16	0	1
5	IF RIGIDITY = *false* AND KINETIC = *true*	EMT	15	0	1
6	IF KINETIC = *true* AND BRADYKINESIA = *false*	EMT	13	0	1
7	IF SPIRO.FREE.PT.ONLY = *true* AND SPIRO.TEMPLATE.ET.ONLY = *false*	PMT	13	0	1
8	IF HARMONICS = *true*	EMT	12	0	2
9	IF RESTING = *true* AND HARMONICS = *false* AND RIGIDITY = *true*	PMT	12	0	2
10	IF POSTURAL = *true* AND KINETIC = *true* AND RESTING = *false*	EMT	10	0	1
11	IF QUALITATIVE.ASSESSMENT = *PT*	PMT	10	0	1
12	IF RESTING = *false* AND POSTURAL = *true* AND BRADYKINESIA = *false*	EMT	8	0	2
13	IF POSTURAL = *true* AND ANAMNESIS = *positive* AND BRADYKINESIA = *false*	EMT	8	0	2
14	IF SPIRO.FREE.ET.ONLY = *true* AND SPIRO.TEMPLATE.ET.ONLY = *true*	EMT	7	0	1

We compared classification accuracies improvements for ABCN2, Naive Bayes (NB), and kNN.[1] The ABCN2's classification accuracy on the test set improved from the initial 52% (NB: 63%; kNN: 58%) to the final 82% (NB: 81%; kNN: 74%). This result shows that the higher consistency with expert's knowledge is *not* obtained at the expense of classification accuracy.

The overall expert's time involvement was about 20 hours, which is rather low considering the high complexity of the presented domain. The relevance of the critical and

---

[1] In Naive Bayes, the conditional probabilities were estimated by relative frequencies for discrete attributes and by LOESS for continuous attributes. In kNN, the Euclidian distance was selected and k was set to 5.

counter examples shown to the expert is also reflected by the fact that they assisted the expert to spot occasional mistakes in the data.

6   Conclusions

We described a complete knowledge elicitation process with Argument Based Machine Learning (ABML) for a neurological decision support system. A neurological domain, namely differentiating between essential, Parkinsonian, and mixed tremor, served as a case study to demonstrate the following benefits of ABML for knowledge elicitation.

1. It is easier for domain experts to articulate knowledge; the expert only needs to explain a single example at the time.
2. It enables the expert to provide only relevant knowledge by giving him or her critical examples.
3. It helps the expert to detect deficiencies in his or her explanations by providing counter examples.

Our step-by-step presentation of the knowledge elicitation process with ABML can serve the reader as a guideline on how to effectively use the argument-based approach to knowledge elicitation. In addition, the main result of this paper is a quantitative demonstration of how the rules become increasingly more consistent with expert's knowledge during the ABML knowledge elicitation process. It is also interesting to note that ABML loop resulted in a simplification of the original set of attributes.

References

1. Forsyth, R., Rada, R.: Machine learning: applications in expert systems and information retrieval. Halsted Press, New York (1986)
2. Langley, P., Simon, H.A.: Applications of machine learning and rule induction. Commun. ACM 38(11), 54–64 (1995)
3. Martens, D., Baesens, B., Gestel, T.V., Vanthienen, J.: Comprehensible credit scoring models using rule extraction from support vector machines. European Journal of Operational Research 183(3), 1466–1476 (2007)
4. Verbeke, W., Martens, D., Mues, C., Baesens, B.: Building comprehensible customer churn prediction models with advanced rule induction techniques. Expert Systems with Applications 38(3), 2354–2364 (2011)
5. Webb, G.I., Wells, J., Zheng, Z.: An experimental evaluation of integrating machine learning with knowledge acquisition. Mach. Learn. 35(1), 5–23 (1999)
6. Možina, M., Žabkar, J., Bratko, I.: Argument based machine learning. Artificial Intelligence 171(10/15), 922–937 (2007)
7. Možina, M., Guid, M., Krivec, J., Sadikov, A., Bratko, I.: Fighting knowledge acquisition bottleneck with argument based machine learning. In: The 18th European Conference on Artificial Intelligence (ECAI), pp. 234–238 (2008)
8. Groznik, V., Guid, M., Sadikov, A., Možina, M., Georgiev, D., Kragelj, V., Ribarič, S., Pirtošek, Z., Bratko, I.: Elicitation of Neurological Knowledge with ABML. In: Peleg, M., Lavrač, N., Combi, C. (eds.) AIME 2011. LNCS, vol. 6747, pp. 14–23. Springer, Heidelberg (2011)
9. Pahwa, R., Lyons, K.E.: Essential tremor: Differential diagnosis and current therapy. American Journal of Medicine 115, 134–142 (2003)

Min-Max Itemset Trees for Dense and Categorical Datasets*

Jennifer Lavergne, Ryan Benton, and Vijay V. Raghavan

University of Louisiana at Lafayette: The Center for Advanced Computer Studies,
Lafayette LA 70503, USA
jjslavernge686@hotmail.com, rbenton@cacs.louisiana.edu,
raghavan@louisiana.edu

Abstract. The itemset tree data structure is used in targeted association mining to find rules within a user's sphere of interest. In this paper, we propose two enhancements to the original unordered itemset trees. The first enhancement consists of sorting all nodes in lexical order based upon the itemsets they contain. In the second enhancement, called the Min-Max Itemset Tree, each node was augmented with minimum and maximum values that represent the range of itemsets contained in the children below. For demonstration purposes, we provide a comprehensive evaluation of the effects of the enhancements on the itemset tree querying process by performing experiments on sparse, dense, and categorical datasets.

Keywords: data mining, association mining, targeted association mining, itemset tree, min-max itemset tree, dense data, categorical data.

1 Introduction

Association mining is a technique for discovering strong co-occurrences in large sets of data. These datasets are made up of transactions, or rows. Transactions are made up of itemsets, such as {egg, milk, cheese} or {coughing, 101° fever, dizziness, nausea}, and represent objects which occur together. For ease of processing most algorithms map string itemsets to numerical values, i.e. egg = 1, milk = 2, etc.

Examples of transactional datasets can be seen in Fig. 1. Figure 1a shows a sparse dataset that contains highly variable transaction lengths and has the potential for a low average transaction length. Figure 1b shows a dense dataset, which contains less variation in transactions size and has a higher average transaction length than a sparse dataset. Figure 1c shows a categorical dataset, or datasets that contain items that can be separated into categories.

An example of a categorical dataset is one which contains vehicle information. Example categories are color = {blue, red, green, yellow ... }, windows = {manual, power}, type = {car, truck, van ... } and so on. Each category can

---

* This project was funded in part by the Louisiana Highway Safety Commission.

L. Chen et al. (Eds.): ISMIS 2012, LNAI 7661, pp. 51–60, 2012.

only have one of its members in a given transaction at a time. If a certain item category contains $\{1, 2, \ldots, 5\}$, an itemset $I = \{1,2\}$ would never occur in the dataset. In Fig. 1c, all the values appearing in a column are considered to belong to the same item category.

1	2	3	4	5
3	4			
5	6	9		
2	8			

1	2	3	4	5
3	4	7	8	9
5	6	7	8	
2	5	6	8	9

1	7	10	21	23
3	9	13		25
5	8		16	32
2	7	11	19	29

(a) Sparse transactional dataset (b) Dense transactional dataset (c) Categorical transactional dataset

Fig. 1. The difference between sparse, dense, and categorical transactional datasets

Brute force association mining processes an entire dataset and finds many redundant and useless rules. A well-known algorithm based on the Apriori principle has been proposed, which reduces the number of candidate itemsets to only those that occur frequently, and thereby reduces search complexity. The Apriori principle states that any itemset that is infrequent will only create supersets that are also infrequent and, therefore, should not be included in the list of candidate itemsets [1,2,3]. An itemset is considered frequent when its support is larger than a user-defined threshold minimum support, aka minsup. The support of an itemset I is calculated using the formula: support(I) = count(I)/count(transactions in the dataset). If support(I) $\geq$ minsup, I is considered frequent.

For example, given the table in Fig. 1a, assume minsup = 30%. If I (the itemset of interest) is $\{3,4\}$, then the support of I = support($\{3,4\}$) = 2/4 = 50%. Since 50% $\leq$ 30%, I is frequent and is included as a candidate itemset. Conversely if we look at the itemset I = $\{1\}$, support($\{1\}$) = 1/4 = 25% < 30%. This itemset and any supersets containing it will be infrequent; therefore, it is not included in the candidate itemset list [2].

When generating rules based upon the list of candidate itemsets, the minsup threshold may generate a very large list of results. In order to reduce the number of candidate itemsets a number of "constraint-based" techniques [4,5,6] have been proposed. These techniques reduce the search space by imposing constraints upon the association mining process.

Another technique proposed for reducing the number of candidate itemsets is the Maximal Frequent Itemset Tree [7] which prunes the itemset lattice generated using the Apriori principle [2]. This data structure finds all maximal frequent itemsets, which are frequent itemsets that have no supersets that are also frequent, and returns those as candidate itemsets.

Traditional association mining returns all strong rules from all frequent itemsets in the dataset regardless of whether they are within the user's sphere of interest or not. Another technique, targeted association mining, proposed only generating rules from itemset that are of interest to the user [8,9,10]. An example of this is a user who is only interested in rules containing $\{cheese\}$ or $\{101°,$ fever, nausea$\}$. Ideally, only frequent supersets of these itemsets will be used for generating rules. Targeted association mining uses an index tree structure called an itemset tree to find the desired candidate itemsets, which results in a significantly smaller set of rules than traditional mining [8].

An inspection of the rule mining process of the itemset tree algorithm presented by Kubat et al. [9] revealed two issues. These issues resulted in numerous search paths being traversed unnecessarily for each query, thereby incurring extra processing costs.

This paper proposes two enhancements to reduce the overall complexity of the mining process on an itemset tree. First, the Ordered itemset tree sorts the nodes of the tree in lexical order. This will allow a query to terminate early when a larger itemset is reached. Second, the Min-Max itemset tree proposes the inclusion of a range, min and max for each node, of the itemsets contained in the children below. If a query is not within the range the query process does not need to search this subtree. Finally, a comprehensive evaluation of the effects of the min-max itemset tree on the tree querying process is presented. Experiments are performed on three different dataset types: sparse, dense, and categorical. By exploring these dataset types, this paper plans to test the robustness of the min-max itemset tree enhancement.

This paper is organized in the following manner. Section 2 discusses the itemset tree data structure and its itemset querying process. Section 3 describes the Ordered and Min-max itemset tree enhancements. Section 4 presents the experiments completed on a sparse synthetic dataset. Section 5 discusses the dense dataset experiments and results. Section 6 covers the conclusions and future work.

2   Itemset Trees

Definition 1: An *itemset tree*, T, consists of a root pair [I, f(I)], where I is an itemset and f(I) is I's frequency in the dataset, and a possibly empty set $\{T_1, T_2, \ldots, T_k\}$, each element of which is an itemset tree. If a query itemset Q is in the root itemset I, then it will also be in the children of the root and if Q is not in the root, then there is a possibility that it may be in the children of the root if: first_item(I) $\leq$ first_item(Q) and last_item(I) $<$ last_item(Q) [8]. For the latter case, notation I refers to the itemset associated with the root of one of the children itemset trees, recursively. The item IDs of both sets I and Q are stored in increasing order.

Itemset trees were created to replace a transactional database with an index tree structure. It is possible to recreate the entire database from an itemset tree, losing only the original transaction insertion order. Each item in the transactional dataset is mapped to an integer value to facilitate the tree mining process. The resulting tree will inhabit less space than the original dataset [9]. With the itemset tree, it is possible to not only find all rules which are within the user's sphere of interest, but also all those strong association rules which are contained in the dataset [10].

The first issue contributing to the excess costs is that when an itemset tree is built, the nodes are not placed in lexical order. This results in missed opportunities for early termination of a query. The second issue allows a query to potentially search an entire tree to find an itemset that was either never in the tree, or was not in most of the subtrees. The third issue is when building an itemset tree on a dense dataset, the resulting tree will possess deep paths with many leaf nodes. Querying such a tree, especially considering the first two issues, can result in many unneeded paths traversed and higher processing costs.

3   Enhancements

In order to address the issues discussed in the previous section, we propose two itemset tree enhancements. The first enhancement, which is referred as an Ordered Itemset Tree (OIT), requires that the nodes in the tree be ordered in a lexical fashion. This enhancement permits the query function to terminate a traversal early by negating the requirement to check nodes that, by definition, violate the lexical constraints. The second enhancement, which is referred to as the Min-Max Itemset Tree (MMIT), augments each node with two fields and, at its most efficient, is also an ordered tree. These fields are the minimum and maximum 1-itemset value contained under any given parent node. This will allow a traversal to decide whether or not the query in question could potentially be in the subtree.

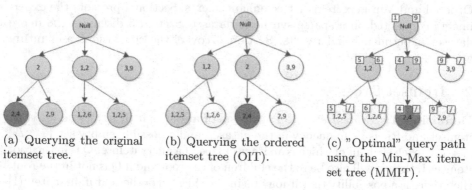

(a) Querying the original itemset tree.

(b) Querying the ordered itemset tree (OIT).

(c) "Optimal" query path using the Min-Max itemset tree (MMIT).

Fig. 2. Comparison for querying an itemset tree for Q = {2,4}

A demonstration of the traversal on the original unordered itemset tree can be seen in Fig. 2a. Traversing an ordered itemset tree, seen in Fig. 2b, eliminates node itemsets larger than the search itemsets from the search path. The Min-Max itemset tree, seen in Fig. 2c, further reduces the search path by comparing the range to the search itemset. Nodes skipped are represented as white and the nodes in the traversal path are represented as gray. The large circles in the trees contain the node's itemset and the smaller circles represent the support of that node.

Definition 2: An *Ordered Itemset Tree*, T, consists of a root pair [I, f(I)], where I is an itemset and f(I) is I's frequency in the dataset, and a possibly empty set {T_1,

$T_2, \ldots, T_k$}, each element of which is an ordered itemset tree. If a query itemset Q is in the root itemset I, then it will also be in the root's children and if Q is not in the root, then there is a possibility that it may be in the root's children if: first_item(I) $\leq$ first_item(Q) and last_item(I) < last_item(Q). For any two child trees T_i and T_j, the first_item(root_itemset(T_i)) < first_item(root_itemset(T_j)) must hold for all i < j.

The Ordered Itemset Tree is built incrementally by placing each transaction read in from the dataset in a lexical order with respect to the other nodes of the tree.

Definition 3: An *Ordered Min-Max Itemset Tree*, T, consists of a root pair [I, f(I), min, max], and a possibly empty set $T_1, T_2, \ldots, T_k$, each element of which is an ordered min-max itemset tree. I is an itemset, f(I) is a frequency, min and max represent the range of 1-items contained in the node's children. All items in the tree must be lexically ordered. For any two child trees T_i and T_j, the first_item(root_itemset(T_i)) < first_item(root_itemset(T_j)) must hold for all i < j. For each node T_i, the max and min values are included. The max is the largest 1-itemset contained in the set of all T_i's children. The min is the smallest 1-itemset in T_i's children excluding T_i's items.

The Min-Max Itemset Tree introduces the min and max fields, seen as boxes attached to the nodes in Fig. 2c. As a tree is built each transaction makes its way through the tree to its proper place in the tree. As parent nodes are passed, their min and max fields are updated, if needed, to reflect the new transaction's placement.

3.1   Query Algorithm Comparison

The original itemset tree querying algorithm, seen in Fig. 3a, traversed the tree using minimal knowledge about the itemsets contained in the nodes. It also lacked knowledge about the sibling and children nodes of the node currently being searched.

```
Let R' be the top null root node.
Let ci be the current search node.
Count(Q, ci):
  For each ci s.t. first_item(ci) ≤ first_item(Q)
    Case 1: If Q ⊆ ci, return(F(Q)+F(ci))
    Case 2: If Q ⊄ ci, and last_item(ci) > last_item(Q)
      return Count(F(Q)+ Count(Q, T(ci)))
```

```
Let T be a tree rooted at R and Q be a query.
Let R' be the top null root node.
Let ci be the current search node.
Count(Q, ci):
  For each ci s.t. first_item(ci) ≤ first_item(Q)
    If min(R') > min(Q) and/or max(R') ≥ max(Q),
      return(Support(Q));
    If first_item(ci) = first_item(Q) and
      last_item(ci) > last_item(Q), return(Support(Q));
  Else
      Case 1: If Q ⊆ ci return(F(Q) + F(ci))
      Case 2: If Q ⊄ ci, and last_item(ci) > last_item(Q)
        and min(R') ≤ min(Q) and max(R') > max(Q).
        return Count(F(Q) + Count(Q, T(ci)))
```

(a)   Querying   the   original itemset tree.

(b) Querying the ordered min-max tree.

Fig. 3. Querying algorithms for the original itemset tree and the ordered min-max itemset tree

The Min-Max Itemset Tree querying algorithm, seen in Fig. 3b, includes the information missing from the unordered itemset tree querying algorithm. Knowledge of the siblings inherent in an ordered tree says that all sibling nodes before will be smaller, whereas all nodes after will be larger. This allows the traversal to terminate early. Knowledge of the range of itemsets contained by the children node is found using the Min-Max Itemset Tree algorithm. If a query itemset does not fall within these bounds, the traversal can skip the entire subtree. Including these small bits of knowledge can greatly reduce the number of nodes searched to find an itemset with little overhead.

3.2   Preliminary (Sparse Dataset) Results

The preliminary experiments were executed on three datasets created by the IB-MQuest synthetic dataset generator. This generator created by the IBMQuest research group at IBM generates sparse market basket style datasets which mimic real world transactional datasets with high co-occurrences [2,11]. The three datasets contained 30,000, 40,000, and 50,000 transactions and were queried using four different query sets designed. Based upon the results, the number of nodes traversed followed a distinct pattern of: UIT(nodes) $\geq$ Unordered-MMT(nodes) > OIT(nodes) $\geq$ OMMT(nodes). Overall, the OMMT outperformed all other tree types by requiring the least amount of nodes to find the queries.

4   Expanded Experiments

In the preliminary results, it was demonstrated that the Ordered Min-Max Itemset Tree outperformed the other three tree types. Therefore for the following experiments, we limited the trees to only the original unordered itemset tree and the Ordered Min-Max Itemset Tree. For each tree generated, a set of queries was also created to conform to these four query types: deep path high support queries, deep path low support queries, shallow path high support queries, and shallow path low support queries. In addition, queries which did not appear in the dataset were included to simulate the possibility that the user is looking for itemsets that have not occurred.

4.1   Tree Generation Experiments

In order to better understand the impact of varying the number of transactions and the number of unique 1-itemsets in a dataset, 16 datasets were created. Transactions sizes was also varied to obtain datasets with number of 1-itemsets varying from 25,000 - 200,000. The number of unique 1-itemsets were also doubled for each dataset from 50 - 400. Each dataset is a subset of the previous dataset mimicking daily store transactions. An analysis of the dynamics of the created trees can be seen in Table 1 and Fig. 4. Table 1 details the number of nodes in each tree created using the above combinations. In Fig. 4, the effects

Table 1. Experimental dataset description

Dataset	Number of nodes in the tree			
Transactions	50 1-items	100 1-items	200 1-items	400 1-items
25,000	31,993	31,383	30,247	29,917
50,000	62,265	61,483	59,555	58,151
100,000	62,265	61,483	59,555	58,151
200,000	62,265	61,483	59,555	58,151

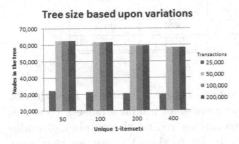 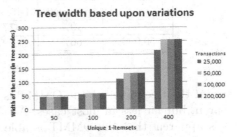

(a) Tree size variations.   (b) Tree width variations.

Fig. 4. Tree size and width as the number of transactions and unique 1-itemsets varies

of varying the number of transactions and unique 1-itemsets has on the size and the width of the trees can be seen.

Two observations can be made from the results in Table 1 and Fig. 4. First, as the number of transactions increases, the number of nodes in the tree remains static past 50,000. This occurs because increasing the number of transactions only repeats the same subset of itemsets, merely increasing their support in the tree. Second, as the number of unique 1-itemsets increases, the width of the tree increases. Hence, for dense datasets, experiments are limited to the 25,000 and 50,000 transactional datasets and all four levels of unique 1-itemsets.

4.2   Dense Dataset Experiments and Results

In order to discern the search savings when using the ordered min-max tree instead of the original itemset tree, a query set was created for each dataset. The query sets generated included all four query types listed previously. Queries were also separated into those that are in the tree and those that are not in the tree, i.e. support = 0. It is important to note that for all datasets, the average transaction length is 10. Thus, the larger the number of distinct items, the smaller is the density of the dataset.

The number of nodes required to find an itemset were averaged across all queries and the reduction in nodes when comparing the ordered min-max tree to the unordered original itemset tree can be seen in Table 2. It is apparent that despite the density differences of the datasets, the Ordered Min-Max Itemset Tree was still able to outperform the original unordered tree in all cases. As the

Table 2. Percent reduction in nodes needed for search comparing the UIT and OMMT

	Queries with support > 0			
Transactions	50 unique 1-itemsets	100 unique 1-itemsets	200 unique 1-itemsets	400 unique 1-itemsets
25,000	17%	36%	54%	50%
50,000	20%	11%	21%	18%
	Queries with support = 0			
Transactions	50 unique 1-itemsets	100 unique 1-itemsets	200 unique 1-itemsets	400 unique 1-itemsets
25,000	19%	77%	96%	99%
50,000	66%	29%	99%	99%

size of the tree increases the OMMT is able to reduce the number of nodes for search, but far less in most cases. As the number of unique itemsets increases it is apparent that the OMMT is able to further reduce the number of nodes needed for a search, since there are more nodes to skip. For queries that were not contained within the tree, the min-max tree was able to reduce the number of search nodes by more than half in most cases.

4.3   Categorical Dataset Experiments and Results

Categorical variables are created to structure data by organizing items into groups, i.e. colors = {blue, red, green, orange}. For these experiments, three datasets have been obtained from the Louisiana Highway Safety Database; these were used to create itemset trees. All datasets contained 238 unique 1-itemsets, all of which are included in seven distinct categorical groups. The specifications of each dataset can be seen in Table 3.

Table 3. Experimental results based upon categorical data with 238 unique 1-itemsets and 7 groups

Transactions	Number of nodes in the tree	Transaction length range
39,483	33,892	6 - 12 items
51,290	34,919	6 - 11 items
76,815	56,690	6 - 12 items

An UIT and OMMT were generated for each dataset and were queried using the same query-set. This query-set includes queries having the support and path characteristics mentioned in previous experiments. The number of nodes required to find each query using the UIT and OMMT were averaged and the percent difference is presented in Table 4.

Table 4. Average number of nodes searched to find queries using categorical datasets

Queries with support > 0			
Transactions	Avg nodes for UIT	Avg nodes for OMMT	% difference
39,483	248.38	75.01	69%
51,290	245.70	74.41	70%
76,815	249.75	77.14	69%
Queries with support = 0			
Transactions	Avg nodes for UIT	Avg nodes for OMMT	% difference
39,483	574.35	64.09	89%
51,290	158.27	63.3	60%
76,815	757.13	175.58	77%

Upon inspection of the results in Table 4, two observations can be made. First, when compared to the preliminary results, it appears that the OMMT's ability to reduce the number of nodes for categorical data is higher than with non-categorical data. Second, the OMMT is able to significantly reduce the number of nodes needed to be searched for queries that do not appear in the tree by a significant amount.

5   Conclusions and Future Work

In this paper, two critical weaknesses of the itemset tree, proposed by Kubat et al. [9], were presented. In order to address these weaknesses, the Ordered Itemset Tree and the Min-Max Itemset Tree were proposed. These enhancements reduced the number of nodes searched for any given query on an itemset tree by permitting early termination of the search and the skipping of fruitless subtrees. Initial experiments on sparse datasets, which showed significant reductions in the number of nodes searched for all three tree types, were summarized.

For the expanded results, it is apparent that not only was the Ordered Min-Max Tree able to outperform the original itemset tree for queries within the tree, but was especially well suited to the queries which were not contained in the tree. As the number of unique 1-itemsets increased, the OMMT was able to decrease the number of nodes searched to find an itemset. As the tree size increased, the number of nodes searched to find an itemset using the OMMT has less of a difference from the UIT. This occurs due to the fact that a tree with more nodes, but the same number of 1-itemset is more likely to have a min or max included. When applied to categorical datasets, the OMMT appears to be able to reduce the number of nodes searched to find, or not find, any given query than with non-categorical transactional datasets.

Future work will follow three distinct paths. First, as noted in [11], the IBM QUEST system may not be representative of many market basic (sparse) database sets. Hence, we plan to investigate the impact of the proposed enhancements using other generators, including Cooper and Rozsypal [11,12]. Second, we will improve upon the theoretical understanding of itemset trees, with a particular focus on investigating dynamic query sets. Third, itemset trees have been adopted to handle full association mining [10]. However, their applicability to temporal mining/rare rule mining has never been investigated. We plan to enhance the data structure to permit association mining within those realms.

References

1. Srikant, R., Agrawal, R.: Mining Generalized Association Rules. In: Proceedings of the 21st International Conference on Very Large Data Bases, VLDB 1995, pp. 407–419 (1995)
2. Agrawal, R., Srikant, R.: Fast Algorithms for Mining Association Rules in Large Databases. In: Proceedings of the 20th International Conference on Very Large Data Bases, pp. 487–499 (1994)
3. Srikant, R., Agrawal, R.: Mining quantitative association rules in large relational tables. SIGMOD Rec., 1–12 (1994)
4. Cheung, Y., Fu, A.: Mining frequent itemsets without support threshold: with and without item constraints. IEEE Trans. on Knowledge and Data Engineering, 1069–2004 (1994)
5. Pei, J., Jiawei Han, J.: Constrained frequent pattern mining: a pattern-growth view. SIGKDD Explor. Newsl., 31–39 (2002)
6. Pei, J., Jiawei Han, J., Lakshmanan, L.: Mining frequent itemsets with convertible constraints. In: Proceedings of Data Engineering, pp. 433–442 (2001)
7. Burdick, D., Calimlim, M., Gehrke, J.: MAFIA: a maximal frequent itemset algorithm for transactional databases. In: Data Engineering, pp. 443–452 (2001)
8. Hafez, A., Deogun, J., Raghavan, V.V.: The Item-Set Tree: A Data Structure for Data Mining. In: Mohania, M., Tjoa, A.M. (eds.) DaWaK 1999. LNCS, vol. 1676, pp. 183–192. Springer, Heidelberg (1999)s
9. Kubat, M., Hafez, A., Raghavan, V., Lekkala, J., Chen, W.: Itemset trees for targeted association querying. IEEE Transactions on Knowledge and Data Engineering, 1522–1534 (2003)
10. Li, Y., Kubat, M.: Searching for high-support itemsets in itemset trees. Intell. Data Anal., 105–120 (2006)
11. Cooper, C., Zito, M.: Realistic Synthetic Data for Testing Association Rule Mining Algorithms for Market Basket Databases. In: Kok, J.N., Koronacki, J., Lopez de Mantaras, R., Matwin, S., Mladenič, D., Skowron, A. (eds.) PKDD 2007. LNCS (LNAI), vol. 4702, pp. 398–405. Springer, Heidelberg (2007)
12. Rozsypal, A., Kubat, M.: Association Mining in Gradually Changing Domains. In: Proceedings of the Sixteenth International Florida Artificial Intelligence Research Society Conference, pp. 366–370 (2003)

TRARM-RelSup: Targeted Rare Association Rule Mining Using Itemset Trees and the Relative Support Measure

Jennifer Lavergne, Ryan Benton, and Vijay V. Raghavan

University of Louisiana at Lafayette: The Center for Advanced Computer Studies,
Lafayette LA 70503, USA
jjslavernge686@hotmail.com, rbenton@cacs.louisiana.edu,
raghavan@louisiana.edu

Abstract. The goal of association mining is to find potentially interesting rules in large repositories of data. Unfortunately using a minimum support threshold, a standard practice to improve the association mining processing complexity, can allow some of these rules to remain hidden. This occurs because not all rules which have high confidence have a high support count. Various methods have been proposed to find these low support rules, but the resulting increase in complexity can be prohibitively expensive. In this paper, we propose a novel targeted association mining approach to rare rule mining using the itemset tree data structure (aka TRARM-RelSup). This algorithm combines the efficiency of targeted association mining querying with the capabilities of rare rule mining; this results in discovering a more focused, standard and rare rules for the user, while keeping the complexity manageable.

Keywords: data mining, association mining, targeted association mining, itemset tree, rare association rule mining.

1 Introduction

For years, organizations have sought to make sense of the large repositories of data collected from their day to day workings. Hidden within these datasets are potentially interesting correlations. An example is a department store that wishes to know which articles of clothing, jewelry, and/or shoes are selling well together. They utilize this information and advertise slower moving items with more desirable ones in magazines and stores. Prediction of new items that may sell well together can also be drawn from this information.

The process by which these gems of information are discovered is referred to as association mining [11,2,12]. Association mining processes large quantities of transactional data and returns rules based upon strong co-occurrences found. This process can be broken up into two separate subtasks: finding frequent co-occurrences and generating implications. An itemset is considered frequent if the frequency of its occurrence in a dataset is greater than some user-defined threshold. Examples of itemsets from a department store transactional dataset

L. Chen et al. (Eds.): ISMIS 2012, LNAI 7661, pp. 61–70, 2012.

would be I = {shoes, shirt, tie} and I = {necklace, earrings}. Unfortunately, using brute force association mining to find frequent co-occurences can be highly complex and require significant amount of processing time. Agrawal et al. [2] observed that if an itemset (a set of items from the dataset) is infrequent, then any superset containing the itemset will also be infrequent. This is known as the Apriori principle. Using the Apriori principle, we are able to prune large amounts of candidate itemsets from the mining process, reducing complexity and processing time [2].

The first association mining subtask scans the dataset for itemsets I that occur frequently together. Support is found using the formula support = count(I)/ number of transactions in the dataset [2]. The Apriori principle uses a user-defined threshold minimum support, or minsup, to determine if an itemset is considered frequent. Once the set of all frequent itemsets is generated, the second subtask uses this set to generate association rules. An association rule is an implication X $\Rightarrow$ Y, where I = X$\cup$Y, X and Y are disjoint itemsets. In order to determine if an association rule is potentially interesting, Apriori uses the conditional probability, i.e. how confident are we that Y will occur given X. Confidence is found using the formula confidence = Support(X$\cup$Y)/Support(X) [2]. Like minsup, we use a user-defined threshold minimum confidence, or minconf, to determine a rule's potential interestingness. If a rule's confidence $\geq$ minconf and it is derived from a frequent itemset, then the rule is considered strong and is returned to the user as output.

Association mining is designed to return all strong rules. This can result in prohibitively large rule sets returned to the user. A user may not be interested in most of the rules returned. For example, a department store owner may not be interested in correlations containing only their last season inventory, in favor of correlations concerning their newest products.

In 1999, Hafez et al. [6] proposed the concept of targeted association mining. Targeted association mining discovers rules based upon the interests of users and, therefore, can generate more focused results. The itemset tree data structure was developed to facilitate the targeted rule mining process. The use of the tree structure reduces the complexity of rule mining when compared to the Apriori style algorithms and generates more focused results [6,8]. Additional work has been completed showing that the itemset tree structure can also be used for efficiently finding all possible frequent association rules in a dataset [9].

Association mining and targeted association mining both focus on finding frequent itemsets based upon the minsup threshold. Unfortunately, not all high-confidence rules have a high support. By using the Apriori principle, these high confidence/low support rules are lost. This is referred to as the rare rule problem [1,13]. One can lower the minsup value in order to find more rare rules, but this greatly increases complexity and enlarges the rule set. Another study proposed the use of more than one minimum support value in order to separate itemsets into different frequency zones and then mined the rules therein. This method can potentially lose rules that may occur outside the frequency zones [3].

Thus, on one side, the rare rule mining permits the discovery of high confidence/low support rules, but at great computational cost. On the other, targeted mining currently allows for the efficient discovery of rules containing items of interest to the user, but it lacks the ability to find high-confidence/low-support rules. By merging the targeted mining with rare rule mining, the ability to mine targeted standard and rare rules, in a computationally efficient manner could be achieved. This is the premise of TRARM-RelSup algorithm that is being proposed in this paper. TRARM-RelSup combines the efficiency of the itemset tree for targeted association mining with the strategy of using relative support in order to efficiently find targeted rare association rules.

This paper is organized in the following manner. Section 2 discusses work related to the TRARM-RelSup algorithm. Section 3 will cover the proposed approach for combining targeted and rare association mining in the TRARM-RelSup algorithm. In Section 4, we discuss the experimental setup and the queries and comparisons used therein. Section 5 discusses experimental results. Finally, Section 6 presents our conclusions and planned future work.

2 Related Work

This section covers background information that will be used in the TRARM-RelSup algorithm. First the rule mining process implemented by the itemset tree algorithm and data structure will be covered [6,8]. Then, a discussion on the relative support measure used for finding rare rules within a dataset will be presented. Relative support finds all rules with a potentially high confidence value using a user-defined threshold, minRelSup, for pruning low confidence candidate itemsets. This measure will be adopted by the TRARM-RelSup algorithm for incorporating rare rule mining capabilities to the basic itemset tree querying method.

2.1 Targeted Association Mining

The itemset tree data structure was developed by Hafez et al. [6] and facilitates the targeted association mining process. Users specify query itemsets or areas of interest and the itemset tree generates all rules containing those specific items of interest to the user. All itemsets within the transaction set are mapped to integer values to expedite the mining process. The rule mining process has also been extended such that an itemset tree can be searched efficiently to find all rules in a dataset [9].

Definition 1: An *itemset tree*, T, consists of a root pair [I, f(I)], where I is an itemset and f(I) is I's frequency in the dataset, and a possibly empty set $\{T_1, T_2, \ldots, T_k\}$, each element of which is an itemset tree. If a query itemset Q is in the root itemset I, then it will also be in one or more children itemset trees of the root and if Q is not in the root, then there is a possibility that it may be in the children of the root if: first_item(I) $\leq$ first_item(Q) and last_item(I)

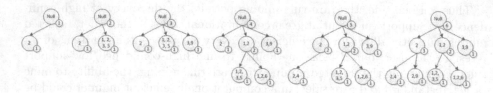

Fig. 1. Building an itemset tree

< last_item(Q) [6]. Note that in the latter case, notation I refers to the itemset associated with the root of one of the children itemset trees, recursively. The items in both sets I and Q are stored in increasing order of item IDs.

Example: The following steps detail the tree generation process seen in Fig. 1 and using the transactions: {2}, {1,2,3,5}, {3,9}, {1,2,6}, {2,4}, {2,9}.

- Step 1: Add transaction = {2} as the null root's first child.
- Step 2: {1,2,3,5} has no lexical overlap with {2}, so it becomes the root's second child.
- Step 3: {3,9} has no lexical overlap with {2}, so it becomes the root's third child.
- Step 4: {1,2,3,5} and {1,2,6} have a common lexical overlap of {1,2}. A node is created for the overlap with {1,2,3,5} and {1,2,6} as its children.
- Step 5: {2,4} becomes the first child of {2}.
- Step 6: {2,9} becomes the second child of {2}.

There are two ways to query an itemset tree. The first query type finds the support of a given query itemset Q. This algorithm takes advantage of the knowledge that if a query itemset Q is in the root itemset I, then it will also be in the children itemset trees of the root. Also if Q is not in the root, then there is a possibility that it may be in the children of the root if: first_item(I) ≤ first_item(Q) and last_item(I) < last_item(Q). Using this logic, one can traverse the tree and find all nodes that contribute support to Q and skip all subtrees that fail both conditions [6,8].

The second query type finds all itemsets in the tree that contain an itemset Q and is used as a preprocessing step before generating association rules. This algorithm is similar to query type 1. However, when a node whose itemset contains Q is found, not only is that itemset returned, but also all the itemsets contained in the subtree below. Only those itemsets whose support > minsup will be added to the results set [6,8].

In order to generate rules using this algorithm, the first step is to combine all itemsets into L = {l_1, ... ,l_n} where no l_i is a proper subset of and l_j, for i ≠ j, and n = |L|. The condition that l_i should not be a subset of l_j is applied only in the case that the support of l_i is equal to the support of l_j. Second, for each l_j the confidence is for the rules with LHS equal to Q and RHS equal to all subsets of l_j - Q is confidence = support((subset of (l_j - Q)) ∪ Q) / Support(Q). If confidence ≥ minconf, the rule is considered potentially interesting.

2.2   Rare Rule Mining

Recent work in rare rule mining has explored the ever present issue of trading accuracy for speed and vice versa. Reducing minsup to include all possible high confidence rules causes an undesirable increase in complexity and results in many duplicate and useless rules [1]. The MSApriori algorithm uses multiple minsups for creating frequent zones reduces complexity when compared to the brute force rule mining. Bu,t overall, it still generates large result sets. In addition this method also loses the itemsets outside of the frequency zones [3]. In 2010, Gedikli and Jannach [5] proposed adapting MSApriori [3] for creating weighted association rules for recommenders. This new recommendation system generates proposed personalized rule sets based upon the rules sets of the user's surrounding neighbors. This method works best when mined upon "very sparse" datasets [5]. The concept of a support difference (SD) for assigning a minimum supports to an itemset was proposed by Kiren et al [7]. This difference considers the maximum support deviation allowable when comparing the itemset's support to that of its corresponding supersets. If a superset's support is within this bound, it can be considered "frequent". This method also dynamically applies an additional minsup for each itemset in the transaction set based upon their current support [7].

A confidence-like measure, relative support, attempts to narrow the search field by reducing the number of candidate itemsets. This measure uses knowledge of rule confidence to prune itemsets that do not have the potential for a high confidence value. A method for finding rare and frequent rules with high support proposes the use of both minsup and the user-defined minRelSup threshold for finding candidate itemsets. The relative support of an itemset I can be found using the formula $RelSup = supp(I)/supp(X)$, where X is a subset of I with the highest support. This method finds all candidate itemsets with the potential for high confidence rules. All rules generated from this method will have confidence $\geq$ minRelSup. This reduces complexity for finding candidate itemsets, but increases complexity for the rule generation process due to the increased number of candidate itemsets [14]. Also the resulting rules sets are still the same as an Apriori-like algorithm and contain all the same redundant and uninteresting rules as before.

3   Combining Itemset Trees and Rare Rule Mining: TRARM-RelSup Algorithm

The previous sections have covered three different types of association mining: frequent, targeted, and rare. Frequent association mining has complexity issues and fails to find all high confidence rules [1,13,10]. Also, not all rules generated are of interest to the user, and the result set it large and unfocused. Targeted association mining reduces some of the complexity of the mining process, but still fails to find all high confidence rules. Rare rule mining is said to have higher complexity than frequent association mining due to the addition of frequent and rare candidate itemsets. This also results in large and unfocused rule sets.

This paper proposes a rare rule mining algorithm, TRARM-RelSup, which combines the efficiency of targeted association mining with the capabilities of rare rule mining. By combining these concepts, we can allow users the ability to choose the areas they want to search in for a more focused results set and we can not only find strong rules, but also rare rules that contain items of user interest. We will begin by determining the user-defined interest itemsets, the UDII query set Ω.

3.1 TRARM-RelSup

TRARM-RelSup finds both frequent rules as well as rare rules above a certain confidence level. This algorithm combines the efficiency of the itemset tree with that of the rare rule mining relative support method. Rules will be generated using this method from itemsets, I, that are above either minsup and minrelsup.

TRARM-RelSup algorithm: In order to find strong association rules containing UDII set = Ω using itemset trees and relative support:

- Step 1: Build an itemset tree based upon the transactional dataset T.
- Step 2: For each query Q_i contained in the UDII set Ω, find all nodes N which contain Q_i in the tree, their supports, and the support of Q_i.
- Step 3: For each N_i in N, return the set C of all itemsets whose support $\geq$ minsup OR whose relSup $\geq$ minRelSup.
- Step 4: For each C_i a subset of C_j in C, remove C_i if support(C_i) = support(C_j).
- Step 5: For each C_i in C, generate rules with LHS = Q_i and RHS = all subsets of C_i - Q_i whose confidence $\geq$ minconf.

Using this algorithm, TRARM-RelSup is capable of discovering high support rules as well as rare rules with high confidence that would be missed by the original itemset tree algorithm. Using the transactional dataset in Example 1 and minsup = 40%, the rule $\{1\} \Rightarrow \{2\}$ with supp = 33% and confidence = 100% would not appear in the results set. TRARM-RelSup would find this rule, and return it as a part of its high confidence rare rules set.

4 Experimental Setup

This section outlines the experiments carried out for testing the TRARM-RelSup algorithm. First, there will be a discussion of the IBM Quest synthetic data generator [4] and the datasets created for the experiments. Second, this section will cover the UDII query sets generated to test various user interest patterns. Finally, the experiments carried out for testing the TRARM-RelSup algorithm will be outlined and comparisons will be discussed.

4.1 IBM Quest Data Generator

IBM QUEST was created by IBM's Quest research group and creates two sets of data D and T. The former is the database of transactions itself and the latter is a set of n candidate frequent itemsets. T populates D by using a Poisson distribution to determine a size for the transactions. Each transaction is created using elements existing in T and are either already existing in the database, or are randomly chosen from T's unused elements. This models market basket style data which contains many co-occurrences [2]. The system is able to create realistic data in large quantities but predictions on algorithm behavior cannot be made. Due to its wide range of variation, it is hard to generalize via empirical analysis [2,4].

The datasets generated for the following experiments can be seen in Table 1.

Table 1. Experimental dataset description

Dataset Transactions	50 1-item datasets		100 1-item datasets	
	Number of nodes in the tree	Width of the itemset tree	Number of nodes in the tree	Width of the itemset tree
25,000	31,993	47	31,383	56
50,000	62,265	47	61,483	58
100,000	62,265	47	61,483	58
200,000	62,265	47	61,483	58

Each of these datasets has an average transaction size of 10 items and contains 50 or 100 unique 1-items. The number of transactions was varied in order to discern whether a larger tree will result in less rare rules due to its denser nature or vice versa. By varying the number of unique itemsets we are able to mimic the diversity found in a transactional dataset but also increase potential for both high and low support rules.

4.2 UDII Query Set

The UDII elements of Ω used for the following experiments range between $1 \leq |Q_i| \leq 3$ and mimic a user's interest in specific items. Ω was populated with itemsets such that their supports ranged between 0.005% and 0.15%. Approximately 50% of the Q_i were selected to contain 1 - 2 itemsets in common with other Q_j, where $Q_i \neq Q_j$. The other 50% are mutually exclusive.

4.3 Experiments

The following experiments are designed to test the ability of the proposed algorithm to discover rare rules that may potentially be interesting for the user.

Experiments for testing the TRARM-RelSup algorithm:

- Build itemset trees on all 8 IBM Quest datasets.
- Query all trees with the 8 UDII query sets using the original itemset tree and TRARM-RelSup algorithm.
- Generate rules from the resulting set $\mathcal{C}$ using the rule generation algorithm.
- Compare the Itemset Tree and TRARM-RelSup results sets.

The goal of these experiments is to demonstrate that TRARM-RelSup is capable of finding not only the strong rules found by the original itemset tree, but also a significant number of the rare rules the original algorithm would miss. Comparisons of the number of rules generated will be made for each of the 8 datasets with emphasis on strong rules discovered vs rare rules discovered. For each experiment, the minSup measure is varied in order to discern the effect of that parameter on rare rules set size and the strong rules set. For both experiment sets, the minRelSup and minConf remain constant as minsup varies. The variations of minsup as well as the values of minRelSup and miconf for the two experiment sets can be seen below:

- Experiment set 1: (minSup = 0.15%, minSup = 0.1%, minSup = 0.05%), minRelSup = 0.005%, minConf = 0.005%.
- Experiment set 2: (minSup = 0.15%, minSup = 0.1%, minSup = 0.05%), minRelSup = 0.02%, minConf = 0.02%.

All parameters were selected based upon the characteristics of the generated datasets.

5   Results

For each tree the number of rules generated ranged between 3,000 - 24,000 depending upon tree size and the number of unique 1-items. Tables 2 and 3 show the number of rules generated by the original itemset tree algorithm and TRARM-RelSup. The strong rules listed were found by both the original itemset tree algorithm and TRARM-RelSup. The rare rules listed were only found by the TRARM-RelSup algorithm.

Table 2. Experiment set 1: number of rules found with minRelSup = 0.005% and minconf = 0.005%

Transactions	1-items	minSup = 0.15%		minSup = 0.1%		minSup = 0.05%	
		strong	rare	strong	rare	strong	rare
25,000	50	293	23,745	478	23,560	1,198	22,840
50,000	50	260	19,096	440	18,916	987	18,369
100,000	50	122	1,846	133	1,835	152	1,816
200,000	50	101	1,024	104	1,021	106	1,019
25,000	100	111	14,457	179	14,389	545	14,023
50,000	100	110	13,085	177	13,018	463	12,732
100,000	100	110	11,324	177	11,257	458	10,976
200,000	100	110	13,085	177	13,018	463	12,732

The number of rules generated by each experiment can be seen in Tables 2 and 3. Several observations can be made based upon these results. First, as minRelSup and minconf decreases, the total number of rules discovered increases

Table 3. Experiment set 2: number of rules found with minRelSup = 0.02% and minconf = 0.02%

		minSup = 0.15%		minSup = 0.1%		minSup = 0.05%	
Transactions	1-items	strong	rare	strong	rare	strong	rare
25,000	50	293	4,786	478	4,601	1,198	3,881
50,000	50	260	3,492	440	3,312	987	2,765
100,000	50	122	1,123	133	1,112	152	1,193
200,000	0	101	1,016	104	1,013	106	1,011
25,000	100	111	2,888	179	2,820	545	2,454
50,000	100	110	2,158	177	2,091	463	1,805
100,000	100	110	1,991	177	1,924	458	1,643
200,000	100	110	2,158	177	2,091	463	1,805

dramatically. Second, decreasing minsup displays a correspondingly steady increases in the number of strong rules, but not as dramatic as the decrease in the number of rare rules using minRelSup. Third, as the number of 1-items in a transaction set increases from 25 to 50, the number of rules decreases marginally. This occurs because as the number of 1-items increases, the tree becomes more varied and the overall supports are reduced to compensate. Finally, as the numbers of transactions increase but the number of 1-items remain constant, the dataset repeats the same items resulting in a more concentrated smaller tree and vice versa. This results in a decrease in the number of rules discovered by the algorithms.

6 Conclusions and Future Work

Based upon the experiments performed for TRARM-RelSup, several observations can be made. First, for all experiments, TRARM-RelSup demonstrated a drastic increase in the number of rare rules discovered. Second, as minsup increases, the size of the strong rules set decreases as rules from the stronger are rule set are transferred to the rare rules set. Finally, given the large number of rare rules discovered by TRARM-RelSup when compared to just the number of strong rules discovered by both the original itemset algorithms, it is apparent that there are many rules missed using only minsup to discover potentially interesting itemsets. Even with an increase in the number of discovered itemsets, the linear nature of querying an itemset tree (where n is the number of transactions) [8] results in a manageable increase in rule mining complexity. Also the resulting rules sets are tailored to the user's specific interests.

Future work involving TRARM-RelSup will include exploration of the possibility of reducing a dataset T to T', where T' has only transactions containing the UDIIs. This is based upon the ARM-PDI-RT method proposed by Sha and Chen in 2011 for finding rare rules using Apriori [10]. Also explorations of removing minsup and only using minRelSup when querying an itemset tree are being contemplated.

References

1. Adda, M., Wu, L., Feng, Y.: Rare Itemset Mining Machine Learning and Applications. In: ICMLA 2007, pp. 73–80 (2007)
2. Agrawal, R., Srikant, R.: Fast algorithms for mining association rules in large databases. In: Proceedings of the 20th International Conference on Very Large Data Bases, pp. 487–499 (1994)
3. Bing, L., Wynne, H., Yiming, M.: Mining association rules with multiple minimum supports. In: Proceedings of the Fifth ACM SIGKDD International Conference on Knowledge Discovery and Data Mining, pp. 337–341 (1999)
4. Cooper, C., Zito, M.: Realistic synthetic data for testing association rule mining algorithms for market basket databases. In: Proceedings of the 11th European Conference on Principles and Practice of Knowledge Discovery in Databases, pp. 398–405 (2007)
5. Gedikli, F., Jannach, D.: Neighborhood-Restricted Mining and Weighted Application of Association Rules for Recommenders. In: Chen, L., Triantafillou, P., Suel, T. (eds.) WISE 2010. LNCS, vol. 6488, pp. 157–165. Springer, Heidelberg (2010)
6. Hafez, A., Deogun, J., Raghavan, V.V.: The Item-Set Tree: A Data Structure for Data Mining. In: Mohania, M., Tjoa, A.M. (eds.) DaWaK 1999. LNCS, vol. 1676, pp. 183–192. Springer, Heidelberg (1999)
7. Kiran, R., Reddy, P.: An improved multiple minimum support based approach to mine rare association rules. In: Computational Intelligence and Data Mining, CIDM 2009, pp. 340–347 (2009)
8. Kubat, M., Hafez, A., Raghavan, V., Lekkala, J., Chen, W.: Itemset trees for targeted association querying. IEEE Transactions on Knowledge and Data Engineering, 1522–1534 (2003)
9. Li, Y., Kubat, M.: Searching for high-support itemsets in itemset trees. Intell. Data Anal., 105–120 (2006)
10. Sha, Z., Chen, J.: Mining association rules from dataset containing predetermined decision itemset and rare transactions. In: Seventh International Conference on Natural Computation, vol. 1, pp. 166–170 (2011)
11. Srikant, R., Agrawal, R.: Mining generalized association rules. In: Proceedings of the 21st International Conference on Very Large Data Bases, VLDB 1995, pp. 407–419 (1995)
12. Srikant, R., Agrawal, R.: Mining quantitative association rules in large relational tables. SIGMOD Rec., 1–12 (1994)
13. Szathmary, L., Napoli, A., Valtchev, P.: Towards rare itemset mining. Tools with Artificial Intelligence, 305–312 (2007)
14. Yun, H., Ha, D., Hwang, B., Ryu, K.: Mining association rules on significant rare data using relative support. Journal of Systems and Software 67, 181–191 (2003)

Incremental Rules Induction Method
Based on Three Rule Layers

Shusaku Tsumoto and Shoji Hirano

Department of Medical Informatics, School of Medicine, Faculty of Medicine
Shimane University
89-1 Enya-cho Izumo 693-8501 Japan
{tsumoto,hirano}@med.shimane-u.ac.jp

Abstract. This paper proposes a new framework for incremental learning based on rule layers constrained by inequalities of accuracy and coverage. Since the addition of an example is classified into one of four
possibili- ties, four patterns of an update of accuracy and coverage are
observed, which give two important inequalities of accuracy and coverage for induction of probabilistic rules. By using these two inequalities,
the proposed method classifies a set of formulae into three layers: the
rule layer, subrule layer and the non-rule layer. Then, the obtained rule
and subrule layers play a central role in updating rules. If a new example contributes to an increase in the accuracy and coverage of a formula
in the subrule layer, the formula is moved into the rule layer. If this
contributes to a decrease of a formula in the rule layer, the formula is
moved into the subrule layer. The proposed method was evaluated on
datasets regarding headaches and meningitis, and the results show that
the proposed method outperforms the conventional methods.

Keywords: incremental rule induction, rough sets, accuracy, coverage,
subrule layer.

1 Introduction

Several symbolic inductive learning methods have been proposed, such as induction of decision trees [1,2,3], and AQ family [4,5,6]. These methods are applied
to discover meaningful knowledge from large databases, and their usefulness is
in some aspects ensured. However, most of the approaches induces rules from all
the data in databases, and cannot induce incrementally when new samples are
derived. Thus, we have to apply rule induction methods again to the databases
when such new samples are given, which causes the computational complexity
to be expensive even if the complexity is n^2.

Thus, it is important to develop incremental learning systems to manage large
databases [7,8]. However, most of the previously introduced learning systems
have the following two problems: first, those systems do not outperform ordinary learning systems, such as AQ15 [6], C4.5 [9] and CN2 [4]. Secondly, those
incremental learning systems mainly induce deterministic rules. Therefore, it is

L. Chen et al. (Eds.): ISMIS 2012, LNAI 7661, pp. 71–80, 2012.

indispensable to develop incremental learning systems which induce probabilistic rules to solve the above two problems.

Extending concepts of rule induction methods based on rough set theory, we introduce a new approach to knowledge acquisition, called PRIMEROSE-INC3 (Probabilistic Rule Induction Method based on Rough Sets for Incremental Learning Methods), which induces probabilistic rules incrementally.

This method first calculates all the accuracy and coverage values of attributes and induces rules. Then, it classifies attributes into rule layers and subrule layers. Then when an additional example is given, the method is classified into one of the four cases. When it updates rule layers and subrule layers and induce rules. The method repeats this process.

This paper is organized as follows: Section 2 briefly describe rough set theory and the definition of probabilistic rules based on this theory. Section 3 discusses problems in the incremental learning of probabilistic rules. Section 4 provides formal analysis of incremental updates of accuracy and coverage, where two important inequalities are obtained. Section 5 presents an induction algorithm for incremental learning based on the above results. , which is then evaluated in Section 6. Finally, Section 7 concludes this paper.

2   Probabilistic Rules

2.1   Rough Sets

In the following sections, we use the following notation introduced by Grzymala-Busse and Skowron [10], based on rough set theory [11]. Let U denote a nonempty finite set called the universe and A denote a nonempty, finite set of attributes, i.e., $a : U \to V_a$ for $a \in A$, where V_a is called the domain of a, respectively. Then a decision table is defined as an information system, $A = (U, A \cup \{d\})$. The atomic formulas over $B \subseteq A \cup \{d\}$ and V are expressions of the form $[a = v]$, called descriptors over B, where $a \in B$ and $v \in V_a$. The set $F(B, V)$ of formulas over B is the least set containing all atomic formulas over B and closed with respect to disjunction, conjunction, and negation.

For each $f \in F(B, V)$, f_A denotes the meaning of f in A, i.e., the set of all objects in U with property f, defined inductively as follows:

1. If f is of the form $[a = v]$, then $f_A = \{s \in U | a(s) = v\}$.
2. $(f \wedge g)_A = f_A \cap g_A; (f \vee g)_A = f_A \vee g_A; (\neg f)_A = U - f_a$.

2.2   Classification Accuracy and Coverage

Definition of Accuracy and Coverage. By use of the preceding framework, classification accuracy and coverage, or true positive rate are defined as follows.

Definition 1. *Let R and D denote a formula in $F(B, V)$ and a set of objects that belong to a decision d. Classification accuracy and coverage(true positive rate) for $R \to d$ is defined as:*

$$\alpha_R(D) = \frac{|R_A \cap D|}{|R_A|} (= P(D|R)), \tag{1}$$

$$\kappa_R(D) = \frac{|R_A \cap D|}{|D|} (= P(R|D)), \tag{2}$$

where $|S|$, $\alpha_R(D)$, $\kappa_R(D)$, and $P(S)$ denote the cardinality of a set S, a classification accuracy of R as to classification of D, and coverage (a true positive rate of R to D), and probability of S, respectively.

2.3 Probabilistic Rules

By use of accuracy and coverage, a probabilistic rule is defined as:

$$R \overset{\alpha,\kappa}{\to} d \quad s.t. \qquad R = \wedge_j [a_j = v_k],$$
$$\alpha_R(D) \ge \delta_\alpha \text{ and } \kappa_R(D) \ge \delta_\kappa. \tag{3}$$

If the thresholds for accuracy and coverage are set to high values, the meaning of the conditional part of probabilistic rules corresponds to the highly overlapped region. This rule is a kind of probabilistic proposition with two statistical measures, which is an extension of Ziarko's variable precision model (VPRS) [12].

3 Problems in Incremental Rule Induction

The most important problem in incremental learning is that it does not always induce the same rules as those induced by ordinary learning systems [1], although an applied domain is deterministic. Furthermore, since induced results are strongly dependent on the former training samples, the tendency of overfitting is larger than in the ordinary learning systems.

The most important factor of this tendency is that the revision of rules is based on the formerly induced rules, which is the best way to suppress the exhaustive use of computational resources. However, when induction of the same rules as ordinary learning methods is required, computational resources will be needed, because all the candidates of the rules should be considered.

Thus, for each step, computational space for deletion of candidates and addition of candidates is needed, which causes the computational speed of incremental learning to be slow. Moreover, in the case when probabilistic rules should be induced, the situation becomes much severer, since the candidates for probabilistic rules become much larger than those for deterministic rules.

4 Theory for Incremental Learning

[Question] **How will accuracy and coverage change when a new sample is added to the dataset ?**

---

[1] Here, ordinary learning systems denote methods that induce all rules by using all the samples.

Usually, datasets will monotonically increase. Let $[x]_R(t)$ and $D(t)$ denote a supporting set of a formula R in given data an a target concept d at time t.

$$[x]_R(t+1) = \begin{cases} [x]_R(t) + 1 & \textit{an additional example satisfies } R \\ [x]_R(t) & \textit{otherwise} \end{cases}$$

$$D(t+1) = \begin{cases} D(t) + 1 & \textit{an additional example belongs} \\ & \textit{to a target concept } d. \\ D(t) & \textit{otherwise} \end{cases}$$

Thus, from the definition of accuracy (Eqn.(1) and coverage (Eqn. (2)), accuracy and coverage may nonmonotonically change. Since the above classification gives four additional patterns, we will consider accuracy and coverage for each case as shown in Table 1. in which $|[x]_R(t)|$, $|D(t)|$ and $|[x]_R \cap D(t)|$ are denoted by n_R, n_D and n_{RD}.

Table 1. Four patterns for an additional example

t:	$[x]_R(t)$	$D(t)$	$[x]_R \cap D(t)$
original	n_R	n_D	n_{RD}

t+1	$[x]_R(t+1)$	$D(t+1)$	$[x]_R \cap D(t+1)$
Both negative	n_R	n_D	n_{RD}
R: positive	$n_R + 1$	n_D	n_{RD}
d: positive	n_R	$n_D + 1$	n_{RD}
Both positive	$n_R + 1$	$n_D + 1$	$n_{RD} + 1$

4.1 Both: Negative

The first case is when an additional example does not satisfy R and does not belong to d. In this case,

$$\alpha(t+1) = \frac{n_{RD}}{n_R} \quad \textit{and} \quad \kappa(t+1) = \frac{n_{RD}}{n_D}.$$

Table 2. An additional example neither satisfies R nor d

t+1	$[x]_R(t+1)$	$D(t+1)$	$[x]_R \cap D(t+1)$
Both negative	n_R	n_D	n_{RD}

Table 3. An additional example only satisfies R

t+1	$[x]_R(t+1)$	$D(t+1)$	$[x]_R \cap D(t+1)$
R: positive	$n_R + 1$	n_D	n_{RD}

4.2   R: Positive

The second case is when an additional example satisfies R, but does not belong to d.

In this case, accuracy and coverage become:

$$\Delta\alpha(t+1) = \alpha(t+1) - \alpha(t) = \frac{n_{RD}}{n_R+1} - \frac{n_{RD}}{n_R} = \frac{-\alpha(t)}{n_R+1}$$

$$\alpha(t+1) = \alpha(t) + \Delta\alpha(t+1) = \frac{\alpha(t)n_R}{n_R+1}.$$

4.3   d: Positive

The third case is when an additional example does not satisfy R, but belongs to d.

Table 4. An additional example only satisfies d

t+1	$[x]_R(t+1)$	$D(t+1)$	$[x]_R \cap D(t+1)$
d: positive	n_R	n_D+1	n_{RD}

$$\Delta\kappa(t+1) = \kappa(t+1) - \kappa(t) = \frac{n_{RD}}{n_D+1} - \frac{n_{RD}}{n_D} = \frac{-\kappa(t)}{n_D+1}$$

$$\kappa(t+1) = \kappa(t) + \Delta\kappa(t+1) = \frac{\kappa(t)n_D}{n_D+1}.$$

4.4   d: Positive

Finally, the fourth case is when an additional example satisfies R and belongs to d.

$$\alpha(t+1) = \frac{\alpha(t)n_R+1}{n_R+1} \quad and\kappa(t+1) = \frac{\kappa(t)n_D+1}{n_D+1}.$$

Thus, in summary, Table 6 gives the classification of four cases of an additional example.

Table 5. An additional example satisfies R and d

t+1	$[x]_R(t+1)$	$D(t+1)$	$[x]_R \cap D(t+1)$
Both positive	n_R+1	n_D+1	$n_{RD}+1$

Table 6. Summary of change of accuracy and coverage

Mode				$\alpha(t+1)$	$\kappa(t+1)$
Both negative	n_R	n_D	n_{RD}	$\alpha(t)$	$\kappa(t)$
R: positive	n_R+1	n_D	n_{RD}	$\frac{\alpha(t)n_R}{n_R+1}$	$\kappa(t)$
d: positive	n_R	n_D+1	n_{RD}	$\alpha(t)$	$\frac{\kappa(t)n_D}{n_D+1}$
Both positive	n_R+1	n_D+1	$n_{RD}+1$	$\frac{\alpha(t)n_R+1}{n_R+1}$	$\frac{\kappa(t)n_D+1}{n_D+1}$

4.5   Updates of Accuracy and Coverage

From Table 6, updates of Accuracy and Coverage can be calculated from the original datasets for each possible case. Since rules is defined as a probabilistic proposition with two inequalities, supporting sets should satisfy the following constraints:

$$\alpha(t+1) > \delta_\alpha \ \kappa(t+1) > \delta\kappa$$

Then, the conditions for updating can be calculated from the original datasets: when accuracy or coverage does not satisfy the constraint, the corresponding formula should be removed from the candidates. On the other hand, both accuracy and coverage satisfy both constraints, the formula should be included into the candidates. Thus, the following inequalities are important for inclusion with an additional positive example.

$$\alpha(t+1) = \frac{\alpha(t)n_R+1}{n_R+1} > \delta_\alpha,$$
$$\kappa(t+1) = \frac{\kappa(t)n_D+1}{n_D+1} > \delta\kappa.$$

The other two inequalities are important for deletion with an additional negative example.

$$\alpha(t+1) = \frac{\alpha(t)n_R}{n_R+1} < \delta_\alpha,$$
$$\kappa(t+1) = \frac{\kappa(t)n_D}{n_D+1} < \delta\kappa.$$

Thus, the following inequalities are obtained for accuracy and coverage.

Theorem 1. *If accuracy and coverage of a formula R to d satisfies one of the following inequalities, then R may include into the candidates of formulae for probabilistic rules, whose update indices will satisfy the constraints for accuracy and coverage with addition of a positive example.*

$$\frac{\delta_\alpha(n_R+1)-1}{n_R} < \alpha_R(D)(t+1) < \delta_\alpha, \tag{4}$$

$$\frac{\delta_\kappa(n_D+1)-1}{n_D} < \kappa_R(D)(t+1) < \delta_\kappa. \tag{5}$$

□

Theorem 2. *If accuracy and coverage of a formula R to d satisfies one of the following inequalities, then R may include into the candidates of formulae for probabilistic rules, whose update indices will not satisfy the constraints for accuracy and coverage with addition of a negative example.*

$$\delta_\alpha \le \alpha_R(D)(t+1) < \frac{\delta_\alpha(n_R+1)}{n_R}, \tag{6}$$

$$\delta_\kappa \le \kappa_R(D)(t+1) < \frac{\delta_\kappa(n_D+1)}{n_D}. \tag{7}$$

It is notable that the lower and upper bounds can be calculated from the original datasets.

Select all the formulae whose accuracy and coverage satisfy the above inequalities They will be a candidate for updates. A set of formulae which satisfies the inequalities for probabilistic rules is called a *rule layer* and a set of formulae which satisfies Eqn (4) and (5) is called a *subrule layer*. Figure 1 illustrates the relations between a rule layer and a sublayer.

5   An Algorithm for Incremental Learning

5.1   Algorithm

To provide the same classificatory power to incremental learning methods as ordinary learning algorithms, we introduce an incremental learning method PRIMEROSE-INC3 (Probabilistic Rule Induction Method based on Rough Sets for Incremental Learning Methods)[2]. Figure 2 illustrates an algorithm of the rule induction method.

6   Experimental Results

PRIMEROSE-INC3 [3] was applied to meningitis [14], which has 198 examples with three classes and 25 attributes.

The proposed method was compared with the former version PRIMEROSE-INC, the non-incremental versions: PRIMEROSE [15] and PRIMEROSE0 [4], and the other three conventional learning methods: C4.5, CN2 and AQ15. The experiments were conducted using the following three procedures. First, these samples randomly split into pseudo-training samples and pseudo-test samples. Second, using the pseudo-training samples, PRIMEROSE-INC3, PRIMEROSE-INC, PRIMEROSE, and PRIMEROSE0 induced rules and the statistical measures [5]. Third, the induced results were tested by the pseudo-test samples. These

---

[2] This is an extended version of PRIMEROSE-INC[13].

[3] The program is implemented by using SWI-prolog.

[4] This version is given by setting δ_α to 1.0 and δ_κ to 0.0.

[5] The thresholds δ_α and δ_κ are set to 0.75 and 0.5, respectively in these experiments.

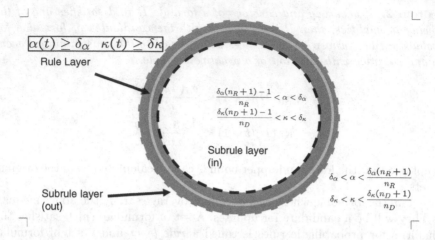

Fig. 1. Intuitive Diagram of Rule and Subrule Layers

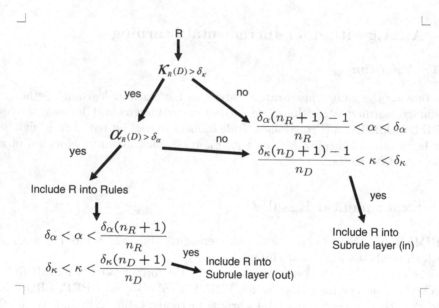

Fig. 2. Algorithm

procedures were repeated 100 times and each accuracy is averaged over 100 trials. Table 7 gives the comparison between PRIMEROSE-INC2 and other rule induction methods with respect to the averaged classification accuracy and the number of induced rules. These results show that PRIMEROSE-INC3 outperformed all the other non-incremental learning methods.

Table 7. Experimental Results: Accuracy and Number of Rules (Meningitis)

Method	Accuracy	No. of Rules
PRIMEROSE-INC3	$89.8 \pm 3.4\%$	77.2 ± 2.2
PRIMEROSE-INC	$89.5 \pm 5.4\%$	67.3 ± 3.5
PRIMEROSE	$89.5 \pm 5.4\%$	67.3 ± 3.0
PRIMEROSE0	$72.1 \pm 2.7\%$	12.9 ± 2.1
C4.5	$85.8 \pm 2.4\%$	16.3 ± 2.1
CN2	$87.0 \pm 3.9\%$	19.2 ± 1.7
AQ15	$86.2 \pm 2.6\%$	31.2 ± 2.1

7 Conclusion

By extending concepts of rule induction methods based on rough set theory, called PRIMEROSE-INC3 (Probabilistic Rule Induction Method based on Rough Sets for Incremental Learning Methods), we have introduced a new approach to knowledge acquisition, which induces probabilistic rules incrementally,

The method classifies elementary attribute-value pairs into three categories: a rule layer, a subrule layer and a non-rule layer by using the inequalities obtained from the proposed framework. This system was evaluated on clinical datasets regarding headache and meningitis. The results show that PRIMEROSE-INC2 outperforms previously proposed methods.

This is a preliminary work on incremental learning based on rough set theory. Our future work will be to conduct further empirical validations and to establish a theoretical basis of this method.

Acknowledgements. This research is supported by Grant-in-Aid for Scientific Research (B) 24300058 from Japan Society for the Promotion of Science(JSPS).

References

1. Breiman, L., Freidman, J., Olshen, R., Stone, C.: Classification and Regression Trees. Wadsworth International Group, Belmont (1984)
2. Cestnik, B., Kononenko, I., Bratko, I.: Assistant 86: A knowledge-elicitation tool for sophisticated users. In: EWSL, pp. 31–45 (1987)
3. Quinlan, J.R.: Induction of decision trees. Machine Learning 1(1), 81–106 (1986)
4. Clark, P., Niblett, T.: The cn2 induction algorithm. Machine Learning 3 (1989)
5. Michalski, R.S.: A theory and methodology of inductive learning. Artif. Intell. 20(2), 111–161 (1983)
6. Michalski, R.S., Mozetic, I., Hong, J., Lavrac, N.: The multi-purpose incremental learning system aq15 and its testing application to three medical domains. In: AAAI, pp. 1041–1047 (1986)
7. Shan, N., Ziarko, W.: Data-based acqusition and incremental modification of classification rules. Computational Intelligence 11, 357–370 (1995)
8. Utgoff, P.E.: Incremental induction of decision trees. Machine Learning 4, 161–186 (1989)

9. Quinlan, J.: C4.5 - Programs for Machine Learning. Morgan Kaufmann, Palo Alto (1993)
10. Skowron, A., Grzymala-Busse, J.: From rough set theory to evidence theory. In: Yager, R., Fedrizzi, M., Kacprzyk, J. (eds.) Advances in the Dempster-Shafer Theory of Evidence, pp. 193–236. John Wiley & Sons, New York (1994)
11. Pawlak, Z.: Rough Sets. Kluwer Academic Publishers, Dordrecht (1991)
12. Ziarko, W.: Variable precision rough set model. Journal of Computer and System Sciences 46, 39–59 (1993)
13. Tsumoto, S.: Incremental rule induction based on rough set theory. In: [16], pp. 70–79
14. Tsumoto, S., Takabayashi, K.: Data mining in meningoencephalitis: The starting point of discovery challenge. In: [16], pp. 133–139
15. Tsumoto, S., Tanaka, H.: Primerose: Probabilistic rule induction method based on rough sets and resampling methods. Computational Intelligence 11, 389–405 (1995)
16. Kryszkiewicz, M., Rybinski, H., Skowron, A., Raś, Z.W. (eds.): ISMIS 2011. LNCS, vol. 6804. Springer, Heidelberg (2011)

RPKM: The Rough Possibilistic K-Modes

Asma Ammar[1], Zied Elouedi[1], and Pawan Lingras[2]

[1] LARODEC, Institut Supérieur de Gestion de Tunis, Université de Tunis
41 Avenue de la Liberté, 2000 Le Bardo, Tunisie
asma.ammar@voila.fr, zied.elouedi@gmx.fr
[2] Department of Mathematics and Computing Science, Saint Marys University
Halifax, Nova Scotia, B3H 3C3, Canada
pawan@cs.smu.ca

Abstract. Clustering categorical data sets under uncertain framework is a fundamental task in data mining area. In this paper, we propose a new method based on the k-modes clustering method using rough set and possibility theories in order to cluster objects into several clusters. While possibility theory handles the uncertainty in the belonging of objects to different clusters by specifying the possibilistic membership degrees, rough set theory detects and clusters peripheral objects using the upper and lower approximations. We introduce modifications on the standard version of the k-modes approach (SKM) to obtain the rough possibilistic k-modes method denoted by RPKM. These modifications make it possible to classify objects to different clusters characterized by rough boundaries. Experimental results on benchmark UCI data sets indicate the effectiveness of our proposed method i.e. RPKM.

1 Introduction

Clustering is an unsupervised learning technique where its main aim is to discover structure of unlabeled data by grouping together similar objects. There are two main categories of clustering methods. They consist of hard (or crisp) methods and soft methods. Crisp approaches cluster each object of the training set into a particular cluster. In contrast to the hard clustering methods, in soft approaches, objects belong to different clusters. Actually, clustering objects into separate clusters presents a difficult task because clusters may not necessarily have precise boundaries. In order to deal with this imperfection, many theories of uncertainty have been proposed. We can mention the fuzzy set, the possibility and the Rough set theories that have been used with different clustering methods to handle uncertainty [1] [5] [6] [11] .

In this work, we develop the rough possibilistic k-modes method denoted by RPKM. This proposed approach is based on the standard k-modes (SKM) and it uses possibility and rough set theories to handle uncertainty in the belonging of the objects to several clusters. Hence, it forms clusters with rough limits. The use of these uncertainty theories provides many advantages. They can express the degree of belongingness of each object to several clusters using possibilistic membership values and they allow the detection of peripheral objects (i.e. an object belongs to several clusters) using the upper and lower approximations.

L. Chen et al. (Eds.): ISMIS 2012, LNAI 7661, pp. 81–86, 2012.
© Springer-Verlag Berlin Heidelberg 2012

2   The K-Modes Method

The k-modes method (SKM) [9] [10] deals with large categorical data sets. It is based on the k-means [7] and it uses the simple matching dissimilarity measure and the frequency-based function to cluster the objects into k clusters.

Assume that we have two objects X_1 and X_2 with m categorical attributes defined respectively by $X_1=(x_{11}, x_{12}, ..., x_{1m})$ and $X_2=(x_{21}, x_{22}, ..., x_{2m})$. The simple matching method denoted by d $(0 \leq d \leq m)$ is described in Equation (1):

$$d(X_1, X_2) = \sum_{t=1}^{m} \delta(x_{1t}, x_{2t}) \ .$$ (1)

Note that $\delta(x_{1t}, x_{2t})$ is equal to 0 if $x_{1t} = x_{2t}$ and equal to 1 otherwise. Moreover, $d=0$ if all the values of attributes relative to X_1 and X_2 are similar. However, if there are no similarities between them, $d=m$.

Generally, if we have a set of n objects $S = \{X_1, X_2, ..., X_n\}$ with its k-modes $Q = \{Q_1, Q_2, ..., Q_k\}$ for set of k clusters $C = \{C_1, C_2, ..., C_k\}$, we can aggregate it into k clusters with $k \leq n$. The minimization of the clustering cost function is min $D(W, Q) = \sum_{j=1}^{k} \sum_{i=1}^{n} \omega_{i,j} d(X_i, Q_j)$, where W is an $n \times k$ partition matrix and $\omega_{i,j} \in \{0, 1\}$ is the membership degree of X_i in C_j.

3   Possibility and Rough Set Theories

3.1   Possibility Theory

Possibility Distribution. Let us consider $\Omega = \{\omega_1, \omega_2, ..., \omega_n\}$ as the universe of discourse where ω_i is an element (an event or a state) from Ω [12]. The possibilistic scale denoted by L is defined in the quantitative setting by $[0, 1]$.

A fundamental concept in possibility theory is the possibility distribution function denoted by π. It is defined from the set Ω to L and associates to each element $\omega_i \in \Omega$ a value from L. Besides, we mention the normalization illustrated by $max_i \{\pi(\omega_i)\} = 1$, the complete knowledge defined by $\exists \omega_0$, $\pi(\omega_0) = 1$ and $\pi(\omega) = 0$ otherwise and the total ignorance defined by $\forall \omega \in \Omega, \pi(\omega) = 1$.

3.2   Rough Set Theory

Information System. Data sets used in RST are presented through a table known as an information table. Generally, an information system (IS) is a pair defined such that $S = (U, A)$ where U and A are finite and nonempty sets. U is the universe and A is the set of attributes. The value set of a also called the domain of a is denoted by V_a and defined for every $a \in A$ such that $a : U \to V_a$.

Indiscernibility Relation. Assume that $S = (U, A)$ is an IS, the equivalence relation $(IND_S(B))$ for any $B \subseteq A$ is defined in Equation (2):

$$IND_S(B) = \{(x, y) \in U^2 | \forall a \in B \ a(x) = a(y)\} \ .$$ (2)

Where $INDS(B)$ is B- indiscernibility relation and $a(x)$ and $a(y)$ denote respectively the value of attribute a for the elements x and y.

Approximation of Sets. Suppose we have an IS: $S = (U, A)$, $B \subseteq A$ and $Y \subseteq U$. The set Y can be described through the attribute values from B using two sets called the B-upper $\overline{B}(Y)$ and the B-lower $\underline{B}(Y)$ approximations of Y.

$$\overline{B}(Y) = \bigcup_{y \in U} \{B(y) : B(y) \cap Y \neq \phi\}. \tag{3}$$

$$\underline{B}(Y) = \bigcup_{y \in U} \{B(y) : B(y) \subseteq Y\}. \tag{4}$$

By $B(y)$ we denote the equivalence class of B identified by the element y. Equivalence class of B describes elementary knowledge called granule.

The B-boundary region of Y is described by: $BN_B(Y) = \overline{B}(Y) - \underline{B}(Y)$.

4   Rough Possibilistic K-Modes

The aim of the RPKM is to deal with uncertainty in the belonging of objects to several clusters based on possibilistic membership degrees and to detect peripheral objects in the clustering task using rough sets.

There are several cases where an object can be similar to different clusters and it can belong to each cluster with a different degree. This fact can be caused by the high similarities between the values of the modes and objects. Clustering such objects to exactly one cluster is difficult and even impossible in some situations. Besides it can make the clustering results inaccurate. In order to avoid this limitation, we propose the RPKM that defines possibilistic membership using possibility theory in order to specify the degree of belongingness of each object to different clusters. Then, the RPKM derives clusters with rough boundaries by applying the upper and the lower approximations. Thus, an object is assigned to an upper or a lower approximation with respect to its possibilistic membership.

4.1   The RPKM Parameters

1. The simple matching dissimilarity measure: The RPKM deals with categorical and certain attributes' values as the SKM does, so the simple matching method is applied, using Equation (1). It indicates how dissimilar are the objects from the clusters by comparing their attributes' values.
2. The possibilistic membership degree: It presents the degree of belongingness of each object of the training set to the available clusters. It is denoted by ω_{ij} where i and j present respectively the object and the cluster. ω_{ij} expresses the degree of similarity between objects and clusters. To obtain this possibilistic membership, which is defined in $[0, 1]$, we transform the dissimilarity value obtained through Equation (1) to a similarity value such that $similarity = total\ number\ of\ attributes - dissimilarity$. After that, we normalize the obtained result.

3. The update of clusters' modes: It uses Equation (5).

$$\forall j \in k, t \in A, Mode_{jt} = \arg\max_{v} \sum_{i=1}^{n} \omega_{ijtv} \ . \tag{5}$$

Where $\forall i \in n, \max_j (\omega_{ij}) = 1$, ω_{ijtv} is the possibilistic membership degree of the object i relative to the cluster j defined for the value v of the attribute t and A is the total number of attributes.

4. The deriving of the rough clusters from the possibilistic membership: We adapt the ratio [3] [4] to specify to which region each peripheral object belongs. In fact, after specifying the final ω_{ij} for each object, we compute the ratio defined by Equation (6).

$$ratio_{ij} = \frac{\max \omega_i}{\omega_{ij}}. \tag{6}$$

After that, the ratio relative to each object is compared to a threshold≥ 1 [3] [4] denoted by T. If $ratio_{ij} \leq T$ it means that the object i belongs to the upper bound of the cluster j. If an object belongs to the upper bound of exactly one cluster j, it means that it belongs to the lower bound of j. Note that every object in the data set satisfies the rough sets' properties [3].

4.2   The RPKM Algorithm

1. *Select randomly the k initial modes, one mode for each cluster.*
2. *Compute the distance measure between all objects and modes using Equation (1) then precise the membership degree of each object to the k clusters.*
3. *Allocate an object to the k clusters using the possibilistic membership.*
4. *Update the cluster mode using Equation (5).*
5. *Retest the similarity between objects and modes. Reallocate objects to clusters using possibilistic membership degrees then update the modes.*
6. *Repeat step 5 until all objects are stable.*
7. *Derive the rough clustering through the possibilistic membership degrees by computing the ratio of each object using Equation (6) and assigning each object to the upper or the lower bound of the cluster.*

5   Experiments

5.1   The Framework

In order to test the RPKM, we have used several real-world data sets taken from UCI machine learning repository [8]. They consist of Shuttle Landing Control (SLC), Balloons (Bal), Post-Operative Patient (POP), Congressional Voting Records (CVR), Balance Scale (BS), Tic-Tac-Toe Endgame (TE), Solar-Flare (SF) and Car Evaluation (CE).

5.2   Evaluation Criteria

The evaluation criteria consist of the accuracy (AC), the iteration number (IN) and the execution time (ET). The AC$=\frac{\sum_{l=1}^{k} a_C}{n}$ is the rate of the correctly classified objects, where n is the total number of objects and a_C is the number of objects correctly classified in C. It can be verified that the objects with the highest degree are in the correct clusters. The IN denotes the number of iterations needed to classify the objects into k rough clusters. The ET is the time taken to form the k rough clusters and to classify the objects.

5.3   Experimental Results

In this section, we make a comparative study between the RPKM, the SKM and the KM-PM (the k-modes method based on possibilistic membership) proposed in [2] which is an improved version of the SKM where each object is assigned to all the clusters with different memberships. This latter specifies how similar is each object to different clusters. However, it cannot detect the boundary region computed using the set approximations as the RPKM does.

Table 1. The evaluation criteria of RPKM vs. SKM and KM-PM

	Data sets	SLC	Bal	POP	CVR	BS	TE	SF	CE
SKM	AC	0.61	0.52	0.684	0.825	0.785	0.513	0.87	0.795
	IN	8	9	11	12	13	12	14	11
	ET/s	12.43	14.55	17.23	29.66	37.81	128.98	2661.63	3248.61
KM-PM	AC	0.63	0.65	0.74	0.79	0.82	0.59	0.91	0.87
	IN	4	4	8	6	2	10	12	12
	ET/s	10.28	12.56	15.23	28.09	31.41	60.87	87.39	197.63
RPKM	AC	0.67	0.68	0.77	0.83	0.88	0.61	0.94	0.91
	IN	4	4	8	6	2	10	12	12
	ET/s	11.04	13.14	16.73	29.11	35.32	70.12	95.57	209.68

As shown in Table 1, the RPKM has improved the clustering task of both the SKM and the KM-PM. Both of the KM-PM and the RPKM allow objects to belong to several clusters, in contrast to the SKM which forces each object to belong to exactly one cluster. The difference in the behaviors leads to different clustering results. Generally, the RPKM and the KM-PM provide better results than the SKM based on the three evaluation criteria. Furthermore, we observe that the RPKM gives the most accurate results for all data sets. Moving to the second evaluation criterion (i.e. the IN), the KM-PM and the RPKM need the same number of iterations to cluster the objects, to give the final partitions and to detect the peripheral objects. However, the last evaluation criterion i.e. the execution time relative to the RPKM is higher than the execution time of the KM-PM since, our proposed approach needs more time to detect boundary regions and to specify to which bound (upper or lower) each object belongs. We can observe that the ET of the RPKM is lower than the ET relative to the SKM. This result is due to the time taken by the SKM to cluster each object to distinct cluster which slows down the SKM algorithm. Moreover, in the SKM it is possible to obtain several modes for a particular cluster which leads to random choice, this latter may affect the stability of the partition and as a result, increases the execution time.

Generally, the RPKM has improved the clustering task by providing more accurate results through the detection of clusters with rough limits.

6 Conclusion

In this paper, we have highlighted the uncertainty in the clustering task by combining the SKM with the possibility and the rough set theories. This combination has been addressed in the RPKM which successfully clustered objects using possibilistic membership degrees and detected objects that belong to rough clusters.

The RPKM has been tested and evaluated on several data sets from UCI machine learning repository [8]. Experimental results on well-known UCI data sets have proved the effectiveness of our method compared to the SKM and the KM-PM.

References

1. Ammar, A., Elouedi, Z.: A New Possibilistic Clustering Method: The Possibilistic K-Modes. In: Pirrone, R., Sorbello, F. (eds.) AI*IA 2011. LNCS, vol. 6934, pp. 413–419. Springer, Heidelberg (2011)
2. Ammar, A., Elouedi, Z., Lingras, P.: K-Modes Clustering Using Possibilistic Membership. In: Greco, S., Bouchon-Meunier, B., Coletti, G., Fedrizzi, M., Matarazzo, B., Yager, R.R. (eds.) IPMU 2012, Part III. CCIS, vol. 299, pp. 596–605. Springer, Heidelberg (2012)
3. Joshi, M., Lingras, P., Rao, C.R.: Correlating Fuzzy and Rough Clustering. Fundamenta Informaticae (2011) (in press)
4. Lingras, P., Nimse, S., Darkunde, N., Muley, A.: Soft clustering from crisp clustering using granulation for mobile call mining. In: Proceedings of the GrC 2011: International Conference on Granular Computing, pp. 410–416 (2011)
5. Lingras, P., West, C.: Interval Set Clustering of Web Users with Rough K-means. Journal of Intelligent Information Systems 23, 5–16 (2004)
6. Lingras, P., Hogo, M., Snorek, M., Leonard, B.: Clustering Supermarket Customers Using Rough Set Based Kohonen Networks. In: Zhong, N., Raś, Z.W., Tsumoto, S., Suzuki, E. (eds.) ISMIS 2003. LNCS (LNAI), vol. 2871, pp. 169–173. Springer, Heidelberg (2003)
7. MacQueen, J.B.: Some methods for classification and analysis of multivariate observations. In: Proceeding of the 5th Berkeley Symposium on Math., Stat. and Prob., pp. 281–296 (1967)
8. Murphy, M.P., Aha, D.W.: Uci repository databases (1996), http://www.ics.uci.edu/mlearn
9. Huang, Z.: Extensions to the k-means algorithm for clustering large data sets with categorical values. Data Mining and Knowledge Discovery 2, 283–304 (1998)
10. Huang, Z., Ng, M.K.: A note on k-modes clustering. Journal of Classification 20, 257–261 (2003)
11. Pal, N.R., Pal, K., Keller, J.M., Bezdek, J.C.: A possibilistic fuzzy c-means clustering algorithm. IEEE Transactions on Fuzzy Systems, 517–530 (2005)
12. Zadeh, L.A.: Fuzzy sets as a basis for a theory of possibility. Fuzzy Sets and Systems 1, 3–28 (1978)

Exploration and Exploitation Operators
for Genetic Graph Clustering Algorithm

Jan Kohout[1] and Roman Neruda[2,*]

[1] Faculty of Mathematics and Physics, Charles University in Prague, Czech Republic
[2] Institute of Computer Science, Academy of Sciences of the Czech Republic, Prague,
Czech Republic

Abstract. In this paper, two evolutionary algorithms for clustering in the domain of directed weighted graphs are proposed. Several genetic operators are analyzed with respect to maintaining the balance between exploration and exploitation properties. The approach is extensively tested on medium-sized random graphs.

1 Introduction

The goal of a clustering task in general is to divide a given set of objects into subsets (clusters) such that objects from the same subset are more similar to each other than objects that come from different subsets. In case of graphs, the objects to be organized into homogeneous clusters are nodes (vertices) of the graph. The measure of similarity of the nodes is then given by edges of that graph and their weights. Unlike in the case of metric spaces, it can be difficult to say how much similar each two nodes are. So, the algorithms designed for clustering in metric spaces can't be easily modified to work properly on graph nodes. Graph clustering can have many applications in area of data analysis and exploration, especially when the analyzed data have naturally a structure of a graph. For example, it can be utilized for communities detection in social networks or for organizing world wide web pages into groups for improving searching results among them.

Our attention is focused on *directed weighted graphs* (i.e., graphs where each edge is an ordered pair of nodes and has assigned a real number as its weight). We further assume that the range of the weight function is a subset of interval $[0, 1]$, and we study non-overlapping clusterings where the number of clusters is not known in advance.

As a measure of similarity we will take the connectedness of nodes by the edges and their weights (the higher the weight of the edge, the more similar the nodes are). A sufficient clustering quality measure should take to account both intra-cluster density of edges and inter-cluster sparsity of edges, as expressed by $Coverage$, mentioned in [1]:

$Coverage(C) = \frac{\sum_{e' \in I_C(E)} w(e')}{\sum_{e \in E} w(e)}$, where $I_C(E) \subseteq E$ is a set of graph edges that are intra-cluster edges in clustering C.

However, this is not a good quality measure for optimization, because the maximum is always reached for the clustering composed of just one cluster containing all nodes,

* This research has been supported by the Czech Republic Science Foundation project no. P202/11/1368.

independently on the edges. The *Modularization Quality (MQ)*, was proposed in [2], here we modify it for weighted graphs. Let C is a clustering composed of k clusters $V_1, ..., V_k$. Denote $I_i(E)$ a set of intra-cluster edges of cluster V_i, $Ex_{ij}(E)$ a set of inter-cluster edges between clusters V_i and V_j. Define $A_i = \frac{\sum_{e \in I_i(E)} w(e)}{|V_i|^2}$, $B_{ij} = 0$ for $i = j$, and $\frac{B_{ij} = \sum_{e \in Ex_{ij}(E)} w(e)}{2|V_i||V_j|}$ for $i \neq j$. Finally, we can define $MQ(C)$ as follows:

$$MQ(C) = \begin{cases} A_1 & k = 1 \\ \frac{\sum_{i=1}^{k} A_i}{k} - \frac{\sum_{i,j=1}^{k} B_{ij}}{\frac{k(k-1)}{2}} & k > 1 \end{cases} \tag{1}$$

In [3], the authors present an alternative function for measuring clustering quality, the $Performance(C) = 1 - \frac{2|E|(1 - 2Coverage(C)) + \sum_{i=1}^{k} |V_i||V_i - 1|}{|V||V-1|}$.

2 Algorithms Proposal

Two evolutionary clustering algorithms for graphs domain were proposed and tested. The algorithms differ in genetic operators used and also in the way how the evolutionary search process is guided.

The *Simple Genetic Clusterizer (SGC)* is the basic version of the genetic algorithm used for clustering graph nodes. The main new characteristics of this algorithm are the following:

Integer encoding is used as mentioned in [4]. Each individual is represented by a vector of $|V|$ integers, the number j at the i-th position identifies that the node i belongs to cluster number j. The function *Performance* is used for fitness. The *Tournament* selection is used in this algorithm, with tournament size 8 and $p = 0.75$. Moreover, the best individual is preserved for the next generation in each cycle.

Several crossover operators were implemented that work with the encoding of solutions into integer vectors, such as the standard *Uniform* and *Twopoint* crossovers. Another crossover operator which is more cluster-oriented is called *Clusterwise* crossover. This crossover is similar to the *Uniform* crossover with the difference that the exchange of information between parents is made at the level of clusters. The operator works in the following manner: the clusters in both solutions are numbered such that the cluster containing the node number 1 has number 0, cluster number 1 is the cluster which contains node with the lowest number that does not belong to the cluster 0 etc. By this process, the clusters in both parental solutions are numbered from 0 to $k_1 - 1$ ($k_2 - 1$ respectively), where k_1, k_2 are the cluster counts from the solutions. Then, clusters with matching numbers are randomly exchanged like genes in the *Uniform* crossover, the unaffected genes by the clusters exchange are copied from parents into offsprings.

We have implemented several mutation operators including the ones that support exploiting by refining the solutions. The *Basic* mutation is an implementation of the uninformed value changing mutation. A specified number of nodes is randomly chosen and for each of them the assignment of the cluster number is randomly changed. The *Swapping* mutation interchanges cluster numbers in one or more randomly selected pairs of nodes. The *Splitting* mutation operator selects a cluster at random and splits it into two clusters. Nodes from the original cluster are divided into new clusters randomly. The *Joining* mutation randomly selects two clusters and merges them together. The *MinDensity splitting* operator supports exploiting. It is similar to *Splitting*, but the cluster S_i selected for splitting is the cluster with the lowest intra-cluster edge density, which is defined as follows: $Density(S_i) = \frac{\sum_{e \in I_i(E)} w(e)}{|S_i|(|S_i|-1)}$. Nodes from the original cluster are not distributed randomly to new clusters, instead of this, one node is selected from which the process of building the new cluster is started. During the building of the new cluster, nodes from the original cluster are added to the new cluster while density of the new cluster increases. Nodes that did not increase density of the new cluster remain in the original cluster. The *MaxCut joining* is the more exploiting version of the *Joining* mutation. Two clusters selected for joining are chosen such that the total sum of all edge weights between them is maximal among all possible pairs of clusters from the solution. Finally, the *Cluster refining* mutation selects one cluster from the solution which will be refined. Refinement of the cluster is performed utilizing the cluster density, as defined above. The cluster for refinement is selected according to the following rule: with probability p, the cluster with the minimal density is selected for refinement, otherwise, with probability $1 - p$, the cluster with the maximal density selected. The refinement of the cluster works subsequently: For each edge $e = (v, u)$ which crosses the border of the cluster (i.e., one node of the edge lies inside the cluster and one node lies outside), examine all of the three following possibilities and continue with the one with the highest density of the refined cluster: (i) exclude both nodes u and v from the refined cluster, (ii) include both nodes in the refined cluster (iii) leave the original state (e crosses the border). This process is repeated for each border-crossing edge.

A summarization of SGC using pseudo-code can be viewed in the listing of Algorithm 1.

Algorithm 1. Simple genetic clusterizer (SGC)

1: Set mutations to be used in the *Master mutation*
2: **while** Maximal number of evolution cycles not reached **do**
3:     Select parents from the current generation, preserve the best individual
4:     Create offsprings by the crossover
5:     Mutate individuals using the *Master mutation*
6: **end while**
7: Return the best individual

The *Two Phase Genetic Clusterizer (TPGC)* is based on the previous one, but tries to deal with the problem of exploration versus exploitation. The evolution process runs in two phases - the exploring phase and the refining phase. These two phases are regularly alternated during the running of the algorithm. The phases differ in a set of genetic operators used and, furthermore, a refining procedure is repeatedly called the in refining phase. Lengths of both phases are set via parameters of the algorithm. In the exploring phase, the following mutations are used:*Basic*, *Splitting*, and *Joining*. On the other hand, in the refining phase, mutations in the list are changed to *Cluster Refining*, *MinDensity Splitting*, and *MaxCut Joining*.

Moreover, a special procedure that refines one selected individual is applied in each iteration of the refining phase. The individual for refining is selected randomly from the population. Depending on the algorithm's settings, more (all of them randomly selected) individuals can be refined in each cycle. The refining procedure takes each node of the graph and tries to add it to each of the clusters encoded in the individual which is being refined. At the end of the algorithm, the refining procedure is called for the last time to improve the best solution, just before it is returned as the result. The search process of TPGC is illustrated by pseudo-code in the listing of Algorithm 2.

Algorithm 2. Two phase genetic clusterizer (TPGC)

 1: **while** Maximal number of evolution cycles not reached **do**
 2:    **if** Refining phase is on **then**
 3:       Select individuals for the next generation
 4:       Mutate individuals with refining phase mutations
 5:       Refine randomly selected individual(s) by the refining procedure
 6:    **else**
 7:       Select parents from the current generation
 8:       Create offsprings by the crossover
 9:       Mutate individuals with exploring phase mutations
10:    **end if**
11:    **if** terminal condition satisfied **then**
12:       Escape from the evolution loop
13:    **end if**
14: **end while**
15: Refine the best individual by the refining procedure and return it

---

3   Experiments

In this section we present several experiments performed with the implemented cluster-izers to assess their performance and estimate proper values for their settings. Due to the limited space, only the main results can be included here, for the complete report one can cf. [5]. For these experimental purposes, six random graphs $G_1, ..., G_6$ were generated, each with 50 nodes but with different number of edges. The weights of all edges were randomly chosen from range [0,1]. Numbers of edges in the graphs were: G_1 : 293 edges, G_2 : 225 edges, G_3 : 316 edges, G_4 : 172 edges, G_5 : 331 edges, G_6 : 251 edges.

The first pair of experiments was evaluating the performance of mutation and cross-over operators. The results suggest that there is a nontrivial benefit from usage of *Clusterwise* and *Uniform* crossover operators (with both of them the quality reached is roughly the same). Likewise, the *Basic* mutation tend to be more successful than in the case when the *Swapping* mutation is used too often. The third experiment was test-ing the influence of lengths of exploring and refining phases on the quality of the final solution in Two Phase Genetic Clusterizer. The set of values tested for the exploring phase length was $\{5, 9, 13, 17\}$ while the set of values for refining phase length was $\{1, 2, 3, 4, 5\}$. TPGC can use the same genetic operators as SGC and so it is in this case, except the mutation operators. The configuration of the algorithm for this test was following: The results of this experiment show that, on the one hand, exact lengths of both phases have not dramatic impact on the quality of the solution but, on the other hand, it seems that shorter lengths have a little bit greater chance to find better solution. This is demonstrated by the graphs in the figure 1.

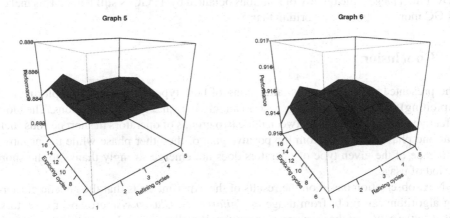

Fig. 1. *Performance* of the returned solutions depending on the refining phase and the exploring phase lengths

The final experiment was comparing the SGC and TGC algorithms. This experiment has two goals: to test usability of a fitness function where two clustering quality measures are combined together and to compare results of SGC and TPGC when this new fitness function is used. In the previous experiments, the fitness function was realized by *Performance* quality measure. Because the graph clustering problem is quite complex, there is not exactly one function which would be the best one to be used as a subject of optimization , so we have decided to combine two measures: $Fitness(C) = \frac{1+MQ(C)}{2} + Performance(C)$.

Both of the tested clusterizers were executed on all graphs $G_1, ..., G_6$ for fifty times to obtain average values for results that are presented in the table 1.

Table 1. Comparison of TPGC (70 cycles) and SGC (190 cycles), both used with combined fitness function

Graph	MQ	Performance	Coverage	Time (s)
G_1	0.366 (0.348)	0.912 (0.906)	0.528 (0.508)	1.14 (2.30)
G_2	0.321 (0.315)	0.937 (0.929)	0.581 (0.517)	1.07 (2.08)
G_3	0.287 (0.269)	0.898 (0.886)	0.443 (0.407)	1.17 (2.16)
G_4	0.296 (0.289)	0.947 (0.938)	0.468 (0.425)	1.17 (2.06)
G_5	0.328 (0.300)	0.881 (0.874)	0.314 (0.281)	1.28 (2.22)
G_6	0.274 (0.248)	0.910 (0.899)	0.342 (0.304)	1.32 (2.15)

The experiment has proven that TPGC algorithm is definitely more successful than SGC by means of all clustering quality measures used on all graphs. TPGC is a little bit more time consuming than SGC (when executed for the same number of cycles), but but the letter one is able to find solutions of higher quality, and moreover, even if the number of iterations of SGC is significantly higher (which makes the runtime of SGC much longer) the quality of solutions obtained by TPGC is still better. This makes TPGC more perspective algorithm than SGC.

4   Conclusion

The presented results show that alternating of both types of operators (exploring and exploiting) during the searching process can improve quality of the solutions. The most effective way is to alternate between these two groups of operators in short periods such that one phase can profit from the positive gain of the other phase while the negative influence of the given type of operators does not emerge strongly thanks to the short period of usage.

Next observation based on the results of the experiments is that the genetic clustering algorithm can profit from usage of *Uniform* and *Clusterwise* crossover operators, because the nature of these crossover operators is well compatible with this type of task and with the encoding used. The clusterizers proposed in this work are included in the *Graph Clusterizer* application which is under continuous development [6].

References

1. Gaertler, M.: Clustering. In: Brandes, U., Erlebach, T. (eds.) Network Analysis. LNCS, vol. 3418, pp. 178–215. Springer, Heidelberg (2005)
2. Dias, C.R., Ochi, L.S.: Efficient evolutionary algorithms for the clustering problems in directed graphs. In: Proc. of the IEEE Congress on Evolutionary Computation, pp. 983–988. IEEE Computer Press (2003)
3. Brandes, U., Gaertler, M., Wagner, D.: Experiments on Graph Clustering Algorithms. In: Di Battista, G., Zwick, U. (eds.) ESA 2003. LNCS, vol. 2832, pp. 568–579. Springer, Heidelberg (2003)
4. Hruschka, E., Campello, R., Freitas, A., de Carvalho, A.: A survey of evolutionary algorithms for clustering. IEEE Transactions on Systems, Man, and Cybernetics, Part C: Applications and Reviews (2009)
5. Kohout, J., Neruda, R.: Two genetic graph clustering algoritms. Technical Report V-1161, Institute of Computer Science, Academy of Sciences of the Czech Republic (2012)
6. Kohout, J.: Graph clusterizer web page (2012),
   http://sourceforge.net/projects/gclusterizer/

Local Probabilistic Approximations
for Incomplete Data

Patrick G. Clark[1], Jerzy W. Grzymala-Busse[1,2], and Martin Kuehnhausen[1]

[1] Department of Electrical Engineering and Computer Science, University of Kansas,
Lawrence, KS 66045, USA
{pclark,jerzy,mkuehnha}@ku.edu
[2] Department of Expert Systems and Artificial Intelligence, University of
Information Technology and Management, 35-225 Rzeszow, Poland

Abstract. In this paper we introduce a generalization of the local approximation called a local probabilistic approximation. Our novel idea is associated with a parameter (probability) α. If $\alpha = 1$, the local probabilistic approximation becomes a local lower approximation; for small α, it becomes a local upper approximation. The main objective of this paper is to test whether proper local probabilistic approximations (different from local lower and upper approximations) are better than ordinary local lower and upper approximations. Our experimental results, based on ten-fold cross validation, show that all depends on a data set: for some data sets proper local probabilistic approximations are better than local lower and upper approximations; for some data sets there is no difference, for yet other data sets proper local probabilistic approximations are worse than local lower and upper approximations.

1 Introduction

Lower and upper approximations are fundamental concepts of rough set theory [1, 2]. An idea of the local and global approximations was introduced in [3]. Later on, local and global approximations were discussed in [4]. Local approximations are unions of complex blocks which, in turn, are intersections of attribute-value pair blocks. Global approximations are unions of characteristic sets, where the characteristic set is a generalization of the elementary set, well known in rough set theory.

In this paper we introduce a novel idea of the local probabilistic approximation, a generalization of the local lower and upper approximations. The local probabilistic approximation is defined using a parameter α that has an interpretation as a conditional probability of the concept given complex block. If $\alpha = 1$, the local probabilistic approximation is the local lower approximation; if α is small (in this paper 0.001), the local probabilistic approximation is the local upper approximation.

Theoretical properties of global probabilistic approximations, based on an equivalence relation, were studied for many years in variable precision rough set theory, Bayesian rough sets, etc. [5–10]. Global probabilistic approximations

L. Chen et al. (Eds.): ISMIS 2012, LNAI 7661, pp. 93–98, 2012.

based on an arbitrary binary relation were defined in [11]. First results of their practical usefulness were published in [12].

In this paper we will distinguish two interpretations of a missing attribute value: lost values and "do not care" conditions. If an attribute value was originally given but now is not accessible (e.g., was erased or forgotten) we will call it *lost*. If a data set consists lost values, we will try to induce rules from existing data. Another interpretation of a missing attribute value is based on a refusal to answer a question, e.g., some people may refuse to tell their citizenship status, such a value will be called a "do not care" condition. In data sets with "do not care" conditions we will replace such a missing attribute value with all possible attribute values.

2   Attribute-Value Pair Blocks

We assume that the input data sets are presented in the form of a *decision table*. Rows of the decision table represent *cases* and columns are labeled by *variables*. The set of all cases will be denoted by U. Some variables are called *attributes* while one selected variable is called a *decision* and is denoted by d. The set of all attributes will be denoted by A.

An important tool to analyze data sets is a *block of an attribute-value pair*. Let (a, v) be an attribute-value pair. For *complete* decision tables, i.e., decision tables in which every attribute value is specified, a block of (a, v), denoted by $[(a, v)]$, is the set of all cases x for which $a(x) = v$, where $a(x)$ denotes the value of the attribute a for the case x. For incomplete decision tables the definition of a block of an attribute-value pair is modified.

- If for an attribute a there exists a case x such that $a(x) = ?$, i.e., the corresponding value is lost, then the case x should not be included in any blocks $[(a, v)]$ for all values v of attribute a,
- If for an attribute a there exists a case x such that the corresponding value is a "do not care" condition, i.e., $a(x) = *$, then the case x should be included in blocks $[(a, v)]$ for all specified values v of attribute a.

A special block of a decision-value pair is called a *concept*. For a case $x \in U$ the *characteristic set* $K_B(x)$ is defined as the intersection of the sets $K(x, a)$, for all $a \in B$, where the set $K(x, a)$ is defined in the following way:

- If $a(x)$ is specified, then $K(x, a)$ is the block $[(a, a(x))]$ of attribute a and its value $a(x)$,
- If $a(x) = ?$ or $a(x) = *$ then the set $K(x, a) = U$.

Characteristic set $K_B(x)$ may be interpreted as the set of cases that are indistinguishable from x using all attributes from B and using a given interpretation of missing attribute values.

3   Local Probabilistic Approximations

For incomplete data sets, a set X will be called B-*globally definable* if it is a union of some characteristic sets $K_B(x)$, $x \in U$. A set T of attribute-value pairs, where all attributes belong to set B and are distinct, will be called a B-*complex*. We will discuss only *nontrivial complexes*, i.e., such complexes that the intersection of all attribute-value blocks from a given complex is not the empty set. A block of B-complex T, denoted by $[T]$, is defined as the set $\cap\{[t] \mid t \in T\}$.

For an incomplete decision table and a subset B of A, a union of intersections of attribute-value pair blocks of attribute-value pairs from some B-complexes, will be called a B-*locally definable* set. Any set X that is B-globally definable is B-locally definable, the converse is not true.

Let X be any subset of the set U of all cases. Let $B \subseteq A$. In general, X is not a B-definable set, locally or globally. A B-*local probabilistic approximation* of the set X with the parameter α, $0 < \alpha \le 1$, denoted by $appr_\alpha^{local} B(X)$, is defined as follows

$$\cup\{[T] \mid \exists \ a \ family \ \mathcal{T} \ of \ B\text{-}complexes \ T \ of \ X, T \in \mathcal{T}, \ Pr(X \mid [T]) \ge \alpha\}.$$

Due to computational complexity, in our experiments we used a heuristic approach to computing another local probabilistic approximation, denoted by $appr_\alpha^{mlem2}(X)$, since it is inspired by the MLEM2 rule induction algorithm [4]. Using this approach, $appr_\alpha^{mlem2}(X)$ is constructed from A-complexes Y that are the most relevant to X, i.e., with $|X \cap Y|$ as large as possible, if there is more than one A-complex that satisfies this criterion, the largest conditional probability of X given Y is the next criterion to select an A-complex. Note that if two A-complexes are equally relevant, then the second criterion selects an A-complex with the smaller block cardinality.

4   Experiments

In our experiments we used eight real-life data sets taken from the University of California at Irvine *Machine learning Repository*. These data sets were enhanced by replacing 35% of existing attribute values by missing attribute values, separately by *lost* values and by "*do not care*" *conditions*. Thus, for any data set, two data sets were created for experiments, one with missing attribute values interpreted as lost values and the other one as "do not care" conditions.

The main objective of our research was to test whether local probabilistic approximations, different from local lower and upper approximations, are better than local lower and upper approximations in terms of an error rate. Therefore, we conducted experiments of a ten-fold cross validation increasing the parameter α in local probabilistic approximations inspired by MLEM2, with increments equal to 0.1, from 0 to 1.0. For a given data set, in our experiments we used ten-fold cross validation with a random re-ordering of all cases, but during all eleven experiments this order was constant, i.e., all ten pairs of training and testing data subsets were the same. Results of our experiments are presented in Figures 1–4.

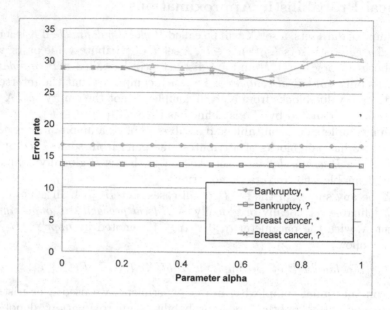

Fig. 1. Error rates for data sets *Bankruptcy* and *Breast cancer* with lost values, denoted by ? and "do not care" conditions denoted by *

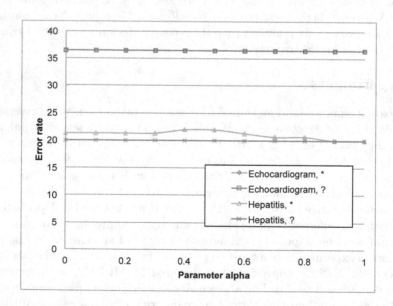

Fig. 2. Error rates for data sets *Echocardiogram* and *Hepatitis* with lost values, denoted by ? and "do not care" conditions denoted by *

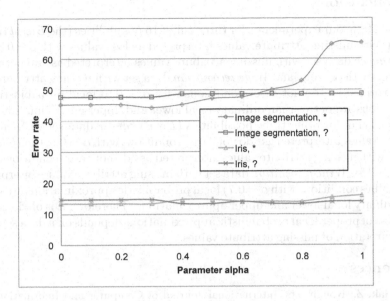

Fig. 3. Error rates for data sets *Image segmentation* and *Iris* with lost values, denoted by ? and "do not care" conditions denoted by *

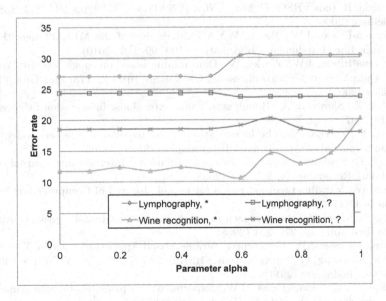

Fig. 4. Error rates for data sets *Lymphography* and *Wine recognition* with lost values, denoted by ? and "do not care" conditions denoted by *

5   Conclusions

As follows from our experiments, for three out of 16 possibilities (the *Breast cancer* data set with missing attribute values interpreted as lost values with α =0.8, the *Breast cancer* data set with missing attribute values interpreted as "do not care" conditions with $\alpha = 0.6$, and *Wine recognition* data set with missing attribute values interpreted as "do not care" conditions with $\alpha = 0.6$) local probabilistic approximations are better than ordinary local lower and upper approximations. On the other hand, for other three possibilities (*Breast cancer* data set with missing attribute values interpreted as "do not care" conditions with $\alpha = 0.9$, the *Hepatitis* data set with missing attribute values interpreted as "do not care" conditions with $\alpha = 0.5$ and for *Wine recognition* data set with missing attribute values interpreted as lost values conditions with $\alpha = 0.7$) local probabilistic approximations are worse than ordinary local lower and upper approximations. Therefore, it is obvious that usefulness of proper local probabilistic approximations depends on a data sets and an interpretation of missing attribute values.

References

1. Pawlak, Z.: Rough sets. International Journal of Computer and Information Sciences 11, 341–356 (1982)
2. Pawlak, Z.: Rough Sets. Theoretical Aspects of Reasoning about Data. Kluwer Academic Publishers, Dordrecht (1991)
3. Grzymala-Busse, J.W., Rzasa, W.: Local and Global Approximations for Incomplete Data. In: Greco, S., Hata, Y., Hirano, S., Inuiguchi, M., Miyamoto, S., Nguyen, H.S., Słowiński, R. (eds.) RSCTC 2006. LNCS (LNAI), vol. 4259, pp. 244–253. Springer, Heidelberg (2006)
4. Grzymala-Busse, J.W., Rzasa, W.: A local version of the MLEM2 algorithm for rule induction. Fundamenta Informaticae 100, 99–116 (2010)
5. Grzymala-Busse, J.W., Ziarko, W.: Data mining based on rough sets. In: Wang, J. (ed.) Data Mining: Opportunities and Challenges, pp. 142–173. Idea Group Publ., Hershey (2003)
6. Pawlak, Z., Skowron, A.: Rough sets: Some extensions. Information Sciences 177, 28–40 (2007)
7. Ślęzak, D., Ziarko, W.: The investigation of the bayesian rough set model. International Journal of Approximate Reasoning 40, 81–91 (2005)
8. Yao, Y.Y.: Probabilistic rough set approximations. International Journal of Approximate Reasoning 49, 255–271 (2008)
9. Ziarko, W.: Variable precision rough set model. Journal of Computer and System Sciences 46(1), 39–59 (1993)
10. Ziarko, W.: Probabilistic approach to rough sets. International Journal of Approximate Reasoning 49, 272–284 (2008)
11. Grzymała-Busse, J.W.: Generalized Parameterized Approximations. In: Yao, J., Ramanna, S., Wang, G., Suraj, Z. (eds.) RSKT 2011. LNCS, vol. 6954, pp. 136–145. Springer, Heidelberg (2011)
12. Clark, P.G., Grzymala-Busse, J.W.: Experiments on probabilistic approximations. In: Proceedings of the 2011 IEEE International Conference on Granular Computing, pp. 144–149 (2011)

On the Relations between Retention Replacement, Additive Perturbation, and Randomisations for Nominal Attributes in Privacy Preserving Data Mining

Piotr Andruszkiewicz

Institute of Computer Science, Warsaw University of Technology, Poland
P.Andruszkiewicz@ii.pw.edu.pl

Abstract. There are several randomisation-based methods in Privacy Preserving Data Mining. In this paper we discuss the additive perturbation and the retention replacement for continuous attributes. We also investigate the randomisations for binary and nominal attributes. We focus on the relations between them, similarities, and differences. We also discuss properties of randomisation-based methods which are important in real applications during implementation and the usage of particular randomisations. We have proven that the retention replacement can be implemented with the randomisation for nominal attributes. We have also shown that the additive perturbation can be approximated with the aforementioned solution for nominal attributes.

1 Introduction

In Privacy Preserving Data Mining there are several randomisation-based methods (perturbation). The first proposed randomisation-based method for continuous attributes was the additive perturbation [1], which adds randomly drawn (from a given distribution) noise to original values of a continuous attribute. An original value of an attribute can be also multiplied by a random value drawn from a given distribution. This approach is used in the multiplicative perturbation [2]. One type of the multiplicative perturbation is the rotation perturbation [3], where a vector of values of attributes for a given object is multiplied by a rotation matrix. Another randomisation-based method is the retention replacement [4], which retains an original value of a continuous attributes with a given probability and replaces an original value with a value drawn from the given distribution otherwise. There are also randomisation-based methods for binary and nominal attributes [5,1,6].

In this paper we discuss the relations, similarities, and differences between the randomisation-based methods, especially the additive perturbation, the retention replacement, and the randomisations for nominal attributes because in literature the randomisation-based methods were investigated separately without taking into account the relations between them. Moreover, we stress, based on our experience, some properties of the randomisation-based methods that

L. Chen et al. (Eds.): ISMIS 2012, LNAI 7661, pp. 99–104, 2012.

are important during implementation, usage and have not been discussed in literature.

The remainder of this paper is organized as follows: In Section 2, we discuss the relations between the additive perturbation and the retention replacement. Then, in Section 3, we prove that the retention replacement randomisation can be implemented with the randomisations for nominal attributes. Section 4 shows that the randomisation for nominal attributes approximates the additive perturbation. Finally, in Section 5, we summarise the conclusions of our study and outline future avenues to explore.

2   Additive Perturbation versus Retention Replacement

In this section, we stress the properties of the considered randomisations which we have learnt during implementations and usage of these randomisations. The presented properties have not been discussed in literature.

One of the main drawbacks of the additive perturbation is that the distortion of a continuous attribute changes, in general, the domain of an attribute. Furthermore, this process may generate values which are nonnatural for a given attribute, e.g., negative values for an attribute with only nonnegative values. Moreover, the higher level of privacy, that is, the higher range of the domain of the distortion distribution, the "more nonnatural" values we may obtain. For a normal distortion distribution the drawback is even worse because we may obtain value high below 0 for small levels of privacy (a normal distribution is not strictly bounded contrary to a uniform distribution).

This drawback generates next difficulties which are connected with discretisation of distorted values of an attribute. In order to reconstruct original distribution of an attribute (e.g., in the process of building a decision tree classifier over distorted data), we need to discretise distorted values (for details about reconstruction please refer to [1]). The range of a distorted attribute depends on the level of incorporated privacy, thus we need to bear in mind that discretisation parameters need to be updated for different levels of privacy we would like to incorporate. We can retain the same number of intervals in the discretisation and obtain different intervals lengths or change the number of intervals to maintain the same length of intervals. Moreover, increasing level of privacy we need to consider higher range of distorted values of an attribute although the usable range of original values of an attribute is significantly smaller, thus we obtain smaller resolution of reconstructed values of an attribute assuming a constant number of intervals.

In the case of a normal distortion distribution we need to tackle the problem of unlimited range of possible distorted values during the discretisation. The solution is to bound values to lower and higher limits of considered range of a distorted attribute, but this should be taken into account in the process of reconstruction and calculation of probabilities of changing a values from a given interval I_i to interval I_j.

3 Retention Replacement and Randomisations for Nominal Attributes

Both, the retention replacement and the nominal attributes distortion with $\mathbf{P}$ matrix[1] are similar because with a given probability an original value is kept, otherwise a value is drawn according to a continuous distribution or a probability distribution over possible values of a nominal attribute. The difference is that we can obtain an original value of an attribute in the retention replacement randomisation even when we draw a new value because a distortion distribution has the same domain as an original attribute. In the nominal distortion with $\mathbf{P}$ matrix, we cannot obtain an original value if we change a value of an attribute.

In this section, we will prove that assuming discretisation the retention replacement randomisation can be implemented as the nominal distortion. Moreover, we will prove that increasing number of intervals up to ∞ the retention replacement can be represented by the nominal distortion.

Let us assume that we have an original attribute X with continuous values. We distort the values of X by means of the retention replacement randomisation-based method with the probability of retaining an original value equal to p and the probability density function, used to draw a distorted value when an original value of the attribute is changed, equal to $h()$. We obtain the attribute Z with distorted values. Let C be a nominal attribute obtained by discretising Z. A is a nominal attribute obtained by discretising X. The matrix $\mathbf{P}$ is the representation of the probabilities of changing/retaining original values of the attribute A during the distortion. In other words, we assume that the matrix $\mathbf{P}$ is used to distort (transform) values of the attribute A and obtain distorted values of the attribute C. In real applications, values of the attribute X and A are not known because only distorted continuous values are stored. However, the parameters of the retention replacement randomisation-based method used in the process are known. We assume that the attribute X and attribute Z were discretised in the same way into k intervals, $I_1, \ldots, I_k$, thus the attribute A and C are nominal attributes with k values, $v_1, \ldots, v_k$.

In order to calculate probabilities $P(v_i \rightarrow v_j), i, j = 1, \ldots, k$ in $\mathbf{P}$ matrix, we need to calculate probability that an original value v_i of an attribute A will be changed to a value v_j:

$$P(v_i \rightarrow v_j) = P(C = v_j | A = v_i). \tag{1}$$

Assuming that the attributes A and C are the discretised attributes X and Z, respectively, we can write:

$$P(C = v_j | A = v_i) = P(Z = z \in I_j | X = x \in I_i). \tag{2}$$

When $X \in I_j$, with the probability p we have a chance that the original value x of the attribute X will be retained. Moreover, with probability $(1 - p)$ a new value will be drawn and we have a chance that this value will lie in the interval

---

[1] For details about $\mathbf{P}$ matrix please refer to [7].

I_j with probability $\int_{I_j} h(t|X)dt$. When $X \notin I_j$, the only chance is that with probability $(1-p)$ a new value will be drawn and it will lie in the interval I_j with probability $\int_{I_j} h(t|X)dt$. Let $\mathbf{1}(condition)$ be an indicator function which takes 1 when *condition* is met and 0 otherwise. Using the indicator function we can write:

$$P(Z = z \in I_j | X = x \in I_i) = (p \cdot 1 + (1-p) \int_{I_j} h(t|X)dt)\mathbf{1}(X \in I_j)$$
$$+(p \cdot 0 + (1-p) \int_{I_j} h(t|X)dt)\mathbf{1}(X \notin I_j). \tag{3}$$

Since $h()$ is independent of X, we omit this attribute and simplify the equation:

$$P(Z = z \in I_j | X = x \in I_i) = p\mathbf{1}(X \in I_j) + (1-p) \int_{I_j} h(t)dt. \tag{4}$$

For the uniform perturbation and intervals with the same length, $\int_{I_j} h(t)dt = \frac{1}{k}$.
P matrix for the discretised attribute with k values will look as follows:

$$\mathbf{P} = \begin{pmatrix} p + (1-p) \int_{I_1} h(t)dt & (1-p) \int_{I_1} h(t)dt & \cdots & (1-p) \int_{I_1} h(t)dt \\ (1-p) \int_{I_2} h(t)dt & p + (1-p) \int_{I_2} h(t)dt & \cdots & (1-p) \int_{I_2} h(t)dt \\ \vdots & \vdots & \ddots & \vdots \\ (1-p) \int_{I_k} h(t)dt & (1-p) \int_{I_k} h(t)dt & \cdots & p + (1-p) \int_{I_k} h(t)dt \end{pmatrix}.$$

We check whether the probabilities in columns sum to 1 [7].
$p + (1-p) \int_{I_1} h(t)dt + (1-p) \int_{I_2} h(t)dt + \ldots + (1-p) \int_{I_k} h(t)dt = p + (1 - p)(\int_{I_1} h(t)dt + \int_{I_2} h(t)dt + \ldots + \int_{I_k} h(t)dt) = p + (1-p)(\int_{I_1 \cup I_2 \cup \ldots \cup I_k} h(t)dt) = p + (1-p) = 1$, thus we obtained a proper matrix **P**.

Increasing the number of intervals up to ∞ the nominal distortion transforms into the retention replacement randomisation-based method because the probability distribution function for the nominal attribute is based on the continuous distortion distribution $h(t)$ and the length of an intervals comes to 0 or to the precision of the digital representation of a number.

By defining $\mathbf{P}'$ matrix which contains probabilities $P(v_i \to v_i | change\ the\ ori-ginal\ value)$ that an original value of the nominal attribute A is transformed given that we change the original value with p probability (as in the retention replacement), we obtain:

$$\mathbf{P}' = \begin{pmatrix} \int_{I_1} h(t)dt; & \int_{I_1} h(t)dt; & \cdots & \int_{I_1} h(t)dt \\ \int_{I_2} h(t)dt; & \int_{I_2} h(t)dt; & \cdots & \int_{I_2} h(t)dt \\ \vdots & \vdots & \ddots & \vdots \\ \int_{I_k} h(t)dt; & \int_{I_k} h(t)dt; & \cdots & \int_{I_k} h(t)dt \end{pmatrix}.$$

The columns of $\mathbf{P}'$ constitute the probability distribution function which is the continuous distortion distribution in the retention replacement randomisation when the length of intervals $\to 0$. This ends the proof. $\qquad\square$

To sum up, assuming a discretisation of a continuous attributes in the retention replacement, as it is assumed in the additive perturbation [1], a distorted continuous attribute can be transformed to a nominal attribute and processed according to available classification algorithms [7]. Moreover, when the number of intervals comes to ∞, the retention replacement can be represented by the nominal distortion.

4 Additive Perturbation and Randomisations for Nominal Attributes

As well as the retention replacement randomistaion, the additive perturbation can be represented as the nominal distortion.

We assume that we have the attributes A, C, X, and Z, as in Section 3. Let f_Y be a distorting probability density function for the additive perturbation and Y represents random values drawn from this distribution and added to original values of X. As for the retention replacement randomisation, we calculate the following probability:

$$P(C = v_j | A = v_i) = P(Z = z \in I_j | X = x \in I_i, X + Y = Z). \tag{5}$$

Assuming that $f_Z(z)$ is a probability density function of Z, we write:

$$P(Z = z \in I_j | X = x \in I_i, Z = X + Y) = \int_{I_j} f_Z(z | X \in I_i, X + Y = Z)dz. \tag{6}$$

Since Y is independent of X:

$$P(Z = z \in I_j | X = x \in I_i, Z = X + Y) = \int_{I_j} f_Y(z - x | X \in I_i, X + Y = Z)dz. \tag{7}$$

The value of the above probability depends on the exact value of X and can be different for values of X that lie in the same interval. Thus, we introduce approximation, as in the process of reconstruction of the original distribution for continuous attributes distorted by means of the additive perturbation [1]. Thus, distance between two points is approximated as the distance between mid-points of the corresponding intervals. Hence:

$$P(Z = z \in I_j | X = x \in I_i, Z = X + Y) = \int_{I_j} f_Y(m(z) - m(x) | X \in I_i, X + Y = Z)dz, \tag{8}$$

where $m(x)$ denotes the mid-point of the interval in which x lies.

To sum up, assuming discretisation the additive perturbation can be approximated with the nominal distortion and increasing the number of intervals up to ∞, we approach to the additive perturbation.

5   Conclusions and Future Work

In this work, we have shown the relations between the randomisation-based methods in Privacy Preserving Data Mining. We have discussed the properties of the additive perturbation and the retention replacement that we have learnt during application of these methods. One of the main drawbacks of the additive perturbation is that this perturbation changes the domain of a continuous attribute and may lead to distorted values that are nonnatural for the attribute.

We have also proven that the retention replacement can be implemented as the nominal distortion when we assume a discretisation of a continuous attribute. Moreover, for infinity number of intervals in the discretisation, the nominal distortion can represent the retention replacement. The additive perturbation has been also shown to be approximated by the nominal distortion. The approximation is more accurate for higher number of intervals.

The difference between the additive perturbation and the retention replacement is that a distorted value of an attribute for the case when a value is changed depends on an original value in the additive perturbation, contrary to the retention replacement. As we have shown in Sections 3 and 4, the nominal distortion covers both cases. Moreover, both randomisations cannot differentiate a distortion distribution based on an original value of an attribute. However, in the nominal distortion it is possible because for each possible value of a nominal attribute we can have a different probability distribution, not only shifted by a given number of positions. Considering these properties, we can say that, in general, the nominal distortion cannot be represented by the additive perturbation nor the retention replacement.

In future, we plan to investigate the relations between the multiplicative, the rotation perturbation and the remaining randomisation-based methods.

References

1. Agrawal, R., Srikant, R.: Privacy-preserving data mining. In: Chen, W., Naughton, J.F., Bernstein, P.A. (eds.) SIGMOD Conference, pp. 439–450. ACM (2000)
2. Kim, J.J., Winkler, W.E.: Multiplicative noise for masking continuous data. Technical report, Statistical Research Division, US Bureau of the Census, Washington, D.C. (2003)
3. Chen, K., Liu, L.: Privacy preserving data classification with rotation perturbation. In: ICDM, pp. 589–592. IEEE Computer Society (2005)
4. Agrawal, R., Srikant, R., Thomas, D.: Privacy preserving olap. In: SIGMOD 2005: Proceedings of the 2005 ACM SIGMOD International Conference on Management of Data, pp. 251–262. ACM, New York (2005)
5. Rizvi, S.J., Haritsa, J.R.: Maintaining data privacy in association rule mining. In: VLDB 2002: Proceedings of the 28th International Conference on Very Large Data Bases, pp. 682–693. VLDB Endowment (2003)
6. Andruszkiewicz, P.: Privacy preserving data mining on the example of classification. Master's thesis, Warsaw University of Technology (2005) (in Polish)
7. Andruszkiewicz, P.: Privacy preserving classification for continuous and nominal attributes. In: Proceedings of the 16th International Conference on Intelligent Information Systems (2008)

Decision Support for Extensive Form Negotiation Games

Sujata Ghosh[1], Thiri Haymar Kyaw[2], and Rineke Verbrugge[2,*]

[1] Indian Statistical Institute, Chennai Centre
SETS Campus, MGR Film City Road
Chennai 600113, India
sujata@isichennai.res.in
[2] Department of Artificial Intelligence
University of Groningen
The Netherlands
thirihaymarkyaw@gmail.com, rineke@ai.rug.nl

Abstract. This paper presents a tool, NEGEXT, for finding individual and group strategies to achieve certain goals while playing extensive form negotiation games. NEGEXT is used as a model-checking tool which investigates the existence of strategies in negotiation situations. We consider sequential and parallel combinations of such games also. Thus, it may aid students of negotiation in their understanding of extensive game-form negotiation trees and their combinations, as well as in their learning to construct individual and group strategies.

1 Introduction

Negotiation may be found everywhere: From mundane conversations between partners about who will fetch the children from school and who will cook dinner, to the sale of an apartment whilst the seller is trying to hide from the buyer that she has bought a new house already, and to fully-fledged international multi-party multi-issue negotiations about climate control, and so forth. Negotiation is a complex skill, and one that is not learnt easily. Thus, many negotiations are broken off, even when they have potential for a win-win solution. Moreover, in many negotiations that do result in an agreement, one or more participants "leave money on the table": they could have done better for themselves [1]. Thus, it is no wonder that several scientific fields have made contributions to analyzing, formalizing, and supporting negotiation.

Kuhn [2] highlighted the importance of using extensive form games in modeling negotiation situations in an objective way, by focusing on the temporal and dynamic nature of the negotiations. For a general overview on negotiation games, see [3]. Researchers in multi-agent systems investigate many aspects of negotiations: some design negotiation mechanisms [4], some analyze negotiation as a form of dialogue [5], whilst others build software to simulate and support

---

* This research was supported by the Netherlands Organisation of Scientific Research grant 600.065.120.08N201 and Vici grant NWO 227-80-001.

L. Chen et al. (Eds.): ISMIS 2012, LNAI 7661, pp. 105–114, 2012.

negotiation [6,7]. In summary, the fast-growing body of research on negotiation provides varied sophisticated models for negotiations.

This paper reports the development of a simple tool, NEGEXT (http://www.ai.rug.nl/~sujata/negext.html), written in the platform-independent Java language. NEGEXT has been constructed to aid students of negotiation. This toolkit, based on extensive form games [8], will aid in under-standing how to combine negotiations, and in planning one's strategic moves in interaction situations when the opponents' possible moves can be approximated. Even though some visualization tools for extensive form game trees already exist,[1] as well as software for negotiation support,[2] we believe we are the first to make a tree-based negotiation toolkit that incorporates the possibility of representing learning from game to game, by sequential and parallel composition (cf. [9]). Moreover, the toolkit has a model-checking component which computes whether and how an individual or a specific coalition can achieve a given objective.

2   Analyzing Negotiation Situations Using Game Trees

A finite extensive form game can be represented by a finite tree where the nodes correspond to players' positions and edges correspond to moves of the players. The terminal nodes of the tree are the end-points of the game, which are generally termed as leaves of the tree. A strategy for a player i is a subtree of the finite game tree, which consists of single edges from player i nodes, and all possible edges from the other players' nodes. A strategy for a group of players K is a subtree consisting of single edges from player k's nodes, where $k \in K$, and all possible edges from player k''s nodes, where $k' \notin K$. Note that all the players have complete knowledge of the whole game. Let us briefly describe the notion of sequential and parallel combinations of extensive form games and players' strategies in such games, providing an application of each in negotiations.

2.1   Sequential Composition

Any two extensive form games can be *sequentially composed* by plugging in the second game at each leaf node of the first game, each leaf node of the first game becoming the root node of the second game [9]. One can extend this idea to define sequential composition of a game with a set of games. Here, at each leaf node of the first game, some game from a given set of games is plugged in.

Story-I: Sequential Cooperation between Two Companies. We now de-scribe a situation which can be dealt with by combining two trees sequentially. Suppose that two Research & Development companies on biotechnology, Biocon and Wockhard, have just entered into a joint project concerning the discovery of biomarkers. They need to hire two types of specialists, a genomic expert and a proteomic expert, one specialist per company. Unfortunately the two compa-nies did not discuss beforehand which of them should hire which type of expert.

---

[1] See for example http://www.gametheory.net/Mike/applets/ExtensiveForm/ ExtensiveForm.html and http://www.gambit-project.org

[2] See for example http://www.negotiationtool.com/

The hiring process can be represented by the simple game trees in Figure 1a,1b, where the nodes B and W stand for Biocon and Wockhard, respectively. The action "hire g" stands for "hire a genomic expert" and action "hire p" stands for "hire a proteomic expert". The propositional atoms have the following meanings:

| g_B: genomics is covered by Biocon | g_W: genomics is covered by Wockhard |
| p_B: proteomics is covered by Biocon | p_W: proteomics is covered by Wockhard |

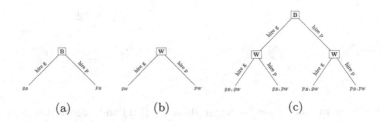

(a)   (b)   (c)

Fig. 1. Hiring game trees: 1(c) is the sequential combination of 1(b) after 1(a)

Wockhard notices that it is a good idea to await Biocon's hiring decision. In order to analyze the different possibilities, they combine the two games sequentially (Figure 1c), plugging in the game of Figure 1b at each leaf node of the game of Figure 1a. Figure 1c clearly shows that hiring one expert after another, in a perfect information game, is much better than the imperfect information game represented by the two game trees in Figure 1a, 1b. Wockhard *learns* who was hired by Biocon, and hires someone with complementary expertise. The two 'middle branches' in Figure 1c provide win-win solutions.

2.2 Parallel Composition

An *interleaving parallel combination* of two extensive form games, as defined in [9], gives rise to a set of games. The main idea of the interleaving parallel operator is as follows: A play of such a game basically moves from one game to the other. One player can move in one of the games, and in the next instant some other player or even the same player may move in the other game. A parallel game takes care of such interleaving. The different orders in the way moves of the players can be interleaved give rise to different games in a parallel combination of two games. Interleaving parallel games allows for copy-cat moves, and in general enables the transfer of strategies in between games. For the formal details, see [9].

Story-II: Analysis of Job Application in Lockstep Synchrony. This example has been inspired by the trick regarding *how not to lose while playing chess with a grandmaster* [9], but with additional aspects of translation. Suppose that Prof. Flitwick (FL) applied for a position in the research group of Prof. Sprout (SP), and he turns out to be the favored candidate. Job negotiations N_1 between Flitwick and Sprout are on the verge of starting. In the same period, Prof. Sprout herself applied for a professorship in the department of Prof. Quirrell (QR), and has just been selected by Prof. Quirrell. The time arises to discuss the particulars of this position as well, in negotiation N_2.

Prof. Sprout is not a very savvy negotiator herself, but she knows that both Prof. Flitwick and Prof. Quirrell are. Moreover, Prof. Sprout notices that the issues of negotiation are quite similar in both cases: The prospective employer can offer either a much higher salary than the standard one, or a standard salary. On the employee's turn, he or she can choose to offer teaching subjects of their own choice, or whatever the department demands (Figure 2).

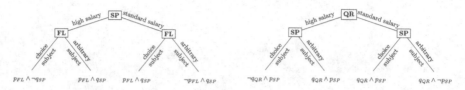

Fig. 2. Appointment game trees for negotiation N_1 (left) and negotiation N_2 (right)

The main pay-offs of the negotiators can be given in terms of propositional letters as follows, where:

- p_i: the new employee i gets more satisfaction in the job (for i is FL or SP);
- q_j: the new employer j is happy with the terms (where j is SP or QR).

In both negotiations, the goal of the new employee can be formulated conditionally: either she procures a higher salary, and then he or she is ready to do whatever he or she is asked; or, if she gets a standard salary, then she would like to teach her favorite subjects only. The goal of the employer would be to have a settlement favorable for the group, i.e. favoring paying a standard salary and / or the new employee teaching according to demand. The salaries and subjects in question are different in both negotiations, but there is a reasonable one-one map of possible offers from one negotiation to the other, represented by the actions of offering "high salary" versus "low salary" and "choice subject" versus "arbitrary subject". The general idea for Prof. Sprout is to be a copy-cat:

- wait for QR to make an offer in negotiation N_2 about the salary;
- translate that offer to the terms of negotiation N_1, and propose a similar offer to FL;
- await the counteroffer from FL about subjects to teach, translate it to the terms of negotiation N_2, and make that offer to QR.

To model this, one needs to compose the trees in Figure 2 in a parallel fashion with interleaving moves, see Figure 6. It is clear that winning proposals are possible in both subtrees, namely on the two 'inside branches' of each, and therefore also in the parallel copy-cat negotiation.

3   NEGEXT Toolkit

We have developed the NEGEXT toolkit to aid students in strategic interactions during negotiations, with negotiations represented as extensive form games and their sequential and parallel combinations. NEGEXT has been written in Java version 1.7.0_02 using the Eclipse editor. It can run on any system that has a Java Virtual Machine (JVM) or Java-enabled web browsers. NEGEXT uses an applet to display the graphical user interface. The user can draw game trees using a menu of input nodes. NEGEXT can also generate the sequential combination of two game trees and all possible parallel combinations. Another feature of the software is that for both individual players and coalitions, it can check whether a strategy to achieve a proposition exists, and if so, point to such a winning strategy in the tree. We used a modified form of binary tree data structure for organizing the tree nodes.

The main frame of the program consists of two panels. In the menu panel on the left, the user can make choices for drawing input trees in a step-by-step fashion. The right panel has two canvas areas for displaying the game trees (Figure 4).

3.1   Drawing a Tree Using NEGEXT

For negotiation problems that can be represented as extensive form games, one can draw a tree using NEGEXT. The user draws a tree from the root node to the terminal leaf nodes, one after another. For a current node, he needs to define the current level of the tree, the parent node, and the player whose turn it is. For the example described in Figure 1, the user may draw the tree as follows:

Level 0 for root node and no need to define the parent (as it has none). Select player i on the left panel (i =1 to 10). Players are represented by colors; in this case red represents Biocon.

Level 1 nodes: no need to define the parent (default parent is root node). Select the player i (i =1 to 10) for the left child and right child nodes; in this case both are green, representing Wockhard.

Level 2: terminal nodes with blue color; parent = 1 (level 1, left child) or parent = 2 (level 1, right child). Define the proposition and add the terminal nodes one after another.

Checking strategy for the coalition of player 1 and player 2 and for formula $g_B \wedge p_W$. The red line shows their joint strategy.

3.2 Checking for Strategies That Achieve Goals in NEGEXT

After the user has drawn a tree, NEGEXT can check whether a strategy exists for either one player or a coalition of players by applying Algorithm 1. The basic idea behind the algorithm for finding strategies for an individual player i is to observe every edge from the root to the terminal nodes, considering a single edge from any node representing this player's turn and all possible edges from the nodes representing other players' turns. If there is no alternate strategy for a different player that prevents the player under consideration from having a winning strategy, then there exists a strategy for this player to achieve the proposition (see **Algorithm 1**).

Algorithm 1. Algorithm1 Finding strategy for Player(s)

Input: A single player i or set of players Sp, Formula φ
for all TreeNode **do**
  if ISTERMINALNODE(TreeNode) and ISTRUE(**Formula** φ, **Node** TreeNode)
  then
    if Input of Player == Player i **then**
      if PARENTOF(**Node** TreeNode) == Player i **then**
        if ISROOT(Player i) **then**
          Print Player i has a winning strategy for Formula;
        else
          Print Player i has no winning strategy for Formula;
        end if
      else if (PARENTOF(PARENTOF(**Node** TreeNode))) == Player i **then**
        Check both left child and right child of ParentOf(**Node** TreeNode);
        if (Check is OK) **then**
          Print Player i has a winning strategy for Formula;
        else
          Print Player i has no winning strategy for Formula;
        end if
      end if
    else if Input of Player == set of players Sp **then**
      while (PARENTOF(**Node** TreeNode) != null) **do**
        if PARENTOF(**Node** TreeNode) == Player $i \in Sp$ **then**
          $Sp = Sp \setminus \{Player\ i\}$;
          Node = (PARENTOF(**Node** TreeNode));
        else
          Print set of players Sp has no winning strategy for Formula.
        end if
      end while
      if $Sp == \emptyset$ **then**
        Print set of players Sp has a winning strategy for Formula
      end if
    end if
  end if
end for

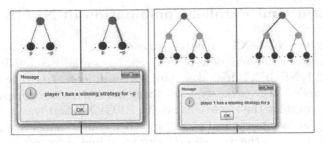

Fig. 3. Strategy checking for an individual player

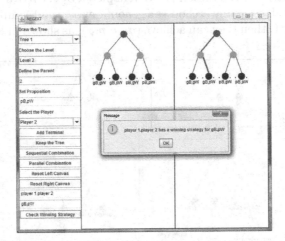

Fig. 4. Strategy checking for a group of players

For the user to find a strategy to achieve a particular formula, he needs to input the player's name ("player 1") or the names of the coalition of players ("player 1, player 2") and input the goal formula corresponding to a proposition at a terminal node. Then NEGEXT checks all tree nodes starting from the root to the terminal nodes: If a proposition of any terminal node t matches the input formula, NEGEXT checks the parent node of t. If the parent node of t is matched with player i and it is also the root node, then player i has an individual strategy for this proposition. For example, in the left part of Figure 3, player 1 (red) has a strategy to achieve $\neg p$. If it is a node at level 1 at which it is another's turn, then the algorithm needs to check whether all its children have p. For example, in the right part of Figure 3, player 1 (red) has a strategy to achieve p. Otherwise, player i has no strategy to achieve this proposition.

To find a group strategy for a formula, NEGEXT also checks all tree nodes: If a proposition of any terminal node t matches the input formula, NEGEXT checks the parent node of t recursively up to the root. If all predecessor nodes of t are matched with all players in the group, then the group has a strategy for the formula, otherwise it does not. In the toolkit, the input tree is shown in the left canvas area, and the output tree with a strategy path (red line) is shown in the right canvas area (Figure 4).

4   Sequential and Parallel Combination in NEGEXT

Let us consider how NEGEXT can be used to analyze the stories in Subsections 2.1 and 2.2 which involve combinations of trees. For sequentially combining the trees of Subsection 2.1, NEGEXT first takes as inputs the different trees to combine, one after another, and then gives the sequential combination tree in a separate window (Figure 5). To combine the trees sequentially, NEGEXT first traverses all nodes of tree 1 from the root to the leaf nodes, and then concatenates tree 2 to all leaf nodes of tree 1. Suppose the user asks the system to find one possible strategy for achieving $(g_B \wedge p_W)$ by the coalition $\{B, W\}$, depicted by the red and green players in the screenshots. If company B chooses a genomic expert and company W chooses a proteomic expert, then they can achieve $(g_B \wedge p_W)$, as shown in Figure 5.

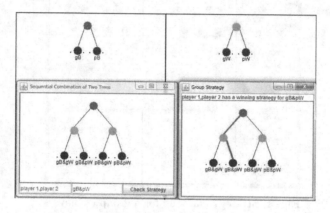

Fig. 5. Sequential combination and strategy checking

Now let us consider the parallel combination of trees, corresponding to the situation described in Section 2.2. As in the previous case, the different trees that are to be combined are taken as separate inputs, one after another, as given in Figure 6. Here, agents QR, SP and FL are represented by pink, red, and green, respectively. Let us define abbreviations as follows: a for $p_{FL} \wedge \neg q_{SP}$; b for $p_{FL} \wedge q_{SP}$; c for $\neg p_{FL} \wedge q_{SP}$; d for $\neg q_{QR} \wedge p_{SP}$; e for $q_{QR} \wedge p_{SP}$; and f for $q_{QR} \wedge \neg p_{SP}$.

Before combining these trees in an interleaving way, we should note here that the parallel combination of two trees will give rise to a bunch of possible trees. These trees will appear in an enumeration, from which the user selects one combination as the final tree. In this case, the particular tree as depicted by the story is given in Figure 6. Figure 7 presents a copy-cat strategy that the common red player (Sprout) can follow in order to end up in a winning situation in the parallel game. Note that formally this strategy is a $\langle QR, SP, FL \rangle$-strategy, which may have been elicited by the user's question as to whether the set $\{QR, SP, FL\}$ can achieve the goal $p_{FL} \wedge p_{SP} \wedge q_{QR} \wedge q_{SP}$. Thus, using NEGEXT enables students to see clearly how players QR, FL, and SP can jointly achieve an intuitive goal.

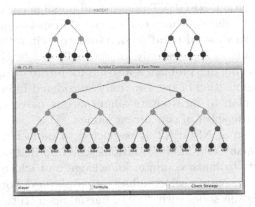

Fig. 6. An interleaving parallel combination

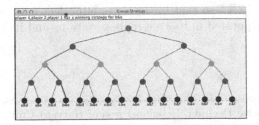

Fig. 7. A group strategy for *QR*, *SP*, and *FL* in the combined tree

5   Conclusions and Future Work

In this work, we have presented a toolkit to represent negotiations which span over a finite time, and we consider the actions of the negotiators one after another in response to each other. Two examples have been provided to advocate the fact that some real-life negotiations can be aptly described by perfect information extensive form games. The NEGEXT toolkit can help students of negotiation to learn how to respond in order to achieve their goals, including situations where it is not easy to compute the optimal response. The current version of the toolkit has not been tested for learnability and usability yet, so a first step in future research will be to improve it on the basis of a usability study, in which subjects will be asked to use the tool in order to find strategies given particular negotiation trees.

We used a binary tree data structure for drawing trees in the NEGEXT toolkit, implemented on a pure Java applet. In real life, players often have more than two options. In addition, current NEGEXT still allows only at most level 2 trees (root plus intermediate level plus leaves) for sequential or parallel combination, because of the node placing in the current graphical user interface. We aim to relax both restrictions in future work so that NEGEXT will be able to represent various types of branching at different nodes, as well as deeper trees.

In parallel combination, NEGEXT presents all possible parallel combinations to the user. NEGEXT allows the user to click the "Parallel Combination" button repeatedly in order to view all parallel combined trees in different interleaving ways. If the user wants to know of only an optimally combined tree, NEGEXT may confuse him. Solving this problem is also left for future work. Note that while defining parallel combination of trees we only considered interleaving moves in between trees. We plan to incorporate simultaneous moves as well, bringing NEGEXT closer to the spirit of concurrent games.

The current version of NEGEXT is restricted to perfect information situations. However, in many real-life negotiations, the information dilemma looms large: Which aspects to make common knowledge and which aspects to keep secret or to divulge to only a select subset of co-players? For example, Raiffa distinguishes negotiation styles with "full or partial open truthful exchange" [1]. Such aspects of imperfect or incomplete information cause far-reaching asymmetries between parties, sometimes with grave consequences [10]. It will be future work to extend NEGEXT so that incomplete, imperfect, and asymmetric information can be incorporated in its tree representations and its strategic advice.

References

1. Raiffa, H., Richardson, J., Metcalfe, D.: Negotiation Analysis: The Science and Art of Collaborative Decision. Belknap Press of Harvard Univ. Press, Cambridge (2002)
2. Kuhn, H.: Game theory and models of negotiation. The Journal of Conflict Resolution 6(1), 1–4 (1962)
3. Brams, S.J.: Negotiation Games: Applying Game Theory to Bargaining and Arbitration. Routledge, London (2003)
4. Rosenschein, J., Zlotkin, G.: Rules of Encounter: Designing Conventions for Automated Negotiation Among Computers. MIT-Press, Cambridge (1994)
5. Parsons, S., Wooldridge, M., Jennings, N.: Agents that reason and negotiate by arguing. Journal of Logic and Computation 8, 261–292 (1998)
6. Lin, R., Kraus, S., Wilkenfeld, J., Barry, J.: Negotiating with bounded rational agents in environments with incomplete information using an automated agent. Artificial Intelligence Journal 172(6-7), 823–851 (2008)
7. Hindriks, K.V., Jonker, C., Tykhonov, D.: Towards an Open Negotiation Architecture for Heterogeneous Agents. In: Klusch, M., Pěchouček, M., Polleres, A. (eds.) CIA 2008. LNCS (LNAI), vol. 5180, pp. 264–279. Springer, Heidelberg (2008)
8. Osborne, M., Rubinstein, A.: A Course in Game Theory. MIT Press, Cambridge (1994)
9. Ghosh, S., Ramanujam, R., Simon, S.: Playing Extensive Form Games in Parallel. In: Dix, J., Leite, J., Governatori, G., Jamroga, W. (eds.) CLIMA XI. LNCS, vol. 6245, pp. 153–170. Springer, Heidelberg (2010)
10. Oakman, J.: The Camp David Accords: A case study on international negotiation. Technical report, Princeton University, Woodrow Wilson School of Public and International Affairs (2002)

Intelligent Alarm Filter Using Knowledge-Based Alert Verification in Network Intrusion Detection

Yuxin Meng[1], Wenjuan Li[2], and Lam-for Kwok[1]

[1] Department of Computer Science, College of Science and Engineering,
City University of Hong Kong, Hong Kong, China
ymeng8@student.cityu.edu.hk
[2] Computer Science Division, Zhaoqing Foreign Language College,
Guangdong, China
wenjuan.anastatia@gmail.com

Abstract. Network intrusions have become a big challenge to current network environment. Thus, network intrusion detection systems (NIDSs) are being widely deployed in various networks aiming to detect different kinds of network attacks (e.g., Trojan, worms). However, in real settings, a large number of alarms can be generated during the detection procedure, which greatly decrease the effectiveness of these intrusion detection systems. To mitigate this problem, we advocate that constructing an alarm filter is a promising solution. In this paper, we design and develop an intelligent alarm filter to help filter out NIDS alarms by means of knowledge-based alert verification. In particular, our proposed method of knowledge-based alert verification employs a rating mechanism in terms of expert knowledge to classify incoming NIDS alarms. We implemented and evaluated this intelligent knowledge-based alarm filter in a network environment. The experimental results show that the developed alarm filter can accurately filter out a number of NIDS alarms and achieve a better outcome.

Keywords: Intelligent System, Alarm Filtration, Alert Verification, Knowledge Representation and Integration, Network Intrusion Detection.

1 Introduction

Network threats (e.g., Trojan, malware, virus) are now becoming a big problem with the rapid development of networks [2]. In order to address this issue, network intrusion detection systems (NIDSs) [3,4] have been widely deployed in current network environments (e.g., an insurance company, a bank) to defend against different kinds of network attacks. The network intrusion detection systems can be roughly classified into two types: signature-based NIDS and anomaly-based NIDS. For the signature-based NIDS [5,6], it detects an attack by comparing incoming packet payloads with its stored signatures. The signature (or called *rule*) is a kind of descriptions for a known attack. On the other hand, the anomaly-based NIDS [7,8] identifies an intrusion by detecting the great deviations between network events and its pre-defined normal profile. The normal profile is used to represent a normal behavior or network connection.

L. Chen et al. (Eds.): ISMIS 2012, LNAI 7661, pp. 115–124, 2012.

Problem. For the network intrusion detection systems, a big suffering problem is that a lot of alarms (e.g., false alarms, non-critical alarms[1]) can be produced during their detection process, which greatly decrease the detection effectiveness and heavily increase the burden of analyzing NIDS alarms [10,11]. The reasons for this issue are described as below:

- For the signature-based NIDS, its detection ability is heavily depending on its stored signatures. However, these signatures are weak in representing multi-step attacks. Thus, it is hard for this kind of NIDSs to decide the situation of an attack attempt (i.e., whether it is a successful attack or not). In this case, it has to report and alert all detected attack attempts aiming to reduce potential security risks [9].
- Whereas, for the anomaly-based NIDSs, it is very hard for them to establish an accurate normal profile so that many unwanted alarms (e.g., non-critical alarms) will be generated [14]. For instance, traffic accidents (e.g., a sudden increase of the network traffic) can easily violate and crack the normal profile, causing a lot of false alarms.

Contributions. Overall, the large number of non-critical alarms including the false alarms is a key limiting factor to encumber the development of NIDS [1]. To mitigate this big challenge, we advocate that designing an alarm filter is a promising solution. In this paper, we therefore attempt to design an intelligent alarm filter by using a method of knowledge-based alert verification (named *intelligent KAV-based alarm filter*), aiming to reduce the unwanted alarms based on expert knowledge. In this work, we define the *unwanted alarms* as either false alarms, non-critical alarms and unwanted true alarms. The contributions of our work can be summarized as follows:

- Alert verification is a method to help determine whether an attack is successful or not. In this work, we developed a method of knowledge-based alert verification to determine whether a NIDS alarm is critical or not by means of expert knowledge.
- To implement the method of knowledge-based alert verification, we developed a dependent component called *Rating Measurement* to classify and rate NIDS alarms by means of a KNN-based classifier. This classifier is trained with some rated-alarms which have been rated by using expert knowledge. During the filtration, the filtration rate is determined by a pre-decided *rating threshold*.
- We implemented and evaluated the *intelligent KAV-based alarm filter* in a real network environment. During the experiment, the filter can intelligently rate incoming alarms and filter out unwanted alarms. The experimental results show that this filter can positively and accurately reduce a large number of NIDS alarms (e.g., from 85.9% to 90.6%) in the deployed network environment.

The remaining parts of this paper are organized as follows: in Section 2, we introduce some research work on constructing alarm filter in network intrusion detection; Section 3 presents the architecture of *intelligent KAV-based alarm filter* and describes each component in detail; Section 4 shows the experimental network environment and analyzes the experimental results; finally, we conclude our work with future directions in Section 5.

---

[1] A non-critical alarm is either a false alarm or a non-critical true alarm.

2 Related Work

A lot of solutions have been proposed aiming to reduce the number of unwanted alarms in the field of network intrusion detection. The method of alert verification is used to help filter out NIDS alarms by determining whether an attack is successful or not. For instance, Zhou et al. [16] described an approach to verify intrusion attempts by using lightweight protocol analysis. This approach tracked responses from network applications and verified the results by analyzing the header information. Several other work regarding to this method can be found in [15], [20] and [17].

To filter out alarms for a NIDS, another widely used method is to construct an alarm filter by using computational intelligent methods (e.g., machine learning algorithms). Pietraszek [19] proposed and developed a system of adaptive alert classifier which used both the analysts' feedback and machine learning techniques to help reduce false positives. Their classifier could drop alerts in terms of their classification confidence. Law and Kwok [21] designed a false alarm filter by using KNN (k-nearest-neighbor) classifier and achieved a good filtration rate. Then, Alharbt et al. [23] constructed an alarm filter by using continuous and discontinuous sequential patterns to detect abnormal alarms. Later, Meng et al. [12] presented an adaptive false alarm filter to help NIDS filter out a large number of false alarms. By adaptively selecting the most appropriate machine learning algorithm, the filter can keep a good filtration rate. Meng and Li [22] further designed a non-critical alarm filter to detect and refine non-critical alarms based on contextual information such as application and OS information.

In real settings, we find that expert knowledge is very crucial in deciding whether an alarm is critical or not. In this work, we therefore attempt to design an intelligent alarm filter to filter out NIDS alarms by using a method of knowledge-based alert verification. That is, we use expert knowledge to determine whether an alarm is critical or not. The proposed method of knowledge-based alert verification also clearly distinguishes our work from other work. Specifically, the filter refines NIDS alarms by employing a rating mechanism in which each NIDS alarm will be classified and rated. In the evaluation, the filter achieved good performance in both alarm classification and alarm filtration.

3 Intelligent KAV-Based Alarm Filter

In this section, we first describe the architecture of the *intelligent KAV-based alarm filter* and we then give an in-depth description of each component.

3.1 Architecture

The high-level architecture of the *intelligent KAV-based alarm filter* is illustrated in Fig. 1. The alarm filter mainly consists of three components: *Alarm Database*, *Rating Measurement* and *Alarm Filter*. The *Alarm Database* is responsible for storing rated alarms and training the machine learning classifier (e.g., KNN-based classifier). The component of *Rating Measurement* is responsible for classifying incoming NIDS alarms and rating different alarms by giving them relevant scores. Finally, the component of *Alarm Filter* will filter out NIDS alarms in terms of the pre-decided *rating*

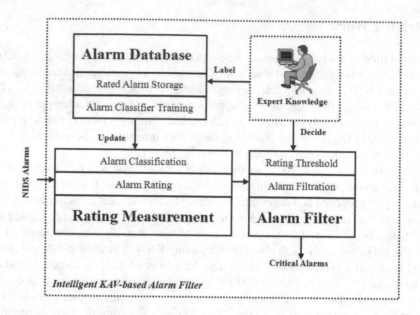

Fig. 1. The high-level architecture of intelligent KAV-based alarm filter

threshold. The *expert knowledge* is very important in the architecture and is responsible for rating alarms in the training process and deciding the *rating threshold*.

In real deployment, there are two phases in the filter: *Preparation Phase* and *Filtration Phase*. Experts first label and rate several NIDS alarms (e.g., one hundred rated alarms), and then store them into the component of *Alarm Database*. This component will train the machine learning classifier by using these rated alarms. In addition, experts also should decide the *rating threshold* in the component of *Alarm Filter*. We denoted the above procedures as *Preparation Phase*.

After the *Preparation Phase*, the filter will enter into the *Filtration Phase*. In this phase, NIDS alarms first arrive at the component of *Rating Measurement*. This component will intelligently give score to each incoming NIDS alarm by using the machine learning classifier. In general, a higher score or rating value means that relevant alarm is more important. Later, these rated alarms will be forwarded to the component of *Alarm Filter*, and this component conducts alarm filtration in terms of the pre-decided *rating threshold*. For example, if the score of an alarm is smaller than the *rating threshold*, then this alarm will be filtered out as unwanted alarm.

3.2   The Component of Alarm Database

This component is mainly used to store rated NIDS alarms and train a machine learning classifier by using these labeled alarms. Thus, it contains two parts: *Rated Alarm Storage* and *Alarm Classifier Training*. In the part of *Alarm Classifier Training*, we use a KNN-based classifier to classify and rate incoming alarms. The selection of this classifier is based on the following two points:

Table 1. The scores and meanings for different rate-values

Classification	Rate (4.0, 5.0]	Rate (3.0, 4.0]	Rate (2.0, 3.0]	Rate (1.0, 2.0]	Rate (0, 1.0]
Meaning	Very Critical	Critical	Important	Not Important	Non-Critical

- The KNN (k-nearest neighbor) algorithm is a method for classifying objects based on closest training examples in the feature space. That is, an object is classified in terms of its distances to the nearest cluster. Therefore, this classifier is good at clustering the rated alarms and classifying incoming alarms.
- Based on our previous work [12], the KNN-based classifier can achieve a very high filtration rate with high classification accuracy. In addition, this classifier has a fast speed in the phases of both training and classification, which is a desirable property when deployed in a resource-limited platform (e.g., a mobile phone, an agent-based network).

In Fig. 1, experts will give scores to a number of NIDS alarms and store them in the *Alarm Database*. Then, this component can train the KNN-based classifier to establish a KNN-based model by using these rated alarms. In addition, this component can update the KNN-based model in the component of *Rating Measurement* periodically.

3.3   The Component of Rating Measurement

This component is responsible for classifying incoming NIDS alarms by using KNN-based classifier, and rating these alarms by giving a score to each of them. The rating classification and meanings of the rating mechanism are described in Table 1. It is easily visible that the rate value of 5 is the highest score which means the alarm is very critical for security experts and networks, while the rate value of 0 is the lowest score. Generally, higher score indicates that an alarm is more important. The numerical accuracy is 0.1 so that each rate has a range. For example, for a *very critical* alarm, its rating value can be ranged from 4.0 to 5.0. For an *important* alarm, its rating value can be ranged from 2.0 to 3.0. This numerical accuracy can improve the effectiveness of training and decrease the classification errors.

To better illustrate the alarm classification by using KNN-based classifier, we give a case in Fig. 2. The white-point is an incoming alarm waiting for classification. The black-points are rated alarms and are gathered into three clusters rated as 4.3, 3.2 and 2.7. To classifier the white-point, the KNN-based classifier will calculate the Euclidean distance (e.g., D1, D2, D3) between the white-point and the other three clusters respectively. The distance can be used to represent the similarity between the white-point and the clusters. The shorter the distance, the more similar they are. The calculation of the Euclidean distance is presented as below:

$$[Distance\ (P1,\ P2)]^2 = \sum_{0}^{N}(P1_i - P2_i)^2 \qquad (1)$$

$P1_i$ and $P2_i$ are the values of the ith attribute of points $P1$ and $P2$ respectively. If an alarm is classified into one cluster, then this alarm will be given the rating value the

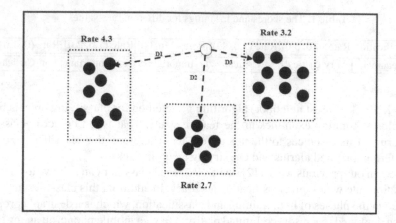

Fig. 2. A case of classifying alarms using KNN-based classifier in the component of Rating Measurement. The white-point is an alarm waiting for classification while the other black-points are rated alarms and are gathered into three clusters.

same as that cluster. For instance, if an alarm is classifier to the cluster rated as 3.2, then this alarm will be rated as 3.2 accordingly.

3.4 The Component of Alarm Filter

This component is responsible for filtering out NIDS alarms by means of a pre-decided *rating threshold*. Thus, it has two parts: *Rating Threshold* and *Alarm Filtration*. In the part of *Rating Threshold*, the rating threshold is pre-decided by expert knowledge. Then, in the part of *Alarm Filtration*, the filtration procedure is shown as follows:

– If the rating value of an alarm is smaller than the *rating threshold*, then this alarm will be filtered out as unwanted alarm.
– If the rating value of an alarm is higher than the *rating threshold*, then this alarm will be output by the filter.

For the *rating threshold*, experts should pre-decide it in the *Preparation Phase* by considering the network structure and deployment. In this case, the expert knowledge is a key factor to affect the filtration performance of the alarm filter. In the field of network intrusion detection, a security expert has the capability to suggest and determine which alarms are useful for them, so that the use of expert knowledge can indeed help improve the filtration performance [19]. Due to the complexity, we leave the extraction of expert knowledge as an open problem in our future experiments.

4   Evaluation

In this section, we implement and evaluate the *intelligent KAV-based alarm filter* in a constructed network environment by using Snort [13] and Wireshark [18]. The experimental deployment is presented in Fig. 3.

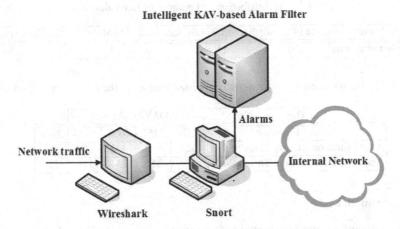

Fig. 3. The deployment of experimental network environment

The Snort is an open-source signature-based NIDS, so that we deploy it in front of *Internal Network* to detect network attacks. The *intelligent KAV-based alarm filter* is deployed close to Snort in order to filter out NIDS alarms. The Wireshark is implemented in front of Snort and is responsible for recording network packets. In real settings, network traffic passes through Wireshark and arrives at Snort. The Snort examines network packets and generates alarms. At last, all the generated alarms will be forwarded into the *intelligent KAV-based alarm filter* for alarm filtration.

In the remaining parts of this section, we begin by describing the experimental methodology. Then, we present and analyze the experimental results.

4.1 Experimental Methodology

In the evaluation, we mainly conduct two experiments (named *Experiment1* and *Experiment2*) to investigate the performance of the *intelligent KAV-based alarm filter*.

- *Experiment1:* In this experiment, we first rated 100 Snort alarms by using expert knowledge with the purpose of training the KNN-based classifier (k=1). The used 100 Snort alarms were randomly extracted from the historic alarm datasets which were collected in the same network environment. Then, we evaluated the *intelligent KAV-based alarm filter* in the experimental environment for 5 days.
- *Experiment2:* In this experiment, we used the same rated Snort alarms to train the KNN-based classifier (k=1). We then evaluated the *intelligent KAV-based alarm filter* by using the same 5-day alarms that were collected in the *Experiment1*. But the difference is that, in the end of each day, we re-trained the KNN-based classifier with 50 new rated Snort alarms.

The feature extraction of the Snort alarms can be referred to our previous work [12]. The *Experiment1* attempts to explore the initial performance of the *intelligent KAV-based alarm filter* with *one-time training*, while the *Experiment2* attempts to investigate the performance of the *intelligent KAV-based alarm filter* with *continuous training* and explore the effect of *continuous training* on the filter.

Table 2. The distribution and number of rated alarms

Rate Value	Rate (4.1, 5.0]	Rate (3.0, 4.0]	Rate (2.0, 3.0]	Rate (1.0, 2.0]	Rate (0, 1.0]
The number of alarms	10	11	9	33	37

Table 3. The filtration results with two rating threshold values in the *Experiment1* for 5 days

Day	DAY1	DAY2	DAY3	DAY4	DAY5
Before Filtration	1376	2456	1654	1209	2098
After Filtration (Rating Threshold: 3)	327	656	432	209	560
After Filtration (Rating Threshold: 4)	172	256	234	114	278

4.2   Experiment1

In this experiment, we used the method of one-time training to explore the performance of the *intelligent KAV-based alarm filter*. The distribution of rated alarms is presented in Table 2. This distribution is similar to the real deployment that the number of non-critical alarms is far higher than that of critical alarms [11,12].

The 5-day experimental results are described in Table 3, it is easily visible that the *intelligent KAV-based alarm filter* can greatly reduce the number of NIDS alarms. By using different rating threshold values, the filtration rate varies accordingly. Based on the security experts' suggestions, the alarms with the rating values of 3 or above are more interesting to them. Thus, we set the rating threshold values to 3 and 4 respectively. For instance, in *DAY2*, the filtration rate is 73.3% for the rating threshold of 3 while the filtration rate is increased to 89.6% if the rating threshold is set to 4. Moveover, by analyzing the data collected by Wireshark, we find that the classification accuracy in this experiment is ranged from 92.5% to 94.8%, which is applicable in real settings.

Overall, for the rating threshold of 3, the filtration rate is ranged from 73.3% to 82.7%, whereas for the rating threshold of 4, the filtration rate can be higher which is ranged from 85.9% to 90.6%. The experimental results show that our designed *intelligent KAV-based alarm filter* is promising in filtering out NIDS alarms.

4.3   Experiment2

In this experiment, we used a method of continuous training to explore the performance of the filter. That is, in the end of each day, we re-trained the KNN-based classifier by using 50 new rated Snort alarms. The experiment was performed on the same alarms collected in the *Experiment1* and the experimental results are described in Table 4.

In the table, the filtration rate for *DAY1* is the same since the continuous training was started in the end of *DAY1*. We find that the filter performed a little better filtration rate than that in the *Experiment1*. For example, in *DAY2*, the filtration rate is increased to 79.3% for the rating threshold of 3 while the filtration rate is further increased to 91.1% for the rating threshold of 4. We also find that the classification accuracy of the filter is increased and ranged from 95.2% to 96.7%

In this experiment, from *DAY2* to *DAY5*, the filtration rate is ranged from 77.1% to 85.4% for the rating threshold of 3, whereas the filtration rate is ranged from 87.4%

Table 4. The filtration results with two rating threshold values in the *Experiment2*

Day	DAY1	DAY2	DAY3	DAY4	DAY5
Before Filtration	1376	2456	1654	1209	2098
After Filtration (Rating Threshold: 3)	327	507	378	176	413
After Filtration (Rating Threshold: 4)	172	218	208	87	223

to 92.8% for the rating threshold of 4. The experimental results show that by using the continuous training, both of the filtration rate and the classification accuracy of the filter can be further improved.

5   Concluding Remarks

Network intrusion detection systems are very essential and widely used in current network environment to detect network attacks. But it is a big problem that a large number of alarms can be generated during their detection. To mitigate this issue, in this paper, we propose and design an intelligent alarm filter by using the method of knowledge-based alert verification (named *intelligent KAV-based alarm filter*). In particular, this filter consists of three major components: *Alarm Database*, *Rating Measurement* and *Alarm Filter*. The expert knowledge is used in both *Alarm Database* and *Rating Measurement* to rate alarms and decide the rating threshold respectively. In the evaluation, we conducted two experiments to investigate the performance of the *intelligent KAV-based alarm filter*. The experimental results showed that by using one-time training, the filter could achieve a very high rate in both alarm classification and alarm filtration. The results also showed that by using a method of continuous training, the filtration rate and classification accuracy could be further improved.

The filter achieved a good result in a network environment. Future work could include evaluating the impact of different k values on the classification results and using larger and more alarm datasets to validate our conclusion. In addition, future work could also include investigating how to accurately and efficiently extract expert knowledge.

References

1. Axelsson, S.: The Base-Rate Fallacy and the Difficulty of Intrusion Detection. ACM Transactions on Information and System Security, 186–205 (August 2000)
2. Symantec Corp., Internet Security Threat Report, vol. 16, http://www.symantec.com/business/threatreport/index.jsp (accessed on May 26, 2012)
3. Paxson, V.: Bro: A System for Detecting Network Intruders in Real-Time. Computer Networks 31(23-24), 2435–2463 (1999)
4. Scarfone, K., Mell, P.: Guide to Intrusion Detection and Prevention Systems (IDPS), pp. 800–894. NIST Special Publication (2007)
5. Vigna, G., Kemmerer, R.A.: NetSTAT: a Network-based Intrusion Detection Approach. In: Proceedings of Annual Computer Security Applications Conference (ACSAC), pp. 25–34. IEEE Press, New York (1998)
6. Roesch, M.: Snort: Lightweight Intrusion Detection for Networks. In: Proceedings of 13th Large Installation System Administration Conference (LISA), pp. 229–238. USENIX Association Berkeley, CA (1999)

7. Valdes, A., Anderson, D.: Statistical Methods for Computer Usage Anomaly Detection Using NIDES. Technical report, SRI International (January 1995)
8. Ghosh, A.K., Wanken, J., Charron, F.: Detecting Anomalous and Unknown Intrusions Against Programs. In: Proceedings of Annual Computer Security Applications Conference (ACSAC), pp. 259–267 (1998)
9. Ptacek, T.H., Newsham, T.N.: Insertion, Evation, and Denial of Service: Eluding Network Intrusion Detection. Technical Report, Secure Networks (January 1998)
10. McHugh, J.: Testing Intrusion Detection Systems: a Critique of the 1998 and 1999 Darpa Intrusion Detection System Evaluations as Performed by Lincoln Laboratory. ACM Transactions on Information System Security, 262–294 (2000)
11. Lippmann, R.P., Fried, D.J., Graf, I., Haines, J.W., Kendall, K.R., McClung, D., Weber, D., Webster, S.E., Wyschogrod, D., Cunningham, R.K., Zissman, M.A.: Evaluating Intrusion Detection Systems: the 1998 DARPA off-line Intrusion Detection Evaluation. In: Proceedings of DARPA Information Survivability Conference and Exposition, pp. 12–26 (2000)
12. Meng, Y., Kwok, L.-F.: Adaptive False Alarm Filter Using Machine Learning in Intrusion Detection. In: Wang, Y., Li, T. (eds.) Practical Applications of Intelligent Systems. AISC, vol. 124, pp. 573–584. Springer, Heidelberg (2011)
13. Snort-The Open Source Network Intrusion Detection System, http://www.snort.org/ (accessed on April 25, 2012)
14. Sommer, R., Paxson, V.: Outside the Closed World: On using Machine Learning for Network Intrusion Detection. In: Proceedings of IEEE Symposium on Security and Privacy, pp. 305–316 (2010)
15. Kruegel, C., Robertson, W.: Alert Verification: Determining the Success of Intrusion Attempts. In: Proceedings of Workshop on Detection of Intrusions and Malware and Vulnerability Assessment (DIMVA), pp. 25–38 (July 2004)
16. Zhou, J., Carlson, A.J., Bishop, M.: Verify Results of Network Intrusion Alerts Using Lightweight Protocol Analysis. In: Proceedings of Annual Computer Security Applications Conference (ACSAC), pp. 117–126 (December 2005)
17. Mu, C., Huang, H., Tian, S.: Intrusion Detection Alert Verification Based on Multi-level Fuzzy Comprehensive Evaluation. In: Hao, Y., Liu, J., Wang, Y.-P., Cheung, Y.-M., Yin, H., Jiao, L., Ma, J., Jiao, Y.-C. (eds.) CIS 2005. LNCS (LNAI), vol. 3801, pp. 9–16. Springer, Heidelberg (2005)
18. Wireshark, Homepage, http://www.wireshark.org (accessed on April 10, 2012)
19. Pietraszek, T.: Using Adaptive Alert Classification to Reduce False Positives in Intrusion Detection. In: Jonsson, E., Valdes, A., Almgren, M. (eds.) RAID 2004. LNCS, vol. 3224, pp. 102–124. Springer, Heidelberg (2004)
20. Kruegel, C., Robertson, W., Vigna, G.: Using Alert Verification to Identify Successful Intrusion Attempts. Journal of Practice in Information Processing and Communication 27(4), 220–228 (2004)
21. Law, K.H., Kwok, L.-F.: IDS False Alarm Filtering Using KNN Classifier. In: Lim, C.H., Yung, M. (eds.) WISA 2004. LNCS, vol. 3325, pp. 114–121. Springer, Heidelberg (2005)
22. Meng, Y., Li, W.: Constructing Context-based Non-Critical Alarm Filter in Intrusion Detection. In: Proceedings of International Conference on Internet Monitoring and Protection (ICIMP), pp. 75–81 (2012)
23. Alharby, A., Imai, H.: IDS False Alarm Reduction Using Continuous and Discontinuous Patterns. In: Ioannidis, J., Keromytis, A.D., Yung, M. (eds.) ACNS 2005. LNCS, vol. 3531, pp. 192–205. Springer, Heidelberg (2005)

Clustering-Based Media Analysis for Understanding Human Emotional Reactions in an Extreme Event

Chao Gao[1] and Jiming Liu[2],*

[1] College of Computer and Information Science,
Southwest University, Chongqing, China
[2] Department of Computer Science,
Hong Kong Baptist University, Kowloon Tong, Hong Kong
jiming@comp.hkbu.edu.hk

Abstract. An extreme event such as a natural disaster may cause social and economic damages. Human beings, whether individuals or society as a whole, often respond to the event with emotional reactions (e.g., sadness, anxiety and anger) as the event unfolds. These reactions are, to some extent, reflected in the contents of news articles and published reports. Thus, a systematic method for analyzing these contents would help us better understand human emotional reactions at a certain stage (or an episode) of the event, find out their underlying reasons, and most importantly, remedy the situations by way of planning and implementing effective relief responses (e.g., providing specific information concerning certain aspects of an event). This paper presents a clustering-based method for analyzing human emotional reactions during an event and detecting their corresponding episodes based on the co-occurrences of the words as used in the articles. We demonstrate this method by showing a case study on Japanese earthquake in 2011, revealing several distinct patterns with respect to the event episodes.

1 Introduction

In a human society, extreme events, such as natural and technological disasters (e.g., earthquakes and industrial accidents), economic crises and regional conflicts, are inevitable and sometime unpredictable [1]. The occurrences of extreme events may cause serious social and economic damages [2]. In the face of such an event, human beings will experience different emotional reactions at different times [3], and may adjust their behaviors in order to recover from the impacts of the event [4]. For example, people may feel anxious about the degree of a certain damage and may be easily influenced by rumors [5]. At some point, people may be fearful about the severity of the situation and may prefer to evacuate [6], such as the spontaneous evacuation when facing a perceived radiation threat at Three Mile Island [4]. The anxious or fearful emotion may arouse people to take actions to avoid potential harms, whereas the angry emotion may drive people to overcome some encountered difficulties [4][7]. These emotions may also lead to irrational crowd behaviors [4][6].

Throughout the different stages of an extreme event, human emotional reactions are, to a certain extent, revealed in news articles or reports [8][9]. Thus, it would be desirable if such reactions can be timely and accurately captured by analyzing the contents

---

* Prof. Jiming Liu is the corresponding author of this paper.

L. Chen et al. (Eds.): ISMIS 2012, LNAI 7661, pp. 125–135, 2012.

from creditable mass media and professional relief-related media. Understanding human emotional reactions based on these media can readily help us identify social needs at different times and thus plan effective relief missions accordingly (e.g., deploying rescue teams to providing specific information concerning certain aspects of the event).

An extreme event often consists of a series of episodes [2]. Here, an episode could be seen as an associated subevent within an entire event (e.g., tsunami may be associated with earthquake) [4]. These episodes may draw different public concerns and result in different emotional reactions at different times. While existing methods for sentiment analysis that are aimed at detecting human emotions from certain contents [9][10] do not address which episode of an event plays a dominant role in the change of human emotional reactions, our present work utilizes clustering-based analysis, a widely-used method [11], to identify episode-specific contents from all event-related news articles. By doing so, we analyze human emotional reactions to each event episode based on LIWC [3][12], a text-analysis dictionary containing more than 4500 words, grouped into 80 categories (e.g., emotional, perceptual and social processes).

The remainder of this paper is organized as follows: Section 2 introduces the related work. Section 3 presents the basic ideas of our method. Section 4 analyzes human reactions during Japanese earthquake in 2011. Section 5 highlights our major contributions.

2   Related Work

Currently, there exist two types of methods for detecting human reactions to an extreme event from media; they are: content-based analysis [5][13][14] and keyword-based analysis [15][16][17].

Content-based analysis is essentially a manual processing method that provides qualitative interpretations of extracted texts. These texts are classified by coders into different categories based on certain coding schemes, such as RIAS (including 14 categories) [5] or Taylor's IUE model (including 8 categories) [13]. Utilizing this method, Mechel et al. [18] have analyzed the interrelated reports that cover the BP Oil Spill in 2010. They have found that traditional media focus more on political news and facts, whereas social media are often driven by rumors and human-interested stories. Oh et al. [5] and Bollen et al. [14] have examined human emotional changes during a stock crisis and an earthquake in Haiti, respectively. Although content-based analysis has good theoretical foundations in social sciences, some limitations restrict its widespread use: (1) coders may not agree with each other on how to rate the type of an emotional word even though they have been well trained; (2) it is time-consuming if rating an article by multiple coders; (3) coders' emotions may affect the results, especially when reading sad news about a catastrophe [3].

In order to perform real-time analysis, some computer-aided approaches have been proposed for detecting human reactions to an extreme event. One of the most effective methods is keyword-based analysis [15][16][17]. For example, SATO et al. [15] have compared the changes of keywords over different times, and the chronological changes of different keywords from mass media. They have found various human activities at the different stages of an earthquake, such as *saving life* in the first 100 hours and *reconstruction* after 1000 hours. Signorini et al. [16] have tracked the public sentiment with respect to H1N1 in USA. Nevertheless, it remains to be inadequate to depict human emotional reactions based only on a few manually-selected words, as people often use different emotional words to describe the same event [19]. Also, keyword-based analysis can only provide a macroscopic view of human emotional reactions, rather than a microscopic representation showing where each event episode may play a distinct role in emotional reactions. Moreover, some irrelevant contents in the news may affect the quality of the results. To eliminate the irrelevant words and analyze the effect of an event episode on human emotional reactions, in this paper we present a clustering-based analysis method that evaluates the co-occurrences of the words, as used in news articles, and utilizes a sentimental analysis tool (LIWC [3][12]). We take Japanese earthquake in 2011 as a case study to reveal distinct patterns of emotional reactions with respect to earthquake, tsunami and nuclear crisis by analyzing the contents of a typical mass medium (i.e., BBC) and a professional relief-related medium (i.e., ReliefWeb).

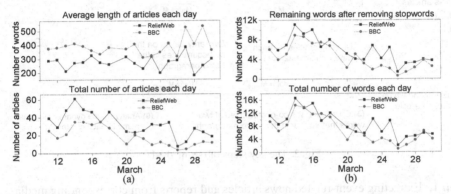

Fig. 1. The numbers of articles and words extracted from BBC and ReliefWeb over time

3   Clustering-Based Analysis

3.1   Datasets

The content data used in this work are extracted from BBC and ReliefWeb, which correspond to two typical public media, i.e., a general mass medium and a special website for relief. Generally speaking, the BBC materials are based on the stories of common people (e.g., victims or refugees in an extreme event) and/or the press releases from government offices, whereas the articles and reports on ReliefWeb mainly come from the special relief agencies (e.g., international organizations and NGOs such as OCHA).

We retrieve all "Japan" related news from BBC[1] and ReliefWeb[2] between March 11, 2011 to March 30, 2011. Fig. 1 shows the numbers of articles and words being released each day.

---

[1] http://www.bbc.co.uk/search/news/japan?start_day=11
&start_month=03&start_year=2011&end_day=30
&end_month=03&end_year=2011&sort=reversedate

[2] http://reliefweb.int/disaster/
eq-2011-000028-jpn?search=&sl=environment-term_listing

3.2   Clustering-Based Co-occurrence Analysis

We first extract all the words from each article, and construct a word-matrix for each day. Second, we measure the asymmetric similarity of two words based on their co-occurrences in the same paragraph. The more they are related, the closer they will appear [20]. Then, we cluster the similarity matrix by maximizing the modularity measure Q [21]. The words in each cluster are divided into two parts: (1) episode-specific words and (2) emotion-related words based on the LIWC categories [3]. By doing so, we analyze the patterns of human emotional reactions with respect to distinct event episodes by counting the number of emotional words in each cluster. The key steps of our method are illustrated in Fig. 2.

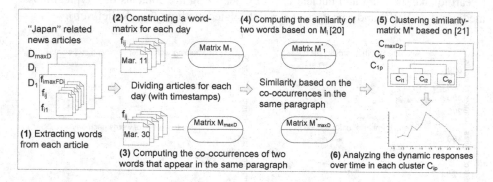

Fig. 2. The key steps of our method

Step 1. Extracting event-related news articles and reports from corresponding media

The news released on the same day are included into a folder D_i, where $i \in [1, maxD]$. Each D_i has a number of articles f_{ij}, i.e., $D_i = \{f_{i1}, ..., f_{ij}\}$, where $j \in [1, maxFD_i]$, and $maxFD_i$ denotes the total number of articles in D_i.

Step 2. Constructing a word-matrix M_i for each D_i

Based on a *stopwords* list[3], we remove redundant words (such as prepositions) from D_i. The rest of the words in D_i form a matrix $M_i[N_{D_i}][N_{D_i}]$, where N_{D_i} is the total number of different words having removed stopwords.

Step 3. Computing the co-occurrences of words in M_i

For each article f_{ij} in D_i, if two words (e.g., n_x, n_y) appear in the same paragraph, we perform: $M_i[x][y] + +$; $M_i[x][x] + +$; and $M_i[y][y] + +$.

Step 4. Computing the similarity of words in M_i

A *pseudo-inclusion* measure [20] is used to define the asymmetrical relationship between two words. The asymmetric similarity matrix M_i^* is defined as: $M_i^*[x][y] = (\frac{M_i[x][y]}{M_i[x][x]})^\alpha (\frac{M_i[x][y]}{M_i[y][y]})^{1/\alpha}$, where α denotes the coefficient parameter. If $\alpha=1$, M_i^* is the classical proximity index in scientometrics. Particularly, if $M_i^*[x][y]$ is low and $M_i^*[y][x]$ is high for $\alpha \gg 1$, it means that y is general relative to x, and x belongs to a

---
[3] http://www.lextek.com/manuals/onix/stopwords1.html

specific subdomain relative to y. Take two terms from ReliefWeb as an example. On March 11, "kill" and "earthquake" are in the 684th and 3rd line in M_1^*, respectively. For α=10, $M_1^*[3][684]$=0 and $M_1^*[684][3]$=0.8603. It means that "kill" is general relative to "earthquake", i.e., "kill" is always accompanied by "earthquake". We evaluate it in the original texts. There are 8 articles containing "kill" on March 11, in which "kill" and "earthquake" simultaneously appear in the same paragraph.

Step 5. Clustering asymmetric similarity matrix $M_i^*[x][y]$

We aim to find what types of emotional words are clustered together in describing an episode of an extreme event. That is to say, we want to identify the effect of each event episode on human reactions. Newman has proposed a fast community detection algorithm for a matrix based on the modularity Q [21], defined as $Q = \sum_k (c_{kl} - a_k^2)$. At the beginning, each word is regarded as a single community. c_{kl} is defined as the fraction of edges that connect words in group C_k to others in group C_l, and $a_k = \sum_l c_{kl}$. This greedy algorithm aims to maximize Q at each step through merging different small clusters into a big one, i.e., maximizing $\triangle Q = 2(c_{kl} - a_k a_l)$ at each step. When Q reaches to the maximum value, the best division is obtained. Each word is labeled with a cluster ID and grouped into a different cluster C_k based on the similarity matrix $M_i^*[x][y]$.

Step 6. Analyzing the changes of human reactions over time through counting emotional words in the same type of clusters

After step 5, the words in M_i is divided into different clusters, i.e., D_i can be further denoted as $\{C_{i1}, ..., C_{ip}\}$, where there are p clusters on i^{th} day. Some tightly-connected words, which depict the same episode, are clustered together. Specifically, some emotional words can be extracted in each C_{ip} based on LIWC [3]. Through computing the number of emotional words of a certain type, or the ratio of a certain type of emotional words to all emotional words in the same type of cluster in each D_i, we can observe human emotional reactions to an event episode. For example, we can reveal human emotional reactions to the nuclear crisis by extracting related words from $C_{i\{nuclear\}}$ of D_i.

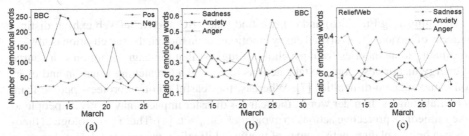

Fig. 3. (a) Dynamic changes of Pos and Neg emotions over time during Japanese earthquake in 2011. (b)(c) Subcategories of negative emotions extracted from BBC and ReliefWeb, respectively.

4 Results and Discussion

In this section, we provide some results on revealing human emotional reactions during Japanese earthquake in 2011.

4.1   General Results without Clustering-Based Analysis

Based on the LIWC categories, Fig. 3(a) reveals the distinct patterns of human negative and positive emotions with respect to Japanese earthquake in 2011. We can further analyze the detailed subcategories of negative emotions (i.e., sadness, anxiety and anger); the resulting patterns are plotted in Figs. 3(b) and (c) for BBC and ReliefWeb, respectively. From these plots, it is difficult for us to explain which event episode (e.g., earthquake or nuclear crisis) affects the change of human emotional reactions. In other words, we cannot capture what concerns people during such an event based on the whole contents of news articles.

Figure 4 further presents the ratios of the words used specifically for describing human perceptual and social processes according to the subcategories of LIWC [3]. Fig. 4(a) shows that the information that people acquire during the earthquake is mainly told by others. And, people are interested in acquiring more information through communicating with others, as shown in Fig. 4(b). Based on the uncertainty reduction theory (URT), under the conditions of stress and uncertainty, people tend to communicate with others in order to know what has happened, and further to reduce perceived uncertainty and alleviate their anxious emotions about an unknown event [22].

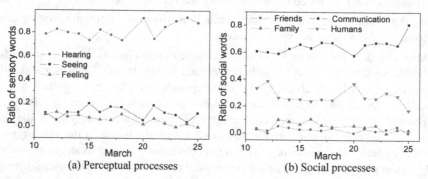

Fig. 4. Dynamic changes of human reactions during Japanese earthquake in 2011 extracted from BBC news

By comparing Figs. 3(b) and (c), we find that BBC and ReliefWeb exhibit distinct patterns of anxious (fearful) and angry emotions, even though the sad emotion is always dominant. Some theories from psychology suggest that the angry emotion is a source of energy that drives people to take actions to control a threatening situation and overcome encountered difficulties [7]. Whereas, the fearful emotion protects people from potential dangers. In order words, the fear for disaster impacts may motivate people to take immediate protective actions to avoid these impacts [4]. These psychological theories can help explain why two types of media exhibit different patterns during Japanese earthquake; ReliefWeb, as a professional relief-related medium, provides more supports and encourages people to take actions to recover from this event, whereas BBC, as a public medium, pays more attention to how to protect ourselves from a danger.

Meanwhile, we also notice two phases in Fig. 3(c). In the first phase before March 20 in 2011, the angry emotion is higher than the anxious emotion. After that, the angry emotion is lower than the anxious emotion. We cannot further explain this based only on the current method. In the next section, we apply clustering-based analysis to identify and understand human reactions to each event episode during Japanese earthquake.

4.2 Clustering-Based Analysis

Based on the asymmetric similarity of words, clustering-based analysis identifies episode-specific words from all the articles of each day; this is similar to detecting hot topics from news articles [23]. With this method, we aim to reveal the underlying reasons for distinct patterns of human emotional reactions

As described in Section 3, through clustering analysis at Step 5, some words about event episodes (e.g., earthquake, tsunami and nuclear crisis) and their corresponding impacts will be clustered together, respectively. A cluster is composed of both description-related (e.g., magnitude, fukushima and meltdown) and psychology-related words (e.g., cry, confuse and kill). Fig. 5 shows an example of the hierarchal structure of D_1. Some irrelevant events, such as "illegal political donations of PM Naoto Kan" and "London Olympiad", are clustered separately. These irrelevant contents may affect our analysis results. In order to accurately capture the changes of human emotional reactions over time, we extract and compare the ratio of emotional words in the same type of cluster on each day (e.g., the nuclear-episode on i^{th} day, $C_{i\{nuclear\}}$), rather than the all words on i^{th} day (i.e., D_i). Based on these episode-specific clusters, we can uncover how an episode (i.e., subevent) affects the changes of human reactions.

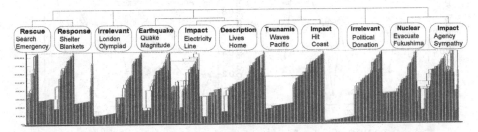

Fig. 5. An example to illustrate the hierarchal structure of clustering results. There are 11 clusters on March 11, 2011 (D_1).

Figure 6 presents three human emotional reactions to the "earthquake" and "nuclear" episodes as extracted from the two media. By comparing Figs. 6(a) and (b) with (c) and (d), respectively, we find that people feel more anxious about the nuclear crisis and sad about the earthquake. That is because the nuclear crisis could have more potential risk and uncertainty than the earthquake, while people could more readily learn figures about the damages of the earthquake from TV or other media. The anxious emotion would become especially high when people lack situation-specific information about the impacts of the event [4]. This also explains the higher anxious emotion as shown in Fig. 3. Under such a situation, rumors could fast spread (such as the nuclear fallout map[4] and the radiation rain of Philippines[5]), and may result in some irrational behaviors

---

[4] http://www.youtube.com/watch?v=RBye93OVkLU
[5] http://www.bbc.com/news/technology-12745128

(e.g., panic buying of iodide in USA[6] and salt in China[7]). In this case, the emergency management should quickly release more specific information about the radiation impact in order to reduce such an anxious emotion. Moreover, ReliefWeb, as a special medium for diaster relief, exhibited a relatively calm and professional attitude about the nuclear crisis if we compare Figs. 6(c) with (d). As shown in Fig. 6(d), the anxious emotion increased sharply in ReliefWeb only after more evidences about radiation leakage were confirmed on March 20, 2011, which could also be used to further explain the change of the anxious emotion in Fig. 3(c).

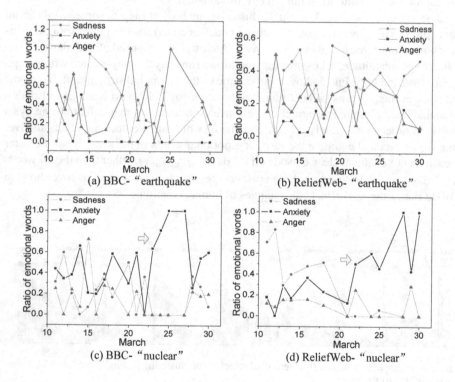

Fig. 6. Human emotional reactions to the "earthquake" and "nuclear" episodes during Japanese earthquake extracted from BBC and ReliefWeb, respectively

Figure 7 presents the number of words used specifically for describing human perceptual processes concerning different episodes at different times. In the face of a catastrophe, BBC and other news agencies provide more live pictures on the damages. Therefore, nature disasters could arouse more "seeing" related perceptual processes than a nuclear crisis. Towards to an unseen risk, people would exhibit more "hearing" related perceptual processes. On the other hand, we can see that there is a distinct peak in Fig. 7(b). That is because Japan government and IAEA published many reports to

---

[6] http://www.abc.net.au/news/2011-03-18/
worldwide-reports-of-iodine-overdoses-who/2649816
[7] http://af.reuters.com/article/worldNews/idAFTRE72G1LZ20110317

explain the nuclear situation (e.g., IAEA continuously released six reports about the nuclear crisis on ReliefWeb on March 15, 2011). Based on clustering analysis, most of the comments from authorities are clustered into a single and basic group (i.e., at the bottom of Fig. 5) as the reactions of the society to Japanese earthquake. However, only the cluster of nuclear crisis is associated to these comments, and merged into a bigger cluster (i.e., at the top of Fig. 5). From this point of view, we may infer how serious the nuclear crisis was.

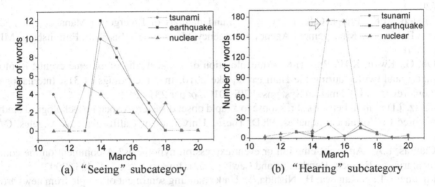

(a) "Seeing" subcategory   (b) "Hearing" subcategory

Fig. 7. A comparison of the perceptual processes among three different event episodes during Japanese earthquake

5   Conclusion

This paper has presented a clustering-based method for analyzing human emotional reactions based on news articles released during an extreme event. Different from the existing work, our method (1) eliminates the noises of irrelevant contents by means of clustering episode-specific words; (2) identifies the relationship between emotion-related words and episode-specific words based on an asymmetric similarity matrix; (3) provides reasonable results based on a set of episode-specific words, rather than a few manually-selected keywords. In addition, our method allows us not only to observe human emotional reactions (e.g., sadness, anxiety and anger) over an entire event, but also to find their underlying reasons, i.e., why a specific emotional reaction plays a dominant role at a certain stage. By taking Japanese earthquake in 2011 as an example, we have revealed some distinct patterns of human emotional reactions to the natural disasters and to the nuclear crisis. Our results show that professional relief-related media contain more supportive emotions than general public media, and people exhibit more anxious emotions about the nuclear crisis than the natural disasters. Our finding also shows that during Japanese earthquake, there was a need to provide more specific information about the radiation impact, so as to reduce the anxious level of the society.

Acknowledgment. This project was supported primarily by Hong Kong Baptist University and in part by the Fundamental Research Funds for the Central Universities (XDJK2012B016) and the PhD fund of Southwest University (SWU111024). The authors would like to thank Benyun Shi for his contribution on this paper.

References

1. Coyle, D., Meler, P.: New Technologies in Emergencies and Conflicts: The Role of Information and Social Networks, pp. 1–52. United Nations Foundation and Vodafone Foundation (2009)
2. Mendonca, D., Wallace, W.A.: A cognitive model of improvisation in emergency management. IEEE Transactions on Systems, Man and Cybernetics: Part A 37(4), 547–561 (2007)
3. Tausczik, Y.R., Pennebaker, J.W.: The psychological meaning of words: LIWC and computerized text analysis methods. Journal of Language and Social Psychology 29(1), 24–54 (2010)
4. Lindell, M.K., Prater, C.S., Perry, R.W.: Fundamentals of Emergency Management. Federal Emergency Management Agency Emergency Management Institute, Emmitsburg, MD (2006)
5. Oh, O., Kwon, K.H., Rao, H.R.: An exploration of social media in extreme events: Rumor theory and twitter during the Haiti earthquake 2010. In: Proceedings of 31st International Conference on Information Systems (ICIS 2010), paper 231 (2010)
6. Goltz, J.D.: Initial behavioral response to a rapid onset disaster: A social psychological study of three California earthquakes, Ph.D. Thesis, University of California, Los Angeles, CA (2006)
7. Camras, L.A.: An event-emotion or event-expression hypothesis? A comment on the commentaries on Bennett, Bendersky, and Lewis (2002). Infancy 6(3), 431–433 (2002)
8. Fukuhara, T., Nakagawa, H., Nishida, T.: Understanding sentiment of people from news articles: Temporal sentiment analysis of social events. In: International Conference on Weblogs and Social Meida (ICWSM 2007), p. 2 (2007)
9. Calvo, R.A., Mello, S.D.: Affect detection: An interdisciplinary review of models, methods, and their applications. IEEE Transactions on Affective Computing 1(1), 18–37 (2010)
10. Neviarouskaya, A., Prendinger, H., Ishizuka, M.: SentiFul: A lexicon for sentiment analysis. IEEE Transactions on Affective Computing 2(1), 22–36 (2011)
11. Rosvall, M., Bergstrom, C.T.: Mapping change in large networks. PLoS ONE 5(1), e8694 (2010)
12. Golder, S.A., Macy, M.W.: Diurnal and seasonal mood vary with work, sleep, and daylength across diverse cultures. Science 333(6051), 1878–1881 (2011)
13. Sreenivasan, N.D., Lee, C.S., Goh, D.H.: Tweet me home: Exploring information use on Twitter in crisis situations. In: Proceedings of the 14th International Conference on Human-Computer Interaction (HCI 2011), pp. 120–129 (2011)
14. Bollen, J., Mao, H., Zeng, X.J.: Twitter mood predicts the stock market. Journal of Computational Science 2(1), 1–8 (2011)
15. Sato, S., Hayashi, H., Maki, N., Inoguchi, M.: Development of automatic keyword extraction system from digitally accumulated newspaper articles on disasters. In: Proceedings of the 2nd International Conference on Urban Disaster Reduction, 6 p. (2007)
16. Signorini, A., Segre, A.M., Polgreen, P.M.: The use of Twitter to track levels of disease activity and public concern in the U.S. during the influenza a H1N1 pandemic. PLos ONE 6(5), e19467 (2011)
17. Thelwall, M., Buckley, K., Paltoglou, G.: Sentiment in twitter events. Journal of the American Society for Information Science and Technology 62(2), 406–418 (2011)
18. Meckel, M.: A spill and a spin: How traditional and social media are interrelated in covering the BP oil spill 2010. In: Proceedings of 61st Annual Conference of the International Communication Association, pp. 1–6 (2011)
19. Thelwall, M., Wilkinson, D., Uppal, S.: Data mining emotion in social network communication: Gender differences in MySpace. Journal of the American Society for Information Science and Technology 61(1), 190–199 (2010)

20. Chavalarias, D., Cointet, J.P.: Bottom-up scientific field detection for dynamical and hierarchical science mapping, methodology and case study. Scientometrics 75(1), 37–50 (2008)
21. Newman, M.E.J.: Fast algorithm for detecting community structure in networks. Physical Review E 69(6), 066133 (2004)
22. Lachlan, K.A., Westerman, D.K., Spence, P.R.: Disaster news and subsequent information seeking: Exploring the role of spatial presence and perceptual realism. Electronic News 4(4), 203–217 (2010)
23. Chen, K.Y., Luesukprasert, L., Chou, S.T.: Hot topic extraction based on timeline analysis and multidimensional sentence modeling. IEEE Transactions on Knowledge and Data Engineering 19(8), 1017–1025 (2007)

An Event-Driven Architecture
for Spatio-temporal Surveillance of Business Activities

Gabriel Pestana[1], Joachim Metter[2], Sebastian Heuchler[2], and Pedro Reis[3]

[1] INOV/IST, Lisbon, Portugal
gabriel.pestana@inov.pt
[2] BIJO-DATA GmbH, Sesslach, Germany
{jmetter,sheuchler}@bijodata.de
[3] ANA-Aeroportos de Portugal, Lisbon, Portugal
pereis@ana.pt

Abstract. The SECAIR project provides an event-driven architecture for critical environments. It deals with context awareness issues and surveillance of moving objects in real-time. The proposed architecture adopts a data fusion process to handle with location based data. It is therefore well-suited for business activity monitoring, supporting managers at analysing and processing complex event streams in real-time. A prototype applied to the surveillance of situation awareness for airport environments is presented. The main goal is to monitor for events in very congested areas for indoor and outdoor areas. The SECAIR system provides a collaborative environment for a better management of ground handling operations, in compliance with existing business rules. An advanced Graphical-User Interface with geographical and analytical capabilities is also presented.

Keywords: Situation Awareness, event-driven architecture, spatio-temporal data, spatial Data Warehouse, Spatial Dashboard, Control Services.

1   Introduction

In today's highly competitive service-oriented business environment, it is essential to keep decision makers informed about which events are affecting business-critical operations. This is the case of an airport, usually classified as a critical infrastructure, with a set of business rules which need to be continuously checked. It also needs to encompass functionalities for an unambiguous surveillance of surface traffic caused by aircrafts and vehicles, without reducing the number of operations or the airport safety level [1].

In the airport environment, to efficiently coordinate ground movements caused by aircraft, passengers and cargo, decision makers must be able to respond to an increasingly complex range of threats. But to reach the required level of coordination for decision-making and simultaneously to respond to every airport surveillance demands an informational cockpit with the following functionalities is required:

L. Chen et al. (Eds.): ISMIS 2012, LNAI 7661, pp. 136–142, 2012.

capacity to geographically communicate large amounts of relevant information with techniques to visualize multiple business data such as operational metrics and descriptive data (i.e., metadata). A feature typically assigned to corporate spatial dashboards.

This is particularly true when the visualization of key performance indicators (KPI), describing how business is performing, are correlated with a geographical representation of on-going events [2].

In this paper an innovative event-driven approach for a continuous monitoring of airport events is presented based on the prototype that is being designed within the SECAIR project [3]. The project makes use of a spatial data warehouse (SDW) and deals with complex event streams using multiple localisation technologies, to reach the required level of data integration for decision-making.

The remainder of the paper is organized as follows. Section 2 provides a short analysis of related work, outlining technologies used to continuously collect data about airport surface movements. Section 3 and 4 presents the system overall description, emphasizing the main points which make the proposed solution different from usual tracking systems. Section 5 presents the conclusions and future work.

2   Related Work

The range of technologies required to provide the early detection and intervention in an airport surveillance environment typically include a mix of many different systems. For instance, surface movement radar (SMR) and secondary surveillance radar (SSR) systems are quite common in large airports. However these solutions are extremely expensive to purchase and operate, and are subject to masking and distortion in the vicinity of airport buildings, terrain or plants [4].

The extensive deployment of satellite system and air-to-ground data links results in the emergence of complementary means and techniques. Among these, ADS-B (Automatic Dependent Surveillance-Broadcast) and MLAT (Multilateration) techniques may be the most representative [5]. However, current radar based systems have many problems to track surface targets, especially in very dense traffic areas, such as the apron area. But since most of the aircrafts turn-off their transponders after landing, there is a strong demand for a new sensing technology. Some solutions include near-range radar networks, Mode 3/A, S or VHF multilateration, magnetic flux sensors [6], CCTV systems with video analytics, or D-GPS installed in vehicles [4]. However, none of these localisation technologies is individually able to meet all the user's requirements for airport surveillance.

Although, it is already possible to find algorithms addressing the very stringent integrity requirements to support aircraft surface movement [7]. Most of existing systems still operate independently from each other, limiting the opportunity for providing automated decision support [8].

In the literature it is possible to find related research projects (e.g., 2004: Airnet and ISMAEL, 2006: EMMA, 2008: AAS and LOCON), that have successfully tested for ground movements surveillance, presenting actionable data with a high degree of certainty or as a cost-effective solution. Within the SECAIR project, the differentiating approach is that the localisation technologies are coherently integrated using advanced data-fusion techniques in order to reduce installation costs and to address multipath effects reduction. The project advanced fusion techniques operates with high-performance Global Navigation Satellite System (GNSS) and improved radio based tracking, combined with video based technology to accomplish an automatic and reliable prediction of safety events. This integration extends the state-of-the-art for the surveillance of airport surface traffic, enabling unique automated decision support capabilities, with context aware services, that have not thus far been tested.

3   System Description and Architecture

The SECAIR system is designed to detect Safety and Security events caused by ground vehicle and aircraft movement at airports, using commercially available localisation technologies. For instance, to track handling operations at congested areas for indoor and outdoor environments (e.g., boarding gates at the passenger terminal and aircraft parking areas at the apron).

The following data sources for vehicle or person positions, coming from different localization technologies that provide at least one position per second, are used:

- A standalone GNSS, collecting positions each second and transmitting it to the central system via wireless data communication (e.g., Wi-Fi);
- An indoor-outdoor tracking system based on radio frequency localization (IOTS);
- A video surveillance and tracking system (VSTS);
- IEEE 802.15.4a Ultra Wide Band (UWB) standard for a low-rate wireless personal area network.

As many as possible all target objects (i.e., vehicles, and staff) are equipped with at least one localisation technology. To allow for fusion of data from different systems all location-based data are converted to WGS84 coordinates at the server side. As illustrated in Fig.1, these data are continuously transmitted (Data Capture layer) to a central processing Data Fusion module at the Business Data Processing Layer. The SECAIR advanced fusion techniques operate with improved radio based tracking and video based technology enabling the surveillance of non-cooperative resources (i.e., aircraft, vehicle, or personnel not equipped with a localisation device).

Instead of using a simple median-based approach like most commercially available solutions, SECAIR uses an advanced data fusion algorithm that makes use of Quality-of-Location (QoL) values as well as of environmental context information, for instance the position of walls or obstacles. This is further augmented by information on the object in question, such as its type, size or maximum speed.

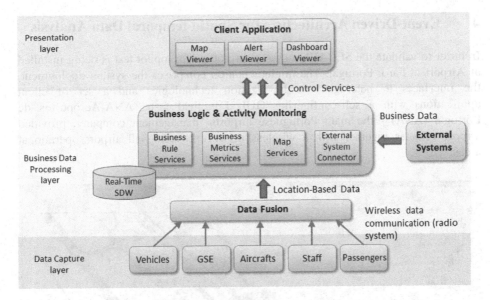

Fig. 1. Block diagram with a high level view of the SECAIR system

Constrains associated to the type of localisation technology are taken into account as well. For instance, video data may have issues with conglomeration of objects (which are difficult to distinguish), GNSS will probably give bad positional data inside a building, and radio signals can suffer from reflection from walls inside buildings. Such approach is especially noticeable in border cases such as the transition from inside a building to the outside, which typically means several localisation systems that were reliable before suddenly send bogus data, while other systems that may not even have received a signal start to provide excellent location based data.

The Business Data Processing layer includes software modules such a role-classified multi-view of business rules, customized business metrics and the segmentation of the airport into multiple operational areas interacting with each other over a common stream of location-based data. All modules use Windows Communication Foundation (WCF), part of the .NET framework 4, to transmit messages via TCP. WCF doesn't define a required host, allowing creating clients that access services running in different context environments.

At the Presentation layer, the surveillance capability of the SECAIR system is presented in three different ways. The Map Viewer represents moving objects as colour coded point features with a timestamp and a set of descriptive data (see Fig.2); including metadata about aircrafts, vehicles, drivers, flight data, or airport operations. The Alert Viewer lists alert messages with start and end time plus additional descriptive data. In Fig.2, the three vehicles visualized with a colour coded label outline different levels of alert messages presented at the Alert Viewer. The Dashboard Viewer graphically displays spatio-temporal business indicators to provide a clear picture of the airport status, using the Squarified Treemap algorithm [9].

4   Event-Driven Architecture for Spatio-temporal Data Analysis

In order to validate the SECAIR, a system prototype for a pilot test is being installed at Airport of Faro, Portugal. The implementation comprises the system deployment, the interfaces to heterogeneous localization technologies and a set of client applications with a self-configuring GUI. For field tests, ANA-Aeroportos de Portugal (ANA) - the main Portuguese airports' management company, provided airport vehicles together with a wireless network covering all airport operational areas.

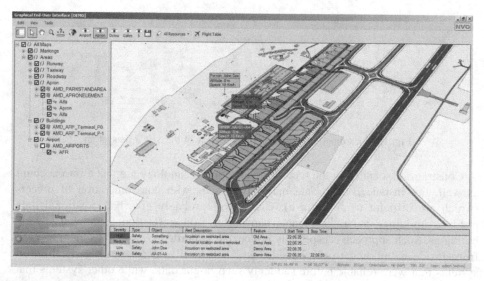

Fig. 2. Prototype version of the graphical user interface (GUI) layout

The system deployment also comprises interoperability with existing airport systems, for instance, to collect flight information and data about Staff. Table 1 presents the multidimensional database structure of the Spatial Data Warehouse (SDW) that will hold the analytical processing capabilities of the SECAIR system. The matrix lists at the columns the dimensions used to describe business logic and at the rows the events (facts) related to the different domains of surface surveillance. Any dimension (column) with more than one "X" implies that this dimension must be conformed across multiple fact tables, forming a constellation.

The Airport Layout dimension is a spatial dimension with metadata about each operational area stored in thematic layers characterizing the airport layout in conformity to the ED119 std. [10]. For instance, airport circulation constraints on areas related to ground traffic movements include speed limits for different types of moving objects (e.g., operational vehicles and A/C), constrains for specific vehicles categories (e.g., auto-stairs, high-loaders, passenger busses), or data related to the airport operational status (e.g., normal or low visibility operations). The other dimensions are non-spatial in the sense that they only store business data obtained through interoperability with existing airport systems.

The three fact tables store specific events related to surface traffic, therefore classified as spatial facts with a new position stored for each object being monitored. The Ground Movements is the most granular and detailed fact table, classifying each movement in relation to any business rule infringement. The Safety & Security Events aggregates data related only to safety and security events. Finally the Vehicle Services fact table is particularly adjusted for fleet management.

Table 1. The SDW Matrix of the SECAIR Data Structure

Dimensions Granular Facts (ROLAP)	Airport Layout	Aircrafts	Vehicles	Stakeholders	Flights	Staff	Tasks	Alerts	Time Hour	Time Day
Ground Movements	X	X	X	X	X	X	X	X	X	X
Safety & Security Events	X	X	X	X		X		X		X
Vehicle Services	X		X	X	X	X	X		X	X

Especially at higher levels in the architecture, some streams are opt-in, so clients need to register first. This reduces the amount of unnecessary network traffic. Messages also feature a priority property which is used throughout the system to identify important circumstances, making sure they are handled first.

Client applications can subscribe to different events, receiving also in an event-driven way all information, which can consist of location data and other business or device related data. For instance, whenever a vehicle protection area intersects with another moving object or infrastructure, a collision avoidance event is triggered informing the driver to move to a safety distance. This scenario requires the vehicle to be equipped with an onboard unit which immediately informs the driver of the danger. For ease of use SECAIR offers a touch screen display and a radiofrequency (RFID) reader for an automatic login procedure that uses the airport ID card of the driver to validate if the diver is authorized to operate the identified vehicle.

Within the SECAIR project, WCF provides an explicit support for service-oriented development with a unified programming model for rapidly build service-oriented interfaces. The communication interface is mainly event driven, with the business logic and the external systems continuously sending events to update the GUI after a successful login.

5   Conclusions

The paper presents an event-driven architecture for the surveillance and tracking of Safety and Security events in critical areas, with functionalities to support spatio-temporal data processing. This means that any occurrences are shown immediately (within fractions of a second) and, as far as possible, dealt with automatically. It is therefore well-suited for business activity monitoring, supporting business users at analysing and processing complex event streams in real-time.

The architecture adopts a multi-layer approach with heterogeneous localization technologies and a data fusion process to handle with event streams emitting continuously location-based data for each surveyed object.

The proposed system is being designed to deal with spatio-temporal requirements, including scalability and security of data. A SDW was specified to hold a very high volume of fine-grained events, which must be processed and analyse individually before taking appropriate control actions. The innovative mix between geovisualization functionalities and KPIs for the spatial dashboard also introduces new challenges, contributing to improve situation awareness and coordination of ground handling operations.

References

[1] ICAO, Safety Management Manual (SMM), 2nd edn. Doc 9859 AN/474 (2009) ISBN 978-92-9231-295-4
[2] Wyman, O.: Guide to Airport Performance Measures. Airports Council international, ACI (2012)
[3] Metter, J. (Project coordinator): Security on airport and other critical large infrastructures (SECAIR). EUREKA-Eurostars project ref. E!6030 (2011-2014)
[4] Hu, T.T.W., Wang, L., Maybank, S.: A survey on visual surveillance of object motion and behaviors. IEEE Transacions on Systems Man, and Cybernetics-part C: Applications and Reviews, 334–351 (2004)
[5] Soto, A., Merino, P., Valle, J.: ADS-B integration in the SESAR surface surveillance architecture. In: Proc. of Enhanced Surveillance of Aircraft and Vehicles (ESAV), pp. 13–18 (2011)
[6] Gao, H., et al.: Safe airport operation based on innovative magnetic detector system. Institution of Engineering and Technology, Transp. Syst. 3(2), 236–244 (2009)
[7] Schuster, W., Bai, J., Feng, S., Ochieng, W.: Integrity monitoring algorithms for airport surface movement. SpringerLink 16(1), 65–75 (2012)
[8] Mehta, V., et al.: Decision Support Tools for the Tower Flight Data Mananger System. In: Integrated Communications, Navigation and Surveillance Conference, ICNS, pp. I4-1–I4-12 (2011)
[9] Chintalapani, G., Plaisant, C., Shneiderman, B.: Extending the Utility of Treemaps with Flexible Hierarchy. In: Proc. of Int. Conf. on Information Visualisation, London (2004)
[10] EUROCAE, ED-119B - Interchange Standards For Terrain, Obstacle, and Aerodrome Mapping Data (2011)

Multiple Model Text Normalization
for the Polish Language

Łukasz Brocki, Krzysztof Marasek, and Danijel Koržinek

Polish-Japanese Institute of Information Technology, Warsaw, Poland

Abstract. The following paper describes a text normalization program for the Polish language. The program is based on a combination of rule-based and statistical approaches for text normalization. The scope of all words modelled by this solution was divided in three ways: by using grammar features, lemmas of words and words themselves. Each word in the lexicon was assigned a suitable element from each of the aforementioned domains. Finally, the combination of three n-gram models operating in the domains of grammar classes, word lemmas and individual words was combined together using weights adjusted by an evolution strategy to obtain the final solution. The tool is also capable of producing grammar tags on words to aid in further language model creation.

1 Introduction

In the field of Natural Language Processing there has always been a grave demand for the employment of large quantities of textual data [2,3]. This is especially true for Automatic Speech Recognition (ASR), or Machine Translation (MT). To make software as effective as possible, such texts need to undergo several stages of preparation, one of the most essential being normalization.

During the development of Language Models (LM), which are often used in the ASR and MT tasks, corpora of over 100 million words are frequently utilized. Manual processing of such gigantic amounts of texts, even by a large team of people would be at best ineffective and expensive, if not simply impossible. The only feasible and working solution is the utilization of computer programs which can perform the same task in a reasonable amount of time.

Text normalization is the process of converting any abbreviations, numbers and special symbols into corresponding word sequences. The procedure must produce texts consisting exclusively of words from a given language. In particular, normalization is responsible for:

1. expansion of abbreviations in the text into their full form
2. expansion of any numbers (e.g. Arabic, Roman, fractions) into their appro priate spoken form
3. expansion of various forms of dates, hours, enumerations and articles in contracts and legal documents into their proper word sequences

This task, although seemingly simple, is in fact quite complicated - especially in languages like Polish which, for example contains 7 cases and 8 gender forms for

L. Chen et al. (Eds.): ISMIS 2012, LNAI 7661, pp. 143–148, 2012.

nouns and adjectives, with additional dimensions for other word classes. That is why most abbreviations have multiple possible expansions and each number notation over a dozen outcomes. It is worth noting that in Polish the dictionary of only the words related to numbers contains almost 1000 entries. This is because each number can be declined by both the case and gender/number which in Polish most often changes the suffix of a word, thus producing a completely new word (if unique letter sequences are treated as unique words). The amount of possible outcomes of normalization of word sequences containing abbreviations and numbers grows exponentially with the number of words that need to be expanded. It is also worth mentioning that sentences in Polish follow a strictly grammatical structure and the adjoining words in the sequence have to be grammatically correct. Example: *Wypadek był na sto dziewięćdziesiątym piątym kilometrze autostrady.* (translation: The accident happened on the hundred and fiftieth kilometer of the highway.)

The authors needed a program to normalize texts in Polish in order to prepare corpora used for training language models for use in Automatic Speech Recognition and Machine Translation. This work describes the technical aspects of the text normalization tool designed specifically for the Polish language.

2   Text Acquisition

The first step in building the software was the acquisition of a large quantity of textual data. The authors managed to obtain text corpora surpassing 1 billion words in size. They originated from various online newspapers (~48%), specialized newspapers (~9%), legal documents (~18%), wikipedia (~9%), parliament transcripts (~9%), radio, tv, usenet, subtitles, etc...

Most of the data was gathered directly from Internet sources using custom software. It was later balanced during the training phase to avoid overfitting to certain domains and styles of speech. Most of these texts (excluding the finished corpora) were acquired during a span of several months. The task workload is estimated at around 6-12 man-months.

3   Text Preparation

All the acquired texts were initially processed to remove any unnecessary data like tags, formatting and words out of context (e.g. values from tables or contents of ads). This processing was done by a single person during a span of about 2 months. Following that, texts that were suspected to contain too much garbage were removed from the dataset. This was done using a program that counted words from a Polish spelling word list. The discarded texts had either a large amount of typos or non-Polish words. Finally, all the data was gathered and saved into a simple and manageable text format. It's worth noting, however, that at that point, the most important and most difficult task - text normalization - has still not yet been performed.

The authors have decided to use both rule-based mechanisms and statistical language models in the text normalization software. Paradoxically, to build even the simplest statistical language model for use in this tool, already normalized texts were required. To that end, a small group of linguistics students were trained to create a small balanced manually normalized corpus. This was later used to build the initial language model for the first iteration of the text normalization software. A collection of carefully chosen texts amounting to 2.5 million words were used for this manual corpus. A group of 8 people performed this task in the span of 4 months.

The word list was split into several smaller sub-lists. Each sub-list was assigned to two independent linguistic students that didn't know each other or had any way of communicating. The results were then merged by a program that also generated a list of all inconsistencies between the two lists. This list of inconsistencies was finally analyzed and corrected by a third, independent person, with a PhD in linguistics and most experienced of the group. This stage took another 4 months to complete. The final result was a dictionary of most frequent Polish words, their grammatic description and lemmas.

In Polish, many unique letter sequences can be derived from more than one word lemma. That is why it was necessary to reevaluate the whole normalized corpus once again in order to disambiguate all the words given their actual context. This was made easier thanks to a program that looked for such words within the corpus and allow for appropriate alterations. This whole stage took about a month.

The outcome of the 9 month work of the whole group was a balanced, manually normalized text corpus with disambiguated word lemmas and grammatic features. A dictionary of the most frequent words, abbreviations and all common number forms with lemmas and grammar features was also created.

4   Synthetic Texts

After a few iterations of the software it was observed that most errors appear in sequences that occur least frequently in the training set. These were sequences that had a generally clear grammatical structure, but the used text corpus was too small for the language models to reflect that structure. A common technique used to improve the statistical language models is to generate synthetic texts with a previously established grammatical structure [13].

Synthetic texts were generated to contain context-free sequences of several words from the domains including numbers (with various units of measurment), dates and times. All the words contained within the synthetic texts were also placed in a separate dictionary. These texts obviously didn't require any normalization or disambiguation because this was already included in the generation process. The texts were constantly generated and added throughout the second half of the project. The final list contained around 3 million words.

5 Development of Language Models

Many experiments were done in order to create the best working language model for use in the text normalization software. The best found configuration will be described in this section.

The word lexicon contained 39362 words, 14631 word lemmas and 603 grammar classes. Given that the word context during normalization of texts in Polish is often more than 5 words (e.g. while expanding long numbers) it was established that the best course of action in this case would be to create long range n-gram models. A model with a range of n=3 was used for the individual words, n=5 for word lemmas and n=7 for grammar classes. Since the manually normalized corpus was rather small, the experiments showed that expanding the range of the individual word model didn't drastically improve its performance (as witnessed by perplexity measures). A different result was observed with word lemma and grammar models however. Because both the number of lemmas and grammar classes is considerably smaller than the number of individual words, the former were better modelled in the text corpus and this allowed for increasing the context length of their language models. The ratio 3-5-7 was established as optimal for the given corpus, dictionary and domain. Larger contexts didn't decrease the perplexity significantly. It is worth noting that the 7-gram range for grammar classes has a significant advantage to lower ranges because in Polish a word at the start of the sentence determines the case for the entire sentence and thus can affect the morphology of words also at the end of the sentence.

To develop the language model, only the normalized corpus consisting of 2.5 million words and synthetic corpus of 3 million words were used. The training data consisted of 10 collections, each containing texts from a certain domain. Synthetic texts comprised 4 different domains and the manually normalized texts had 6 different domains amounting to 2 million words. 250 thousand words each were chosen from the manual corpus for a testing and development set. All the models were linearly interpolated. A $(\mu + \lambda)$ Evolution Strategy [4,5] that minimized the perplexity of the final model (consisting of 3 smaller models: word, lemma and grammar) on the development set was used.

The Evolution Strategy optimized hundreds of parameters, specifically:

1. weights of 30 text domain sets (10 parameters for each model)
2. linear interpolation weight for all n-grams in all models. The weights depended on the frequency of occurrence of given n-gram - there were 5 ranges of frequency
3. linear interpolation weights for the word, lemma and grammar classes models (combining the smaller models into one larger)

After preparing all the data and tools in previous stages, it takes about a week to generate a new language model. The best model trained on the normalized data the perplexity of 376 on an independent test set (around 400 thousand words).Perplexity [2] is a common benchmark used in estimating the quality of language modeling. The lower values are generally preferable, although they obviously depend on both the effectivness of the model, as well as the complexity of the test data. In our experiment, the value isn't too small, but it's worth noting that the test set contained texts from varying domains and not a single domain, as is often the case in the majority of domain experiments, where state-of-the-art models achieve values well below 100.

6 Text Normalizer Architecture

The most important components of the software for text normalization are the decoder, language model and a set of expansion rules. The expansion rules are used in the expansion of commonly used abbreviations and written date and number forms. A synchronous Viterbi style decoder that generated a list of hypotheses ordered by the values retrieved from the language model was used. Each time the text contained a word sequence that could be expanded, all the possible expansions were fed into the decoder. Because the expansion of long numbers or some abbreviations expects that several words need to be added at once, hypotheses of varying lengths may end up competing against each other. This was remedied by the normalization of hypotheses' probabilities to their lengths. Such a normalization was equivalent to the addition of a heuristic component commonly used in asynchronous decoders like A^*. The decoding process is generally quite fast, but word sequences that contain many abbreviations and numbers can severely slow it down. For this reason, a maximum number of hypothesis was set to 2000. Over 1500 different rules of abbreviation expansion were manually created including a number of algorithms for parsing of date and hour formats, converting Roman numerals to Arabic, parsing real numbers and improving the quality of the normalizer in case the source texts missed decimal points (marked by commas in Polish) or if they were replaced by spaces or periods.

7 Experiment Results

Because the fragments sometimes expanded into a sequence of words, a simple word error rate seemed inappropriate for this purpose. Instead a fragment error rate was evaluated. To that end, a special test set was chosen from independent data. The source of this data was the same as the training data. This was processed by the normalizer and then manually corrected by a linguist. Each fragment that needed normalization was analyzed by the expert and marked either as correct or incorrect. The test set contained 1845 normalized fragments, 1632 of which were normalized correctly and 213 incorrectly, giving 88.5% accuracy.

8 Conclusion

This work described the development of software used for the normalization of texts in Polish language. The development took considerable effort seeing as it was necessary to manually build a training corpus, dictionary and set of rules for abbreviation expansion. The project was assisted by 8 students of linguistics. The result of this work is a program that is able to normalize a 100 million word corpus on a modern computer in 2-3 days.

This work represents one on the few efforts in normalization of large quantities of textual data for Polish [1] and arguably the first used for ASR and MT purposes. The results clearly show it is possible to create a reasonably accurate working system for any domain. It is worth noting that normalization of domain independent data can be problematic. For example, systems trained on news data tend to produce many errors when used on legal documents and vice versa. More experiments on domain constraints and adaptation are necessary.

Acknowledgments. The work was sponsored by a research grant from the Polish Ministry of Science, no. N516 519439.

References

1. Filip, G., Krzysztof, J., Agnieszka, W., Mikołaj, W.: Text Normalization as a Special Case of Machine Translation. In: Proceedings of the International Multiconference on Computer Science and Information Technology, Wisła, Poland, vol. 1 (2006)
2. Jelinek, F.: Statistical Methods for Speech Recognition. MIT Press, Cambridge (1998)
3. Dumke, R.R., Abran, A. (eds.): IWSM 2000. LNCS, vol. 2006. Springer, Heidelberg (2001)
4. Michalewicz, Z.: Genetic algorithms + Data Structures = Evolution Programs. Springer (1994)
5. Michalewicz, Z., Fogel, D.B.: How to Solve It: Modern Heuristics. Springer (1999)
6. Przepiórkowski, A.: Korpus IPI PAN. Wersja wstępna / The IPI PAN Corpus: Preliminary version. IPI PAN, Warszawa (2004)
7. Savary, A., Rabiega-Wiśniewska, J., Woliński, M.: Inflection of Polish Multi-Word Proper Names with Morfeusz and Multiflex. In: Marciniak, M., Mykowiecka, A. (eds.) Aspects of Natural Language Processing. LNCS, vol. 5070, pp. 111–141. Springer, Heidelberg (2009)
8. http://sgjp.pl/morfeusz/
9. Bilmes, J.A., Kirchhoff, K.: Factored language models and generalized parallel backoff. In: Proceedings of HLT/NACCL, pp. 4–6 (2003)
10. Chen, S.F., Goodman, J.T.: An empirical study of smoothing techniques for language modeling. Computer, Speech and Language 393, 359–393 (1999)
11. Katz, S.M.: Estimation of Probabilities from Sparse Data for the Language Model Component of a Speech Recognizer. IEEE Transactions on Acoustics, Speech and Signal Processing 3, 400–401 (1987)
12. Kneser, R., Ney, H.: Improved backing-off for n-gram language modeling. In: International Conference on Acoustics, Speech and Signal Processing, pp. 181–184 (1995)
13. Chung, G., Seneff, S., Wang, C.: Automatic Induction of Language Model Data for A Spoken Dialogue System. In: 6th SIGdial Workshop on Discourse and Dialogue Lisbon, Portugal, September 2-3 (2005)

Discovering Semantic Relations
Using Prepositional Phrases

Janardhana Punuru and Jianhua Chen

Computer Science Division
School of Electrical Engineering and Computer Science
Louisiana State University
Baton Rouge, LA 70803-4020
{jpunur1,cschen}@lsu.edu

Abstract. Extracting semantical relations between concepts from texts is an important research issue in text mining and ontology construction. This paper presents a *machine learning-based* approach to semantic relation discovery using prepositional phrases. The semantic relations are characterized by the prepositions and the semantic classes of the concepts in the prepositional phrase. WordNet and word sense disambiguation are used to extract semantic classes of concepts. Preliminary experimental results are reported here showing the promise of the proposed method.

1   Introduction

Learning semantic relations from domain texts is a fundamental task in automatic ontology construction and text mining. Non-taxonomical semantic relations tend to be more difficult to discover simply because there are too many such relations in contrast to the taxonomical ("is- a") relation. Researchers have developed various approaches [2] [5] [13] to the task of non-taxonomical relation extraction. Methods presented in [2], [5] and [13] exploit the syntactic structure and dependencies between the words for relations extraction and exploit statistical tests to verify the statistical significance on the occurrence of concept pair and the verb together. In [10], we proposed the SVO (Subject-Verb-Oobject) approach that starts with triplets of the form (C_1, V, C_2) such that C_1, V, and C_2 occur as subject, verb and object in a sentence, and utilizes the log-likelihood ratio measure to select valid semantical relations. It was shown that our SVO approach produces very good precision in experiments using the *Electronic Voting* data set.

Even though the SVO method is able to identify relations with high accuracy, the count of relations obtained does not represent the whole domain. To improve the coverage of non-taxonomic relations, we present in this paper a supervised learning technique to find the semantic relations using prepositional phrases.

2   Using Prepositional Phrases

Intuitively, prepositional phrases often indicate semantic relations between concepts. Consider the phrases "management of company" and "revolt of the

L. Chen et al. (Eds.): ISMIS 2012, LNAI 7661, pp. 149–154, 2012.

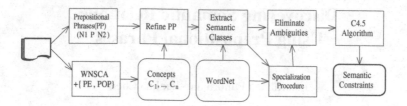

Fig. 1. Architecture for Learning Semantic Constraints

workers" as an example. Clearly the phrases indicate that company relates to management via relationship possess, and revolt relates to worker via relationship performed by. In this paper we present a supervised learning technique for finding semantic relationship between concepts occurring in prepositional phrases. The detailed architecture for obtaining the training data and learning the semantic constraints is presented in the Figure 1. The supervised approach for learning semantic constraints for semantic relations can be divided into the following four steps.

1. Extraction of unambiguous prepositional phrases.
2. Selection of attributes and their values for training instances.
3. Eliminating inconsistencies in the training data.
4. Learning rules for labeling the relations.

2.1 Ambiguity in Prepositional Phrases

One of the major difficulties in identifying semantic relations from prepositional phrases is the ambiguity in preposition attachment. For example, from the observation of the following two sentences,

1. I bought the shirt with pockets
2. I washed the shirt with soap

It is clear that in sentence 1, *with pockets* describes the *shirt*. However, in sentence 2, *with soap* modifies the verb *wash*. Prepositional phrase disambiguation in natural language processing takes (N_1, V_1, P, N_2) as input and to identify whether the given prepositional phrase $[P, N_2]$ to be attached to the noun(N_1) or the verb(V_1). In this paper we focus on unambiguous prepositional phrases for learning the constraints.

Unambiguous prepositional phrases are extracted from the part of speech tagged text using a simplified chunker and the extracted concepts list obtained by applying our method [8] and indicated as "WNSCA+{PE, POP}" in Figure 1. The input text is initially processed using Brill's part of speech tagger [1]. Words occurring as determiners, adjectives, and cardinal numbers are removed from the part of speech tagged text. From the filtered text, concepts are extracted, and then for each occurrence of the prepositional phrase(i.e., preposition and the following noun in the form (P, N_2)), the preceding text(up to 5 words) is searched for the occurrence of noun, verb, or both. If both noun and verb are present then such prepositional phrase is considered as an ambiguous one. We only keep triplets of the form (N_1, P, N_2) found such that either N_1 or N_2 must occur in the extracted concepts list.

2.2 Relationship Labeling

For each of the unambiguous prepositional phrases selected, the relationship between the concepts occurring in the phrase needs to be labeled. We define a fixed set of candidate relation labels as shown in Table 1. The relation labels are collected observing several semantic relations listed in [12], [4], and [7].

Table 1. Semantic Relations for Prepositional Phrases $(A, Preposition, B)$

No.	Semantic Relation	$A \to B$	$A \leftarrow B$
1	subtype		transaction of purchase
2	part of		
3	attribute	model of computer	
4	procedure	construction of plant	project in production
5	perform by	authorization from director	company for distribution
6	cause		debt for investment
7	measurement	billion in investment	amount in dollar
8	use		
9	location		bank in city
10	require		knowledge of negotiation
11	produce	processor of product	
12	antonym	breakup of conglomerate	
13	synonym		
14	performed on	marketing of satellite	market for acquisition
15	source		support from Board
16	member	head of planning	
17	possess	person with deposit	tax on income
18	recipient		share by affiliate
19	constraint		
20	associated with	bid after trading	
21	temporal	Working through weekend	budget on schedule
22	collection	syndicate of investor	

Each unambiguous prepositional phrase extracted from the text is manually labeled with one of the relations in Table 1. The assigned label indicates the semantic relation between the concepts in the prepositional phrase along with the direction of the relationship.

2.3 Training Data Construction and Inconsistency Elimination

Each manually labeled prepositional phrase is considered as an instance for the supervised learning algorithm. We used the C4.5 [11] decision tree algorithm for the learning task. To make use of the C4.5 algorithm, attributes or features need to be selected and their possible values determined so that each prepositional phrase instance is converted to a record for decision tree learning. Three attributes namely, *Source Class*, *Preposition*, and *Target Class* are defined. For a given prepositional phrase, the *Source Class* attribute takes the semantic class of the noun preceding the preposition, *Preposition* attribute takes the actual preposition in the phrase, and *Target Class* attribute takes the semantic class of the noun following the preposition. For example, for the phrase "maker of refrigerator", attribute values are *Source Class* = entity#1, *Preposition* = of, and *Target Class* = entity#1. Attributes values for *Source Class* and *Target Class* attributes are extracted from the WordNet [6] by identifying the senses of concepts. Weighted sense disambiguation approach [9] is used to find the senses of the concepts. As shown in the above example, semantic classes of the concepts appear as class name#sense number. Here, class name is the top-level class in the WordNet and sense number is the sense index in the WordNet for class name. In the above example entity#1 indicates the top-level class as entity with sense number 1.

Training data obtained this way may still contain inconsistencies. Namely, there may exist two or more instances with same attribute values but with different relation labels. To resolve ambiguities a specialization procedure is applied. For each set of ambiguous examples, *Source Class* attribute value is replaced with immediate hyponym of its current value in the hierarchy of the noun preceding the preposition. If ambiguity is not eliminated, *Target Class* attribute value is replaced with the immediate hyponym of its current value in the hierarchy of noun following the preposition. This process is repeated until the ambiguity is resolved or no more specialization can be done. Similar specialization procedure is also used in [3] for learning constraints for part-whole relations.

3   Empirical Evaluations

The proposed semantic relation discovery method is experimented with *Electronic Voting* and *Tenders, Offers, and Mergers (TNM)* corpora. The Electronic Voting domain text consists of 15 documents (with a total of more than 10,000 words) extracted from New York Times website in 2004. These documents describe issues in using electronic voting machines for elections.. The TNM Corpus is collected from TIPSTER Volume 1 corpus distributed by NIST. The TIPSTER Volume 1 corpus consists of news articles from Wall Street Journal in 1987 through 1989. In TIPSTER corpus data each news article is labeled with its topic. The TNM Corpus is obtained by collecting news articles with the topic label *Tender offers, mergers, and acquisitions*. The TNM corpus consists of 270 articles with a total size of 29.9 MB.

3.1   Two Classification Schemes

Due to space limitations we omit detailed discussions on experiments for the *Electronic Voting* domain text, and mainly focus on the studies on the TNM corpus. For both corpora, we tried two alternative classification schemes, as described below.

Frequent Relations Approach. The top four frequent relations in the training data are labeled by 3, 14, 9, and 17 in Table 1. That is "attribute", "performed on", "location", and "possess" are the most common semantic relations in the training data. The remaining relations do not have much representatives. So we lump the these relations into one new relation "0" and construct one decision tree using the revised training data. Here we have a classifier with 5 outcome classes.

Multiple Binary Classifiers Approach(MBCA). Another approach we implemented for learning rules for semantic relation labeling is MBCA. In this approach, C4.5 algorithm is used for learning rules for each of the top four relations separately from the training data. Training data is constructed in such a way that all instances belong to either the target relation or not. Modified training data is fed to C4.5 algorithm to learn the rules for the target relation. This procedure is repeated for each of the four relations. The learned rules for each of the relations are combined and sorted based on their accuracies. The sorted rule set is used for classification of unknown instances.

3.2 Experiments on TNM Data

After initial success with the *Electronica Voting* data, experiments are conducted on *Tenders Offers, Mergers and Acquisitions(TNM)* data to further confirm the validity of the proposed method. After filtering out ambiguous prepositional phrases, we got 116350 triplets of the form (N_1, P, N_2). It is difficult to manually label all of the 116350 phrases. To obtain the training data, we randomly selected a subset of 2000 prepositional phrases and manually labeled each of them with one of the relationship labels. Among the 2000 phrases, 692 phrases are either ill-formed, or contain terms not defined in the WordNet. Thus the remaining 1308 prepositional phrases are used for learning with C4.5.

The resultant 1308 prepositional phrases, when manually labeled, got 28 distinct relation labels. Maximum possible number of relation labels are 22(from $A \rightarrow B$) +22 (from $A \leftarrow B$) = 44 as listed in Table 1. Among the 28 distinct relationships, there exists only 3 to 5 instances for most of the relations. As described earlier, we focus on the top frequent relations and label the instances of the remaining relations as class "0".

Frequent Relations Approach. Among the 1308 examples labeled above from TNM data, relationships with labels 3, 14, 9, and 17 constituted 386 examples. For the remaining 922(1308-386) instances, label 0 is assigned indicating that given instance does not hold any of the four relation labels mentioned above. The modified training data with 5 target class labels $(0 - 4)$ is used to learn the rules for each relations. Using four-fold cross validation, average accuracy of the C4.5 decision tree on test data is $(61.1 + 50.3 + 53.1 + 51)/4) = 53.8\%$. From the obtained decision tree, high accuracy rules are extracted.

The accuracy of the frequent relations approach on *Electronic Voting* and *TNM* domains is summarized in the Table 2. The MBCA is also applied to the *TNM* data. The accuracy of the MBCA on *Electronic Voting* and *TNM* domains is summarized in the Table 3.

Table 2. The FR Approach Results

Domain Text	Accuracy(%)
Electronic Voting	57.8
TNM	53.8

Table 3. The MBCA Approach Results

Domain Text	Precision(%)	Recall(%)	Accuracy(%)
Electronic Voting	42.9	38.4	49.1
TNM	45.7	57.7	47.8

4 Conclusions

We present a method for learning semantic constraints for labeling the relations between the concepts occurring prepositional phrases. This approach learns the semantic constraints using C4.5 algorithm. The learned constraints are represented in terms of the semantic classes of the concepts in the prepositional phrases. Semantic classes of the concepts are extracted from the WordNet using sense disambiguation technique. This approach is experimented with *Electronic Voting* and *TNM* data. The experimental results indicate the presented method is useful for learning semantic constraints. Experimental results emphasize that further experiments need to be conducted on larger training data.

References

1. Brill, E.: A simple rule-based part-of-speech tagger. In: Third Conference on Applied Natural Language Processing, pp. 152–155 (1992)
2. Ciaramita, M., Gangemi, A., Ratsch, E., Saric, J., Rojas, I.: Unsupervised learning of semantic relations between concepts of a molecular biology ontology. In: Proc. of International Joint Conference on Artificial Intelligence (IJCAI 2005), pp. 659–664 (2005)
3. Girju, R., Badulescu, A., Moldovan, D.: Learning semantic constraints for the automatic discovery of part-whole relations. In: Human Language Technologies and North American Association of Computational Linguisitics, pp. 80–87 (2003)
4. Girju, R., Moldovan, D., Tatu, M., Antohe, D.: On the semantics of noun compounds. Computer Speech and Language 19, 479–496 (2005)
5. Kavalec, M., Maedche, A., Svátek, V.: Discovery of Lexical Entries for Non-taxonomic Relations in Ontology Learning. In: Van Emde Boas, P., Pokorný, J., Bieliková, M., Štuller, J. (eds.) SOFSEM 2004. LNCS, vol. 2932, pp. 249–256. Springer, Heidelberg (2004)
6. Miller, G.A.: Wordnet: An on-line lexical database. International Journal of Lexicography 3(4), 235–312 (1990)
7. O'Hara, T., Wiebe, J.: Classifying functional relations in factotum via wordnet hypernym associations. In: Int. Conf. on Computational Linguistics, pp. 347–359 (2003)
8. Punuru, J., Chen, J.: Automatic Acquisition of Concepts from Domain Texts. In: Proceedings of IEEE Int. Conf. on Granular Computing, pp. 424–427 (2006)
9. Punuru, J., Chen, J.: Learning Taxonomical Relations from Domain Texts using wordNet and Word Sense Disambiguation. Proceedings of IEEE Int. Conf. on Granular Computing, August 2012 (to appear)
10. Punuru, J., Chen, J.: Learning Non-Taxonomical Semantic Relations from Domain Texts. Journal of Intelligent Information Systems 38(1), 191–207 (2012)
11. Quinlan, R.J.: C4.5: Programs for Machine Learning. Morgan Kaufmann (1993)
12. Rosario, B., Hearst, M.: Classifying the semantic in noun compounds via a domain-specific lexical hierarchy. In: EMNLP 2001, pp. 82–90 (2001)
13. Schutz, A., Buitelaar, P.: *RelExt*: A Tool for Relation Extraction from Text in Ontology Extension. In: Gil, Y., Motta, E., Benjamins, V.R., Musen, M.A. (eds.) ISWC 2005. LNCS, vol. 3729, pp. 593–606. Springer, Heidelberg (2005)

DEBORA: Dependency-Based Method for Extracting Entity-Relationship Triples from Open-Domain Texts in Polish

Alina Wróblewska[2] and Marcin Sydow[1,2]

[1] Polish-Japanese Institute of Information Technology, Warsaw, Poland,
[2] Institute of Computer Science, Polish Academy of Sciences, Warsaw, Poland
alina@ipipan.waw.pl, msyd@poljap.edu.pl

Abstract. This paper describes DEBORA – a dependency-based approach to the extraction of relations between named entities from Polish open-domain texts. The presented method designed for the purpose of the conducted experiment is adapted to morpho-syntactic properties of Polish. Results show that the method is applicable for Polish, even if there is a room for improvement. The extraction approach may be applied to the problem of graphical entity summarisation.

Keywords: knowledge graphs, information extraction, dependency parsing, graphical entity summarisation.

1 Introduction

Amount of information available in textual resources on the Web is huge and is still growing. For this reason, *Information Extraction* (IE), which aims at automatic or semi-automatic collection of structured data from textual corpora of given domain, is in the mainstream of academic and industrial research. More recently, the scientists' attention is paid to *Open Domain* IE (ODIE), in which information is automatically extracted from textual resources not restricted to any particular domain.

This paper describes a subtask of ODIE – the automatic extraction of *entity-relationship (ER) triples* from Polish texts. ER-triples are instances of semantic relations between pairs of named entities (NE), e.g., (*Warszawa, jest w, Polsce*), Eng. (*Warsaw, is located in, Poland*). Triples extracted from a corpus are regarded as candidates for being facts. Thus, after applying some validation techniques to filter out invalid or unreliable facts, available ER-triples may be used to build a large semantic knowledge base. Since building a knowledge base is a complex process, its first stage consisting in the extraction of triples from Polish open-domain texts is in the scope of this paper.

Most of already proposed relation extraction techniques are based on pre-defined extraction rules or manually annotated training corpora. As the manual development of extraction patterns or the manual corpus annotation are expensive and time-consuming processes, systems based on these techniques are usually limited to one extraction domain. One of the first successful systems for the fast and scalable fact extraction from the Web is the domain-independent system, *KnowItAll* [4]. *KnowItAll* starts with the extraction of entities of pre-defined entity types (e.g., CITY, MOVIE) and then discovers

L. Chen et al. (Eds.): ISMIS 2012, LNAI 7661, pp. 155–161, 2012.

instances of relations between extracted entities using handwritten patterns. Another system called *TextRunner* [2] applies a technique of extracting all meaningful instances of relations from the Web. The system *ReVerb* [5], in turn, overcomes some limitations of the mentioned systems using a novel model of the verb-based relation extraction.

Although many efficient triple-extraction models exist for English and few other languages, this research field is still not explored in a large group of inflecting languages with relatively free word order, such as Polish. The direct application of many extraction techniques designed for English, which is an isolating language with topological argument marking, seems to be not suitable for Polish, which is an inflecting languages with the morphological argument marking[1] and free word order. Topology-based extraction rules defined for English may not be applicable for Polish.

This paper presents experiments on extracting ER-triples from Polish Web documents using a dependency-based method. The paper is structured as follows. Section 2 outlines the dependency-based method of extracting triples. The prototype implementation of the entire extraction procedure is described in Section 3. Section 3.1 gives an overview of experimental results. Finally, a novel application of the extracted triples in graphical entity summarisation is presented in Section 5.

2 DEBORA – A Dependency-Based Method of Triple Extraction

Although existing triple extraction techniques may be efficient for English, they may not be applicable for Polish. English is a language with relatively restrictive word order used to convey grammatical information. Polish, in turn, is characterised by rather flexible word order. Thus, a single fact may have numerous surface representations in a text. Because of this, the iterative pattern induction as in *DIPRE* [3] or extraction of meaningful facts defined as token chains between entities as in *TextRunner* [2] might be difficult or even inapplicable for Polish. As no manually specified domain-independent extraction patterns or seed instances of relations enabling the extraction of further facts are available, triples are discovered using a dependency-based method.

A triple is defined as a tuple $t = (ne1_{subj}, r, ne2)$, where $ne1_{subj}$ denotes a noun phrase recognised as a named entity (NE) and fulfilling the subject function, $ne2$ represents another recognised NE, and r denotes an instance of relation between these NEs. Instances of relations are discovered only between recognised NEs, one of which fulfils the subject function $(ne1_{subj})$. This decision is motivated by the property of Polish, which allows pro-drop pronouns with the subject function. At the current stage of our work, relations between implicitly realized entities are not modelled. Sentences without a subject are ruled out, in order to avoid the coreference resolution problem.

In an idealistic scenario, $ne2$ is realised as a noun phrase that depends on the sentence predicate. However, $ne2$ may also be realised as a noun phrase depending on a

---

[1] In Polish, there is a partial adequacy between the case of a noun and the argument this noun may fulfil, e.g., a noun phrase marked for nominative (NPNOM) typically fulfils the subject function. However, NPNOM may also fulfil the predicative complement function (e.g., Pol. *Jan*.NOM *to artysta*.NOM Eng. 'John is an artist.').

preposition[2] or another noun phrase (e.g., apposition) which are governed by the sentence predicate. These NEs are also involved in the extraction of instances of relations.

In order to extract meaningful instances of relations, identification of grammatical functions seems to be essential. In our approach, only elements of the predicate-argument structure selected from a dependency structure may build instances of relations. An instance of relation between two NEs consists of a sentence predicate and arguments subcategorised by this predicate, excluding arguments fulfilled by two NEs. Currently, the field that triples are extracted from is restricted to a simple sentence or a matrix clause in a complex sentence.

3   Experiments

The dependency-based triple extraction technique has been implemented and integrated with ExPLORER – a currently developed experimental platform for extracting fact database from an open-domain Polish corpus. The system takes textual resources (e.g., web documents) as input and outputs a set of extracted triples. ExPLORER can be generally viewed as a chain of configurable modules (corpus creator, NLP-module, ER-triple extractor). The extraction procedure starts with crawling web documents. Then, a text corpus is extracted and annotated with external publicly available NLP-tools. The best Polish part-of-speech tagger – *Pantera* [1] divides the entire text into sentences and tokens, performs a thorough morphological analysis and augments tokens with their lemmas, part-of-speech tags and morpho-syntactic features. Morpho-syntactically annotated texts are given as an input to a named-entity recogniser – *Nerf*[3] [7], which annotates dates and personal, organisation and place names. The corpus annotated with morpho-syntactic features and NEs constitutes an input to the Polish dependency parser[4] [9]. Finally, ER-relations are automatically extracted with a module based on the heuristic described in Section 2.

3.1   Experimental Results

The extraction method described in section 2 is applied to a set of Polish web news articles (188,415 texts) taken from [6]. Raw texts are split into 6,303,794 sentences with 20.3 tokens per sentence on average. As the goal of the experiment is to discover instances relating NEs, only these sentences with at least two recognised NE (3,265,817 sentences) are parsed with the Polish dependency parser. The morpho-syntactically annotated and dependency-parsed sentences are given to the triple extractor that discovers 58,742 instances of relations between pairs of NEs. The extracted triples concerned 26,469 unique NEs fulfilling the subject function. In order to evaluate the quality of extracted triples and the extraction procedure itself, two evaluations are carried out.

---

[2] If a noun phrase recognised as a NE is governed by a preposition, the preposition constitutes a part of the instance of relation.

[3] According to [7], *Nerf* achieves the general recognition performance of 79% F-score.

[4] According to [9], the Polish dependency parser using the system MaltParser achieves the parsing performance of 71% LAS (*labelled attachment score*).

First Evaluation Experiment. Because the total number of extracted triples is quite large in our web-based experiment, the straightforward computation of *precision* is not a trivial task. Furthermore, the exact computation of *recall* in case of a large web-based input text corpus is completely infeasible, since it would involve counting all valid triples contained in this corpus. Instead, an approximation of precision is computed by sampling 100 random triples and manually examining their validity. This computation is repeated three times and the average precision of 54% is achieved. We also select all triples which represent some particularly interesting relations concerning people and places. The precision of selected relations is manually computed: *bornIn* (occurrences in total: 154) – 98.7%, *died* (228) – 80%, *livedIn* (16) – 87.5%, *isLocatedIn* (7) – 100%.

The results are very promising. Despite the early stage of our work and difficulty with the open-domain extraction task (especially for Polish), the majority of the examined extracted triples are correctly formed and represent interesting facts about entities.

Second Evaluation Experiment. The second evaluation is performed in order to check impact of linguistic processing on the quality of triples extracted with the DEBORA algorithm. For reasons of this evaluation, a small Polish test corpus (64 simple sentences, 9.25 tokens/sentence on average) is manually annotated. In the first step of the evaluation experiment (baseline), 62 triples are automatically extracted from the corpus and 46 of them are correct (precision: 74.2%, recall: 46%). Since Nerf has not recognised any of six alone occurred last names, sentences with these NEs are not taken into account while extracting triples. That is why, input given to the triple extractor is manually corrected (part-of-speech tags and dependency structures[5] are amended and some missing NE labels are added) in the second step of the performed experiment. Manual corrections of input increase the number of extracted triples (92 correct triples, precision: 95.8%, recall: 92%). The evaluation against the gold standard corpus of simple sentences suggests that DEBORA performs quite well (56.8% of F-score) and that the extraction accuracy significantly increases if input is manually corrected in terms of morpho-syntax (93.9% of F-score). A cursory error analysis shows that the poor quality of some extracted triples is mainly due to error-prone linguistic processing (e.g., missing triples caused by unrecognised NEs, partially identified relation and incorrectly composed relation caused by errors in dependency structures). It indicates that better NLP-tools are required in order to further improve the extraction performance.

4   Potential Applications to Graphical Entity Summarisation

Extracted triples, after additional post-processing (e.g., NE normalisation and disambiguation) may be used to automatically build large semantic knowledge bases that can be viewed as large repositories of facts automatically extracted from the open-domain sources like www. Such repositories can be further processed or queried. As a demonstration of such future possibilities, we present an example of an application of DEBORA to compute *graphical entity summarisations* on semantic knowledge graphs [8].

---

[5] A NE subcategorised by the sentence predicate may be incorrectly annotated as a dependent of any other element, or a NE may be incorrectly annotated as an argument of the sentence predicate, even if it is not subcategorised by this predicate.

Figure 1 presents a graphical summary of the Polish 19th century poet *Adam Mickiewicz* automatically created with a diversified summarisation tool developed in the DIVER-SUM project [8] applied to the set of triples concerning the poet automatically extracted by DEBORA from the Polish web corpus described in Section 3.1. The presented example of graphical summary is of surprisingly high quality, especially when one takes into account that it is based on open-domain web articles.

It would be very interesting to further integrate the crawling, extracting, summarising and visualising modules into one coherent platform. One may imagine two operational modes of such platform. In the off-line mode, the user first specifies the web sources to be automatically collected off-line by an intelligent focused web crawler. The crawled corpus is subsequently processed by DEBORA in the off-line manner to build a large knowledge graph that contains extracted facts on NEs from a given domain. Finally, such a knowledge base can be interactively queried by users with the tool similar to the one presented on Figure 1. In the on-line mode, user provides the system with a medium-sized passage of text concerning some domain or entity (similarly to the biographical text used in the second evaluation experiment in Section 3.1). The text is immediately processed by DEBORA and user can interactively use the system to produce graphical summarisations of the entities concerned with the input text. Our

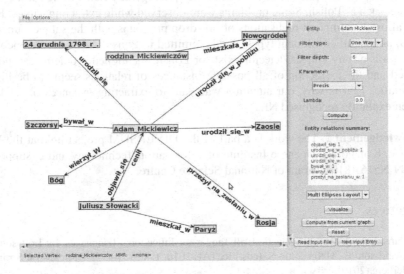

Fig. 1. Graphical entity summarisation obtained with a visualisation tool described in [8] concerning the Polish poet Adam Mickiewicz based on automatically extracted facts from Polish web texts. (NEs were normalised manually for this example). Extracted ERs: *Adam_Mickiewicz urodził_się_w Zaosiu* (Eng. "born in" (location)), *Mickiewicz urodził_się 24_grudnia_1798_r_.* (Eng. "born on" (date)), *Adam_Mickiewicz urodził_się_W pobliżu_Nowogródka* (Eng. "was born close to"), *Mickiewicz przeżył_na_zesłaniu_w Rosji* (Eng. "survived the exile to Russia"), *Mickiewicz wierzył_W Boga* (Eng. "believed in God"), *Adam_Mickiewicz bywał_w Szczorsach* (Eng. "used to be in" (location)).

experimental platform under development, presented in this section, seems to be a promising prototype to achieve the described functionality for Polish.

5   Conclusions

DEBORA – the dependency-based method for extracting ER-triples from Polish open-domain texts has been presented. The method was implemented, integrated within the ExPLORER platform and evaluated. According to the achieved results, the extraction method quite successfully extracts triples, so it seems to be appropriate for Polish. An approximate precision of about 54% was achieved in the evaluation based on samples of triples randomly selected form the open-domain corpus of web documents. Furthermore, an average precision of about 90% characterised selected triples representing some favourable relations such as *bornIn, died, livedIn isLocatedIn*. The second evaluation based on the small gold standard corpus confirmed that the quality of linguistic processing has a huge impact on the accuracy of extracted triples. If better NLP-tools were at hand, the better extraction results might be achieved.

According to [2], deploying a deep linguistic parser to extract relations between entities is not practicable at Web scale. It was shown that it can be a reasonable solution to extract facts from open-domain texts in a morphologically rich and free word order language, such as Polish. Some problems were observed while extracting triples. First, Polish allows for implicit realisations of pro-drop pronouns with the subject function. Second, almost all constituent types may be omitted in Polish. As no coreference resolution tools or any ellipsis detector exist for Polish, mentioned problems can not be managed and the extraction of all possible instances of relations seems to be highly problematic. That is why, our attention was paid to extracting instances of relations between explicitly recognised NEs.

Acknowledgements. The work is a part of the DIVERSUM project hold at the Web Mining Lab of Polish-Japanese Institute of Information Technology and is supported by the N N516 481940 grant of National Science Centre.

References

1. Acedański, S.: A Morphosyntactic Brill Tagger for Inflectional Languages. In: Loftsson, H., Rögnvaldsson, E., Helgadóttir, S. (eds.) IceTAL 2010. LNCS, vol. 6233, pp. 3–14. Springer, Heidelberg (2010)
2. Banko, M., Cafarella, M., Soderland, S., Broadhead, M., Etzioni, O.: Open information extraction from the web. In: Proceedings of the 20th International Joint Conference on Artifical Intelligence, pp. 2670–2676 (2007)
3. Brin, S.: Extracting Patterns and Relations from the World Wide Web. In: Atzeni, P., Mendelzon, A.O., Mecca, G. (eds.) WebDB 1998. LNCS, vol. 1590, pp. 172–183. Springer, Heidelberg (1999)
4. Cafarella, M.J., Downey, D., Soderland, S., Etzioni, O.: KnowItNow: Fast, Scalable Information Extraction from the Web. In: Proceedings of Human Language Technology Conference and Conference on Empirical Methods in Natural Language Processing (HLT/EMNLP), pp. 563–570. Association for Computational Linguistics (2005)

5. Etzioni, O., Fader, A., Christensen, J., Soderland, S., Mausam: Open Information Extraction: The Second Generation. In: Proceedings of the Twenty-Second International Joint Conference on Artificial Intelligence, pp. 3–10 (2011)
6. Korpus Rzeczpospolitej. Corpus of articles published in the Polish newspaper *Rzeczpospolita*, http://www.cs.put.poznan.pl/dweiss/rzeczpospolita
7. Savary, A., Waszczuk, J.: Narzędzia do anotacji jednostek nazewniczych. In: Przepiórkowski, A., Bańko, M., Górski, R.L., Lewandowska-Tomaszczyk, B. (eds.) Narodowy Korpus Języka Polskiego. Wydawnictwo Naukowe PWN, Warszawa (2012)
8. Sydow, M., Pikuła, M., Schenkel, R.: To Diversify or Not to Diversify Entity Summaries on RDF Knowledge Graphs? In: Kryszkiewicz, M., Rybinski, H., Skowron, A., Raś, Z.W. (eds.) ISMIS 2011. LNCS (LNAI), vol. 6804, pp. 490–500. Springer, Heidelberg (2011)
9. Wróblewska, A.: Polish Dependency Bank. Linguistic Issues in Language Technology 7(1) (2012), http://elanguage.net/journals/index.php/lilt/article/view/2684

Lexical Ontology Layer – A Bridge between Text and Concepts*

Grzegorz Protaziuk[1], Anna Wróblewska[1], Robert Bembenik[1],
Henryk Rybiński[1], and Teresa Podsiadły-Marczykowska[2]

[1] Institute of Computer Science, Warsaw University of Technology
Nowowiejska 15/19, 00-665 Warsaw, Poland
{G.Protaziuk,A.Wroblewska,R.Bembenik,H.Rybinski}@ii.pw.edu.pl
[2] Institute of Biocybernetics and Bioengineering,
Trojdena 4, 02-109 Warszawa, Poland
tpodsiadly@ibib.waw.pl

Abstract. Intelligent methods for automatic text processing require linking between lexical resources (texts) and ontologies that define semantics. However, one of the main problems is that while building ontologies, the main effort is put to the construction of the conceptual part, whereas the lexical aspects of ontologies are usually diminished. Therefore, analyzing texts, it is usually difficult to map words to concepts from the ontology. Usually one should consider various linguistic relationships, such as homonymy, synonymy, etc. However, they are not clearly reflected in the conceptual part. We propose LEXO - a special lexical layer, which is thought as a bridge between text and the conceptual core of the ontology. LEXO is dedicated to storing linguistic relationships along with textual evidence for the relationships (as discovered in the text mining process). In addition, we present an algorithm based on LEXO for determining meaning of a given term in an analyzed text.

Keywords: ontology, lexical layer, ontology localization, disambiguation of meanings of word, context representation.

1 Introduction

Ontologies play an important role with respect to the advancement of established information systems, systems for data and knowledge management, or systems for collaboration and information sharing, as well as, for the development of the revolutionary fields such as semantic technologies and the Semantic Web [1]. In [2] a definition of ontology that consists of two layers, namely the conceptual (CL) and the lexical one (LL) is presented. In our paper this definition is a starting point for introducing LEXO – a special structure of LL, which is thought as a bridge between text and the conceptual core of the ontology. We posit that the

---

* This work is supported by the National Centre for Research and Development (NCBiR) under Grant No. SP/I/1/77065/10 by the Strategic scientific research and experimental development program: "In-terdisciplinary System for Interactive Scientific and Scientific-Technical Information".

L. Chen et al. (Eds.): ISMIS 2012, LNAI 7661, pp. 162–171, 2012.
© Springer-Verlag Berlin Heidelberg 2012

conceptual layer is independent of the lexical one, so many lexical layers may be specified for one core layer. A lot of effort has been put into finding solutions to the problems concerning the conceptual layer of ontology, in particular to ontology representation ([2,3]), ontology engineering ([4]), ontology learning ([5]) and ontology integration ([6]). Problems related to the lexical layer have not been studied so extensively. In the literature one can find a few approaches aiming at ontology localization that have been made quite recently. The most prominent ones include LMF [7], LIR [8], LexInfo [9] and LEMON [10].

LMF is a meta-model that provides a standardized framework, allowing creation and use of computational lexicons. The LMF meta-model is organized into packages. From our perspective the most relevant are: Core Package, Morphology Package, NLP Morphological Patterns Package, NLP Syntax Package, Constraint Expression Package, NLP Semantic and NLP Multilingual Notations Package. The core package contains the basic elements of the model and their dependencies. The central entity in the LMF meta-model is the Lexical Resource, which has an associated Global Information object capturing administrative details and information related to encoding. The Lexical Resource consists of several language-specific Lexicons. A Lexicon then comprises Lexical Entries (i.e. words, multi-word entities such as terms and idioms, etc.) which are realized in different Forms and can have different meanings (Senses). LMF was used as a basis for the construction of other linguistic frameworks.

LIR is a model inspired by LMF; it associates lexical information with OWL ontologies. The main goal of LIR is to provide a model allowing enrichment of ontology with a lexico-cultural layer for capturing the language-specific terminology used to refer to certain concepts in the ontology. The LIR model has focused on multilingual aspects, as well as on capturing specific variants of terms (such as abbreviations, short forms, acronyms, transliterations, etc.) which are all modeled as subclasses of the property hasVariant. To account for multilinguality, the classes LexicalEntry, Lexicalization, Sense, Definition, Source and UsageContext are all associated with a certain Language to model variants of expression across languages. LIR also allows documenting the meaning of certain concepts in different cultural settings.

LexInfo is composed of the three main building blocks: LingInfo, LexOnto and the LMF. LingInfo defines a lexicon model where terms can be represented as objects that include lexical information, morpho-syntactic decomposition and point to semantics as defined by a domain ontology. LingInfo supports the representation of linguistic information, which includes: language-ID (ISO-based unique language identifier), part-of-speech, morphological and syntactic decomposition, and statistical/grammatical context models (linguistic context represented by N-grams, grammar rules, etc.). The goal of LexOnto is to capture the syntactic behavior of words and the relation between that behavior and the ontology. LMF is characterized above.

LEMON is a simplified version of the mentioned above frameworks with a strong focus on usability in information extraction. The main goals of creating LEMON [11] were: (1) conciseness of representation, (2) basing on RDF(S), (3) openness - it does not prescribe the usage of a particular inventory of linguistic categories and properties, (4) support for the reuse of any linguistic ontology, and (5) assignment of semantics to lexical entries by means of references to ontological entities.

Usually the lexical layer may be understood as a mapping. In particular, given a set T of terms (*term* is a word or a phrase from a certain text), and the sets C (concepts), and I (instances) we define LL_map as $LL\_map : T \rightarrow 2^{C \cup I}$. Such a mapping results directly from the definition in [2]. It can be used in an IR system for prompting a user for a clarification in the case when a homonym is used in a query. In our work we consider lexical layer from a perspective of semantic processing of natural language texts. To this end we also incorporate within the layer a contextual information, resulting from text mining, and showing in more detail the relationships between terms, concepts and instances. For a given text corpora we define a set CTX of possible contexts for terms in T, and attempt to build a function LL_func as $LL\_func : T \times CTX \rightarrow (C \cup I)$. Such a function may be used inter *alia* for semantic annotations of texts, for word sense disambiguation, or for translations.

The paper has the following structure: Section 2 presents basic concepts, and the LEXO structure, Section 3 presents a method of using LEXO for word sense disambiguation in an analyzed text. Section 4 presents experiments, and Section 5 draws the conclusions.

2   LEXO Lexical Layer

The proposed lexical layer (LEXO) for an ontology is thought as a bridge connecting words and/or phrases from texts written in a natural language with appropriate notions from the conceptual layer of some ontology. LEXO in contrast to the aforementioned approaches to realizing a lexical layer focuses less on a morphological specification of a given language but rather concentrated on relationships between words or phrases which may be extracted from texts, and which are useful for distinguishing between different meaning of a given word in a given texts. It is dedicated first and foremost to applications associated with semantic analysis of texts, so in the design of the layer we put emphasis on:

- representation of different meanings of words\phrases;
- description of a context of a given meaning in order to make it possible to determine a given meaning of a word or a phrase based on its context taken from text in which that word or phrase occurs.

In the design of LEXO we follow the well-known idea "a word is characterized by the company it keeps" [12] and adapt to ontology needs the approach presented in [13] for finding discriminants indicating various meanings of a word.

2.1   Notation and Definitions

In this section we introduce the notation and define the key constructs which are used in the description of the structure of LEXO (Section 2.2), and in the presentation of a method for determining the most probable meaning of a word

(Section 3). In the sequel *term* is an instance of a word or a phrase from a natural language. Usually, terms are denoted as t, whenever needed for terms being single words (phrases) we denote them by t^w (t^p respectively). In the definitions below we denote by $\mathcal{T}$ a text corpora in a natural language; by u^t we denote a unit u of text from $\mathcal{T}$, which includes a term t; depending of the process, it can be a phrase, paragraph or a document. Following [13], we define a context of t as a termset X for which $sup(t, X) \leq \varepsilon$ in $\mathcal{T}$. A termset X is called a discriminant against termset Y, if we have $relSup(t, X, Y) \leq \delta$, where $relSup(tXY) = \min\{sup(tY)/sup(tZ)), sup(tZ)/sup(tY)\}$, and δ is a user-defined threshold. In other words the termsets X and Y are such contexts of t, that they determine two various meanings of t in $\mathcal{T}$. Obviously, for a term t we may have many discriminants. We denote by $Disc(t)$ a set of all discriminants of t. In other words, for given t we have $d_1 \in Disc(t)$ iff there is $d_2 \in Disc(t)$ and d_1 is discriminant against d_2 for t. The set $Disc(t)$ defines a set of meanings, each meaning represented by a subset of discriminants. By $LMean(t)$ we denote a set of meanings of term t described in LEXO, lm_i^t a single meaning of a term t, $lm_i^t \in LMean(t)$. By $disc(lm_i^t)$ we denote a subset of $Disc(t)$ which defines the meaning lm_i^t of t. In the sequel we denote by $Ctx(t, u^t)$ a context of term t in a given unit u of $\mathcal{T}$ – e.g. set of word/phrases which occur together with t in that unit of text.

A term has a special representation in the LEXO structure. In particular, terms may be interconnected by means of linguistic relations. In LEXO we utilize the following linguistic relations: *synonymy, hyponym, and hypernym*. In LEXO we represent meanings of the terms (e.g. discovered by a TM process) by means of the context:

Definition 1. Given a term t, and a meaning m_i of the term t, the LEXO context $LC(lm_i^t)$ is defined as:

$$LC(lm_i^t) = RCE(lm_i^t) \cup disc(lm_i^t) \cup Vlin(lm_i^t)$$

where

- $RCE(lm_i^t)$ is a set of terms associated with t in the meaning lm_i^t by means of the *linguistic* relations.
- $Vlin(lm_i^t)$ is a set of terms associated with t in the meaning lm_i^t by means of *VicinityLinguistic* relation of LEXO. A term $ct \in VLin(lm_i^t)$ only if it occurs frequently enough in a neighborhood (e.g. in a sentence or a paragraph) of t used in the meaning lm_i^t in texts from the domain of interest. For example for creating such set a *frequent termset* accompanying t may be applied [13].

2.2 The Structure of LEXO

A simplified conceptual structure of a LEXO is presented in Fig. 1. The picture illustrates the main classes with attributes and relations between them that are used in the algorithm introduced in Section 3. Below we provide a short description of this structure (the detailed presentation can be found in [14]).

Word represents a single word from the natural language. That word is stored in the *lexicalForm* attribute. Also, its base form (*baseForm* attribute) and POS tag are stored. *Phrase* represents a phrase from a natural language composed of two or more *Word* objects. *LexicalEntry* (LE) is the basic class of the layer. It represents either a word or a phrase in an ontology. The relations *concerns* are used for connection of LE object with *Word* or *Phrase* object. The different meanings of a given term t are expressed by *MeaningOccurrence* (MO) relations with different *LexicalMeaning* objects. The relation has two attributes: *probability* indicating probability of occurrence in texts of the term t in a given meaning and *priority* showing the position of a given meaning in the sequence of lexical meanings of the term t ordered by their subjective significance. *LexicalMeaning* (LM) represents a single lexical meaning of a given term and indicates a notion from conceptual layer corresponding to this meaning. Such meaning is defined foremost by a context $LC(lm_i^t)$. Also examples of usage represented by *UsageExample* class may be associated with it. ContextEntry (CE) represents a term included in some context $LC(lm_i^t)$. The way of representation is identical as in *LexicalEntry* class.

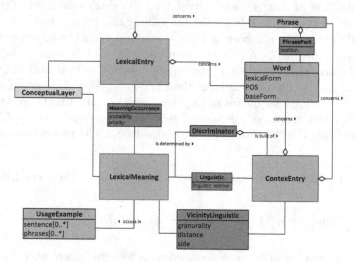

Fig. 1. The simplified structure of LEXO

3   Recognizing the Meaning of a Term Based on LEXO

In this section we present how LEXO can be used for determining meanings of terms while analyzing a text. The selection of the best meaning of a term in a given context is performed based on the adjustment vector which is assigned to each lexical meaning associated with the considered term. The vector consists of the following attributes: $d\#$, $n\#$, $e\#$, *priority*, and *probability*. Given *LexicalEntry le* for the term t and *LexicalMeaning* lm_k^t, the values of these attributes are determined as follows:

- $d\# = |disc(lm_k^t) \cap Ctx(t, u^t)|$ – the number of discriminants connected to lm_k^t that occur in the considered unit of text;
- $n\# = |Vlin(lm_k^t) \cap Ctx(t, u^t)|$ – the number of "neighbors" associated with lm_k^t that occur in the considered unit of text;

- $e\# = |Exmp(u^t, lm_k^t)|$ – the number of examples (phrases or/and sentences) of usage of t in the meaning lm_k included in the analyzed unit of text;
- *priority* and *probability* – the values of attributes of MO relation between le and lm_k^t.

The proposed algorithm *RecoTermMean* determines the meaning of a term t in a given unit u of $\mathcal{T}$ by matching descriptions of meanings in LEXO (their contexts and examples of usage) with a context $Ctx(t, u^t)$. In the sequel we use the following notation: $lm.v$ - an adjustment vector of lm and $lm.v.attrib$ - an attribute of adjustment vector, where *attrib* stands for a name of an attribute.

Here we provide a detailed description of the algorithm. The main function of the algorithm is shown in Listing 1. In line 1 an LE object corresponding to a term t is determined. In line 2 *GetLexMeaning* function is used for getting all LM objects associated with le. Such objects are found by exploring MO relations in which object le participates. Next, for each LM objects of interest an adjustment vector is calculated. Finally, the function *BestMeaning* is applied for selecting the meaning of a term t which is best adjusted to the analyzed unit of a text.

Listing 1. Fuction *RecoTermMean*

```
function RecoTermMean (term t, LEXO lx, Ctx(t,u^t) ctx)):LexicalMeaning
1. LexicalEntry le = FindLE(t, lx);
2. set<LexicalMeanig> LMs = GetLexMeaning(le);
3. for each lm ∈ LMs  (lm.v = calculateCV(lm, ctx));
4. return BestMeaning(LMs);
```

In function *FindLE*, presented in Listing 2, an LE object corresponding to the input term t^w is found. It is done by finding *Word* (or *Phrase*) object (line 1) and using *concerns* relationship to determine the needed LE object (line 8). *Equal* function returns true only if all elements (lexical form, base from, and POS tag) describing t^w and a *Word* object are equal. If no *Word* instance can be matched with the input word the base form of t^w (instead of the lexical form) is considered. In case of the phrase an instance of *Phrase* is looked for in line 1.

Listing 2. Function *FindLE*

```
function FindLE (term t^w, LEXO lexo): LexicalEntry
1. wo:Word = {wo | wo instanceof Word in lexo, equal (wo, t^w)};
2. if (wo is null)
3. begin
4.   t^w = t^w.baseFrom;
5.   repeat search from line 1;
6. end
7. if (wo not null) return le = {le | concerns(wo, le) holds};
8. else return null;
```

Function *BestMeaninig*, presented in Listing 3, selects the best meaning by comparing values of successive attributes of adjustment vectors assigned to the considered LM objects and choosing object of the highest value of the currently analyzed attribute. The order of the attributes is the following: $d\#$, $e\#$, $n\#$, *probability*, and *priority*. If these attributes are not enough to determine only one meaning the meaning is drawn using *rand* function (line 6). The selected meaning is returned only if at least one element of its context in LEXO or example of usage can be found in the analyzed unit of text. Otherwise, the special meaning "other" is returned.

Listing 3. Function *BestMeaning*

```
function BestMeaning(set<LexicalMeanig> LMS): LexicalMeaning
1. set<LexicalMeanig> bLM = {lm | lm ∈ LMS, lm.v.d# = Max_disc(LMS)}
2. if (|bLM| > 1) bLM = {lm|lm ∈ bLM, lm.v.e# > 0};
3. if (|bLM| > 1) bLM = {lm|lm ∈ bLM, lm.v.n# = Max_neigh(bLM)};
4. if (|bLM| > 1) bLM = {lm|lm ∈ bLM, lm.v.probabilty = Max_prob(bLM)};
5. if (|bLM| > 1) bLM = {lm|lm ∈ bLM, lm.v.priority = Max_prio(bLM)};
6. if (|bLM| > 1) bLM = rand(bLM);
7. if (bLM ≠ ∅)
        return {lm|lm ∈ bLM, lm.v.d# > 0 or lm.v.e# > 0 or lm.v.n# > 0};
8. return "other";
```

4   Experiments

We have carried out several experiments in order to verify the approach proposed in Section 3 to meaning recognition of a term from texts written in a natural language. For this purpose we have selected nine concepts from a conceptual layer of an ontology concerning IT domain and have prepared a corresponding fragment of LEXO, so that the *RecoTermMean* procedure can be applied. These concepts are:

- Two meanings of the term *procedure*: (1) an algorithm denoted as *proc_alg* and (2) a unit of program code denoted as *proc_code*.
- Three meanings of the term *process*: (1) a general scientific process denoted as *sci*, (2) a stochastic process denoted as *sto*, and (3) execution of a computer program denoted as *prog*.
- Two meanings of the term *relation*: (1) an association denoted as *rel_asc* and a table in database denoted as *rel_tab*.
- Two meanings of the term *server*: (1) a device denoted as *srv_dev and* (2) a type of a computer program denoted as *srv_prog*.

4.1   Populating LEXO with Instances

Texts from a domain of interest written in appropriate natural language are the natural resource of entries to a lexical layer. However, extracting high quality entries from such resource by using automatic or semi-automatic methods is a challenging task. The methods introduced in the literature of subject mostly concern adding new entries to a conceptual layer but they can be adapted relatively easily to the needs of enriching a lexical layer. For example in [15] the grammatical patterns have been used for discovering phrases (compound nouns), in [16] synonyms were discovered based on so called *frequent termsets*, in [17] the procedure for discovering separate contexts of a given term has been proposed.

In order to evaluate practical usefulness of the *RecoTermMean* procedure we need to know a meaning of a term in a given context, so we decided not to adapt the mentioned methods but apply the following procedure for generating required data.

Procedure for populating LEXO with instances.

1. For each selected concept from CL layer create LE object and associate with it LM instances corresponding to specified meaning.
2. For each word of interest (procedure, process, relation, server) create the set of sentences including the word with a given meaning from articles available on the Internet. Label each sentence with the meaning of the word.
3. Assign each word in selected sentences by its part-of-speech tag and its base form.
4. For populating LEXO take the base form of nouns and adjectives occurring in three or more selected sentences. For each word create CE object and associate it with appropriate LM instance by means of VincinityLinguistic relation.

4.2   Results

The data used in the tests consists of more than three hundred sentences. The sentences were manually selected from scientific articles in such a way that: (1) a sentence includes one of the words w of interest, (2) the word w is used in one of the meanings defined in the ontology. Each sentence has been labeled with the appropriate meaning of the word *w*.

We evaluated the practical usefulness of procedure *RecoTermMean* by means of the accuracy measure i.e. we calculated the percentage of sentences for which the meaning of the word was correctly recognized. The overall results are presented in Table 1, whereas the detailed results are provided in Table 2 and Table 3. In these tables phrase "with POS" indicates results obtained by applying the original *equal* function (used in function *FindLE*), whereas the phrase "without POS" indicates results obtained using the modified *equal* function in which the POS tag is not taken into consideration.

Table 1. Overall correct recognition of meaning

mode\ word	Procedure	Process	Relation	Server	ALL
without POS	77%	62%	93%	81%	74%
with POS	76%	64%	86%	79%	73%

The achieved results are promising. Generally, the meaning of the word was correctly recognized in three sentences out of four. Only *sci* meaning of process has not been recognized. The probable reasons of such poor result are:

- In many situations a given type of a process (e.g. process in computing) is a specialization of a scientific process,
- It is very difficult to point out one definition of a scientific process and several different characterizations of such process are used.

Table 2. Recognition of meaning for procedure, relation, and server (without POS / with POS)

meaning\assigned meaning	proc_alg	proc_prog	other
proc_alg	74% / 74%	14% / 2%	12% / 24%
proc_prog	2% / 2%	81% / 79%	17% /17%
meaning\assigned meaning	rel_asc	rel_tab	other
rel_asc	90% / 83%	5% / 12%	5% / 5%
rel_tab	2% / 2%	95% / 93%	2% / 5%
meaning\assigned meaning	srv_dev	srv_prog	other
srv_dev	77% / 68%	18% / 23%	5% / 9%
srv_prog	86% / 90%	14% / 10%	0% / 5%

Table 3. Recognition of meaning for process (without POS / with POS)

meaning\assigned meaning	sto	sci	comp	other
sto	83% / 83%	6% / 6%	11% / 11%	0% / 0%
sci	4% / 0%	9% / 9%	39% / 39%	48% / 52%
comp	8% / 4%	2% / 2%	78% / 82%	12% / 5%

5   Conclusions

In this paper we have investigated a problem of application of ontology lexical layer in semantic processing of texts written in a natural language. We have sketched LEXO – the special structure for LL, which is seen as a bridge between terms from a natural language and concepts defined in an ontology. We have defined formally the key constructs (such as discriminant, context of lexical meaning) which are applied in LEXO. We have introduced the *RecoTermMean* algorithm - a procedure for automatic determination of meaning of a word in text written in a natural language. In the algorithm for indicating the most probable meaning of a word only LEXO and surroundings of that word in text are explored. As the procedure is independent of a conceptual layer it may be used for LL associated with different core layers of ontologies. The performed tests have shown that the proposed method can indicate meaning with high accuracy even in case of a limited number of instances in LEXO. It can be expected that filling the entire structure of LEXO with instances allows essentially improving the achieved accuracy.

References

1. Staab, S., Studer, R.: Handbook on Ontologies, 2nd edn. Springer (2009)
2. Hotho, A., Maedche, A., Staab, S., Zacharias, V.: On Knowledgeable Unsupervised Text Mining. In: Text Mining, pp. 131–152 (2003)
3. Antoniou, G., van Harmelen, F.: Web Ontology Language: OWL. In: Handbook on Ontologies, 2nd edn., Springer (2009)
4. Fernández-López, M., Gómez-Pérez, A., Jurysto, N.: METHONOLOGY: From Ontological Art Towards Ontological Engineering. In: Spring Symposium on Ontological Engineering of AAAI, pp. 33–40. Stanford University, California (1997)
5. Gawrysiak, P., Protaziuk, G., Rybiński, H., Delteil, A.: Text Onto Miner – A Semi Automated Ontology Building System. In: An, A., Matwin, S., Raś, Z.W., Ślęzak, D. (eds.) ISMIS 2008. LNCS (LNAI), vol. 4994, pp. 563–573. Springer, Heidelberg (2008)
6. de Bruijn, J., Ehrig, M., Feier, C., Martin-Recuerda, F., Scharffe, F., Weiten, M.: Ontology Mediation, Merging, and Aligning. John Wiley & Sons, Ltd. (2006)
7. Francopoulo, G., George, M., Calzolari, N., Monachini, M., et al.: Lexical markup framework (LMF). In: Proceedings of the Fifth International Conference on Language Resource and Evaluation, LREC 2006 (2006)
8. Montiel-Ponsoda, E., Aguado-de-Cea, G., Gomez-Perez, A., et al.: Modelling multilinguality in ontologies. In: Proceedings of the 21st International Conference on Computational Linguistics, COLING (2008)
9. Cimiano, P., Buitelaar, P., McCrae, J., Sintek, M.: LexInfo: A Declarative Model for the Lexicon-Ontology Interface. Web Semantics: Science, Services and Agents on the World Wide Web 9(1), 29–51 (2011)
10. McCrae, J., Aguado-de-Cea, G., Buitelaar, P., Cimiano, P., Declerck, T., Gomez-Perez, A., Gracia, J., Hollink, L., Montiel-Ponsoda, E., Spohr, D., Wunner, T.: The Lemon Cookbook, http://www.monnet-project.eu/Monnet/Monnet/English/Navigation/LemonCookbook
11. McCrae, J., Spohr, D., Cimiano, P.: Linking Lexical Resources and Ontologies on the Semantic Web with Lemon. In: Antoniou, G., Grobelnik, M., Simperl, E., Parsia, B., Plexousakis, D., De Leenheer, P., Pan, J. (eds.) ESWC 2011, Part I. LNCS, vol. 6643, pp. 245–259. Springer, Heidelberg (2011)
12. Firth, J.R.: A synopsis of linguistic theory 1930-1955. In: Studies in Linguistic Analysis, pp. 1–32. Philological Society, Oxford (1957)
13. Rybiński, H., Kryszkiewicz, M., Protaziuk, G., Kontkiewicz, A., Marcinkowska, K., Delteil, A.: Discovering Word Meanings Based on Frequent Termsets. In: Raś, Z.W., Tsumoto, S., Zighed, D.A. (eds.) MCD 2007. LNCS (LNAI), vol. 4944, pp. 82–92. Springer, Heidelberg (2008)
14. Wróblewska, A., Protaziuk, G., Bembenik, R., Podsiadły-Marczykowska, T.: LEXO: a Lexical Layer for Ontologies – design and building scenarios. Studia Informatica 33(2B(106)), 173–186 (2012)
15. Protaziuk, G., Kryszkiewicz, M., Rybiński, H., Delteil, A.: Discovering Compound and Proper Nouns. In: Kryszkiewicz, M., Peters, J.F., Rybiński, H., Skowron, A. (eds.) RSEISP 2007. LNCS (LNAI), vol. 4585, pp. 505–515. Springer, Heidelberg (2007)
16. Rybiński, H., Kryszkiewicz, M., Protaziuk, G., Jakubowski, A., Delteil, A.: Discovering Synonyms Based on Frequent Termsets. In: Kryszkiewicz, M., Peters, J.F., Rybiński, H., Skowron, A. (eds.) RSEISP 2007. LNCS (LNAI), vol. 4585, pp. 516–525. Springer, Heidelberg (2007)
17. Nykiel, T., Rybinski, H.: Word Sense Discovery for Web Information Retrieval. In: MCD Workshop 2008, Piza, ICDM, pp. 267–274. IEEE CS (2008)

Using Web Mining for Discovering Spatial Patterns and Hot Spots for Spatial Generalization

Jan Burdziej[1] and Piotr Gawrysiak[2]

[1] Nicolaus Copernicus University of Torun, Gagarina 11, 87-100 Torun, Poland
jan.burdziej@umk.pl
[2] Institute of Computer Science, Faculty of Electronics and Information Technology,
Warsaw University of Technology, Nowowiejska 15/19, 00-665 Warsaw, Poland
p.gawrysiak@ii.pw.edu.pl

Abstract. In this paper we propose a novel approach to spatial data generalization, in which web user behavior information influences the generalization and mapping process. Our approach relies on combining usage information from web resources such as Wikipedia with search engines index statistics in order to determine an importance score for geographical objects that is used during map preparation.

Keywords: web mining, spatial generalization, spatial patterns, geosemantics.

1 Introduction

One of the simplest definitions of a map is that it is a model of the real world [1]. This statement implies an inherent characteristic of all maps: a cettain level of simplification. The process of reducing and/or modifying real world objects and phenomena in terms of their size, shape and numbers within map space is called a generalization [2]. The extent of this reduction depends on multiple factors, of which one of the most important is the map scale. It is a ratio indicating the number of metric units on the ground that are represented by a unit in the space of the map model [3]. For example the depiction of objects such as cities will be different on various map scales [4]. On a large scale map cities can be represented as polygons indicating their exact boundaries. On a medium scale map a modification of the shape can result in the same objects being represented by points. On a small scale map only a subset of biggest cities can be shown in order to avoid information overload and to prevent symbol overlay.

Traditionally, the generalization process was time consuming and required deep knowledge of the theme presented on the map [3],[1]. In the era of ubiquitous cartography and extreme information overload, the need for an automated generalization approach becomes greater than ever before. Although there are already software tools and frameworks that perform automated generalization of spatial data (e.g. Douglas-Peucker algorithm is often used to simplify curves),

L. Chen et al. (Eds.): ISMIS 2012, LNAI 7661, pp. 172–181, 2012.

this problem still remains mostly in the domain of active research [5],[6],[7],[8]. While most of the scientific effort is put into the generalization of the shape of map features, choosing the right objects to be shown on the map, (i.e. selecting objects that are more important than others), is still one of the biggest challenges for cartographers (see [9],[10]). The 'importance' of a particular object is typically related to some numerical characteristics of this object (e.g. the population of the city, the width of the road etc.) or to some other subjective classification, e.g. the administrative category of the city (such as national vs. district capitals) or a road category (e.g. highways vs. local roads). This approach has been successfully applied to modern spatial databases, as all objects within the database can be easily described by some numerical and textual characteristics. It is however by definition mostly static, as the geographical properties change slowly and geospatial databases are updated at even slower rate. For practical purposes the notion of importance of a geographical feature is a much more dynamic variable, influenced by political situation (e.g. as demonstrated by recent conflicts in Northern Africa and Near East), seasonal travel trends (e.g. increased importance of vacation resorts in summer time) or even simply by economic changes.

It has been argued (see e.g. [11]) that the increased popularity of digital media and communication networks will result in a catastrophic information overload. Fortunately the nature of the digital resources created by people, such as web pages, social network graphs or even Internet search engine queries makes analysis and mining of this information feasible. In this paper we present a novel approach towards geospatial generalization, that relies on Internet data such as above in order to assess changes in importance of geographical objects in order to determine if these should be displayed on a map. Specifically, we describe how the information extracted from Wikipedia statistics and search engines indices can be used for creating country maps of varying detail, we present preliminary results of experiments based on this approach and outline the plans for further research that our team is currently carrying out.

This paper is organized as follows. In the first chapter we present background information related to map generalization. Next chapter contains description of sources and methods of data extraction as well as proposals of techniques for determining an importance score for geographical features, based on example of data mining process related to cities in Poland. In the third chapter, we present analysis of this data, together with visualization examples. The results are further evaluated in the last chapter, which contains also ideas for refinements of our approach.

2   Methods

Statistical information related to word usage in the world wide web is a very valuable resource for a variety of natural language processing tasks. It is currently most often used in ontology related research, where search engine index mining has been used for, inter alia, concept mapping and disambiguation (see e.g.

[12]). It is utilized also directly as a web mining method, for extracting bulk information from webpages, by locating relevant resources, or as a means of assessing keyword co-occurrence and similarity (demonstrated in e.g. [13] or [14]). We decided to exploit this information in order to determine the importance of a given concept (i.e. the name of a geographical feature, such as a city) in real life. Obviously the notion of importance is a very elusive one, however current ubiquity of electronic communication means that it is possible to approximate it, for practical purposes, with a concept, or a keyword popularity, in the Internet. Specifically, it is possible to devise two measures, that might be called passive popularity and active popularity, that can be assessed depending on keyword usage patterns in the Internet:

- **passive popularity** score is understood as a measure of a number of Internet resources describing or defining a given concept. It can be approximated by simple keyword[1] counting over entire corpora of web pages, which is done during page indexing by search engine crawlers (see [15]). Additionally, recent rise of social media, especially microblogging, generates large amounts of data that could be also used for such approximation. Specifically analysis of twitter streams (which are changing dynamically and are usually not indexed by general purpose search engines) such as keyword counting in original tweets seems to be especially valuable.
- **active popularity** score is understood as a number of accesses to a given concept definition or description. It primarily can be measured by analysing web traffic, however due to the distributed nature of the Internet such approach is not feasible, at least not for the entire corpus of all web pages and other Internet resources. For controlled, smaller repositories, such analysis is viable. When the repository in question is an authoritative knowledge resource (e.g. Wikipedia for general purpose knowledge or Internet Movie Database for motion picture related knowledge), the score computed by analysing number of accesses for web pages might be a good approximation of real world importance of a concept. Alternatively, one might also treat hyperlink references to a given page as a notion of access (albeit much less dynamic) which could be used here, together with PageRank related analysis pertaining to a page where the hyperlink originated. Social media analysis can be also used here. For example number of retweets, or number of Facebook likes pointing to a page related to a given concept can be utilized. Finally, the search engines query trends are a good indicator of active popularity. Unfortunately these are rarely accessible to third parties (i.e. researchers not affiliated with search engine provider) with notable exception of Google Trends (see [16]).

During our experiments we evaluated the importance of names of all cities in Poland (908 entities as of June 2012) using both passive and active popularity

---

[1] For the purposes of this paper we assume simple unigram language model, as most geographical names, relevant to generalization problem, are either single words or immutable collocations. Obviously for more generic usage the language model used might influence computation of popularity scores quite significantly.

and comparing the results. For the purpose of measuring passive popularity the analysis of search engine indices was selected. Because we do not have direct access to internal databases of search engines, we had to settle on information provided either by their API or by scrapping search page results. Two experiments were performed. The first relied on Microsoft Bing search engine API, which allowed us to automatically query Bing index on number of webpages containing references to city names (the query used was in the form of +city-name which triggers exact keyword match in Bing). Obviously several names of Polish cities are quite common words, such as adjectives (e.g. Biala meaning white in English) or nouns (e.g. Piaski meaning sands in English) so in order to compensate an additional data extraction was performed, where the names of the cities have been combined with names of corresponding voivodeships (e.g. cityname voivodshipname). Similar extraction process was also performed using Google search engine. However, data extraction via API was not practical, as Google search API only allows for 100 free queries per day. We decided therefore to scrap the data returned by Google search engine and process the resulting HTML with our own parser in order to extract number of unique pages containing given keyword. The queries used were similar in syntax to queries used for data extraction from Bing. Resulting data must not be however treated as exact. Our - obviously quite limited - testing clearly indicates that number of pages reported by these two leading search are only approximate, as e.g. subsequent extraction runs resulted in different numbers. The differences were relatively small, however way to significant (around 1 percent) to be a result of search engine reindexing activity. This means that the best way to measure passive popularity would be a direct analysis of the search engine index (clearly possible only for search engine employees) or usage of a corpus generated directly from such index. One of the possible corpora that might be used towards this end is Google 1T 5-gram corpus available from Linguistic Data Consortium [17], that our group will be analyzing in near future (see Discussion)[2].

The purpose of the second data extraction experiment that we performed was to estimate active popularity of the same set of objects as in the previous case - namely the 908 Polish cities. Towards this end we settled on Polish and English Wikipedia, which contains quite extensive information about all these entities (each city has its own dedicated page in both language versions of Wikipedia). Using Wikipedia raw traffic database, available from Wikimedia Foundation ([18]) we extracted information about individual page accesses for entire year 2011 and January - May period of 2012 for all respective pages, describing Polish cities, and for the purpose of further analysis, we created monthly aggregates.

Some other experiments, used mostly as a proof of concept for further research, outlined in the last chapter of this paper, were also done. This includes specifically data extraction from Google Trends system, in order to calculate

---

[2] Unfortunately such corpora are rarely updated e.g. Google 1T 5-gram is a snapshot of entire Internet created in 2006 thus currently almost 6 years old.

active popularity of city names, resulting from search engine query traffic and real-time analysis of Twitter posts; this work is however still underway.

3   Results

The best results, as far as spatial generalization purposes are involved, were obtained by estimating active popularity via analysis of Wikipedia traffic. For each city, the following were computed: a) the average number of accesses for the last 16 months (Jan 2011 - April 2012), b) the average number of accesses divided by the population (i.e. the number of accesses per citizen - ApC), c) the average number of accesses divided by the area (i.e. the number of accesses per square kilometer - ApK). It turns out that top 20 most popular (i.e. with highest ApC) cities are characterized by a more uniform distribution compared to the top 20 most populated cities. This ranking has significant impact on map creation. Assuming that we select only 20 cities to be labelled, some significantly less populated cities have been qualified for the map (such as Rzeszw ranked 21 based on its population but also Zakopane - the most popular mountain resort in Poland, ranked 169 by population). Moreover, the large conurbation of cities in Silesia region, consisting of Katowice, Zabrze, Bytom, Sosnowiec and Gliwice was declustered and only the city of Katowice was listed among the top 20 accessed cities. Direct comparison of ApC with population statistics revealed a significantly different set of cities. Generally, the selected cities were much smaller. Some that were selected were popular travel destinations, such as eba, Hel, or Krynica Morska. Other cities were related to some historic events (e.g Jedwabne, Tykocin), while others were related to recent events (e.g. Szczekociny) see Fig. 1 and Fig 2., grey maps. Additionally, the variation of active popularity in time has been evaluated. For each month current top 20 cities were selected. Moreover, the accesses per citizen values for all cities were interpolated in order to emphasize the spatial variation of this statistics. Generated maps include both the cities with a constantly high ApC value (e.g. eba, Hel or Karpacz) but also allow identification of some short-term spatio-temporal hot spots. For example

Fig. 1. Maps presenting 20 most important cities according to population and popularity score

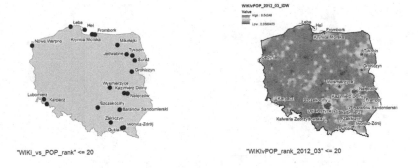

Fig. 2. Direct comparison of population versus popularity

the cities of Nowe Brzesko, Wolbrz and Gocino scored high ApC ranks in January 2011, which was obviously related to the fact that these entities received the status of the town on Jan 1st, 2011. In March 2012 the most significant hot spot was Szczekociny, the place of the most tragic recent rail catastrophe in Poland, which attracted strong media attention (see Fig. 2, color map). Evaluation of these temporal changes (see Fig. 3 Fig. 4) should allow for even more detailed selection of good candidates for mapping, or even creation of interactive maps. The next experiment was to evaluate the passive popularity concept, i.e. analyze Bing and Google search engines' results. It turns out, that the data that we are able to acquire is significantly less useful for mapping purposes. The main problem is amount of linguistic noise, resulting from lack of control over search algorithm of the search engine, and only approximate number of pages in the search engine index, as discussed in Chapter 2. In order to assess the quality of the results and as a way of comparing them with active popularity analysis outlined above we computed Pearson's correlation coefficient between the results and both population and area of cities for Google and Bing. The highest was observed between Bing results (search syntax: +city name voivodeship) and cities area (0.801) with population comparison in the second place (0.787). However

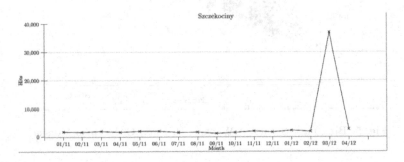

Fig. 3. Wikipedia traffic for Szczekociny (note the peak in March 2012)

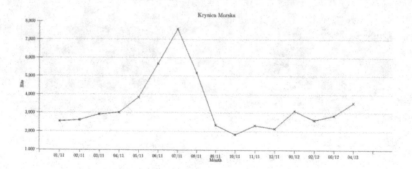

Fig. 4. Wikipedia traffic for Krynica Morska note the vacation traffic peak

the outliers (i.e. cities that would be most interesting for the mapping purposes) included not only popular tourist destinations (such as Krynica Morska or Zakopane), but also those with common names. Two figures below illustrate this by presenting relationship between population and passive (Fig. 5 Bing search case) and active popularity (Fig. 6 Wikipedia) respectively, with sample outliers marked.

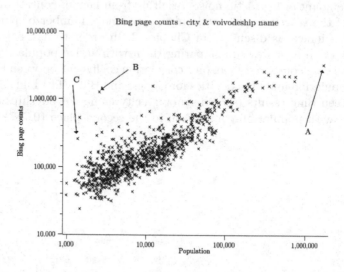

Fig. 5. Bing page counts A:Warsaw, B: Biaa, C: Krynica Morska

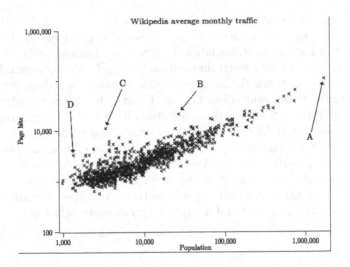

Fig. 6. Wikipedia averages, A: Warsaw, B: Zakopane, C: Kazimierz Dolny, D: Krynica Morska

4 Conclusion

The initial results of our research indicate that both active and passive popularity concepts can be efficiently used as measures of importance. These measures can therefore be incorporated into the web map generalization process for selecting the most important objects that should be shown on a map as well as for defining priorities for object labels. Unlike traditional measures, the active popularity can reflect nearly instant changes of importance, thus can be successfully used to identify short-term hot spots and create news maps. On the other hand, the passive popularity, indicated by the number of Internet resources describing a given object can be more appropriate for reflecting the overall importance. Therefore, the passive popularity can be used additionally to the traditional importance measures, such as the number of population or the area of a city. Our results show that evaluating passive importance involves more uncertainty and it is highly dependent on the particular search engine algorithms. The advantage of web data mining and the proposed concept of active and passive popularity is that these methods can be easily utilized in a web environment, thus can be efficiently incorporated into web maps creation process. The work outlined in this paper is treated by us as a preliminary step for creation of a more robust system, that should eventually allow dynamic creation of geographical map via means of a interactive interface. Towards this end a more reliable methods of estimating active and passive popularity will be needed. We plan especially to improve estimation of passive popularity, first by using Google 1T 5-gram corpus, as a way of computing keyword popularity in the Internet. This corpus, containing information about n-gram counts (from bigrams up to 5-grams) in all web pages indexed by Google by the end of 2006, will be used instead of direct querying the search engines. Eventually, we plan to replace it with our own dedicated Internet crawler, that will be able to use context analysis techniques (see e.g. [19]) in order to perform word sense disambiguation to filter city names similar to common words and include keyword weighting similar to PageRank algorithm. While our active popularity approach yielded good results we plan also to improve it further by incorporating information from Google Trends, that allow to extract search traffic volume information statistics. Additionally we will incorporate seasonal change damping for Wikipedia analysis by comparing traffic to cities pages to baseline traffic for common keywords. We hope that the final system might have important practical utility. Apart from obvious usefulness in traditional cartographic application it might be also used, when combined with data sources with more specialized search engines, such as travel websites, or train and airline commerce portals, to predict and analyze Internet users behaviors.

References

1. Nyerges, T.L.: Representing geographical meaning. In: Buttenfield, B.P., McMaster, R.B. (eds.) Map Generalization, pp. 59–85. Longman Scientific and Technical (1991)
2. Balodis, M.J.: Generalization. In: Anson, R.W. (ed.) Basic Cartography for Students and Technicians. International Cartographic Association, vol. 2, pp. 71–84. Elsevier Applied Science, London (1988)
3. Joao, E.M.: Causes and Consequences of Map Generalisation. In: Research Monographs in Geographic Information Systems. Taylor, Francis (1998)
4. Morehouse, S.: GIS-based map compilation and generalization. In: Muller, J.C., Lagrange, J.P., Weibel, R. (eds.) GIS and Generalization: Methodology and Practice, pp. 21–30. Taylor, Francis, Bristol (1995)
5. Foerster, T., Stoter, J.E., Kraak, M.: Challenges for Automated Generalisation at European Mapping Agencies: a Qualitative and Quantitative Analysis. The Cartographic Journal 47(1), 41–54 (2010)
6. Grunreich, D.: Development of computer-assisted generalization on the basis of cartographic model theory. In: Muller, J.C., Lagrange, J.P., Weibel, R. (eds.) GIS and Generalization: Methodology and Practice, pp. 47–55. Taylor, Francis, Bristol (1995)
7. Vickus, G.: Strategies for ATKIS-related cartographic products. In: Muller, J.C., Lagrange, J.P., Weibel, R. (eds.) GIS and Generalization: Methodology and Practice, pp. 246–252. Taylor, Francis, Bristol (1995)
8. Lee, D.: Experiment on formalizing the generalization process. In: Muller, J.C., Lagrange, J.P., Weibel, R. (eds.) GIS and Generalization: Methodology and Practice, pp. 219–234. Taylor, Francis, Bristol (1995)
9. Duchene, C.: Automated Map Generalisation Using Communicating Agents. In: Proceedings of the 21st International Cartographic Conference (ICC), Durban, South Africa (2003)
10. Ware, J.M., et al.: Automated Map Generalization with Multiple Operators: a Simulated Annealing Approach. International Journal of Geographical Information Science 17(8), 743–769 (2003)
11. Lyman, P., Varian, R.: How Much Information? Journal of Electronic Publishing 6(2) (2000)

12. Gligorov, R., et al.: Using Google distance to weight approximate ontology matches. In: Proceedings of the 16th International Conference on World Wide Web (WWW 2007), pp. 767–776. ACM, New York (2007)
13. Geleijnse, G., Korst, J.: Tagging artists using cooccurrences on the web. In: Proceedings Third Philips Symposium on Intelligent Algorithms (SOIA 2006), pp. 171–182 (2006)
14. Brin, S.: Extracting Patterns and Relations from the World Wide Web. Technical Report. Stanford InfoLab, Stanford University, USA (1999)
15. Brin, S., Page, L.: The anatomy of a large-scale hypertextual Web search engine. In: Enslow, P. (ed.) Proceedings of the Seventh International Conference on World Wide Web, pp. 107–117. Elsevier Science Publishers, Amsterdam (1998)
16. Choi, H., Varian, H.: Predicting the Present with Google Trends, Google Technical Report. Google Inc., USA (2009)
17. 1T5-gram, 2012 Web 1T 5G version 1 corpus. Linguistic Data Consortium (2012), http://www.ldc.upenn.edu/Catalog/CatalogEntry.jsp?catalogId=LDC2006T13
18. Page view statistics for Wikimedia projects (2012), http://dumps.wikimedia.org/other/pagecounts-raw/
19. Kołaczkowski, P., Gawrysiak, P.: Extracting Product Descriptions from Polish E-Commerce Websites Using Classification and Clustering. In: Kryszkiewicz, M., Rybinski, H., Skowron, A., Raś, Z.W. (eds.) ISMIS 2011. LNCS, vol. 6804, pp. 456–464. Springer, Heidelberg (2011)

AGNES: A Novel Algorithm for Visualising Diversified Graphical Entity Summarisations on Knowledge Graphs

Grzegorz Sobczak[1], Mariusz Pikuła[2], and Marcin Sydow[2,3]

[1] Faculty of Mathematics, Informatics and Mechanics, University of Warsaw, Poland
[2] Polish-Japanese Institute of Information Technology, Warsaw, Poland,
[3] Institute of Computer Science, Polish Academy of Sciences, Warsaw, Poland
g.sobczak@students.mimuw.edu.pl,
{msyd,mariusz.pikula}@poljap.edu.pl

Abstract. We present AGNES – a novel algorithm for presenting diversified graphical entity summarisations on semantic knowledge graphs. The main idea is to compute the positions of the vertices in radial system of coordinates based on recursive, in-order traversal of the BST spanning tree of the graph. The implementation of the algorithm is compared on real data with other, existing visualisation tools. We also report a successful user evaluation crowdsourcing experiment that indicates superiority of our algorithm over one of the existing competitors.

Keywords: graphical entity summarisation, semantic knowledge graphs, visualisation, diversity, user evaluation.

1 Introduction

In this paper we present AGNES[1] – a novel visualisation algorithm for *graphical entity summarisation* on semantic knowledge graphs.

In such graphs, nodes represent entities (e.g. writers, books) from some domain (e.g. literature) and directed edges concern binary relations between them (wrote(writer, book), e.g. "Orwell wrote '1984' "). Equivalently, each edge can be interpreted as a single fact concerning the entities. There are possible multiple edges between a given pair of nodes, that makes it a multi-graph.

Example of a small fragment of a semantic knowledge graph is given on Figure 1.

The problem of graphical entity summarisation was originally proposed in [5]:
INPUT: a knowledge base graph B (in the form of semantic knowledge graph, described above), a node q in this graph representing an entity to be summarised and a limit $k \in N$ on number of edges (facts) to be shown
OUTPUT: a connected subgraph $S(B, q, k)$ of B, containing q, and max. of k edges

Intuitively, the summary $S(B, q, k)$ should present in a graphical form a compact and informative summary of the entity q.

The problem of how to automatically *select* the subgraph $S(B, q, k)$ to obtain such informative summary is outside of the scope of this paper and was studied for example

---

[1] AnGle-based NodE Summarisation.

L. Chen et al. (Eds.): ISMIS 2012, LNAI 7661, pp. 182–191, 2012.

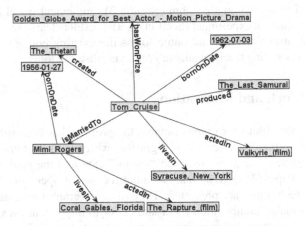

Fig. 1. An example extract from a semantic knowledge base

in [6]. In that paper, two algorithms were proposed and the role of diversity of such summary was experimentally evaluated.

1.1 Aim of the Paper

The task studied in this paper is to *automatically compute the graphical layout* of the given summary $S(B, q, k)$, so that the visualisation is most useful for the user of the summarisation tool. In this paper, we assume that the selection of facts in the summary $S(B, q, k)$ is already computed by an external algorithm. For this task, we use the diversity-aware algorithm presented in [6].

We define the following natural desired properties of the layout for graphical entity summarisation problem:

- the node representing the summarised entity should be placed in the center of the picture
- the other nodes evenly fill the space around the center
- the nodes that are topologically closer to the center should be placed closer to the center, etc.
- the layout should avoid covering any elements of the summary by other elements
- the layout should be adapted to visualise multi-graphs that have quite long labels both on vertices and directed edges

Since it is hard to guarantee that all of the above properties will be satisfied for an arbitrary summary, we additionally assume that our visualisation tool will make it possible for the user to manually introduce some minor corrections after the picture is computed, if needed.

1.2 Contributions

We identify desired properties of a layout algorithm for visualising graphical entity summarisations on knowledge graphs. Taking these properties into account we propose

AGNES, a novel visualisation algorithm (Section 3), implement it and present the results of a user evaluation experiment (Section 4). The results clearly show that AGNES receives positive user feedback and outperforms the existing competitor selected by authors out of the existing freely available graph visualisation tools.

2 Related Work and Motivations

Considering the algorithms for graph drawing, the problem has been studied before in many papers. In [4] there is described an algorithm which works for undirected graphs and weighted graphs. The paper uses the "spring" idea and the goal is to minimise "spring" energy. Paper [3] uses the same idea but focuses on speed of computation. In [7] partitioning techniques are presented. They divide a graph into clusters and draw these clusters. It makes computation faster and easier to implement. Next related work is [1]. In this paper authors focus on five generally accepted aesthetic criteria: distributing the vertices evenly in the frame, minimizing edge crossings, making edge lengths uniform, reflecting inherent symmetry and conforming to the frame.

However, none of them describes an algorithm fully suitable for our situation. The algorithms work for situation when vertices are drawn as single points and edges as lines. For example, in the problem statement we need to take into account the fact that vertices and edges may contain quite long labels.

There are some algorithms like [2] which would be used in our setting but most of them are not entity-oriented. They mostly focus on easthetically nice view of a graph rather than on presenting information about an entity useful for users. For example, algorithm in [2] divides vertices into layers and draws every layer on different height. There is no option to distinguish any vertex, the position of the specific vertex (which corresponds with an entity) depends only on the structure of a graph.

Considering the packages, we examined various existing, freely available software packages for graph visualisation and various built-in layouts available in those packages. The examined graph-visualisation packages include:

- JUNG (http://jung.sourceforge.net/)
- JGraph (http://www.jgraph.com)
- Gephi (http://gephi.org)
- GraphViz (http://www.graphviz.org)
- Walrus (http://www.caida.org/tools/visualization/walrus/)
- GFV (http://gvf.sourceforge.net/)
- Wandora (http://www.wandora.org/wandora/wiki/)
- Welkin (http://simile.mit.edu/welkin/)

A bit surprisingly, it turned out that no built-in layout available in any package found by the authors satisfied all the desired properties listed in Section 1.

The tools that seemed closest to our needs are the first three on the list above. We tried to configure the parameters of the most promising tools to obtain our desired properties, but we always observed some problems. The examples are on Figures 2, 3, 4.

In addition, many existing graph visualisation tools make some assumptions on the visualised graph that do not fit to our setting, for example:

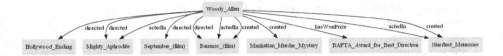

Fig. 2. Example of a summary visualised by the GraphViz tool. The nodes do not center around the summarised node.

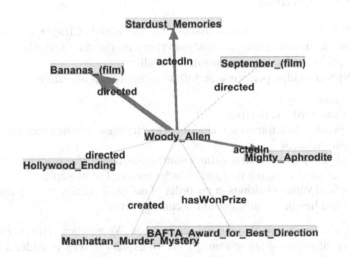

Fig. 3. Example of a summary visualised by the Gephi tool. The nodes overlap, edge labels are always horizontal.

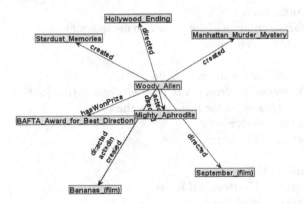

Fig. 4. Example of a summary visualised by the JUNG tool. The nodes overlap with edge labels.

- the visualised graph is a tree
- no multiple edges between a given pair of nodes are allowed
- there are no labels on edges
- the vertices are represented as dots
- the edges are not directed

Interestingly, the problem studied in our work turns out to violate all of the above assumptions. To summarise, it turned out to be necessary to propose a novel, specialised visualisation algorithm for our problem.

3   AGNES Algorithm

In the algorithm that we propose, we assume that the input is a fragment of knowledge graph with one main vertex (summarised entity) and all the other nodes in the summary given in a specified order. It returns positions of all vertices.

The AGNES algorithm performs the following high-level operations:

1. create spanning tree:
   (a) the main vertex is the root
   (b) use breadth-first traversal technique - to minimise the distances from the root to other vertices
2. recursively assign consecutive natural numbers to the vertices (except the main one) using the created spanning tree and in-order traversal of its vertices
3. calculate final vertex positions in the order of assigned numbers taking into account the sizes and heights of subtrees of the central vertex

Every non-main vertex gets its own unique number. We number vertices with the in-order traversal of the resulting spanning tree. To explain it, let's consider a non-main vertex. First we number half of its children (and their descendants), then assign a number to the vertex and assign the numbers to the remaining children.

The pseudo-code of the numbering algorithm can be written as follows:

$I := 0$;
NUMBER(the main vertex)
function NUMBER($vertex$)
    $n :=$ number of children of the $vertex$ in the spanning tree
    $notVisitedChildren := n$;
    $visitedChildren := 0$;
    for $child$ in children of the $vertex$ **do**
        if $notVisitedChildren - visitedChildren \in \{0,1\}$ AND
            $vertex$ is not the main vertex **then**
            $vertex.number := I$;
            $increase(I)$;
        end if
        NUMBER($child$)
        $decrease(notVisitedChildren)$
        $increase(visitedChildren)$
    end for
end function

Note that $notVisitedChildren = n - visitedChildren$. There are many ways in which nodes can be given in the **for** statement. One is to order them by edge labels. Then nodes with similar edges (like "actedIn") are situated close to each other. However, this topic has not been studied and is intended for future work. Simple example of numbering is shown at the Figure 5.

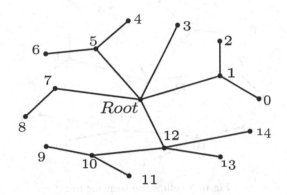

Fig. 5. Numbered spanning tree

Now, when the numbers are assigned to the nodes, each node (except the main one) is assigned its own $angle \in [0, 2\pi)$ and $radius \in (0, 1)$. For every non-main vertex the following holds :

$$angle(vertex) := 2\pi * \frac{vertex.number}{\text{number of non-main vertices}} \tag{1}$$

and $radius(vertex)$ is a function of depth of the vertex and the height of the subtree rooted at this vertex. Below, by distance we mean the (unweighted) length of the path. The general rules are:

- the bigger distance to the main vertex - the bigger radius
- the bigger subtree depth - the smaller radius

Having these two values we can calculate the coordinates of all vertices. Assume that the display space has vertical range from $-l_y$ to l_y and horizontal range from $-l_x$ to l_x. The main vertex is at the central point (position $(0, 0)$). Every other vertex is placed in the position $(x, y) = (l_x r \cos\alpha, l_y r \sin\alpha)$, where r is the $radius$ and α is the $angle$.

For example, consider the following function:

$$radius(vertex) = 0.4 + 0.45 \frac{vertex.depth}{vertex.depth + vertex.subtreeDepth} \tag{2}$$

Vertices' positions in the Figure 5 were calculated by this function. For vertex numbered 12 depth is equal to 1 and subtree depth is equal to 2. Thus radius is equal 0.55. But for vertex numbered 13 its depth is equal to 2 and subtree depth is equal to 0, thus radius is 0.85.

3.1 Discussion of the Basic Properties of the Layout

If some vertices have the same values of depth and subtree height then their radii are equal. Thus the vertices are situated on the same ellipse centered around the main node, like vertices with numbers 1, 5 and 7 on the Figure 6.

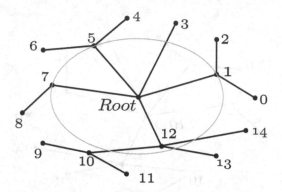

Fig. 6. An ellipse on spanning tree

Assume that vertical and horizontal sizes of layout are equal. Then ellipses are actually circles and angles between every two consecutive nodes (considering assigned numbers) are equal. The way of numbering implies that every subtree is situated in some angle with the endpoint in the main vertex's position. The angle is proportional to the size of the subtree.

The algorithm does not optimise (minimise) the number of edge crossing. A graph is first converted to a spanning tree (to compute the vertex numbers), that's why the temporarily erased edges are not taken into consideration while positions are calculated. But these edges are finally drawn during the visualisation.

3.2 Avoiding Horizontal Edges

After some experimentation, the function *angle* has been additionally improved to avoid the potential problem of vertex and label overlap. This problem is very important in our setting due to the potentially quite long textual labels. We applied the following simple trick. Just before drawing, to avoid any horizontal edges, all the angles are shift to right by a small angle:

$$\frac{1}{3}2\pi * \frac{1}{\text{number of non-main vertices}}$$

Thus the function *angle* was implemented as:

$$angle(vertex) := 2\pi * \frac{vertex.number + \frac{1}{3}}{\text{number of non-main vertices}} \tag{3}$$

Note that the results of subtraction of values returned by function *radius* and $\alpha \in \{0, \pi\}$ are not in interval $(-v, v)$ where $v = \frac{2\pi}{3} \frac{1}{\text{number of non-main vertices}}$.

4   Implementation and User Evaluation

AGNES was implemented by extending one of the freely available graph visualisation packages. We considered JUNG, jGraph and Gephi and finally selected JUNG as not containing native libraries (unlike Gephi) and being better documented than JGraph.

Figure 7 presents an example summary visualised by the resulting prototype implementation of AGNES. The result seems to be quite promising.

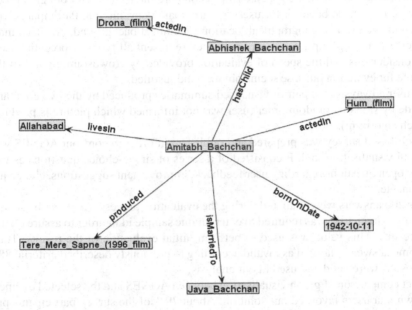

Fig. 7. The example visualisation of graphical entity summary computed on imdb.com data by our implementation of AGNES

We also manually inspected the peformance of our novel algorithm on over 80 other summaries computed on a real dataset concerning imdb.com movie domain. All the inspected visualisations satisfied almost all of the desired properties and avoided most of the negative properties observed in the competing visualisation tools.

The most important minor problem that we regularly observed was partial overlapping of edge label with vertex label but the general impression was much better than in the case of competing tools that we analysed. We believe that this problem will be possible to remove in next versions of our algorithm by additionally controlling the lengths of the edges by taking the label lengths into account. Notice that this issue is almost independent on the angle-placing idea of our algorithm and can be treated separately in our future work.

Most importantly, the general, angle-based positioning of the vertices, that was the main goal of the algorithm seems to be quite successful.

4.1 User Evaluation Experiment

To further evaluate AGNES in a more objective manner we designed and performed a user evaluation experiment of the visualisation algorithm with the CrowdFlower[2] crowdsourcing service.

After examining many existing tools, as described in Section 2, we selected the best (according to our assessment) competitor tool (Gephi), configuring its layout parameters to make it most suitable for the summarisation task and graphically most similar to our implementation. Next, we asked users to compare the results with ours.No predetermined criteria have been set for users to participate in evaluation. Participants were of different age groups, geographical locations, technical background, etc. This made them perfect target group for our survey as they represent all types of potential users. If we consider this and the speed of evaluation provided by crowdsourcing use of this technique for evalution purpose seems obvious and justified.

The user allways saw a pair of visualised summaries produced by the two compared tools side by side in a random order (user was not informed which picture is produced by which algorithm).

A web-based survey was prepared and published in order to confront AGNES with the Gephi visualisation tool. It consisted of a series of single-choice questions as well as some open questions regarding user feedback, improvement suggestions, choice justification, etc.

Over 200 answers were gathered during the evaluation process. As it can be seen in the survey extract, user was required to enter unique sample id in order to assure quality of the results. This value was used to perform initial evaluation of the results. There were some answers qualified as invalid according to previously described criteria. 88% of the results were valid and used in our analysis.

Direct comparison of graph visualisation between AGNES and the selected competitor shown that users favoured our solution. About 79% of the survey participants preferred AGNES. As mentioned before survey also contained open questions so that users could provide more detailed feedback. According to multiple user comments, AGNES was "better looking", "more manageable and easier to read", etc. Arrows showing the direction of the relations made it easier for users to understand presented information and interpret it in appropriate way. Overlapping words, lack of arrows showing relation direction, the "cris-crossed" arrangement of the layout with crossing vertexes were among the points risen by the users as the faults of the visualisation provided by the competing tool. In case of our visualisation, one user suggested a tree structure as more appropriate for data presentation. There was also some suggestion about improving the overall quality of visualisation by using more colours.

Overall, users gave very positive feedback on AGNES, describing it as easy to read, neat and enabling them to quickly read presented information.

5   Conclusions and Further Work

We have presented AGNES – a novel algorithm for visualising diversified graphical entity summarisations on knowledge graphs. The main idea is to compute the positions

---

[2] http://crowdflower.com

of the vertices in radial system of coordinates based on recursive, in-order traversal of the BST spanning tree of the graph. AGNES was implemented and evaluated by real users in a crowdsourcing experiment. The results are very promising and show the superiority over the competing tool.

We observed some minor problems to be corrected in our future work, concerning partial overlapping of vertex labels and edge labels that can be overcome by taking the lengths of labels into account where computing the radius of the vertex.

Another direction of future work would be to add the "spring-like" mechanism to the corners of the rectangular vertices to completely get rid of the potential problem of partial vertex overlaps.

Acknowledgements. The work is a part of the DIVERSUM project held at Web Mining Lab at PJIIT, Warsaw and supported by the N N516 481940 grant of National Science Centre.

References

1. Fruchterman, T.M.J., Reingold, E.M.: Graph drawing by force-directed placement. Software: Practice and Experience 21(11), 1129–1164 (1991)
2. Gansner, E.R., Koutsofios, E., North, S.C., Vo, K.P.: A technique for drawing directed graphs. IEEE Transactions on Software Engineering 19(3), 214–230 (1993)
3. Harel, D., Koren, Y.: A Fast Multi-scale Method for Drawing Large Graphs. In: Marks, J. (ed.) GD 2000. LNCS, vol. 1984, pp. 183–196. Springer, Heidelberg (2001)
4. Kamada, T., Kawai, S.: An algorithm for drawing general undirected graphs. Information Processing Letters 31(1), 7–15 (1989)
5. Sydow, M., Pikuła, M., Schenkel, R.: DIVERSUM: Towards diversified summarisation of entities in knowledge graphs. In: Proceedings of Data Engineering Workshops (ICDEW) at IEEE 26th ICDE Conference, pp. 221–226. IEEE (2010)
6. Sydow, M., Pikuła, M., Schenkel, R.: To Diversify or Not to Diversify Entity Summaries on RDF Knowledge Graphs? In: Kryszkiewicz, M., Rybinski, H., Skowron, A., Raś, Z.W. (eds.) ISMIS 2011. LNCS, vol. 6804, pp. 490–500. Springer, Heidelberg (2011)
7. Walshaw, C.: A Multilevel Algorithm for Force-Directed Graph Drawing. In: Marks, J. (ed.) GD 2000. LNCS, vol. 1984, pp. 171–182. Springer, Heidelberg (2001)

Large Scale Skill Matching
through Knowledge Compilation

Eufemia Tinelli[1], Simona Colucci[2], Silvia Giannini[1], Eugenio Di Sciascio[1],
and Francesco M. Donini[2]

[1] SisInfLab, Politecnico di Bari, Bari, Italy
[2] DISUCOM, Università della Tuscia, Viterbo, Italy

Abstract. We present a logic-based framework for automated skill matching,
able to return a ranked referral list and the related ranking explanation. Thanks to
a Knowledge Compilation approach, a knowledge base in Description Logics is
translated into a relational database, without loss of information. Skill matching
inference services are then efficiently executed via SQL queries. Experimental
results for scalability and turnaround times on large scale data sets are reported,
confirming the validity of the approach.

1 Introduction

We present a logic-based framework for automated skill matching, which combines
the advantages of both semantic-based and database technologies through a Knowledge
Compilation [2] approach. Coherently with it, our contribution makes computationally
efficient the skill matching execution over the information contained in the Knowledge
Base (KB) – modeling intellectual capital according to the formalism of Description
Logics (DLs) – by splitting the reasoning process in two phases: (**i**) *off-line reasoning*
- the KB is pre-processed and stored in a relational database; (**ii**) *on-line reasoning* -
skill matching is performed by querying the data structure coming from the first phase.
Other distinguishing features of the approach include the addition of a fully explained
semantic-based comparison between the job request and the retrieved candidates as well
as the possibility to express both strict requirements and preferences in the job request.
Coherently with this perspective, our approach provides a two-steps matchmaking [5,8]
process: *Strict Match* retrieves candidates fully satisfying all the strict requirements;
Soft Match implements an approximate match by retrieving candidates fully or partially
satisfying at least one user preference.

The approach has been implemented in I.M.P.A.K.T., an integrated system for
automated HR management that provides team composition services [14,6] and Core
Competence extraction [7] (an embryonic I.M.P.A.K.T. version of the retrieval of
candidates ranked referral lists has been presented in [13]).

Among the few semantic-based implemented solutions for HR management, one of
the first ones is –to the best of our knowledge – STAIRS[1], a system still used at US Navy
Department to retrieve referral lists of best qualified candidates w.r.t. a specific task. We

---

[1] http://www.hrojax.navy.mil/forms/selectguide.doc

L. Chen et al. (Eds.): ISMIS 2012, LNAI 7661, pp. 192–201, 2012.

may also cite products offered by Sovren[2], which provide solutions for both CV and job requests parsing starting from several text formats to HR-XML schema. Recently, also *Monster.com*[R], the leading Web job-matching engine, introduced the *Monster Power Resume Search*[TM] service [3]. The product relies on the semantic $6Sense^{(TM)}$ search technology, patented by Monster Worldwide, Inc. . All the previous solutions exploit the semantics of queries– and are able to distinguish between essential and nice-to-have skills– to perform the search process but no ranking explanation is returned. On the other hand, several approaches have been presented, where databases allow users and applications to access both ontologies and other structured data in a seamless way. Das et al. [4] developed a system that stores OWL-Lite and OWL-DL ontologies in Oracle RDBMSs, and provides a set of SQL operators for ontology-based matching. The most popular OWL storage is the recent OWLIM [11], a Sesame plug-in able to add a robust support for the semantics of RDFS, OWL Horst and OWL2 RL. Other systems using RDBMS to deal with large amounts of data are *QuOnto*[4] and *OWLgres*[5], both DL-Lite reasoners providing consistency checking and conjunctive query services. SHER [9] is a highly-scalable OWL reasoner performing both membership and conjunctive query answering over large relational datasets using ontologies modeled in a subset of OWL-DL without nominals. PelletDB[6] provides an OWL 2 reasoning system specifically built for enterprise semantic applications. Although all the previous approaches support languages more expressive than the one we use in our system, they are only able to return either exact matches (*i.e.*, instance retrieval) or general query answering. Instead, we use an enriched relational schema to deal with non-standard inferences and provide effective value-added services.

The rest of this paper is organized as follows. In the next section, the modeling approach translating the KB into the reference relational database is presented. Section 3 introduces the implemented services and Section 4 reports on an experimental evaluation using PostgreSQL 9.1 DBMS showing the effectiveness and the scalability of the proposal. Conclusions and future research directions close the paper.

2 Knowledge Compilation

I.M.P.A.K.T. receives all the information needed to model and manage the domain of human resources from a specifically developed modular ontology $\mathcal{T} = \{M_i | 0 \leq i \leq 6\}$, currently including nearly 5000 concepts. Each ontology module M_i is modeled according to the formalism of $\mathcal{FL}_0(D)$ subset of DLs. In particular, every M_i may include the following items: **i)** a class hierarchy; **ii)** n optional properties $R^i_j, 1 \leq j \leq n$, defined over the classes specifying the module hierarchy; **iii)** optional concrete features p_i, either in the natural numbers or in the calendar dates domain.

---

[2] http://www.sovren.com/default.aspx
[3] http://hiring.monster.com/recruitment/
   Resume-Search-Database.aspx
[4] http://www.dis.uniroma1.it/~quonto/
[5] http://pellet.owldl.com/owlgres/
[6] http://clarkparsia.com/pelletdb/

Hereafter, we shortly describe the content modeled in each ontology module: **Level** models the hierarchy of candidate education and training levels; **ComplementarySkill** models the class hierarchy about complementary attitudes; **Industry** models the hierarchy of company types a candidate may have worked for; **Knowledge** models the hierarchy of possible candidate competence and technical tools usage ability and the related experience role (*e.g.*, developer, administrator, and so on) exploiting the **type** property; **JobTitle** models the hierarchy of possible job positions; **Language** models the hierarchy of possible languages known by the candidate and provides three concrete features for expressing the related level (**verbalLevel**, **readingLevel** and **writingLevel**). Finally, modules **Industry**, **ComplementarySkill**, **Knowledge** and **JobTitle** provide also two predicates: **year**, to specify the experience level in years, and **lastdate**, which represents the last temporal update of work experience. M_0 is the main ontology module: it includes all the previous modules and models a property (called *entry point*) for each imported sub-module.

Thanks to the knowledge modeling outlined so far, it is possible to describe *CV Profiles* in the ABox. The CV classification approach we propose is based on a role-free ABox, which includes only concept assertions of the form $P(a)$, stating that the candidate a (*i.e.*, her CV description) offers profile features P (see Definition 1).

Definition 1 (Profile). *Given the skill ontology $\mathcal{T}$, a profile $P = \sqcap(\exists R_j^0.C)$ is a $\mathcal{ALE}(D)$ concept defined as a conjunction of existential quantifications, where R_j^0, $1 \leq j \leq 6$, is an **entry point** and C is a concept in $\mathcal{FL}_0(D)$ modeled in the ontology module M_j.*

As hinted before, our knowledge compilation approach aims at translating the skill knowledge base into a relational model, without loss of information and expressiveness, in order to reduce on-line reasoning time. Relational schema modeling is therefore the most crucial design issue and it is strongly dependent on both knowledge expressiveness to be stored and reasoning to be provided over such a knowledge base. We recall that $\mathcal{FL}_0(D)$ concepts can be normalized according to the *Concept-Centered Normal Form* (CCNF), [1, Ch.2]. The availability of a finite normal form turns out to be very useful and effective, since all non-standard reasoning services performed by I.M.P.A.K.T. process the atomic information making up the knowledge descriptions, rather than the concept as a whole. Thus, we map the KB to the database according to the following design rules:

1) a table CONCEPT is created to store all the atomic information managed by the system: i) concept and role names; ii) the CCNF atoms of all the $\mathcal{FL}_0(D)$ concepts defined in modules M_j, with $1 \leq j \leq 6$; **2)** two tables mapping recursive relationships over the table CONCEPT, namely PARENT and ANCESTOR; **3)** a table PROFILE including the profile identifier (*profileID* attribute) and the so called *structured information*: extra-ontological content, such as personal data (*e.g.*, last and first name, birth date) and work-related information (*e.g.*, preferred working hours, car availability); **4)** a table $R_j(X)$ is created for each entry point R_j^0, $1 \leq j \leq 6$ where X is the set of attributes $X = \{profileID, groupID, conceptID, value, lastdate\}$. Once the

$CCNF(P) = \sqcap(\exists R_j^0.CCNF(C))$ of a profile P (see Definition 1) has been computed, the assertion $P(a)$ is stored in the database. I.M.P.A.K.T. produces a unique identifier for candidate a and assigns it to attribute $profileID$ in table PROFILE. Then, for each conjunct $\exists R_j^0.C$ belonging to $P(a)$, it adds one tuple for each atom of the $CCNF(C)$ to the related table $R_j(X)$. Thus, all features modeled in profile descriptions according to Definition 1 are stored in tables $R_j(X)$ related to the involved entry points. Notice that, thanks to the fourth rule, our model can be easily extended. If the module M_0 in $\mathcal{T}$ is enhanced by a new entry point in order to capture a novel aspect of candidate CV, then the schema can be enriched by adding the corresponding table $R_j(X)$ to it.

3 Skill Matching Services

To evaluate the matching degree between a job request and a candidate profile, we need that both of them share the KB used for representation. Thus, the job requests submitted to I.M.P.A.K.T. have to be represented according to the syntax detailed in Definition 1. In particular, two groups of user requirements (preferences and strict constraints) compose a job request.

Formally, a *Job Request* $\mathcal{F}$ is defined as follows:

Definition 2 (Job Request). *A Job Request $\mathcal{F}$ is a Profile $\mathcal{F} = \sqcap(\exists R_j^0.C)$ (according to Definition 1), defined as a pair of feature sets $\mathcal{F} = \langle \mathcal{FS}, \mathcal{FP} \rangle$ such that:*

- *$\mathcal{FS} = \{fs_i | 1 \leq i \leq s\}$ is a set of s strictly required features fs_i, of the form $\exists R_j^0.C_i$;*
- *$\mathcal{FP} = \{fp_k | 1 \leq k \leq p\}$ is a set of p preferred features fp_k, of the form $\exists R_j^0.C_k$.*

I.M.P.A.K.T. provides two matchmaking processes, namely *Strict Match* and *Soft Match*, detailed in the following. More formally, *Strict Match* is defined as follows:

Definition 3 (Strict Match). *Given the ontology $\mathcal{T}$, a (part of) Job Request $\mathcal{FS}$ and a set $\mathcal{P} = \{P(a_1), \ldots, P(a_n)\}$ of n candidate profiles, modeled according to Definition 1 and stored in the DB according to the schema detailed in section 2, the Strict Match process returns all the candidate profiles $P(a_j)$ in $\mathcal{P}$ providing all the features fs_i in $\mathcal{FS}$.*

We notice that, thanks to the adoption of CCNF, the *Strict Match* can retrieve candidate profiles $P(a)$ also more specific than $\mathcal{FS}$. On the other hand, the *Soft Match* is devoted to implement the approach to approximate matching: the search has to revert also to candidates having some missing features and/or having features *slightly conflicting* w.r.t. $\mathcal{FP}$. We notice that, according to the formalism adopted, inconsistency may happen *e.g.*, when we have a preference $fp_k = \exists R_j^0.C_k$, with $C_k = D \sqcap \geq_n p$, and a candidate profile $P(a)$ with a specified feature $\exists R_j^0.C$, where $C = D \sqcap =_m p$, with $m < n$. In order to satisfy user preferences, candidate profiles modeling concrete features with values in an interval around the required value could represent a good result. We name such concrete features as *slightly conflicting* features (see Definition 4, $MC3$ class).

In order to search for possible approximate matches, *Soft Match* needs to investigate on single atoms of CCNF($\mathcal{FP}$) and compare them with candidate profiles features, which are stored in the DB in their CCNF. Thus, $\mathcal{FP}$ elements need to be further manipulated before the execution of *Soft Match* (notice that for *Strict Match* I.M.P.A.K.T. compares candidates features with the ones in $\mathcal{FS}$ without any preprocessing of $\mathcal{FS}$).

More formally, we define *Soft Match* as:

Definition 4 (Soft Match). *Given the skill ontology $\mathcal{T}$, a (part of) Job Request $\mathcal{FP}$ and a set $\mathcal{P} = \{P(a_1), \ldots, P(a_n)\}$ of candidate profiles, modeled according to Definition 1 and stored in the DB according to the schema detailed in Section 2, the Soft Match process returns a ranked list of candidate profiles $P(a_j)$ in $\mathcal{P}$ belonging to one of the following match classes:*

1. $MC1$ *is the set of profiles $P(a_j)$, such that each $P(a_j)$ provides at least one feature atom corresponding to a concept name in $fp_k \in CCNF(\mathcal{FP})$*[7];
2. $MC2$ *is the set of profiles $P(a_j)$, such that each $P(a_j)$ fully satisfies at least one feature $fp_k \in CCNF(\mathcal{FP})$ combining in C_k both a concept name and a concrete feature*[8];
3. $MC3$ *is the set of profiles $P(a_j)$, such that each $P(a_j)$ partially satisfies one feature $fp_k \in CCNF(\mathcal{FP})$ combining in C_k both a concept name and a concrete feature*[9].

Finally, in the most general case of job request $\mathcal{F}$ containing both $\mathcal{FS}$ and $\mathcal{FP}$, I.M.P.A.K.T. performs a two-step matchmaking approach, namely *Matchmaking*, which starts with *Strict Match* process, computing a set of profiles fully satisfying strict requirements, and then proceeds with *Soft Match* process, trying to approximately match preferences with profiles belonging to the set returned by *Strict Match*.

According to Definition 3, results retrieved by *Strict Match* have a 100% coverage level of the job request $\mathcal{F}$ and thus they do not need to be ranked after retrieval. On the contrary, a ranking process according to a unified measure is necessary for *Soft Match* resulting profiles w.r.t. $\mathcal{F}$. We remind that, among CCNF atoms deriving from features $fp_k \in \mathcal{FP}$, I.M.P.A.K.T. distinguishes between atomic concepts and value restrictions (*i.e.*, qualitative information) and concrete features (*i.e.*, quantitative information), since they need a different manipulation in the ranking process. I.M.P.A.K.T. computes a logic-based ranking by applying the following rules: **(1)** each conjunct in the retrieved candidate profile receives a score on the basis of the number and type (concept name or value restriction or concrete feature) of matched features $fp_k \in \mathcal{FP}$; **(2)** each conjunct ranked according to rule 1 is "re-weighted" based on the relevance of its related entry point R_j^0, $1 \leq j \leq 6$. In rule **(1)**, the score for qualitative information is computed by simply counting the retrieved atoms matching the requested ones. On the other hand, in order to assign a score to each feature specification involving p in a candidate profile, $\mathcal{FP}$ features in the form $\geq_n p$, $=_n p$ and $\leq_n p$ are managed by a different and specifically designed scoring function. Examining the second rule in our score computation strategy, it is easy to notice that a relevance order relation needs to be set among entry

---

[7] See queries $\mathcal{Q}(fp_k)$ and $Q_{NULL}(fp_k)$ in Section 3.1.

[8] See query $Q_n(fp_k)$ in Section 3.1.

[9] See query $Q_{n_{m\%}}(fp_k)$ in Section 3.1.

points (see the following formula (1) for our current implementation). Both *Strict Match* and *Soft Match*, regardless of their different behavior w.r.t. ranking, share the same *Explanation* process of match between a retrieved candidate $P(a)$ and a job request $\mathcal{F}$. Such a process classifies profile features w.r.t. each requirement in $\mathcal{F}$ in the following four groups: **Fulfilled**: $P(a)$ features either perfectly matching or slightly conflicting those requested by $\mathcal{F}$; **Conflicting**: $P(a)$ features slightly conflicting with $\mathcal{FP}$ requirements; **Additional**: $P(a)$ features either more specific than the ones required in $\mathcal{F}$ or not exposed in the user request and belonging to entry point $R_j^0 = hasKnowledge$; **Underspecified**: $\mathcal{F}$ requirements which are not included in $P(a)$ features.

We notice that for a Job Request $\mathcal{F}$ such that $\mathcal{FP} = \emptyset$, that is the case of *Strict Match* only, the explanation of the match related to each returned $P(a)$, is characterized by empty sets of *Underspecified* and *Conflicting* features and by a set of *Fulfilled* features equivalent to $P(a)$ itself.

3.1 SQL-Based Implementation

Coherently with the approach introduced and motivated so far, once our KB has been pre-processed and stored into the DB according to our relational schema, I.M.P.A.K.T. is able to perform all the reasoning services only through standard SQL queries. Notice that we do not use a specific preference language as in [10,3,12] but we exploit a set of standard SQL queries built on-the-fly according to both user requirements (*i.e.*, strict requirements and preferences) and required features (*i.e.*, atoms contained in each feature).

Let us consider a strict requirement fs_i of the form $\exists R_j^0.C_i$. We recall that C_i is a concept description in $\mathcal{FL}_0(D)$ which we can model as a conjunction of concepts defined according to the KB modeling: A – concept name, $\forall R.D$ – universal quantification, $\leq_n p (\geq_n p, =_n p)$ – concrete feature, *i.e.* $fs_i = \exists R_j^0.(A \sqcap \forall R.D \sqcap \geq_n p)$. From database querying point of view, fs_i has to be translated in a set of syntactic elements to search for in the proper $R_j(X)$ table. *Strict Match* asks for a profile to include all of the previous syntactic elements to be retrieved. Since each of these elements fills one tuple of a $R_j(X)$ table, the resulting query, $Q_s(fs_i)$, retrieves a results set by adopting the following conceptual schema: (*set of profiles in* $R_j(X)$ *containing A*) INTERSECT (*set of profiles in* $R_j(X)$ *containing R.D*) INTERSECT (*set of profiles in* $R_j(X)$ *containing* $\geq_n p$)[10].

According to such a schema and the required fs_i, the query $Q_s(fs_i)$ is automatically built on-the-fly considering a number of conditions in WHERE clause defined according to atoms in C_i. In particular, Fig. 1 presents an executable example for the query $fs_i = \exists hasknowledge.(Java \sqcap \forall skillType.Programming \sqcap \geq_3 years)$. We notice that $Q_s(fs_i)$ in Fig. 1 has three conditions in WHERE clause, as expected. On the other hand, *Soft Match* relics on a query schema involving each element $CCNF(fp_k)$, $\forall fp_k \in \mathcal{FP}$. In particular, let $CCNF(fp_k) = \exists R_j^0.CCNF(C_k)$ be a normalized preference;

---

[10] To improve engine performance, I.M.P.A.K.T. exploits, as far as possible, EXISTS operator instead of INTERSECT. Also for performance reasons, conditions in the form conceptID=(SELECT conceptID FROM CONCEPT WHERE name='X') are not executed at run-time but a lookup on a hash table directly assigns the proper conceptID value.

```
SELECT profileID
FROM hasKnowledge as R
WHERE conceptID = (SELECT conceptID
                   FROM concept WHERE name='Java')
AND EXISTS (SELECT *
            FROM hasKnowledge
            WHERE profileid=R.profileid AND groupid=R.groupid
            AND conceptid = (SELECT conceptID
                             FROM concept WHERE name='skillType.Programming'))
AND EXISTS (SELECT *
            FROM hasKnowledge
            WHERE conceptid=(SELECT conceptID
                             FROM concept WHERE name='years')
            AND value >= 3 AND profileid=R.profileid AND groupid=R.groupid)
```

Fig. 1. SQL definition of query $Q_s(fs_i)$ w.r.t. a single feature $fs_i = \exists hasKnowledge.(Java \sqcap \forall skillType.Programming \sqcap \geq_3 years)$.

a single query $Q(fp_k)$ or a set of queries $Q_p(fp_k)$ is built according to the following schema:

- if none of $\{\leq_n p, \geq_n p, =_n p\}$ elements occur in $CCNF(C_k)$, then a single query $Q(fp_k)$ is built which retrieves the profiles containing – w.r.t. the related *entry point* R_j^0– at least one among syntactic element occurring in $CCNF(C_k)$;
- otherwise a set of queries $Q_p(fp_k) = \{Q_n(fp_k), Q_{NULL}(fp_k), Q_{n_{m\%}}(fp_k)\}$ is built retrieving candidate profiles belonging to a different match class –*i.e.*, either profiles fulfill fp_k $(Q_n(fp_k))$ or profiles do not fulfill it $(Q_{n_{m\%}}(fp_k))$ or profiles do not specify p $(Q_{NULL}(fp_k))$. The resulting set of candidate profiles is made up by the UNION of all the tuples retrieved by each of the query in $Q_p(fp_k)$.

As for *Strict Match*, for each $CCNF(fp_k)$ the previous queries are automatically built on-the-fly according to syntactic elements occurring in $CCNF(C_k)$. Here, due to the lack of space, we do not report the SQL definition of both $Q(fp_k)$ query and the set of queries $Q_p(fp_k)$. We only notice that the score for each retrieved atom (*i.e.*, tuple) of candidate feature is computed directly in the SELECT clause of each query implementing the *Soft Match*. In particular, for qualitative information (*i.e.*, atomic concepts and value restrictions) score is equal to 1, whereas for concrete feature p score is an expression computed according to scoring functions aforementioned strategy. Moreover, we notice that, by construction, *Soft Match* retrieves candidate profiles belonging to one of the match classes in Definition 4 for each feature fp_k. Thus, such candidates profiles have to be properly rearranged for defining the final results set. Each retrieved profile is finally ranked according to a linear combination of scores:

$$rank = \sum_{i=1}^{N} w_i * score_{l^i} \tag{1}$$

where w_i are heuristic coefficients belonging to the $(0, 1)$ interval, N is the number of relevance levels defined for the domain ontology and $score_{l^i}$ represents the global score computed summing the score of tuples related to entry points falling in the same relevance level l^i. I.M.P.A.K.T. defines a number $N = 3$ of ontology levels represented by $Level = \{l^1, l^2, l^3\}$ (l^1 is the most relevant one), with *hasKnowledge* set to l^1; *hasIndustry, hasComplementarySkill* and *hasJobTitle* set to l^2 and the remaining entry points set to l^3. Moreover, the following values are assigned to w_i coefficients: $w_1 := 1$, $w_2 := 0.75$ and $w_3 := 0.45$.

4   System Performances

In this section we focus on the evaluation of *data complexity* and *expressiveness complexity* of our knowledge compilation approach and present obtained results. I.M.P.A.K.T. is a client-server application developed in Java. Our current implementation exploits the open source PostgreSQL 9.1 DBMS. In order to prove the effectiveness and efficacy of the proposed approach, we initially created a real dataset by collecting approximately 180 CVs on ICT domain, originated from three different employment agencies. The dataset has been exploited for an iterative refinement phase of both the Skill Ontology development and the setting of the Skill Matching parameters (*i.e.*, entry points levels and weights in scoring strategy). We implemented a synthetic KB instances generator able to automatically build satisfiable profiles according to a given format (*i.e.*, number of features for each relevance level, number of numeric restrictions, etc.). In this way, we generated datasets having different size, ranging from 500 to 5500 profiles, with bigger datasets including the smaller ones. We point out that for the datasets construction we considered a number of features for each candidate comparable to the average value of candidate profiles in the previous mentioned real dataset. In particular, each generated profile has at least: 30 features for *hasKnowledge* entry point, 2 features for *hasLevel* and *knowsLanguage* entry points, and 3 features for *hasJobTitle, hasIndustry* and *complementarySkill* entry points. Tests refer to I.M.P.A.K.T. running on an Intel Dual Core server, equipped with a 2.26 GHz processor and 4 GB RAM and measure the retrieval time calculated as average time over ten iterations. Here we report retrieval times of 9 significant queries, selected among several test queries with a different expressiveness divided into 3 groups: (**A**) only strict requirements represented by either generic concepts (Q_4) or features with an higher specificity (Q_5); (**B**) only preferences again represented by either generic concepts (Q_2) or features with an higher specificity (Q_3); (**C**) a combination of all A) and B) groups features (Q_1,Q_6,Q_7,Q_8,Q_9). We notice that: 1) Q_1 query is a translation in our formalism of a real job request available on http://jobview.monster.co.uk titled *"SQL Developer (Business Intelligence)"* and containing 2 strict and 12 soft requirements for entry point *hasKnowledge* and only one soft request for entry point *complementarySkill*; 2) queries from Q_2 to Q_7 are composed by one feature for each entry point; 3) $Q_6 = Q_2 \cup Q_4$ and $Q_7 = Q_3 \cup Q_5$; 4) Q_8 involves only three entry points, *i.e.*, *hasKnowledge, knowsLanguage* and *hasLevel*; 5) Q_9 involves several features for each entry point.

Table 1 shows the retrieval times together with the number of retrieved profiles ($\#p$) for each dataset and request. In particular, in order to better evaluate matching performances, we differentiate among the request normalization process times (see t_n in Table 1), which is dataset-independent, *Strict Match* retrieval times (see t_{st} in Table 1) and *Soft Match* retrieval times (see t_{sf} in Table 1) including also the ranking calculation times.

Table 1. Retrieval times in milliseconds and number of retrieved profiles ($\#p$) for datasets DS_1, DS_2, DS_3, DS_4 and DS_5 of, respectively, 500, 1000, 2000, 3500 and 5500 profiles

	t_n	DS_1			DS_2			DS_3			DS_4			DS_5		
		t_{st}	t_{sf}	$\#p$	t_{st}	t_{sf}	$\#p$	t_{st}	t_{sf}	$\#p$	t_{st}	t_{sf}	$\#p$	t_{st}	t_{sf}	$\#p$
Q_1	724.4	124.2	210.6	4	246.6	240.8	10	545.6	382.5	20	784.5	402.2	28	2334.8	551.8	144
Q_2	335.8	0	305.7	461	0	456.8	927	0	563.6	1829	0	756.2	3202	0	1158.8	5029
Q_3	474.8	0	440.5	396	0	578.2	740	0	782.2	1560	0	1624.8	2729	0	2775	4270
Q_4	225.9	71.2	0	10	110.4	0	13	212.8	0	23	336.4	0	35	423	0	52
Q_5	224.4	74.1	0	1	115.2	0	1	218	0	1	342	0	1	441.4	0	1
Q_6	240.6	96.7	103.8	10	147.4	128.4	13	227.4	139.4	23	344.4	173	35	485.5	180.4	52
Q_7	538.8	84.8	97.8	1	119.8	133.4	1	219.2	179.6	1	343.8	193.8	1	473.2	208	1
Q_8	347	228.6	96.6	17	456.6	113	44	927	125.2	79	1277.4	132.4	131	2593.4	196.8	226
Q_9	317.8	136.8	163	3	244.2	166.5	3	385.6	168.6	4	671.2	180	5	1245.8	252	7

As we expected, retrieval times of both match procedures linearly increase with datasets size (*e.g.* see Q_5). In particular, *Strict Match* times are also dramatically affected by $\#p$ (see results for DS_5 in Table 1), whereas the *Soft Match* times seem to grow more slowly with $\#p$. We therefore observe that the number of retrieved profiles, though affecting the whole matchmaking process, mostly impacts *Strict Match*, since it involves the SQL intersection of several queries by construction. In particular, profiles returned by Q_5 are dataset-independent, as the *Strict Match* procedure always returns the same profile (*i.e.*, no other profile satisfying strict requirement exists in the datasets). In order to verify the approach expressiveness complexity, we evaluated retrieval times of different test queries on one dataset at a time. It has to be observed that: (i) for queries only expressing preferences (Q_2,Q_3) or only strict requirements (Q_4,Q_5), the retrieval time increases with the query expressiveness; (ii) for the other queries, thanks to preliminary execution of *Strict Match*, the *Soft Match* times are always notably reduced, so confirming the theoretical complexity results. In particular, we notice that for larger data sets and a number of retrieved profiles larger than 3000 (see t_{sf} in Q_2 and Q_3 on DS_4, DS_5), expressiveness of soft requirements has a more relevant impact on retrieval times. Moreover, for each dataset, the real-data query Q_1 has retrieval times comparable to all queries belonging to C group considering also the $\#p$ value. Thus, in the whole matchmaking process, involving both strict requirements and preferences, the query expressiveness does not significantly affect retrieval times.

Summing up, we can claim that I.M.P.A.K.T. is able – with time performances encouraging its application in real-world scenarios – to provide crucial value-added information with respect to typical HR management tasks, even on large datasets.

5   Discussion and Future Work

Motivated by the need to efficiently cope with real-life datasets in semantic-enhanced skill matching, we presented a knowledge compilation approach able to translate a KB into a relational database while retaining the expressiveness of the logical representation. The obtained model allows to perform reasoning services through standard-SQL queries, in the framework of I.M.P.A.K.T.. Performance evaluations on various datasets show an efficient behavior although several optimization techniques have not been implemented yet. Future work aims at testing further devised strategies for score calculation, including the possibility for the user to assign a weight to each preference, along with a full optimization of the database.

Acknowledgments. We gratefully acknowledge support of projects UE ETCP "G.A.I.A." and Italian PON ERMES "Enhance Risk Management through Extended Sensors".

References

1. Baader, F., Calvanese, D., Mc Guinness, D., Nardi, D., Patel-Schneider, P. (eds.): The Description Logic Handbook, 2nd edn. Cambridge University Press (2007)
2. Cadoli, M., Donini, F.M.: A survey on knowledge compilation. AI Commun. 10(3-4), 137–150 (1997)
3. Chomicki, J.: Querying with Intrinsic Preferences. In: Jensen, C.S., Jeffery, K., Pokorný, J., Šaltenis, S., Bertino, E., Böhm, K., Jarke, M. (eds.) EDBT 2002. LNCS, vol. 2287, pp. 34–51. Springer, Heidelberg (2002)
4. Chong, E.I., Das, S., Eadon, G., Srinivasan, J.: An Efficient SQL-based RDF Querying Scheme. In: Proc. of VLDB 2005, pp. 1216–1227. VLDB Endowment (2005)
5. Colucci, S., Di Noia, T., Di Sciascio, E., Donini, F.M., Mongiello, M.: Concept Abduction and Contraction for Semantic-based Discovery of Matches and Negotiation Spaces in an E-Marketplace. In: Proceedings of the 6th Int. Conf. on Electronic Commerce, ICEC 2004, pp. 41–50 (2004)
6. Colucci, S., Di Noia, T., Di Sciascio, E., Donini, F.M., Piscitelli, G., Coppi, S.: Knowledge Based Approach to Semantic Composition of Teams in an Organization. In: Proceedings of the 20th Annual ACM (SIGAPP) Symposium on Applied Computing, SAC 2005, pp. 1314–1319. ACM (2005)
7. Colucci, S., Tinelli, E., Di Sciascio, E., Donini, F.M.: Automating competence management through non-standard reasoning. Engineering Applications of Artificial Intelligence 24(8), 1368–1384 (2011)
8. Di Noia, T., Di Sciascio, E., Donini, F.M.: Extending Semantic-Based Matchmaking via Concept Abduction and Contraction. In: Motta, E., Shadbolt, N.R., Stutt, A., Gibbins, N. (eds.) EKAW 2004. LNCS (LNAI), vol. 3257, pp. 307–320. Springer, Heidelberg (2004)
9. Dolby, J., Fokoue, A., Kalyanpur, A., Kershenbaum, A., Schonberg, E., Srinivas, K., Ma, L.: Scalable semantic retrieval through summarization and refinement. In: Proc. of AAAI 2007 (2007)
10. Kießling, W.: Foundations of Preferences in Database Systems. In: Proc. of VLDB 2002, pp. 311–322. Morgan Kaufmann, Los Altos (2002)
11. Kiryakov, A., Ognyanov, D., Manov, D.: OWLIM – A Pragmatic Semantic Repository for OWL. In: Dean, M., Guo, Y., Jun, W., Kaschek, R., Krishnaswamy, S., Pan, Z., Sheng, Q.Z. (eds.) WISE 2005 Workshops. LNCS, vol. 3807, pp. 182–192. Springer, Heidelberg (2005)
12. Bosc, P., Pivert, O.: SQLf: a relational database language for fuzzy querying. IEEE Transactions on Fuzzy Systems 3(1), 1–17 (1995)
13. Tinelli, E., Cascone, A., Ruta, M., Di Noia, T., Di Sciascio, E., Donini, F.M.: I.M.P.A.K.T.: An Innovative Semantic-based Skill Management System Exploiting Standard SQL. In: Proc. of ICEIS 2009, pp. 224–229 (2009)
14. Tinelli, E., Colucci, S., Di Sciascio, E., Donini, F.M.: Knowledge compilation for automated team composition exploiting standard SQL. In: Proc. of ACM SAC 2012, pp. 1680–1685 (2012)

Evaluating Role Based Authorization Programs

Chun Ruan

School of Computing, Engineering and Mathematics
University of Western Sydney, Penrith South DC, NSW 1797 Australia

Abstract. In this paper, we discuss a role based authorization program and its implementation. A role based authorization program (RBAP) is a logic based framework which enables users to describe complex access control policies in a decentralized system. It supports administrative privilege delegations for both roles and access rights. The program SMOD-ELS is a widely used system that implements the answer set semantics for extended logic programs. In this paper, we show how to use SMODELS to evaluate RBAP. The access control policy is also given.

1 Introduction

Role Based Access Control (RBAC) ([4]) is a well-known paradigm of access control model. The main idea of RBAC is that accesses are associated with roles and users are assigned to appropriate roles thereby acquiring accesses. Classic access control is based on the individual (subject) accessing a resource (object). However, in many situations, access rights are associated with positions other than individuals. Individuals get their access rights due to their roles in an organization. When people leave the organization or change the positions, their access rights will be revoked or changed, too. For example, an employee doing student service in a university can access the students' information. If the employee leaves the university or changes to the IT support development department, his/her capability to access the students' information should be revoked or changed, too. If the number of subjects and objects is large, individual access control becomes difficult. When access rights are indeed assigned to roles other than individual subjects, role-based access control can greatly simplify the administration work.

In a role-based access control, roles are placed between the user and the objects. Users get their access rights indirectly by assigning access rights to roles and roles to users. When people leave or change roles, only the mapping from subjects to roles need to be revoked or changed. On the other hand, if the duties of the roles change, only the mapping from roles to access rights need to be changed. Roles can be organized into hierarchies so that access rights can be inherited, which could further reduce the amount of explicit access rights specification. For example, privileges granted to employee can be inherited by all roles in a university. Roles can also be delegated. One entity may act on behalf of another. Acting Head of School is an example of this relationship.

L. Chen et al. (Eds.): ISMIS 2012, LNAI 7661, pp. 202–207, 2012.

On the other hand, logic based approaches have been developed by many researchers for the purpose of formalizing access control policy specifications and evaluations ([2]). SMODELS is a widely used system that implements the answer set semantics for extended logic programs. It is domain-restricted but supports extensions including built-in functions as well as cardinality and weight constraints. We have developed a role based authorization program (RBAP) which is a logic based framework. It supports administrative privilege delegations for both roles and access rights. In this paper, we briefly discuss some features of RBAP that can be used to assist with specifying role based access control policies. We then show how to evaluate RBAP using SMODELS.

The paper is organised as follows. Section 2 presents an overview of RBAP. Section 3 describes how to evaluate RBAP using SMODELS, while Section 4 concludes the paper.

2   RBAP Overview

2.1   Role Constraints

Role constraints are supported in our framework. Firstly, strong exclusion is an important one which is also called Static Separation of Duty. Two roles are strongly exclusive if no one person is ever allowed to perform both roles. In other words, the two roles have no shared users. For example, full-time staff and part-time staff roles should be exclusive. We believe this exclusion requirement can be generalized to any number of roles; in our current framework, we consider up to 4 roles exclusion, which is adequate for most common systems.

Role cardinality requirement is about the number of members in a role. To make this more general, we will allow the expression of a minimum number and a maximum number for a role at a given time. For instance, the promotion committee consists of at least 5, and at most 7 members. The exact role number requirement is then expressed by making minimum number equal to the maximum number.

Role composition requirement is about a role's relationship with another role. For example, a university promotion committee should include two staff members. We will also generalize this constraint to express that one role should contain at least n and at most m members from another role.

In the real world, it is often required that to be able to perform role 1, one needs to be in role 2. For example, many universities require that Unit coordinators to be full time academic staff. Role dependency requirement is thus introduced in our framework to represent this situation.

2.2   Authorization and Administrative Privilege Propagations

Rule based formalism allows implicit roles and their access rights to be derived from explicit role/access right assignments, and hence this can greatly reduce the size of explicit assignment set. In our framework, we support the implicit role/access right assignment by supporting inheritance. For example, the role Head of School will inherit role staff and therefore has all the access rights granted to staff. If a role is allowed to access student profile then the role should be able to access the profile's components such as the personal information and study record. We apply the authorization propagations along hierarchies of roles and objects represented by the corresponding partial orders. For the same reason, we also support the administrative privileges inheritance along role and object hierarchies. For example, if the role staff is allowed to assign role student or grant access right on student profile, then the role Head of School should also be able to do so.

2.3 Syntax

RBAP is a multi-sorted first order language, with seven disjoint *sorts* $\mathcal{R}, \mathcal{U}, \mathcal{O}, \mathcal{A}, \mathcal{T}, \mathcal{W}$, and $\mathcal{N}$ for role, user, object, access right, authorization type and weight or depth respectively. Variables are denoted by strings starting with lower case letters, and constants by strings starting with upper case letters.

A *rule r* is a statement of the form:

$b_0 \leftarrow b_1, ..., b_k, not\, b_{k+1}, ..., not\, b_m, m >= 0$

where $b_0, b_1, ..., b_m$ are literals, and not is the negation as failure symbol. A *Role Based Authorization Program*, RBAP, consists of a finite set of rules.

The predicate set P in RBAP consists of a set of ordinary predicates defined by users, and a set of system built-in predicates designed for users to express role assignment, role to privilege grant, role/access right administrative privilege delegation, and role constraints etc. Some built-in predicates are briefly presented here.

For role assignment delegation, predicate $(g, r, w, d)canAssign(r', r'')$ means that a user g in role r says that role r' can not only assign users to role r'', but also further delegate this administrative privilege on r'' for the maximum delegation depth of d, and g's trust degree on this delegation to r' is w. For role assignment, predicate $(g, r, w)assign(r', u)$ means that a grantor g in a role r assigns user u to role r'. g's trust degree on this role assignment is w. The users can be individuals, agents or processes. For the role delegation, predicate $deleRole(u, u', R)$ means user u delegates its role R to user u' while $deleAll(u, u')$ with Type $U \times U$ means user u delegates its every role to user u'. The delegatee can perform the delegated role due to the delegation. For the exclusion of roles, we define a predicate $exclusive(r_1, r_2, r_3, r_4)$ to represent up to 4 roles exclusion. Please note that r_2 or r_3 can be empty denoted by _ to denote 2 or 3 role exclusion. Predicate $roleNum(r, n, m)$ is defined for the role cardinality constraint. It means that a role r should have at least n and at most m members. For the role composition constraint, we define the predicate $roleComp(r, r', n, m)$. It means that role r should contain at least n and at most m members from another role r'. For the role dependency constraint, we define the predicate $depend(r, r')$ which means that role r depends on role r'.

2.4   Access Control Policy

We can now define our access control policies. A query is a 4-ary tuple (u, r, o, a) in $\mathcal{U} \times \mathcal{R} \times \mathcal{O} \times \mathcal{A}$, which denotes that a user u as an active member of role r requests access a over object o. The access control policy is a function f from $\mathcal{U} \times \mathcal{R} \times \mathcal{O} \times \mathcal{A}$ to $\{true, false, undecided\}$. Given a request (u, r, o, a), if $f(u, r, o, a) = true$ then it is granted. If $f(u, r, o, a) = false$ then it is denied. Otherwise, $f(u, r, o, a) = undecided$, and it is left to be decided by the implemented access control system.

According to answer set semantics, there may exist several authorization answer sets for a given RBAP, and they may not be consistent with each other in the sense that they may contain conflicting literals. We will adopt an optimistic approach to deal with this problem. Let Π be a RBAP, $A_1, ..., A_m$ be its authorization answer sets. For any query (u, r, o, a), $f(u, r, o, a) = true$ if r is SSO and u is a member of SSO (inRole(u,SSO) appears in all $A_i, 1 \leq i \leq m$); or the role r is granted access right a on o, and u is a member of r in some answer set A_i. Otherwise, $f(u, r, o, a) = false$ if the role r is rejected access right a on o, and u is a member of r in some answer set A_i. Otherwise $f(u, r, o, a) = undecided$. On the other hand, there may exist no authorization answer set for a given RBAP Π. For example, there will be no answer set if the role constraints, such as separation of duty, are not satisfied. In this case, we say Π is not well-defined.

3   Using SMODELS to Evaluate RBAP

3.1   SMODELS

SMODELS is a widely used system that implements the answer set semantics for extended logic programs [1]. It is domain-restricted but supports extensions including built-in functions as well as cardinality and weight constraints.

In SMODELS, constants are either numbers or symbolic constants that starts with a lower case letter. Variables start with a capital letter. A function is either a function symbol followed by a parenthesized argument list or a built-in arithmetical expression. A range is the form of: start..end where start and end are constant valued arithmetic expressions. A range is a notational shortcut for defining numerical domains.

Rules are in the format of:

$h : - l_1, l_2, ... l_n$

Where the head is the part to the left of $: -$, and the body is right to $: -$. A program consists of set of rules.

3.2   Domain-Independent Rules

In this section, we present the SMODELS program for the domain-independent rules R in RBAP [3]. Most of the transformations are straightforward. The special predicates, which are more readable to users, need to be transformed into the normal predicates. For example, the predicate $(g, r, w, d)canAssign(r', r'')$ will be transformed into $canAssign(G, R, R', R'', W, D)$. We also add domain definitions based on the SMODELS syntax requirements. Due to the space limit, we only show the rules for role assignment and constraints.

Rules for domains of roles, persons, objects, access methods, authorization types, trust weights and depths.

$\#domain\, person(P; P1; P2), role(R; R1; R2; R3), depth(D; D1),$
    $file(F; F1; F3), method(M; M2; M1), weight(W; W1), authtype(T; T1).$
$weight(0..10).$
$depth(0..10).$
$authtype(0..1).$
$\#const\, wei = 8.$

Rules for role assignment capability delegation correctness.
$(r_1).\ canAssign1(P, sso, R, R1, W, D) : -$
        $canAssign(P, sso, R, R1, W, D), inRole(sso, P).$
$(r_2).\ canAssign1(P, R, R1, R2, (W * W1)/10, D) : -$
        $canAssign(P, R, R1, R2, W, D),$
        $canAssign1(P2, R3, R, R2, W1, D1), inRole(R, P), D1 > D.$

The above two rules say only eligible delegators' delegations will be accepted, which include *sso* or any roles that are delegated the capability to assign, represented by *canAssign1*.

Rules for role assignment correctness.
$(r_3).\ assign1(P, sso, P1, R, W) : -$
        $assign(P, sso, P1, R, W), inRole(sso, P), P! = P1.$
$(r_4).\ assign1(P, R, P1, R1, W) : -$
        $assign(P, R, P1, R1, W), canAssign1(P2, R2, R, R1, W1, D),$
        $inRole(R, P), P! = P1.$

The above two rules say only eligible assigners' assignments will be accepted, which include *sso* or any roles that have the capability to assign, represented by *canAssign1*.

Rules for calculating the role's assigner's trust degree.
$(r_5).\ trust(sso, R, 10) : -$
$(r_6).\ trusts(R1, R2, W * W1/10) : -canAssign1(P, R, R1, R2, W, D),$
        $trust(R, R2, W1)$
$(r_7).\ existHigherTrusts(R, R1, W) : -trusts(R, R1, W),$
        $trusts(R, R1, W1), W1 > W$
$(r_8).\ trust(R, R1, W) : -trusts(R, R1, W),$
        $not\, existHigherTrusts(R, R1, W)$

Please note that an assigner's trust degree is the product of all the trust degrees along its assignment path. When there are multiple trust degrees for an assigner due to the existence of multiple paths to it, we choose the biggest one as the effective one.

Rules for role assignment acceptance. To accept the role assignment, the assignment's effective trust degree has to be greater than the predefined threshold, represented by wei here.

(r_9). $inRole(R1, P1) : -$
$$assign1(P, R, P1, R1, W), trust(R, R1, W1), W1 * W > wei.$$

Rules for role delegation. A person can delegate all of his/her roles or a specific role to another person.

(r_{10}). $inRole(R, P1) : -deleRole(P, P1, R), inRole(R, P), P! = P1.$
(r_{11}). $inRole(R, P1) : -deleAll(P, P1), inRole(R, P), P! = P1.$

Rules for constraints about role dependency, role exclusion, role cardinality, and role composition.

(s_1). $: -depend(R, R1), inRole(R, P), notinRole(R1, P).$
(s_2). $: -exclusive(R, R1), inRole(R, P), inRole(R1, P).$
(s_3). $: -roleNum(R, N, M), M + 1\{inRole(R, P)\}.$
(s_4). $: -roleNum(R, N, M), \{inRole(R, P)\}N - 1.$
(s_5). $: -roleComp(R, R1, N, M), inRole(R, P), M + 1\{inRole(R1, P)\}.$
(s_6). $: -roleComp(R, R1, N, M), inRole(R, P), \{inRole(R1, P)\}N - 1.$

4   Conclusions

In this paper, we have discussed a logic program based formulation to specify and evaluate complex access control policies that support role based access control. Role constraints such as role dependency, role exclusion, role cardinality and composition etc. are supported. We have discussed how to use SMODELS to implement the logic formulation and provided the corresponding SMODELS rules.

For future work, we are considering to apply the system in a real world web-based teaching system. We also plan to add a user-friendly interface to the system.

References

1. Gelfond, M., Lifschitz, V.: Classical negation in logic programs and disjunctive databases. New Generation Computing 9, 365–385 (1991)
2. Gurevich, Y., Neeman, I.: DKAL: Distributed Knowledge Authorization Language. In: Proceedings of the 21st IEEE Computer Security Foundations Symposium, pp. 149–162. IEEE Computer Society (2008)
3. Ruan, C., Varadharajan, V.: Reasoning about Dynamic Delegation in Role Based Access Control Systems. In: Yu, J.X., Kim, M.H., Unland, R. (eds.) DASFAA 2011, Part I. LNCS, vol. 6587, pp. 239–253. Springer, Heidelberg (2011)
4. Sandhu, R.S., Coyne, E.J., Feinstein, H.L., Youman, C.E.: Role based access control models. IEEE Computer 29(2), 38–47 (1996)

A Comparison of Random Forests and Ferns on Recognition of Instruments in Jazz Recordings

Alicja A. Wieczorkowska[1] and Miron B. Kursa[2]

[1] Polish-Japanese Institute of Information Technology,
Koszykowa 86, 02-008 Warsaw, Poland
alicja@poljap.edu.pl
[2] Interdisciplinary Centre for Mathematical and Computational Modelling (ICM),
University of Warsaw,
Pawińskiego 5A, 02-106 Warsaw, Poland
M.Kursa@icm.edu.pl

Abstract. In this paper, we first apply random ferns for classification of real music recordings of a jazz band. No initial segmentation of audio data is assumed, i.e., no onset, offset, nor pitch data are needed. The notion of random ferns is described in the paper, to familiarize the reader with this classification algorithm, which was introduced quite recently and applied so far in image recognition tasks. The performance of random ferns is compared with random forests for the same data. The results of experiments are presented in the paper, and conclusions are drawn.

Keywords: Music Information Retrieval, Random Ferns, Random Forest.

1 Introduction

The pleasure of listening to music can be very enjoyable, especially if our favorite instruments are playing in the piece of music we are listening to. Therefore, it is desirable to have a tool to find melodies played by a specified instrument. The task of automatic identification of an instrument, playing in a given audio segment, lies within the area of interest of Music Information Retrieval. This area has been broadly explored last years [19], [22], and as a result we can enjoy finding pieces of music through query-by-humming [14], and identify music through query-by-example, including excerpts replayed on mobile devices [21], [24]. However, recognition of instruments in real polyphonic recordings is still a challenging task (see e.g. [4], [6], [7]).

In this paper, we address the recognition of plural instruments in real music recordings of a jazz band, and our goal is to identify possibly all instruments playing in each audio frame; polyphony in these recordings reaches 4 instruments. Identification of instruments is performed in short frames, with no assumption on onset (start) nor offset (end) time, nor pitch etc., which is often the case

L. Chen et al. (Eds.): ISMIS 2012, LNAI 7661, pp. 208–217, 2012.

in similar research, thus our methodology requires no preprocessing nor initial segmentation of the data, and the computation can be fast.

Random ferns are classifiers introduced in 2007 [17] and named as such in 2008 [18]. This classification method combines features of decision trees and Bayesian classifiers. Random ferns have been applied so far in image classification tasks, including video data [1], [16], and they have also been adjusted to be used on low-end embedded platforms, such as mobile phones [25]. Since many audio applications are used in mobile environment, it is advisable to consider such platforms as well. This is why we decided to use random ferns. Additionally, we would like to compare the performance of Random Ferns (RFe) with Random Forests (RFo), which yielded quite good results in our previous research [7], [8], [9]. RFe are simpler and more computationally efficient than RFo [10]. We want to use a simpler algorithm because, as more computationally efficient, it can possibly be applied to be used on mobile devices, with limited computational power (utilizing slower CPUs and working on battery power). We hope that the accuracy of RFe is not much worse, and therefore it is worth using them and possibly implement on mobile devices, to get quick results without communication with a cloud for cloud computing (which is an option which can be chosen for low-end platforms), thus achieving low latency. Also, such a method would be useful for massive calculations for indexing purposes, e.g. in archives, to achieve fast computation and get quick results which are a good approximation of the results that would be obtained using more computationally expensive search.

2 Classifiers

The classifiers applied in our research include random ferns and random forests. RFo performed quite well in the research on instrument identification we performed before [7], but their training is time consuming, whereas the training of RFe is faster. The computational complexity of classification performed using the pre-trained classifiers is similar (linearly proportional to the number of trees/ferns and to their average height), but in the case of ferns there is less branching and memory accesses which should yield faster classification.

2.1 Random Forests

RFo is a classifier consisting of a set of weak, weakly correlated and non-biased decision trees, constructed using a procedure minimizing bias and correlations between individual trees [2]. Each tree is built using a different N-element bootstrap sample of the training N-element set. The elements of the bootstrap sample are drawn with replacement from the original set, so roughly 1/3 (called *out-of-bag*) of the training data are not used in the bootstrap sample for any given tree. For a P-element feature vector, K attributes (features) are randomly selected at each stage of tree building, i.e. for each node of any particular tree in RFo ($K < P$, often $K = \sqrt{P}$). Gini impurity criterion (GIC) is applied to find the best split on these K attributes. GIC measures how often an element would be incorrectly labeled if randomly labeled according to the distribution of labels in the subset; the best split minimizes GIC.

Each tree is grown to the largest extent possible, without pruning. By repeating this randomized procedure N_t times, a collection of N_t trees is obtained, constituting a RFo. Classification of an object is done by simple voting of all trees. In this work, the RFo implementation from the R [13] package randomForest [11] was used.

The computational complexity Ct_{Fo} of training a RFo is

$$Ct_{Fo} = N_t \cdot N_o \cdot \log N_o \cdot K , \qquad (1)$$

where N_o is the number of objects, K is the number of attributes tested for each split and N_t is the number of trees in the forest; the computational complexity

$$Cc_{Fo} = N_t \cdot h_t \qquad (2)$$

where h_t is the average height of a tree in the forest.

2.2   Random Ferns

A fern is defined as a simplified binary decision tree of a fixed height D (called a *depth* of a fern) and with a requirement that all splitting criteria at a certain depth i (C_i) are the same. Each leaf node of a fern stores the distribution of classes over objects that are directed to this node. This way a fern can be perceived as a D-dimensional array of distributions, indexed by a vector of D splitting criteria values, see Figure 1.

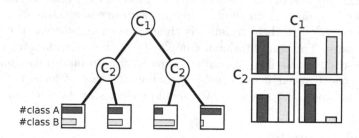

Fig. 1. An example of a fern of depth 2 trained on a binary classification problem (*left*). Splits on each level are based on the same criterion (C_i), thus the fern tree is equivalent to a 2-dim array (*right*). The leaf nodes contain the counts of objects of each class instead of just the names of dominating classes, as in classic decision trees.

The fern forest is a collection of N_f ferns. When classifying a new object, each fern in a forest returns a vector of probabilities that this object belongs to particular decision classes. Ferns are treated as independent, thus all those vectors are combined by simple multiplication and the final classification results for the forest is a class which gets the highest probability, see Figure 2.

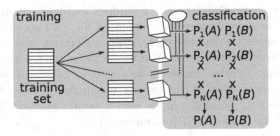

Fig. 2. Training and classification using a fern forest for a binary classification problem. *Bags* are drawn from the training data, and used for building individual ferns, represented here as cubes (*left*). When a new object (represented as an ellipse) is classified, each fern in the forest returns a vector of class probabilities; they are combined by a simple multiplication and the class scoring maximal probability is returned (*right*).

While the original RFe implementation [17,18] was written for a problem of object detection in images, we use the RFe generalization implemented in the R [13] package rFerns [10]; it trains the fern forest model in the following way.

First, N intermediate training sets called *bags* are created by drawing objects with replacement from the training set, each bag being of the same size as the original set. Next, each bag is used to train a fern. All D splits are created purely at random; an attribute is randomly selected and then the splitting threshold is set as a mean of two randomly selected values of this attribute[1]. The distributions of classes in leafs are calculated on a bag with adding 1 for each class (i.e. with a Dirichlet prior); this way the problem of undefined distributions in leafs containing no objects is resolved.

The computational complexity Ct_{Fe} of training a Rfe model is

$$Ct_{Fe} = 2^D \cdot N_f \cdot N_o ,\qquad(3)$$

where D — depth of ferns, N_o — number of objects, N_f — number of ferns; the computational complexity Cc_{Fe} of classifying one sample is

$$Cc_{Fe} = D \cdot N_f .\qquad(4)$$

3   Sound Parameterization

The identification of musical instruments is performed for short frames of audio data, which are parametrized before applying classifiers for training or testing. No assumptions on audio data segmentation or pitch extraction have been made. Therefore, no multi-pitch extraction is needed, thus avoiding possible errors regarding labeling particular sounds in polyphonic recording with the appropriate pitches. The feature vector consists of basic features, describing properties of an

---

[1] In this work we have used only numerical descriptors of sound, thus the description of treating ordinal and categorical attributes is omitted.

audio frame of 40 ms, and additionally difference features, calculated as the difference between the given feature but calculated for a 30 ms sub-frame starting from the beginning of the frame and a 30 ms sub-frame starting 10 ms later. Identification of instruments is performed frame by frame, for consequent frames, with 10 ms hop size. Fourier transform was used to calculate spectral features, with Hamming window. Most of the features we applied represent MPEG-7 low-level audio descriptors, which are often used in audio research [5]. Our feature vector consists of the following 91 parameters [7]:

- *Audio Spectrum Flatness*, $flat_1, \ldots, flat_{25}$ — a multidimensional parameter describing the flatness property of the power spectrum within a frequency bin for selected bins; 25 out of 32 frequency bands were used;
- *Audio Spectrum Centroid* — the power weighted average of the frequency bins in the power spectrum; coefficients are scaled to an octave scale anchored at 1 kHz [5];
- *Audio Spectrum Spread* — RMS (root mean square) value of the deviation of the log frequency power spectrum wrt. *Audio Spectrum Centroid* [5];
- *Energy* — energy (in log scale) of the spectrum of the parametrized sound;
- *MFCC* — a vector of 13 mel frequency cepstral coefficients. The cepstrum was calculated as the logarithm of the magnitude of the spectral coefficients, and then transformed to the mel scale, to better reflect properties of the human perception of frequency. 24 mel filters were applied, and the obtained results were transformed to 12 coefficients. The 13th coefficient is the 0-order coefficient of MFCC, corresponding to the logarithm of the energy [12];
- *Zero Crossing Rate*; a zero-crossing is a point where the sign of the time-domain representation of the sound wave changes;
- *Roll Off* — the frequency below which an experimentally chosen percentage equal to 85% of the accumulated magnitudes of the spectrum is concentrated; parameter originating from speech recognition, where it is applied to distinguish between voiced and unvoiced speech;
- *NonMPEG7 - Audio Spectrum Centroid* — a linear scale version of *Audio Spectrum Centroid*;
- *NonMPEG7 - Audio Spectrum Spread* — a linear scale version of *Audio Spectrum Spread*;
- changes (measured as differences) of the above features for a 30 ms sub-frame of the given 40 ms frame (starting from the beginning of this frame) and the next 30 ms sub-frame (starting with 10 ms shift), calculated for all the features shown above;
- *Flux* — the sum of squared differences between the magnitudes of the DFT points calculated for the starting and ending 30 ms sub-frames within the main 40 ms frame; this feature by definition describes changes of magnitude spectrum, thus it is not calculated in a static version.

Mixes of the left and right channel were taken if the audio signal was stereophonic. Since the recognition of instruments is performed on frame-by-frame basis, no parameters describing the entire sound are present in our feature vector. This feature set was already used for instrument identification purposes using RFo, requiring no feature selection [7], and yielded good results, so we decided to use this feature set in both RFo and RFe classification.

3.1   Audio Data

The audio data we used for both training and testing represent recordings in 44.1kHz/16-bit format. Training was based on three repositories of single, isolated sounds of musical instruments, namely McGill University Master Samples [15], The University of Iowa Musical Instrument Samples [23], and RWC Musical Instrument Sound Database [3]. Clarinet, trombone, and trumpet sounds were taken from these repositories. Additionally, we used sousaphone sounds, recorded by R. Rudnicki in one of his recording sessions [20], since no sousaphone sounds were available in the above mentioned repositories. Training data were in mono format in the case of RWC data and sousaphone, and stereo for the rest of the data. The testing data originate from jazz band stereo recordings by R. Rudnicki [20], and include the following pieces played by clarinet, trombone, trumpet, and sousaphone (i.e., our target instruments):

- Mandeville by Paul Motian,
- Washington Post March by John Philip Sousa, arranged by Matthew Postle,
- Stars and Stripes Forever by John Philip Sousa, semi-arranged by Matthew Postle — Movement no. 2 and Movement no. 3.

To prepare our classifiers to work on larger instrument sets, training data also included sounds of 5 other instruments that can be encountered in jazz recordings: double bass, piano, tuba, saxophone, and harmonica. These sounds were added as additional sounds in training mixes with the target instruments.

4   Methodology of Training of the Classifiers

The goal of training of our classifiers is to identify plural classes, each representing one instrument. We use a set of binary classifiers (RFe or RFo), where each set (which we call a *battery*) is trained to identify whether a target instrument is playing in an audio frame or not. The target classes are clarinet, trombone, trumpet, and sousaphone, i.e. instruments playing in the analyzed jazz band recordings. The classifiers are trained to identify target instruments when they are accompanied by other instruments, and this is why we use mixes of instrument sounds as input data in training.

When preparing training data, we start with single isolated sounds of each target instrument. After removing starting and ending silence [7], each file representing the whole single sound is normalized so that the RMS value equals one. Then we perform parameterization, and train a classifier to identify each instrument — even when accompanied by other sound. Therefore, we perform training on 40 ms frames of instrument sound mixes, mixing from 1 to 4 randomly chosen instruments with random weights and then we normalize it again to get the RMS value equal to one.

The battery of one-instrument sensitive RFo or RFe classifiers is then trained. 3,000 mixes containing any sound of a given instrument are fed as positive examples, and 3,000 mixes containing no sound of this instrument are fed as negative examples. For N instruments we need N binary classifiers (N=4), each one trained to identify 1 instrument. For RFe models, we have been training 1000 ferns of a depth of 10; for RFo, there were 1000 trees and K was set to the default floor of square root of the number of attributes, namely 9.

5   Experiments and Results

The RFo and RFe classifiers, according to the procedure delineated in Section 4, were next used to identify instruments playing in jazz recordings, described in Section 3.1. Ground-truth data were prepared through careful manual labelling [7], based on initial recordings of each instrument track separately.

The accuracy was assessed via precision and recall scores. These measures were weighted by the RMS of a given frame (differently than in our previous work [7], where RMS was calculated for frames taken from instrument channels), in order to diminish the impact of softer frames, which are very hard to perform reasonable identification of instruments, because their loudness is near the noise level. For this reason, our true positive score T_p for an instrument i is a sum of RMS of frames which are both annotated and classified as i. Precision is calculated by dividing T_p by the sum of RMS of frames which are classified as i; respectively, recall is calculated by dividing T_p by the sum of RMS of frames which are annotated as i. As a general accuracy measure we have used F-score, defined as a harmonic mean of such precision and recall.

Table 1. Precision, recall and F-score of the classifiers for jazz band recordings. Each $M \pm S$ data entry represents mean M and standard deviation S over 10 replications of training and testing, accumulated over all target band instruments.

Algorithm	Precision [%]	Recall [%]	F-score [%]
Mandeville			
RFe	88.4±0.6	67±1	76.4±0.6
RFo	92.7±0.2	63±1	75.2±0.7
Washington Post			
RFe	82.36±0.2	73±2	77±1
RFo	87.76±0.3	69±1	77.3±0.5
Stars & Stripes 2			
RFe	79.8±0.4	72±1	76±1
RFo	91±2	68±1	78±1
Stars & Stripes 3			
RFe	94.5±0.2	77±1	84.8±0.7
RFo	94.4±0.3	74±1	83.1±0.9

While in this initial phase of the research we have used PC implementations of the classification algorithms, the timings have been performed on a single core of a Xeon E5620 Linux workstation. R version 2.15.0, rFerns version 0.3 and randomForest version 4.6-6 were used.

Both RFo and RFe are stochastic algorithms, so is the process of creating training sets for the battery. Thus, to assess the stability of the results and make a fair comparison of methods, the whole procedure of creating training sets, training RFe and RFo batteries and testing them on a real recordings has been repeated 10 times.

5.1   Comparison of Random Forests and Random Ferns

The results of performance analysis of RFe and RFo models are given in Table 1. As one can see, for three pieces RFo had superior precision over that of RFe; on the other hand, ferns tend to provide better recall. However, the overall performance of both classifiers measured with the F-score is similar for all pieces.

The detailed comparison of performance analysis of RFe and RFo models for particular instruments is given in Table 2. Sousaphone and trumpet are always quite precisely identified, whereas trombone usually yields lower precision in all pieces, and clarinet in one piece. Recall is lower than precision, but still much improved comparing to our previous results [7]. Again, quite high recall is obtained for sousaphone and is rather good for trumpet, whereas the worst recall is scored by RFo for trombone samples.

Table 2. Precision and recall of both methods on real music; data shown for each instrument independently. The symbol $M \pm S$ denotes that given number has mean M and standard deviation S over 10 replications of training.

	Precision [%]				Recall [%]			
	clarinet	sousaphone	trombone	trumpet	clarinet	sousaphone	trombone	trumpet
	Mandeville							
RFe	91.5±0.2	98.3±0.2	76±2	89.0±0.2	70±2	67±1	71±2	59±2
RFo	91.4±0.2	98.6±0.3	87.3±0.6	90.8±0.2	65±4	80±2	46±2	58±2
	Washington Post							
RFe	80.9±0.4	92.2±0.7	63.6±0.4	92.5±0.5	79±3	76±3	61±2	73±2
RFo	85±1	93.2±0.7	70.3±0.8	96.4±0.6	67±4	88±1	46±2	72±3
	Stars & Stripes 2							
RFe	48.4±0.4	99.4±0.1	78±2	97.8±0.3	81±2	70±4	58±2	77±2
RFo	62±6	99.4±0.1	91±2	99.9±0.1	53±5	94±1	31±3	67±3
	Stars & Stripes 3							
RFo	06.0±0.6	00.2±0.2	88.6±0.6	04.7±0.2	02±2	62±4	61±2	88±1
RFo	95.4±0.4	99.7±0.1	87.8±0.7	94.7±0.1	80±5	83±4	50±2	88±1

On average, RFe and RFo perform classification respectively 25x and 8x faster than the actual music speed; this means RFe offer over 3x speed-up in comparison to RFo, see Table 3.

Table 3. Time taken by each battery of classifiers to annotate the whole testing piece

Piece	Classification time [s]		Piece length [s]
	Random Ferns	Random Forest	
Mandeville	5.7	17.6	139.95
Washington Post	6.0	19.0	148.45
Stars & Stripes 2	2.8	8.3	68.95
Stars & Stripes 3	1.0	3.6	26.2

6   Summary and Conclusions

Experiments presented in this paper show that identification of all instruments playing in real music recordings is possible using both RFo- and RFe-based classifiers, yielding quite good results. We observed improved recall comparing to our previous research [7]; we improved here the RMS weighting, which was previously calculated for separate instrument channels, and in this work, the RMS of all channels together was used for weighting. Our results still are worth improving, but the obtained recall (and precision) are satisfactory, because the task of identification of all instruments playing in a short segment is difficult, and is challenging also for human listeners.

The measured classification speed of RFe suggests that it is a promising method for performing real time annotation, even on low performance devices.

Acknowledgments. This work has been partially financed by the National Science Centre, grant 2011/01/N/ST6/07035. Computations were performed at ICM, grant G48-6. This project was also partially supported by the Research Center of PJIIT, supported by the Polish Ministry of Science and Higher Education. The authors would also like to thank Dr. Elżbieta Kubera from the University of Life Sciences in Lublin for preparing the ground-truth data for initial experiments, and Radosław Rudnicki from the University of York for preparing the jazz band recordings.

References

1. Bosch, A., Zisserman, A., Munoz, X.: Image Classification using Random Forests and Ferns. In: 2007 IEEE 11th International Conference on Computer Vision, pp. 1–8. IEEE (2007)
2. Breiman, L.: Random Forests. Machine Learning 45, 5–32 (2001)
3. Goto, M., Hashiguchi, H., Nishimura, T., Oka, R.: RWC Music Database: Music Genre Database and Musical Instrument Sound Database. In: Proceedings of ISMIR, pp. 229–230 (2003)
4. Herrera-Boyer, P., Klapuri, A., Davy, M.: Automatic Classification of Pitched Musical Instrument Sounds. In: Klapuri, A., Davy, M. (eds.) Signal Processing Methods for Music Transcription. Springer Science+Business Media LLC (2006)
5. ISO: MPEG-7 Overview, http://www.chiariglione.org/mpeg/

6. Kitahara, T., Goto, M., Komatani, K., Ogata, T., Okuno, H.G.: Instrument Identification in Polyphonic Music: Feature Weighting to Minimize Influence of Sound Overlaps. EURASIP J. on Advances in Signal Processing 2007, 1–15 (2007)
7. Kubera, E., Kursa, M.B., Rudnicki, W.R., Rudnicki, R., Wieczorkowska, A.A.: All That Jazz in the Random Forest. In: Kryszkiewicz, M., Rybinski, H., Skowron, A., Raś, Z.W. (eds.) ISMIS 2011. LNCS (LNAI), vol. 6804, pp. 543–553. Springer, Heidelberg (2011)
8. Kursa, M.B., Rudnicki, W., Wieczorkowska, A., Kubera, E., Kubik-Komar, A.: Musical Instruments in Random Forest. In: Rauch, J., Raś, Z.W., Berka, P., Elomaa, T. (eds.) ISMIS 2009. LNCS (LNAI), vol. 5722, pp. 281–290. Springer, Heidelberg (2009)
9. Kursa, M.B., Kubera, E., Rudnicki, W.R., Wieczorkowska, A.A.: Random Musical Bands Playing in Random Forests. In: Szczuka, M., Kryszkiewicz, M., Ramanna, S., Jensen, R., Hu, Q. (eds.) RSCTC 2010. LNCS (LNAI), vol. 6086, pp. 580–589. Springer, Heidelberg (2010)
10. Kursa, M.B.: Random ferns method implementation for the general-purpose machine learning (submitted, 2012), http://arxiv.org/abs/1202.1121v1
11. Liaw, A., Wiener, M.: Classification and Regression by randomForest. R News 2(3), 18–22 (2002)
12. Niewiadomy, D., Pelikant, A.: Implementation of MFCC vector generation in classification context. J. Applied Computer Science 16(2), 55–65 (2008)
13. R Development Core Team: R: A Language and Environment for Statistical Computing (2010), http://www.r-project.org/
14. MIDOMI: Search for Music Using Your Voice by Singing or Humming, http://www.midomi.com/
15. Opolko, F., Wapnick, J.: MUMS — McGill University Master Samples. CD's (1987)
16. Oshin, O., Gilbert, A., Illingworth, J., Bowden, R.: Action Recognition Using Randomised Ferns. In: 2009 IEEE 12th International Conference on Computer Vision Workshops (ICCV Workshops), pp. 530–537. IEEE (2009)
17. Özuysal, M., Fua, P., Lepetit, V.: Fast Keypoint Recognition in Ten Lines of Code. In: 2007 IEEE Conference on Computer Vision and Pattern Recognition. IEEE (2007)
18. Özuysal, M., Calonder, M., Lepetit, V., Fua, P.: Fast Keypoint Recognition using Random Ferns. Image Processing (2008), http://dx.doi.org/10.1109/TPAMI.2009.23
19. Raś, Z.W., Wieczorkowska, A.A. (eds.): Advances in Music Information Retrieval. SCI, vol. 274. Springer, Heidelberg (2010)
20. Rudnicki, R.: Jazz band. Recording and mixing. Arrangements by M. Postle. Clarinet — J. Murgatroyd, trumpet — M. Postle, harmonica, trombone — N. Noutch, sousaphone – J. M. Lancaster (2010)
21. Shazam Entertainment Ltd., http://www.shazam.com/
22. Shen, J., Shepherd, J., Cui, B., Liu, L. (eds.): Intelligent Music Information Systems: Tools and Methodologies. Information Science Reference, Hershey (2008)
23. The University of Iowa Electronic Music Studios: Musical Instrument Samples, http://theremin.music.uiowa.edu/MIS.html
24. TrackID — Sony Smartphones, http://www.sonymobile.com/global-en/support/faq/xperia-x8/internet-connections-applications/trackid-ps104/
25. Wagner, D., Reitmayr, G., Mulloni, A., Drummond, T., Schmalstieg, D.: Real-time Detection and Tracking for Augmented Reality on Mobile Phones. IEEE Transactions on Visualization and Computer Graphics 16(3), 355–368 (2010)

Advanced Searching
in Spaces of Music Information*,**

Mariusz Rybnik[1], Wladyslaw Homenda[1,2], and Tomasz Sitarek[3]

[1] Faculty of Mathematics and Computer Science, University of Bialystok,
ul. Sosnowa 64, 15-887 Bialystok, Poland
MariuszRybnik@wp.pl
[2] Faculty of Mathematics and Information Science, Warsaw University of Technology,
Plac Politechniki 1, 00-660 Warsaw, Poland
homenda@mini.pw.edu.pl
[3] PhD Studies, Systems Research Institute, Polish Academy of Sciences,
ul. Newelska 6, 01-447 Warszawa, Poland
Tomasz.Sitarek@ibspan.waw.pl

Abstract. This study continues construction of specialized grammars covering music information. The proposed grammar is searching-oriented and therefore largely simplified, in order to create easily-searchable structure. Searching is seen here as a particular querying operation in spaces of music information. Searching is discussed for patterns melodically and rhythmically transformed - with transformations in pitch and time dimensions typical for music works. Three operators are proposed to provide convenient meta-data for searching. Searching may be performed in both structured and unstructured musical pieces, regarding voice selection. Proposed methodology may serve for searching of transformed and non-transformed motives, searching of inspirations, comparative analysis of musical pieces, analysis of melodic and rhythmical sequences, harmonic analysis and structure discovery.

Keywords: music information, music representation, syntactic structuring, semantic analysis, querying, searching, knowledge discovery, knowledge understanding.

1 Introduction

The paper is structured as follows. Preliminary information on the subject of the study is included in subsection 1.1. Basic notions of mathematical linguistics are given in subsection 1.2 and concepts of syntax, semantics and understanding are introduced in subsection 1.3. Section 2 presents the proposed simplified

---

* This work is supported by The National Center for Research and Development, Grant no N R02 0019 06/2009.
** Tomasz Sitarek contribution is supported by the Foundation for Polish Science under International PhD Projects in Intelligent Computing. Project financed from The European Union within the Innovative Economy Operational Programme (2007–2013) and European Regional Development Fund.

L. Chen et al. (Eds.): ISMIS 2012, LNAI 7661, pp. 218–227, 2012.

searching-oriented grammar and three operators for detection of melodic and rhythmical transformations. Section 3 discuses advanced searching in spaces of music information with the use of proposed operators and grammar. Section 4 presents an example of advanced searching. Finally, Section 5 concludes the paper and proposes future work.

1.1   The Subject

In this paper we analyze structured spaces of music information. The study is focused on advanced searching operations related to querying. The ideas may be adopted to other methods of describing music information, e.g. Music XML [2], Braille music description [9], MIDI [10] etc. The subject is a continuation of the new issue of "automatic image understanding" raised by R. Tadeusiewicz a few years ago, c.f. [11,12].

1.2   Grammars and Languages

The discussion is based on common definition of grammars and context-free grammars. Let us recall that a system $G = (V, T, P, S)$ is a grammar, where: (a) V is a finite set of *variables* (called also *nonterminals*), (b) T is a finite set of terminal symbols (simply called *terminals*), (c) a nonterminal S is the initial symbol of the grammar and (d) P is a finite set of productions. A pair (α, β) of strings of nonterminals and terminals is a production assuming that the first element α of the pair is a nonempty string. Production is usually denoted $\alpha \to \beta$. Grammars having all productions with α being a nonterminal symbols are context-free grammars.

A derivation in a grammar is a finite sequence of strings of nonterminals and terminals such that: (a) the first string in this sequence is just the initial symbol of the grammar and (b) for any two consecutive strings in the sequence, the latter one is obtained from the former one using a production, i.e. by replacing a substring of the former one equal to the left hand side of the production with the right hand side of it. We say that the last element of the string is *derivable* in the grammar.

For a context-free grammar a derivation can be outlined in a form of derivation tree, i.e. (a) the root of the tree is labeled with the initial symbol of the grammar and (b) for any internal vertex labeled by the left side of a production, its children are labeled by symbols of the right side of the production.

Finally, the set of all terminal strings derivable in a given grammar is the language generated by the grammar. In this paper, terminal strings generated by a grammar are called words, sentences or texts while parts of such units are named phrases. Units of music notation are called scores. Parts of scores have their domain-dependent names (from music theory) that will be used to easily refer subjects, even though they may not be products of the proposed grammar.

1.3 Syntax, Semantics, Understanding

The notion *querying* raises two fundamental aspects: *what (is queried)* and *how (it is queried)*. Firstly, conscious querying (n.b. conscious querying is a case of conscious communication) is associated with understanding *what* is queried, i.e. understanding queried information in terms of its structure and possible meaning. Secondly, queries are expressed and communication is carried out in some language. Consequently, constructions of the language express the aspect *how (it is queried)*.

Understanding, as recognized in this paper, is an ability to identify concepts in the real world, i.e. objects and sets of objects in the world described by constructions of a given language. The term *syntax* is used in the meaning of structuring constructions of the language. *Semantics* is a mapping or relation, which casts constructions of the language on objects and local and global structures of objects of the real world. Therefore, ability to recognize the semantics, i.e. to identify such a mapping or relation, is a denotation of understanding.

Music notation is a language of communication. There is no convincing proof that music notation is or is not a context-free language. Music notation includes constructions of the form *ww* (e.g. repetitions), which are context sensitive ones. Consequently, it seems formally that music notation is a context sensitive language, c.f. [8]. Therefore, a context-free description of music notation is not possible. However, context sensitive methods, which would be utilized in precise description, are not explored enough for practical applications. On the other hand, even if music notation is a context-free language, its complexity does not allow for practical context-free description. For these reasons an effort put in precise formal description of music notation would not be reasonable. Instead we attempt to construct a context-free grammar covering music notation. The term *covering music notation* is used not only in the sense of generating all valid music notation constructions, but also not valid ones. This is why sharp syntactical analysis of music notation is not done.

Usage of a simplified context-free grammar for the purpose of syntactical structuring of music notation is valid in practice. The grammar will be applied in analysis of constructions, which are assumed to be well grounded pieces of music notation. Of course, such a grammar can neither be applied in checking correctness of constructions of music notation, nor in generation of such constructions.

2 Searching-Oriented Grammar

In this paper we modify the *graphically oriented grammar* proposed in our previous work [7] for *paginated music notation*. The simplified grammar is oriented at advanced searching, therefore we call it *searching-oriented grammar*. The searching regards melodic transpositions and rhythmic transformations, typical to music spaces of information, and disregard any graphical aspects of music notation.

Therefore, for the purpose of *searching-oriented grammar* we propose to omit most graphical components typical to *paginated music notation* (e.g. page, stave, barline, clef). Searching is much easier in continuous *notes* space, with no artificial divisions, therefore traditionally fundamental $<measure>$ tag is also disregarded (however in Fig. 2 *measures* are included to maintain readability). For the same reason we propose to convert indispensable music notation symbols connected with *stave* or *measures* into *note* attributes and disregard the unnecessary ones. Symbols from *paginated music notation* referring to a single *note* or a group of *notes* (e.g. articulation: *staccato* dots or *legato* slurs) are maintained as *note* attributes. They could be also possibly treated as a part of searching pattern.

In order to simplify searching in space of music information three new operators are proposed, described in detail in the sections below. We propose to append to *note* the values produced by the operators as additional properties useful in searching.

2.1 Halftone Difference Operator

We introduce an operator designated **halftone difference** $(d_{1/2})$ that defines the difference of *the current note* to *the previous note*, expressed in halftones. The $d_{1/2}$ would be defined for all notes in particular voice except for the first one. Operator $d_{1/2}$ is very similar to *melodic enharmonic interval* in music theory.

The exemplary values (along two other operators) are depicted in Fig. 1. Boxes represent identical melodic patterns (3 notes) for $d_{1/2}$ and d_s and identical rhythmic pattern (4 notes) for q_{rh}.

Fig. 1. Three proposed operators used on a motive from Promenade (Pictures at an Exhibition) by Modest Mussorgsky (please note the lack of measures)

2.2 Scalar Difference Operator

Similarly we introduce an operator designated **scalar difference** (d_s) that defines the difference of *the current note* to *the previous note*, expressed in diatonic scale degrees. The d_s is defined for all notes in particular voice except for the first one, providing that the key (scale) is defined. Please note that the d_s value could be also a fraction for *accidental note* (a note outside of defined scale). Operator d_s is similar to *melodic diatonic interval* in music theory (but more convenient).

2.3   Rhythm Ratio Operator

In order to efficiently process searching for rhythm transformed patterns we propose to introduce an operator designated **rhythm ratio** (q_{rh}) defined as the *ratio* of the current note duration to the previous note duration. The q_{rh} would be defined for all notes in particular voice except for the first one.

2.4   Context-Free Productions of Proposed Grammar

A raw description of the *searching-oriented music notation* could be approximated by the set of context-free productions given below. Components of the grammar $G = (V, T, P, S)$ are as follows. The set of nonterminals includes all identifiers in triangle brackets printed in italic. The nonterminal *<score>* is the initial symbol of G. The set of terminals includes all non bracketed identifiers.

<score>　　　→ *<score_part> <score>* | *<score_part>*
<score_part> → *<voice> <score_part>* | *<voice>*
<voice>　　　→ *<chord> <voice>* | note *<voice>* | rest *<voice>* | *<voice>*
<chord>　　　→ note *<chord>* | note

Grammar Description and Comments: *<score>* may consist of one or several elements of type *<score_part>*, that represent subsequent parts of score in time dimension. It is preferable that *<score_part>* would be maintained in the constant *key*, as searching (when including diatonic transpositions of pattern) may depend on *scale*. Each *<score_part>* consists of one or several elements of type *<voice>*, that represent voices (in the sense of *instrumental parts, polyphonic voices*, leading motive, accompaniment, etc.), performed simultaneously for a given *<score_part>*. *<voice>* may contain terminals **note**, **rest** or non-terminal *<chord>*, that represent subsequent vertical events in time dimension. Definition of *<chord>* cannot be strict in practice, as it may contain notes of various durations and even rests (depending on the strictness of *<voice>* definition). Terminal **rest** contains obligatory attribute *duration*. Terminal **note** contains obligatory attributes *pitch* and *duration* and may contain non-obligatory attributes, for example related to articulation or dynamics. We also propose to include metadata attributes: values of proposed operators: d_s, $d_{1/2}$ and q_{rh}.

On Conversion from Paginated Music Notation to Searching-Oriented Notation: *paginated grammar* is richer than the proposed *searching-oriented grammar*, therefore it is easy to convert *paginated derivation tree* into *searching-oriented derivation tree*. Reverse operation is not directly possible, as it would require formatting of 'raw' voices into its richer graphical representation. In practice it is preferable to link both musical piece representations: paginated for graphical presentation and searching-oriented for potent searches.

　　Music notation can be described by different grammars. Construction of such grammars may reflect various aspects of spaces of music notation, e.g. graphical or logical structuring, c.f. [4]. The above description is constructed regarding raw musical structure, omitting any information unnecessary in the searching context. In order to increase readability some details are skipped. For instance, articulation and ornamentation symbols, dynamics, are not outlined. Expansion of the grammar with these elements (if necessary) is not difficult, we suggest however to include them as **note** attributes, in order to maintain the easy-searchable structure of information.

3   Advanced Searching in Spaces of Music Information

Searching is an operation of locating instance(s) matching a given pattern, i.e. locating instance(s) identical or similar to the pattern. The operation *Search* in the space of music information concerns a *pattern*, which is a structure of music information. Searched pattern is usually a result of another non-trivial operation: *selection*.

Searching in more general context could be seen as a particular *querying* operation. According to the Cambridge Dictionaries Online *query* is *a question, often expressing doubt about something or looking for an answer from an authority*, c.f. [3]. In this paper we assume that *answer from an authority* is also understood as accomplishment of an operation for a given request. *Querying* in spaces of music information could signify operations like: selecting, searching, copying, replacing, pasting, transposing etc. In this work we are interested in advanced searching operations with regard to melody transpositions and rhythm transformations. Please note that the searching for transformed patterns may occur in the same <*voice*>, for different <*voice*> derivation branches and for different <*score*> derivation trees (representing musical pieces).

3.1   Melodic Transformations

This section discuses melodic transformation and searches in this dimension using two introduced operators d_s and $d_{1/2}$. For the discussion we propose the following melodic transformations taxonomy:

1. exact melodic match;
2. *ottava*-type transposition (*all' ottava, all' ottava bassa*, etc.) - a particular case of transposition;
3. chromatic transposition (maintaining the exact number of halftones between the corresponding notes);
4. diatonic transposition (maintaining the diatonic intervals, what could result in slight change of the number of halftones between the corresponding notes);
5. irregular melodic transformations (of variation type).

Exact melody match is detected with a given d_s (or $d_{1/2}$) sequence and at least one corresponding note being identical. Please note that due to lack of artificial divisions (alike measures, staves, pages, etc.) it is relatively easy to query the structure of information. Longer motives could also start in different *moments* of measure, what is natural for the proposed representation.

Chromatic Transposition. With $d_{1/2}$ it is very easy to detect *chromatic transposition*, as sequence of identical values would signify chromatically transposed melody fragments.

Diatonic Transposition. With d_s it is very easy to detect *scalar transposition*, as identical sequences would signify similar melody fragments with regard to the defined *key* (the exact number of halftones can vary however).

Ottava-type Transpositions. are detected with a given d_s (or $d_{1/2}$) sequence and at least one corresponding *note* name (excluding *octave designation*) being identical. This is an alternative to matching of *note names* sequences. It is more potent approach as *ottava*-type transpositions could be detected during general transposition searches.

Irregular melodic transformations (as for example *tonal answer* in fugue) may be detected with the use of similarity measures for the compared sequences. This is however a larger topic out of the scope of this work.

3.2   Rhythm Transformations

This section discuses rhythmic transformation and searches in *duration dimension* using introduced operator $q_r h$. For the discussion we propose the following rhythmic transformations taxonomy:

1. exact match
2. diminution - a melodic pattern is presented in shorter note durations than previously used (usually twice);
3. augmentation - a melodic pattern is presented in longer note durations than previously used (usually twice);
4. irregular rhythmic transformations (of variation type) - out of the scope of this work.

Exact rhythmical match is detected with a given q_{rh} sequence and at least one corresponding duration being identical.

Diminutions and Augmentations. With q_{rh} it is very easy to detect *diminutions* as well as *augmentations* of any kind, as sequence of identical q_{rh} values would signify identical rhythm dependencies.

Irregular rhythm transformations may be detected with the use of similarity measures for the compared sequences. This is however a large topic out of the scope of this work.

3.3 Generalization of Proposed Operators

The three proposed operators are the most potent for a single voice line, however they are applicable as well to chords and even whole musical pieces. Chords sequences could be processed in the following ways:

– **parallel calculations**: assuming the equal number of notes in each chord, operators could be determined in parallel. That would be useful for frequent *thirds* or *sixths* sequences.
– **each-to-each calculations**: each-to-each relations are determined and stored. That increases the amount of data to be generated and could result in arbitrary false searching matches.

Advantage of the *each-to-each calculations*: the whole musical piece could be analyzed at once, disregarding voice selection. It could be very useful for missing, incomplete or erroneous voice data (resulted frequently from automatic transcription or conversion) and may serve as meta-information for discovering structures.

4 Example

In this section we present an example of advanced searching with a short analysis of operators' values and possible continuation (Fig. 2).

Fig. 2. Beginning excerpt of Contrapunctus VII, Art of the Fugue - J.S. Bach, with operators' values

The presented fragment consists of two easily-separable voices. The values of proposed three operators are calculated separately for each voice. It is an example of a polyphonic Baroque form, called fugue. Upper voice presents the *subject* (main theme), and then the lower voice comes with *the tonal answer* (*the subject* slightly altered, in order to maintain harmonic relations) in *diminution*. Operator q_{rh} shows that the rhythmic relations inside both voices are identical. Due to initial alteration of *answer* operator $d_{1/2}$ matches all notes except the first, signifying that the patterns are in large part chromatically transposed. The *accidental note b* in the answer (a result of maintaining the answer in *a-minor* harmonics) cause some of the d_s values to be fractions and match only later part of the patterns.

As one can see the operators can partially detect modified patterns, however in order to fully match them a measure of similarity of the operator-generated sequences should be defined. Such similarity measure could match the similar melody operators' values at the patterns' beginning. It may even employ harmony knowledge in order to relate *fourth* and *fifth* melodic intervals at the beginning, or detect the harmonic relation of *the subject* (d-minor) and *answer* (a-minor).

5   Conclusions and Future Work

In this paper we simplify the *graphically oriented grammar* for paginated music notation, proposed in our previous work [7]. The simplified grammar is oriented at advanced searching operations, that include melodic transpositions and rhythmic transformations. In order to efficiently search in this particular space of information, we propose three operators that describe relations between neighboring notes, regarding melody and rhythm. The resulting values are attached to notes as properties to use in searches. Exemplary search is shown in section 4.

Applications of proposed *searching-oriented grammar*:

- straightforward **searching of *leitmotifs*** regarding melody, rhythm or both, with possible transformations;
- **searching of inspirations**: comparative analysis of musical pieces;
- **rhythmical analysis**: statistical analysis of musical piece regarding used rhythmical figures, with possible transformations;
- **melodic analysis**: statistical analysis of musical piece regarding used melodic sequences, with possible transpositions;
- **harmonic analysis** could benefit from the particular melodic figures (defined and searched easily using operators d_s or $d_{1/2}$), e.g. Chopin chord resolution (third down to base - d_s value equal to -2); or *tonicization* resolution of third into prima (minor second up - $d_{1/2}$ value equal to 1). It is also possible to analyze chords movements and therefore detect harmonic relations.
- **automated searches in whole musical pieces** - disregarding voice selection - could be performed by using generalizations of proposed operators (as described in section 3.3). The melodic and rhythmic analysis using proposed operators could be used to detect the structure of musical work e.g. divide it into separate voices of consistent constitution.

This study continues construction of specialized grammars covering music information. The grammar given in section 2 is *searching-oriented* while other orientation could be more suitable in some applications. Future work in this domain include: a) developing details of another specialized grammars, b) research on semantics, i.e. developing methods of construction of valuation relation, as a key issue in automation of querying, c) development of *Select* and *Replace* operations for music notation, d) development of searching operators in order to detect irregular (but similar to original pattern) transformations, e) studying inherent imperfectness of music information, i.e. incompleteness, uncertainty and incorrectness, hidden under the level of syntax and semantics, f) analysis of rhythmical and melodic sequences for knowledge discovery.

References

1. Bargiela, A., Homenda, W.: Information structuring in natural language communication: Syntactical approach. Journal of Intelligent & Fuzzy Systems 17, 575–581 (2006)
2. Castan, G., et al.: Extensible Markup Language (XML) for Music Applications: An Introduction. In: Hewlett, W.B., Selfridge-Field, E. (eds.) The Virtual Score: Representation, Retrieval, Restoration. The MIT Press (2001)
3. Collins Cobuild English Language Dictionary, http://dictionary.cambridge.org/
4. Homenda, W.: Breaking Accessibility Barriers: Computational Intelligence in Music Processing for Blind People. In: Sordo, M., Vaidya, S., Jain, L.C. (eds.) Advanced Computational Intelligence Paradigms in Healthcare - 3. SCI, vol. 107, pp. 207–232. Springer, Heidelberg (2008)
5. Homenda, W.: Automatic Data Understanding: A Necessity of Intelligent Communication. In: Rutkowski, L., Scherer, R., Tadeusiewicz, R., Zadeh, L.A., Zurada, J.M. (eds.) ICAISC 2010, Part II. LNCS (LNAI), vol. 6114, pp. 476–483. Springer, Heidelberg (2010)
6. Homenda, W., Sitarek, T.: Notes on Automatic Music Conversions. In: Kryszkiewicz, M., Rybinski, H., Skowron, A., Raś, Z.W. (eds.) ISMIS 2011. LNCS (LNAI), vol. 6804, pp. 533–542. Springer, Heidelberg (2011)
7. Homenda, W., Rybnik, M.: Querying in Spaces of Music Information. In: Tang, Y., Huynh, V.-N., Lawry, J. (eds.) IUKM 2011. LNCS, vol. 7027, pp. 243–255. Springer, Heidelberg (2011)
8. Hopcroft, J.E., Ullman, J.D.: Introduction to Automata Theory, Languages and Computation. Addison-Wesley Publishing Company (1979, 2001)
9. Krolick, B.: How to Read Braille Music, 2nd edn. Opus Technologies (1998)
10. MIDI 1.0, Detailed Specification, Document version 4.1.1 (February 1990)
11. Tadeusiewicz, R., Ogiela, M.R.: Automatic Image Understanding a New Paradigm for Intelligent Medical Image Analysis. Bioalgorithms and Med-Systems 2(3), 5–11 (2006)
12. Tadeusiewicz, R., Ogiela, M.R.: Why Automatic Understanding? In: Beliczynski, B., Dzielinski, A., Iwanowski, M., Ribeiro, B. (eds.) ICANNGA 2007, Part II. LNCS, vol. 4432, pp. 477–491. Springer, Heidelberg (2007)

Mood Tracking of Musical Compositions

Jacek Grekow

Faculty of Computer Science, Bialystok University of Technology,
Wiejska 45A, Bialystok 15-351, Poland
j.grekow@pb.edu.pl

Abstract. This paper presents a new strategy for the analysis of emotions contained within musical compositions. We present a method for tracking changing emotions during the course of a musical piece. The collected data allowed to determine the dominant emotion in the musical composition, present emotion histograms and construct maps visualizing the distribution of emotions in time. The amount of changes of emotions during a piece may be different, therefore we introduced a parameter evaluating the quantity of changes of emotions in a musical composition. The information obtained about the emotion in a piece made it possible to analyze a number of pieces, in particular the Sonatas of Ludwig van Beethoven. This analysis has provided new knowledge about the compositions and the method of their emotional development.

Keywords: Emotion detection, Mood tracking, Music visualization.

1 Introduction

Listening to music is a particularly emotional activity [1]. People need a variety of emotions and music is perfectly suited to provide them. However, it turns out that musical compositions do not contain one type of emotion, e.g. only positive or only negative. During the course of one composition, these emotions can take on a variety of shades, change several times with varying intensity. This paper presents a new strategy for the analysis of emotions contained within musical compositions. We present a method for tracking changing emotions during the course of a musical piece. The collected data allowed to determine the dominant emotion in the musical composition, present emotion histograms and construct maps visualizing the distribution of emotions in time.

There are several other studies on the issue of mood tracking. Lu et al. [2], apart from detecting emotions, tracked them, and divided the music into several independent segments, each of which contains a homogeneous emotional expression. Using labels collected through the game MoodSwings, Schmidt et al. [3], [4] tracked the changing emotional content of music. Myint and Pwint [5] presented self-colored music mood segmentation and a hierarchical framework. The use of mood tracking for indexing and searching multimedia databases has been used in the work of Grekow and Ras [6]. The issue of mood tracking is not only limited to musical compositions. The paper by Mohammad [7] is an interesting

L. Chen et al. (Eds.): ISMIS 2012, LNAI 7661, pp. 228–233, 2012.

extension of the issue; the author investigated the development of emotions in literary texts. Also Yeh et al. [8] tracked the continuous changes of emotional expressions in Mandarin speech.

2   System Construction

The proposed system for tracking emotions in a musical composition is shown in Figure 1. It consists of a database of musical compositions, composition segmentation and result presentation module. The segmentation module was combined with classifiers of an external emotion detection system, which was described in a previous paper [9]. The resulting emotion labels were used to designate the consecutive segments of a musical composition. The collected data allowed for analysis of a musical composition in terms of the emotions contained therein.

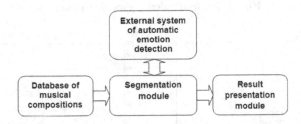

Fig. 1. Construction of the emotion tracking system

3   Mood Tracking

The model we chose in this work is based on Thayer's model [10]. Following its example, we created a hierarchical model of emotions consisting of two levels, L1 and L2 (Fig. 2).

The detection of emotion was conducted in our research on six-second segments. Each consecutive segment was shifted by 2 seconds. In this way, successive segments overlapped at a 2/3 ratio. This allowed to exactly track and detect even the slightest change of emotion in the examined musical composition. For a musical composition lasting $T = 120$ seconds, $N = 60$ segments ($S1, S2, ..., S59, S60$) were analyzed, and for each L1 and L2 level of emotion detection was performed.

4   Results of Mood Tracking

4.1   Emotion Histograms of a Musical Composition

The first method used for presenting the distribution of emotions in a musical composition is emotion histograms (Fig. 3a and Fig. 3b). On the presented graphs, the horizontal axis corresponds to the type of emotion, and the height of the bar indicates how often a specific emotion occurred. Figure 3a presents the histogram of

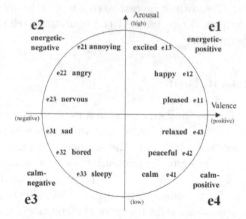

Fig. 2. Arousal-valence emotion plane

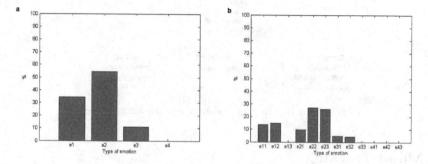

Fig. 3. Histogram of L1 (a) and L2 (b) level emotions in Appassionata sonata, part 1

L1 level emotions in Ludwig van Beethoven's Appassionata sonata. In this musical composition, emotion e2 (energetic-negative) occurs in more than 50% of the segments and is dominant. The second, most significant, emotion is e1 (energetic-positive). Notice that emotion e4 (calm-positive) does not occur at all.

Figure 3b presents the histogram of L2 level emotions in L.v.Beethoven's Appassionata sonata. Analyzing it and comparing it with the L1 level histogram (Fig. 3a), you can see that the emotions of the second quarter of Thayer's model (e2) that occur in this musical composition are e22 (angry) and e23 (nervous), and sub-emotions of the first quarter (e1) are e11 (pleased) and e12 (happy). The percentages of independent L2 level emotions are reduced but also dominant in the piece.

4.2   Emotion Maps

Another method used to analyze the emotion in a musical composition is detailed maps showing the distribution of emotion for the duration of the piece (Fig. 4a and Fig. 4b). The horizontal axis shows the time in seconds and the vertical axis the emotions occurring at a given moment.

On Fig. 4a, presenting a map of L1 level emotions for L.v.Beethoven's Appassionata sonata, you'll notice that e2 is dominant throughout the entire piece with the exception of the central parts (s. 280-310). From the map, you can see when and which emotions occur simultaneously. For example, the beginning of the piece (s. 0-10) is a combination of emotions e2 (energetic-negative) and e3 (calm-negative), and the end (s. 510-540) is a mixture of emotions e2 (energetic-negative) and e1 (energetic-positive). By analyzing the map of L2 level emotions for L.v.Beethoven's Appassionata sonata (Fig. 4b), you can notice not only the detailed distribution of emotions but also the emotional structure of the piece composed of four sections.

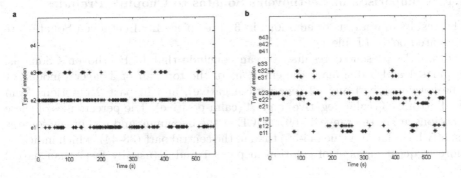

Fig. 4. Map of L1 (a) and L2 (b) level emotions in Appassionata sonata, part 1

4.3   Quantity of Changes of Emotion

Because some pieces may have many emotional changes (e.g., songs of varying moods) while others may be based on a single, dominant emotion (e.g. musical compositions with a steady pace, dynamics and rhythm etc.), we introduced the quantity of changes of emotion (QCE) in a musical composition.

$$QCE = \frac{\sum_{i=1}^{N-1} f(i)}{N} * 100 \tag{1}$$

$$f(i) = \begin{cases} 1, & \text{if } Emotion(i) \neq Emotion(i+1) \\ 0, & \text{if } Emotion(i) = Emotion(i+1) \end{cases} \tag{2}$$

where i is the number of the segment in the piece, N the number of segments in the composition and $Emotion(i)$ represents the emotion of the i segment. The function $f(i)$ indicates whether the adjacent segments have a different (value 1) or same (value 0) emotion. The more changes of emotion in a musical composition, the greater the QCE value.

Table 1. List of L. v. Beethoven's Sonatas with the dominant emotion and the QCE

Piece	QCE	Dominating emotion / in percentage
Appassionata, part 1	53.38	e2 / 55%
Appassionata, part 2	44.28	e3 / 54%
Appassionata, part 3	46.67	e2 / 69%
Waldstein, part 1	36.71	e2 / 54%
Waldstein, part 2	36.89	e3 / 60%
Waldstein, part 3	44.16	e1 / 44%
Pathetique, part 1	50.97	e2 / 45%
Pathetique, part 2	38.51	e4 / 44%
Pathetique, part 3	55.08	e1 / 47%

4.4   Comparison of Beethoven's Sonatas to Chopin's Preludes

The results of analysis of emotions in 3 three-piece L.v.Beethoven Sonatas are presented below (Table 1).

From the presented results, we can conclude that in Beethoven's Sonatas, in parts 1 and 3 dominate the emotions of the top half of Thayer's model: e1 (energetic-positive) and e2 (energetic-negative), and in part 2 emotions from the bottom: e3 (calm-negative) and e4 (calm-positive). The percentages of these emotions are in the range 44-69%. QCE at the beginning and end of the sonatas usually has a larger value (44-55) than in the central part (38-44), which indicates more frequent changes of emotion in parts 1 and 3 (usually fast) than part 2 (slow).

Table 2. List of Chopin's Preludes with the dominant emotion and the QCE

Piece	QCE	Dominating emotion / in percentage
Prelude No.1	27.27	e2 / 77%
Prelude No.5	8.33	e2 / 100%
Prelude No.7	30.00	e4 / 100%
Prelude No.8	21.95	e2 / 97%
Prelude No.15	43.97	e4 / 57%
Prelude No.18	33.33	e2 / 84%
Prelude No.19	3.57	e1 / 100%
Prelude No.22	35.29	e2 / 80%
Prelude No.24	41.07	e2 / 57%

Comparing Chopin's Preludes (Table 2) to Beethoven's Sonatas (Table 1), you can notice that the dominant emotion percentages are much higher in the Preludes (57-100%) than in the Sonatas (44-69%). Also, the quantity of changes of emotion (QCE) in the Preludes posses smaller values than the Sonatas. Based on these results, we can state that Chopin's Preludes are more emotionally homogeneous with a greater dominance of individual emotions and Beethoven's Sonatas are more diverse emotionally.

5   Conclusion

This paper presents a new strategy for the analysis of emotions contained within musical compositions. We present a method for tracking changing emotions during the course of a musical piece. The collected data allowed to determine the dominant emotion in the musical composition, present emotion histograms and construct maps visualizing the distribution of emotions in time. Emotional analysis of musical pieces can be developed in the future through a search for new parameters describing the changes of emotion in a piece, as well as by expanding the collection of studied compositions of various composers.

Acknowledgments. This paper is supported by the S/WI/5/08.

References

1. Pratt, C.C.: Music as the language of emotion. The Library of Congress (1950)
2. Lu, L., Liu, D., Zhang, H.J.: Automatic mood detection and tracking of music audio signals. IEEE Transactions on Audio, Speech and Language Processing 14(1), 5–18 (2006)
3. Schmidt, E.M., Turnbull, D., Kim, Y.E.: Feature Selection for Content-Based, Time-Varying Musical Emotion Regression. In: Proc. ACM SIGMM International Conference on Multimedia Information Retrieval, Philadelphia, PA (2010)
4. Schmidt, E.M., Kim, Y.E.: Prediction of time-varying musical mood distributions from audio. In: Proceedings of the 2010 International Society for Music Information Retrieval Conference, Utrecht, Netherlands (2010)
5. Myint, E.E.P., Pwint, M.: An approach for multi-label music mood classification. In: 2nd International Conference on Signal Processing Systems, ICSPS (2010)
6. Grekow, J., Raś, Z.W.: Emotion Based MIDI Files Retrieval System. In: Raś, Z.W., Wieczorkowska, A.A. (eds.) Advances in Music Information Retrieval. SCI, vol. 274, pp. 261–284. Springer, Heidelberg (2010)
7. Mohammad, S.: From Once Upon a Time to Happily Ever After: Tracking Emotions in Novels and Fairy Tales. In: Proceedings of the ACL 2011 Workshop on Language Technology for Cultural Heritage, Social Sciences, and Humanities, Portland, OR, USA, pp. 105–114 (2011)
8. Yeh, J.-H., Pao, T.-L., Pai, C.-Y., Cheng, Y.-M.: Tracking and Visualizing the Changes of Mandarin Emotional Expression. In: Huang, D.-S., Wunsch II, D.C., Levine, D.S., Jo, K.-H. (eds.) ICIC 2008. LNCS, vol. 5226, pp. 978–984. Springer, Heidelberg (2008)
9. Grekow, J., Raś, Z.W.: Detecting Emotions in Classical Music from MIDI Files. In: Rauch, J., Raś, Z.W., Berka, P., Elomaa, T. (eds.) ISMIS 2009. LNCS (LNAI), vol. 5722, pp. 261–270. Springer, Heidelberg (2009)
10. Thayer, R.E.: The biopsychology arousal. Oxford University Press (1989)

Foundations of Recommender System
for STN Localization during DBS Surgery
in Parkinson's Patients

Konrad Ciecierski[1], Zbigniew W. Raś[2,1], and Andrzej W. Przybyszewski[3]

[1] Warsaw Univ. of Technology, Institute of Comp. Science, 00-655 Warsaw, Poland
[2] Univ. of North Carolina, Dept. of Comp. Science, Charlotte, NC 28223, USA
[3] UMass Medical School, Dept. of Neurology, Worcester, MA 01655, USA
K.Ciecierski@ii.pw.edu.pl, ras@uncc.edu,
Andrzej.Przybyszewski@umassmed.edu

Abstract. During deep brain stimulation (DBS) treatment of Parkinson disease, the target of the surgery is the subthalamic nucleus (STN). As STN is small (9 x 7 x 4 mm) and poorly visible in CT[1] or MRI[2], multi-electrode micro recording systems are used during DBS surgery for its better localization. This paper presents five different analytical methods, that can be used to construct an autonomic system assisting neurosurgeons in precise localization of the STN nucleus. Such system could be used during surgery in the environment of the operation theater. Signals recorded from the micro electrodes are taken as input in all five described methods. Their result in turn allows to tell which one from the recorded signals comes from the STN. First method utilizes root mean square of recorded signals. Second takes into account amplitude of the background noise present in the recorded signal. 3rd and 4th methods examine Low Frequency Background (LFB) and High Frequency Background (HFB). Finally, last one looks at correlation between recordings taken by different electrodes.

Keywords: Parkinson's disease, DBS, STN, FFT, DWT, RMS, LFB, HFB, Hierarchical clustering.

Introduction

Parkinson disease (PD) is chronic and advancing movement disorder. The risk factor of the disease increases with the age. As the average human life span elongates also the number of people affected with PD steadily increases. PD is primary related to lack of the dopamine, that after several years causes not only movement impairments but also often non-motor symptoms like depressions, mood disorder or cognitive decline therefore it has a very high social cost. People as early as in their 40s, otherwise fully functional are seriously disabled

---

[1] Computer Tomography.
[2] Magnetic Resonance Imaging.

L. Chen et al. (Eds.): ISMIS 2012, LNAI 7661, pp. 234–243, 2012.
© Springer-Verlag Berlin Heidelberg 2012

and require continuous additional external support. The main treatment for the disease is pharmacological one. Unfortunately, in many cases the effectiveness of the treatment decreases with time and some other patients do not tolerate anti PD drugs well. In such cases, patients can be qualified for the surgical treatment of the PD disease. This kind of surgery is called DBS[3]. Goal of the surgery is the placement of the permanent stimulating electrode into the STN nucleus. This nucleus is a small – deep in brain placed – structure that does not show well in CT or MRI scans.

Having only an approximate location of the STN, during DBS surgery a set of parallel micro electrodes are inserted into patient's brain. As they advance, the activity of surrounding neural tissue is recorded. Localization of the STN is possible because it has distinct physiology and yields specific micro electrode recordings. It still however requires an experienced neurologist / neurosurgeon to tell whether recorded signal comes from the STN or not [3].

That is why it is so important to provide some objective and human independent way to classify recorded signals. Analytical methods described in this paper have been devised with exactly that purpose. Taking as input recordings made by set of electrodes at subsequent depths they provide information as to which of the electrodes and at which depth passed through the STN.

1   Signal Filtering

Wavelet transforms do not assume stationarity of the signal. It comes from the fact that wavelet base function is also time localized - not stationary. Because of this, the information regarding time in which certain frequencies are present is preserved. From this, looking at specific wavelet transform one can easily identify corresponding sample ranges in raw unfiltered signal [1]. In the case of DWT[4] there are many functions that can be used as a base wave. In [4] authors suggest Daubechies D4 wavelet [1]. This wavelet, having its shape akin to simplified spike shape is especially useful in analyzing neurobiological data. Because of that, in this paper wavelet transforms have been used for filtering.

1.1   DWT

Discrete Wavelet Transform is done both in forward and reverse way in steps. Each step in forward transform gives coefficients corresponding to different frequency ranges. cA (average) contains lower frequencies from input signal, cD (detail) reflect higher ones. If the input signal has $2n$ samples then $card(cA) = card(cD) = n$. cA_j is also an input to the $j + 1$ step of the DWT transform.

2   Signal's Normalization

The need for normalization of the recorded signals comes from several reasons. (1) There is no guarantee that length of the recordings would be the same;

---

[3] Deep Brain Stimulation.
[4] Discrete Wavelet Transform.

(2) While electrodes should have uniform electrical properties, some differences might occur. (3) In different surgeries, different amplification factor can be set in the recording device. First case – acting on a single recording level – implies that method relying on the power of the signal must be normalized in such way that from a given recording they produce power proportional to unit of its length. Last two reasons – acting on electrode level – normalizes amplitude of data calculated from full pass of a single electrode. Normalization must allow one to safely compare signals recorded by different electrodes with different amplification factors. During the PD DBS surgery electrode starts its recording at level $-10000\mu m$ ($10mm$ above predicted STN location) and from there follows its tract for another $15mm$ to the level $+5000\mu m$ ($5mm$ below predicted STN location). First $5mm$ of the tract produces relatively uniform recordings (low amplitude, little or no spiking) [3]. Assume now that a specific method mth taking as input vector of data recorded at subsequent depths produces a vector of coefficients C. Define C_{base} as average of coefficients obtained from the first $5mm$ of the tract. The vector of normalized coefficients is then defined as $C_{NR} = \frac{C}{C_{base}}$. All results from the methods presented in following sections are normalized according to the length of the recording and average from first five depths.

3   Artifacts Removal

Methods described in later sections do not rely on spike detection. They base on data extracted from signal's amplitude, its power for certain frequency ranges or correlations. All methods are highly affected by signal contamination. Because contaminating noise causes increase in both power and amplitude, its removal is especially important to avoid false STN detections. Most simple solution, used by authors in [5], just ignores all contaminated data. Here, solution to salvage uncontaminated portions of such data has been devised.

Artifacts reside mainly in low frequencies (< 375 Hz). Normally proper for such frequencies DWT coefficients have uniform and low amplitude. Looking for coefficients with amplitudes exceeding some threshold should therefore provide information about localization of artifacts in a given recording. Time bound relation between DWT coefficients and signal samples allows for artifacts removal.

If signal reaches maximal allowed amplitude for at least 0.01 % of its samples it is qualified as contaminated, and filtered in a special DWT based way. Six forward steps of DWT are made. Knowing that signal has been sampled with $24KHz$, each coefficient at k^{th} level of DWT corresponds to 2^k samples in the original signal. Also set of cA_6 coefficients corresponding to original's signal in frequency range $0 - 187Hz$. cD_6 coefficients in turn correspond to original's signal in frequency range $187 - 375Hz$. σ_{cA} and σ_{cD} values are calculated from respectively cA_6 and cD_6 sets using formula described in [10].

cA_6 is inspected for values a_j such that $|a_j| > \frac{3}{2}\sigma_{cA}$. cD_6 is inspected for values d_j such that $|d_j| > \frac{3}{2}\sigma_{cD}$. Samples of the original signal that fall in ranges corresponding to found a_j and d_j are set to 0. Later from non zero samples of such modified signal the σ is calculated. Finally, signal is hard thresholded with value 6σ.

4 Analysis of the *RMS* Value

STN is known to produce lots of spikes with high amplitude and also has loud background noise. It can be expected that signals recorded from it would present elevated Root Mean Square Value. This parameter is, among others, also used for Bayesian calculations in [5]. RMS approach is much less computationally demanding than spike detection and can be easier available. It still requires calculation of a sum of squares for all samples from a given recording. Because all samples contribute to the resulting value, this method takes into account both background noise and spikes. If the signal is contaminated by artifacts, the method may produce falsely high RMS values. RMS based method as shown on Fig. 1(a) in a clear way assesses that *anterior*[5] electrode is better than *medial*[6]. RMS method defines clear dorsal[7] border of the STN at -3000 with subsequent increase of the RMS up to depth -3000. Such findings and predicted thickness (6 mm) agree with clinical observations found in [3].

5 Analysis of the Percentile Value – PRC

Spikes amplitude is by far greater than the background noise. Because of that one can find an amplitude value below which no spikes are present or above which spike must rise. This feature is commonly used in many neurological appliances, for example the 50th percentile - *median* of amplitude's module is used for both spike detection and artifact removal process. The approach in which 50th percentile (*median*) of amplitude's module together with other features is used for detecting increased neural activity can be found in [9]. High amplitude ranges that comes from spikes lie after the 95th percentile. Certain percentile value calculated from module of amplitude can thus be used to estimate the amount of background neural activity. Already the 95th percentile shows background activity and discards almost all samples from the spikes. To be however safely independent from any spike activity, even lower percentile can be used. In this paper the 80th percentile is used. Having selected that percentile as a value that can be used for STN distinguishing an electrode comparison chart can be made. Using the same data set that was used for creation of Fig. 1(a), we obtained following results (Fig. 1(b)). Obtained results are in full agreement with those given by RMS method. Both methods state

---

[5] Being most forward, closest to facial plane.
[6] Placed to the left/right side of the central electrode.
[7] Top.

that the best electrode is *anterior* and that STN ranges roughly between -3000 and +3000. Decrease of percentile value is somewhat more steep than in the case of RMS, which a bit more clearly identifies the ventral[8] border of the STN.

6   DWT Based Analysis of the *LFB* Power

It has been postulated in [12] and [11] that background neural activity can be divided into two frequency areas. First, an activity in range below 500 Hz is called Low Frequency Background(LFB). In mentioned paper authors use properties of HFB to pinpoint STN location. Here, in this section a LFB based method for finding the STN is shown. As with Quantile based estimator, here too, it is very important to remove as much of the artifacts as possible. This is due to the fact that most of the power carried by such artifacts resides in range below $375Hz$ that if fully included in the described analysis. Before the transformation can be made, any spikes are removed from the raw MER recorded signal. Later to ensure removal of even highly distorted spikes, signal is also hard thresholded with value of 80th percentile. After artifacts and spikes have all been removed, signal has to be firstly transformed from time to frequency domain. For this DWT is used. DWT is performed fully, i.e. all available forward steps are done. As the result, following list of wavelet coefficient sets is produced: cA_n, cD_n, $cA_{n-1}, \ldots , cD_1$. Each set of coefficients corresponds to specific frequency range. Signal's power is subsequently calculated from all sets representing frequencies below 500 Hz. Once again the LFB power was calculated for the same set of electrodes that were used in previous sections. Results containing LFB power for electrodes and depths are shown on Fig. 1(c). Obtained results are in full agreement with those given by RMS and percentile methods. All methods state that best electrode is *anterior*. LFB based method localizes STN between -3000 and +3000. This yields thickness of about 5mm which is in accordance with brain anatomy [2]. What is especially worth mentioning is the visible division of STN into two different parts. Dorsal part spans -3000 to 0 and has lower power output. Ventral part spans from 0 to +1000 and is definitively more active. This finding agrees with [3].

7   DWT Based Analysis of the *HFB* Power

In this section a modified version of the HFB calculation described in [12] and [11] is presented. Modification comes from the use of DWT based filtering. All principles regarding calculation of HFB and LFB are the same, only difference lies in frequency range. For HFB inspected frequency range is between 500 and 3000 Hz. After calculating HFB for test set of electrodes results shown on Fig. 1(d) were obtained. Ones again, obtained results are in full agreement with those given by RMS, percentile and LFB methods. There is however notable difference, HFB does not show subdivisions of STN nucleus.

---

[8] Bottom.

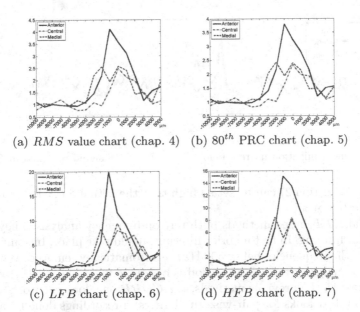

(a) RMS value chart (chap. 4)   (b) 80th PRC chart (chap. 5)

(c) LFB chart (chap. 6)   (d) HFB chart (chap. 7)

Fig. 1. Results of RMS, PRC, LFB and HFB

8   Correlation between Parallel Electrodes Recordings - $CORR$

Authors in paper [6] and [7] suggest that increase in synchronization between neuronal networks in the basal ganglia contributes to clinical observations in PD i.e. rhythmic involuntary muscle movements. Paper [8] also states that phase between different adjacent layers of brain tissue may be inverted. Taking this into account MER data were converted to absolute values prior to DWT calculation. Having done such preprocessing, one can sometimes notice striking similarities between DWT of recordings coming from vicinity of the STN. The synchrony appears to be most evident in frequencies below 23 Hz. Fig. 2 shows a DWT (part of cA_9) transform of 2.7 s long recording from the STN. Synchrony is not a natural phenomenon of the STN and is only present in some cases of PD. It can not be so fully relied upon. Its presence on given depth implies high STN probability. Its lack $DOESNOT$ exclude STN occurrence.

9   Comparison of Methods

Summarizing previous sections, several different methods allowing localization of the STN have been described. Each of the methods have some advantages and disadvantages.

RMS and Percentile(PRC) based methods can be used for classification of single recordings. Their results can be however completely wrong if signal contains any artifacts. If the artifacts are present during the first five recordings,

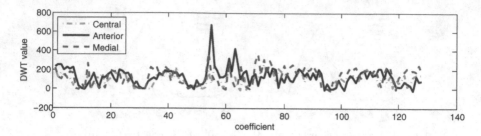

Fig. 2. Synchronization in freq. below 23 Hz for ABS signal recorded in STN

the C_{base} (see section 2) can be set so high that the actual STN location would not be detected at all.

LFB and HFB based methods both rely on frequency analysis. They do not require prior filtering. In fact no prior filtering should take place. In some recording systems the frequencies below 500 Hz are automatically removed. When such recording is analyzed, the LFB method is of course unusable.

In contrast to the previous methods, the $CORR$ one does not work at single recording level. It looks for pair-wise similarities in recordings done by an array of electrodes at the same depth. With certain preprocessing (see section 8) this similarities indeed can be found in frequency range below 23 Hz. As mentioned in [6] and [7] such STN synchrony is a hallmark of the PD. Unless caused by another illness it does not manifest in recording acquired from non-PD patients.

10   Recording Clustering

Using the RMS, PRC, LFB and HFB methods, each recording d made by some electrode at given depth has the set of coefficients described by equation 1.

$$C = \{c_{RMS}(d), c_{PRC}(d), c_{LFB}(d), c_{HFB}(d)\} \tag{1}$$

Having this coefficients, an attempt to obtain meaningful clustering of the recording has been done. Assumption was, that recordings can be divided into three clusters containing respectively:

α – recordings made outside the STN
β – recordings made in areas near the STN
γ – recordings made inside the STN

Clustering has been done using hierarchical cluster tree with Euclidian distance and minimum variance algorithm. As input to the clustering procedure, all possible 2, 3 and 4 – element subsets of C has been tried. Regardless of the coefficients subset taken for clustering, the Spearman rank was greater than 0.86 and in many cases it is above 0.9. This ensures that produced clustering was of good quality. Cluster percentage size also seem to be stable. Having $\mu_\alpha = 76.8$, $\mu_\beta = 17.4$ and $\mu_\gamma = 5.8$ and $\sigma_\alpha = 4.5$, $\sigma_\beta = 4.1$ and $\sigma_\gamma = 2.4$.

10.1   Cluster Cross Comparison

On the following pictures some of the more interesting clustering results are shown. Especially interesting are the results obtained when clustering was made using one subset of coefficients and clustering results are also shown using another subset.

Fig. 3(a) shows results of the clustering done using only coefficients c_{RMS} and c_{HFB}. One can clearly observe that data set has been divided into three sections.

Cluster α – very dense and numerous, containing recordings assumed to be made outside of the STN (6506 recordings). Cluster β – much less dense, containing recordings assumed to be made near the STN (while having area similar to cluster A, it contains only around 1668 recordings). Cluster γ – sparse, containing recordings that are assumed to be coming from the STN (contains only about 893 recordings).

Fig. 3(b) shows results of the same clustering that is shown on Fig. 3(a). It shows however a recording seen from the point of view of other two coefficients (c_{PRC} and c_{LFB}). One can plainly see the cluster α that contains recordings made outside STN. As expected, the recording from that cluster have also small c_{PRC} and c_{LFB} values. Also, as expected, recordings from cluster γ are characterized by largest in population values of c_{PRC} and c_{LFB}. This is a clear evidence that obtained results are stable and regardless of chosen coefficient subset recordings are divided in a similar way. Taking $C_1 \subseteq C$ and $C_2 \subseteq C$ (see equation 1) the resulting α_1 and α_2 clusters in worst case scenario have over 96 % common elements. β_1 and β_2 clusters in worst case have over 75 % common elements. In case of γ clusters, the intersection in never below 78 %. Intersection between α and γ is never bigger than 0.07 %. This allows one to say with good certainty that obtained results are stable and comparable between clusterings. Recording that has been assigned with cluster label α is much, much less likely to come from the STN than another one with cluster label β – even if labels come from different clusterings.

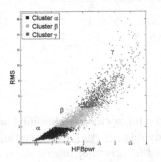 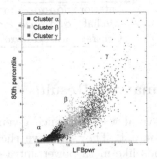

(a) Clustering with RMS and HFB shown on RMS, HFB plane

(b) Clustering with RMS and HFB shown on PRC, LFB plane

Fig. 3. Cluster cross comparison

11   Review of Clustering Results

Let us introduce ranking between clusters. As our goal is to find the STN, the most natural order is that $\alpha < \beta < \gamma$. Having done that, assume that for a given recording, best cluster is the most favorable cluster among all assignments made by different clusterings.

11.1   Sample Case

Let us take the example pass of electrodes that was shown in section 4, 5, 6 and 7. Each electrode produced 16 recordings (from depths ranging -10000 μm to 5000 μm). Labels applied to all subsequent recordings of each electrode are shown in *Clusters* column of Table 1. *Anterior* electrode has five consecutive recordings labeled as γ and largest continuous $\beta+$ sequence (nine depths). *Central* electrode has three subsequent depth labeled as γ, they together with adjacent β recordings form 7 element sequence. *Medial* electrode has five depth labeled as γ, their largest sequence has 4 elements. γ together with adjacent β recordings form 7 element sequence. Basing on obtained results, as the best electrode we select the *Anterior* one – it contains largest sequence of γ labeled recordings. The 2nd best would be the *Medial* electrode, it contains as many γ recordings as the *Anterior* one albeit split into two subsequences.

Having labeled the recordings, we can now offer a way to rank them as to how well did they reached / passed through the STN. For this, we define three measures defined for an electrode: $M_{\gamma 1}$ – count of recordings labeled as γ. $M_{\gamma 2}$ – length of the longest sequence of recordings labeled as γ. $M_{\gamma \beta}$ – length of the longest sequence of recordings labeled as γ or β. Electrodes are then ordered in descending order according to M_{γ}, $M_{\gamma \beta}$ and finally to M_{β}.

Table 1. Electrode rank example

Electrode	Clusters	$M_{\gamma 1}$	$M_{\gamma 2}$	$M_{\gamma \beta}$
Anterior	$\beta\alpha\alpha\alpha\alpha\alpha\beta\gamma\gamma\gamma\gamma\gamma\beta\beta\beta$	5	5	9
Medial	$\alpha\beta\alpha\alpha\alpha\alpha\beta\gamma\gamma\gamma\gamma\beta\gamma\alpha\beta$	5	4	7
Central	$\alpha\alpha\alpha\alpha\alpha\alpha\alpha\alpha\beta\gamma\gamma\gamma\beta\beta\beta$	3	3	7

12   Summary of Results

There are several ways in which given recordings can be classified as coming from STN or not. They all have their advantages and disadvantages. Spike based methods like meta power, intra spike histogram[3] or simple spike count must rely on spike detection. This can be difficult due to the nature of MER recorded signal. Low frequency components or artifacts can lead to incomplete detection of spikes. Also STN is not the only area characterized by high spiking activity. This could led to both false negative and false positive detections of the STN. Method basing on synchrony detection (see chapter 8) can provide some additional insight but are not very reliable. As explained they look for pathological STN condition that does not have to occur.

Last class of methods: RMS, PRC, LFB and HFB described in this paper do not directly rely on spike or synchrony detection. They do require some signal preprocessing. Especially essential is artifact removal. Each of them helps to find the STN location using different approach, still, as pointed out in section 10 they greatly agree in their assessments.

Hierarchical clusterings that base on the described methods proved to be an effective method for STN discrimination. Obtained results are more stable, accurate and less probability dependant then those basing on spike detection. This all, proves that it is possible to construct autonomic and automatic decisive support system for STN detection.

Authors would like to thank Dariusz Koziorowski MD, PhD (Warsaw Medical University) and Tomasz Mandat, MD, PhD (Clinic of Neurosurgery at Warsaw Institute of Psychiatry and Neurology) for providing DBS recordings.

References

1. Jensen, A., Ia Cour-Harbo, A.: Ripples in Mathematics. Springer (2001)
2. Nolte, J.: The Human Brain, Introduction to Functional Anatomy. Elsevier (2009)
3. Israel, Z., et al.: Microelectrode Recording in Movement Disorder Surgery. Thieme Medical Publishers (2004)
4. Alexander, B., et al.: Wavelet Filtering before Spike Detection Preserves Waveform Shape and Enhances Single-Unit Discr. J. Neurosci. Methods, 34–40 (2008)
5. Moran, A., et al.: Real-Time Refinement of STN Targeting Using Bayesian Decision-Making on the RMS Measure. J. Mvmt. Disorders 21(9), 1425–1431 (2006)
6. Levy, R., et al.: Synchronized Neuronal Discharge in the Basal Ganglia of PD Patients Is Limited to Oscillatory Activity. J. Neuroscience, 7766–7775 (2000)
7. Pogosyan, A., et al.: Parkinsonian impairment correlates with spatially extensive subthalamic oscillatory synchronization. J. Neuroscience 171, 245–257 (2010)
8. Buzsaki, G., et al.: High-Frequency Network Oscillation in the Hippocampus. Science 256(5059), 1025–1027 (1992)
9. Gemmar, P., et al.: MER Classification for DBS. 6th Heidelberg Innov. Forum (2008)
10. Ciecierski, K., Raś, Z.W., Przybyszewski, A.W.: Selection of the Optimal Microelectrode during DBS Surgery in Parkinson's Patients. In: Kryszkiewicz, M., Rybinski, H., Skowron, A., Raś, Z.W. (eds.) ISMIS 2011. LNCS (LNAI), vol. 6804, pp. 554–564. Springer, Heidelberg (2011)
11. Novak, P., Przybyszewski, A.W., et al.: Localization of the subthalamic nucleus in Parkinson disease using multiunit activity. J. Neur. Sciences 310, 44–49 (2011)
12. Novak, P., et al.: Detection of the subthalamic nucleus in microelectrographic recordings in Parkinson disease using the high-frequency (> 500 Hz) neuronal background. J. Neurosurgery 106, 175–179 (2007)

From Music to Emotions and Tinnitus Treatment, Initial Study

Deepali Kohli[1], Zbigniew W. Raś[1,2],
Pamela L. Thompson[3,1], Pawel J. Jastreboff[4,1], and Alicja A. Wieczorkowska[5,1]

[1] Univ. of North Carolina, Charlotte, Comp. Science Dept., Charlotte, NC 28223
[2] Warsaw Univ. of Technology, Institute of Comp. Science, 00-665 Warsaw, Poland
[3] Catawba College, Ketner School of Business, Salisbury, NC 28144
[4] Emory Univ. School of Medicine, Dept. of Otolaryngology, Atlanta, GA 30322
[5] Polish-Japanese Institute of Information Technology, 02-008 Warsaw, Poland
{dkohli1,ras}@uncc.edu, pthompso@catawba.edu, pjastre@emory.edu,
alicja@poljap.edu.pl

Abstract. The extended tinnitus database consisting of 758 patients with information repeated from the original database of 555 patients, along with the addition of visits and a new questionnaire, the Tinnitus Function Index and Emotion Indexing Questionnaire, is used to mine for knowledge. New patients in the extended database represent those patients that have completed the Tinnitus Function Index questionnaire (TFI) [10]. The patient visits are separated and used for mining and action rule discovery based on all features and treatment success indicators including several new features tied to emotions (based on a mapping from TFI to Emotion Indexing Questionnaire (EIQ) [14]; EIQ questionnaire is used by our team to build personalized classifiers for automatic indexing of music by emotions). We propose a link between TFI and EIQ leading to a creation of new features in the extended tinnitus database. Then, we extract knowledge from this new database in the form of association action rules to assist with understanding and validation of diagnosis and treatment outcomes.

1 Introduction and Domain Knowledge

Tinnitus, sometimes called "ringing in the ears" affects a significant portion of the population [1]. Some estimates show the portion of the population in the United States affected by tinnitus to be 40 million, with approximately 10 million of these considering their problem significant. Many definitions exist for tinnitus. One definition of tinnitus relevant to this research is ". . . the perception of sound that results exclusively from activity within the nervous system without any corresponding mechanical, vibratory activity within the cochlea, and not related to external stimulation of any kind". Hyperacusis or decreased sound tolerance frequently accompanies tinnitus and can include symptoms of misophonia (strong dislike of sound) or phonophobia (fear of sound). Physiological causes of tinnitus can be difficult or impossible to determine, and treatment

L. Chen et al. (Eds.): ISMIS 2012, LNAI 7661, pp. 244–253, 2012.

approaches vary. It should be mentioned here that we are not dealing with damaged hearing which clearly can not be treated by music.

Tinnitus Retraining Therapy (TRT), developed by Jastreboff [6,7], is one treatment model with a high rate of success and is based on a neurophysical approach to treatment. TRT "cures" tinnitus by building on its association with many centers throughout the nervous system including the limbic and autonomic systems. The limbic nervous system (emotions) controls fear, thirst, hunger, joy and happiness. It is connected with all sensory systems. The autonomic nervous system controls such functions as breathing, heart rate and hormones. When the limbic system becomes involved with tinnitus, symptoms may worsen and affect the autonomic nervous system. TRT combines counseling and sound habituation to successfully treat a majority of patients.

Conceptually, habituation refers to a decreased response to the tinnitus stimulus due to exposure to a different stimulus. Degree of habituation determines treatment success, yet greater understanding of why this success occurs and validation of the TRT technique will be useful. The treatment requires a preliminary medical examination, completion of an Initial Interview Questionnaire for patient categorization, audiological testing, a visit questionnaire referred to as a Tinnitus Handicap Inventory (THI), tracking of medical instruments, and a follow-up questionnaire. The interview collects data on many aspects of the patient's tinnitus, sound tolerance, and hearing loss. The interview also helps determine the relative contribution of hyperacusis and misophonia. A set of questions relate to activities prevented or affected (concentration, sleep, work, etc.) for tinnitus and sound tolerance, levels of severity, annoyance, effect on life, and many others. All responses are included in the database. As a part of audiological testing, left and right ear pitch, loudness discomfort levels, and suppressibility is determined. Based on all gathered information a patient category is assigned. A patient's overall symptom degree is evaluated based on the summation of each individual symptom level, where a higher value means a worse situation. The category is included in the database, along with a feature that lists problems in order of severity (Ex. TH is Tinnitus first, then Hyperacusis).

Patient Categories

- Category 0: *Low Impact on Life, Tinnitus Present*
- Category 1: *High Impact on Life, Tinnitus Present*
- Category 2: *High Impact on Life, Subjective Hearing Loss Present*
- Category 3: *High Impact on Life, Tinnitus not Relevant, Subjective Hearing Loss not Relevant, Hyperacusis Present*
   Category 4: *High Impact on Life, Tinnitus not Relevant, Hyperacusis Present, Prolonged Sound-Induced Exacerbation Present*

The TRT emphasis is on working on the principle of differences of the stimuli from the background based on the fact that the perceived strength of a signal has no direct association with the physical strength of a stimulus, using a functional dependence of habituation effectiveness model. Therefore, once a partial reversal of hyperacusis is achieved, the sound level can be increased rapidly to address tinnitus directly.

As we already mentioned, TRT requires completion of an Initial Interview Questionnaire for patient categorization, audiological testing, and completion of a visit questionnaire referred to as a Tinnitus Handicap Inventory. The Tinnitus Functional Inventory (TFI) is a new visit questionnaire also required by TRT. Its questions (new features) are tied to emotions. In this paper, we define a mapping between TFI features and the set of features used in Thayer's Arousal-valence emotion plane and the mood model for music annotation as described by Grekow and Ras [3]. This way, features related to emotions are used to build emotion-type bridge between tinnitus and music.

Our previous research on tinnitus recommender system was based on knowledge extracted from a dataset without TFI so we did not use emotions as features and the same no reference to music was made [18,23,24].

From the extended tinnitus database (it includes TFI), provided by Jastreboff, we extracted action rules showing that larger positive improvement in emotions yields larger improvement in tinnitus symptoms. The concept of action rule was introduced by Ras and Wieczorkowska [14] and investigated by others [2,5,12,13,16,20,22]. We use Action4ft-Miner Module from Lisp-Miner Project [11,17] to discover action rules.

All of us agree that music invokes emotions in most of us. The general finding of the literature is that the experience of negative emotional states leads people to sharply decrease their exposure to complex, novel and loud music, and simple music at a soft listening level actively soothes negative emotions.

There is a lot of research done in the area of automatic indexing of music by emotions [8,9,15,21,19]. In [13] we introduced the Score Classification Database (SCD) which describes associations between different scales, regions, genres, and jumps. This database was used to automatically index a piece of music by emotions. Also, we have shown how to use action rules extracted from SCD to change the emotions invoked by a piece of music by minimally changing its score. By a score, we mean a written form of a musical composition. In [3], we built hierarchical classifiers for automatic indexing of music by emotions.

Following the approach proposed in [13], we can use action rules to change score of a music piece and the same we can control emotions it invokes. We believe that by applying this strategy to tinnitus patients, we can develop a very successful emotion-based treatment and hopefully control it in real time.

2   Action Rules

An action rule is a rule extracted from a decision system that describes a possible transition of objects from one state to another with respect to distinguished attribute called a decision attribute [14]. It is assumed that attributes used to describe objects in a decision system are partitioned into stable and flexible attributes. Values of flexible attributes can be changed. This change can be influenced and controlled by users. Action rules mining initially was based on comparing profiles of two groups of targeted objects - desirable and undesirable. The concept of an action rule was introduced by Ras and Wieczorkowska and defined as a term $\omega \wedge (\alpha \rightarrow \beta) \Rightarrow (\phi \rightarrow \psi)$, where ω is a conjunction of fixed condition features shared by both groups, $[\alpha \rightarrow \beta]$ represents proposed changes in values of flexible features, and $[\phi \rightarrow \psi]$ is a desired effect of the action. Symbol $\wedge$ is interpreted as logical "and".

When applied to medical data, action rules show great promise; a doctor can examine the effect of treatment choices on a patient's improved state as measured by an indicator that indicates treatment success, such as the Total Score on the Tinnitus Handicap Inventory [24,23]. For example, action rule discovery can be used to suggest a change on a flexible attribute like emotional score in order to see the changes in treatment success as measured by positive change in total score for tinnitus patients in the Tinnitus Handicap Inventory.

3 LISp-Miner for Action Rule Discovery

Ac4ft-Miner procedure is a part of the robust LISp-Miner system developed by Jan Rauch and his colleagues (http://lispminer.vse.cz). LISp-Miner includes an advanced system of software modules that have been developed to implement classification and action rule discovery algorithms on data sets. The 4ft-Miner procedure is used in this research to discover new action rules in the tinnitus data sets covering only new patients (those completing the Tinnitus Functional Index). It has three basic theoretical resources: the GUHA method, association rules and the action rules.

The GUHA method is a method of exploratory analysis with a purpose of providing all interesting facts derived from the analyzed data. Association rules are the rules which express associations or correlation relationships among data items. Action rules express which action should be performed to improve the defined state. Ac4ft-Miner can be thus described as follows: "Ac4ft-Miner finds rules that express which actions should be performed to improve the defined state. It achieves it by examining the dependencies among the data given as an input". Ac4ft-Miner system mines for G-action rules. G-action rules are generalizations of action rules [14]. They may have stable and flexible attributes on antecedent and succedent part of the rule [17]. The input to Ac4ft-Miner is a data matrix and a definition of the set of relevant G-action rules from which true rules are selected [16]. LISp-Miner allows some reduction of the patterns of interest but this requires specific knowledge of the ontology that the dataset satisfies. Attribute values in a rule can be restricted by adding left and right cuts, effectively reducing the values of interest for specific variables. Additionally, variables can be defined as stable and flexible with respect to the decision variable of interest. The desired change in attributes on the left hand side of rule and right hand side of rule can also be defined by using variables "state before" and "state after".

Association rules are mined with the form $\phi \approx \psi$ with ϕ and ψ representing Boolean attributes antecedent and succedent respectively. The association rule

Table 1. Data Matrix

M	ψ	$\neg\psi$
ϕ	a	b
$\neg\phi$	c	d

represented by $\phi \approx \psi$ means that the antecedent and succedent are associated in a way represented by $\approx$ which is called the "4ft-quantifier". This is represented by a quadruple data matrix shown in Table 1:

The a-priori algorithm for association rules discovery is not employed in our research, and the procedure we use follows a complex bit-string method; an explanation of it is provided by Rauch, Simunek, and Nekvapil [11,16]. Let us assume that $Dom(A) = \{a_1, a_2, a_3, .., a_k\}$, where $Dom(A)$ is a domain of the attribute A. For any $i \in \{1, 2, ..., k\}$, the expression $A(a_i)$ denotes Boolean attribute that is true if the value of attribute A is a_i. Assume now that $A_0 \subset Dom(A)$. Similarly, $A(A_0)$ denotes Boolean attribute that is true if there is $a \in A_0$ such that the value of attribute A is a. This way, we can mine for association rules of the form $[A(\_) \wedge B(\_)] \approx C(\_)$ where $(\_)$ is not a single value but a subset of the set of all values of the corresponding attribute. In particular, we can mine for rules of the form $[A(\_) \wedge B(\_)] \rightarrow C(\_)$. The expression $A(\_)$ denotes the Boolean attribute that is true for a particular row of data matrix if the value of A in this row belongs to $(\_)$, and the same is true for $B(\_)$ and $C(\_)$. This approach makes it easy to mine for conditional association rules that are mentioned in [16].

4   Experiments and Results

The tool used in this study is $Ac4FtMiner$ for association action rule discovery with connection to Microsoft Access. This study utilized the Emotion Indexing Questionnaire (see [14]) which consists of two parts. In its first part, users are asked to answer a number of questions including their musical preferences (what formal musical training they have, what kind of music they listen to when they are happy, sad, angry, calm), and a group of questions asking about their current mood (emotions listed in Table 2). In its second part, users are asked to annotate a number of music pieces by emotions listed in Thayer's arousal-valence emotion plane [3]. The second part of the questionnaire is used to build classifiers for automatic indexing of music by emotions. Each emotion in the first part of the questionnaire (Table 2) is represented by a rating scale of 0 to 4 with 0 meaning emotion was absent and 4 meaning emotion was extreme, as measured over the previous one week period including the day the questionnaire was completed.

Table 3 shows the mapping between the terms (emotions) used in the Tinnitus Functional Index (TFI) and emotions listed in Table 2 (the first part of the Emotion Indexing Questionnaire (EIQ)). The terms used in TFI (first column in Table 3) were normalized to be between 0 to 4 as the TFI questions use rating scale of 0 to 10 with 0 as absent and 10 as the worst case. Questions one

Table 2. List of emotions used in EIQ

Tense	Grouchy
Angry	Energetic
Worn − out	Unworthy
Lively	Uneasy
Confused	Fatigued
Shaky	Annoyed
Sad	Discouraged
Active	Muddled
Exhausted	Efficient

Table 3. Mapping between TFI and EIQ

TFI	EIQ
In_control	Active
Annoyed	Annoyed
Cope	Efficient
Ignored	Unworthy
Concentrated	Efficient
Think_clearly	Efficient
Focused_attention	Efficient
Fall/stay_asleep	Fatigued
As_much_sleep	Fatigued, Exhausted, Worn − out
Sleeping_deeply	Exhausted
Social_activities	Energetic
Enjoyment_of_life	Lively
Work_on_other_tasks	Efficient
Anxious, worried	Uneasy
Bothered, upset	Tensed, Angry

and four on the TFI are rated as percentages and hence they were normalized
to fit the range 0 to 4. Next, in the extended tinnitus database, we replaced
the terms used in TFI by emotions used in EIQ. Pearson correlation coefficient
was calculated between these terms using the function $corrcoef(X)$ provided in
$Matlab$. Emotions like Tensed and Angry are perfectly correlated with Pearson
correlation coefficient equal 1.

Then $Ac4FtMiner$ action rule discovery software was applied to the new extended tinnitus database to analyze if changes in emotional scores positively affect the total score from Tinnitus Handicap Inventory. The total score is a sum of emotional score, functional score and catastrophe score in THI [7].

Before the rules are established, we need to analyze the data further. Every patient can be characterized by more than one emotional state. If two attributes (emotions) are strongly correlated, then we must be able to characterize the patient using only one of them. Weaker the correlation between the attributes on the antecedent part of the rule, stronger is the need to use both of them in characterizing the patient.

Assume that two patients in the extended Tinnitus database are represented by vectors $\alpha_1 = [a_1, b_1, ...]$ and $\alpha_2 = [a_2, b_2, ...]$, where $a_1, a_2 \in Dom(a)$, $b_1, b_2 \in Dom(b)$. Distance $\rho(\alpha_1, \alpha_2)$ with respect to attributes a, b is calculated by the following formula:

$$([dist(a_1, a_2) + dist(b_1, b_2)]/2 + Q(a(a_1, a_2), b(b_1, b_2)) * [dist(a_1, a_2) + dist(b_1, b_2)]/2$$

where $Q(a(a_1, a_2), b(b_1, b_2))$ is computed from the Pearson correlation coefficient $P(a, b)$ with respect to all tuples having minimum one of the properties b_1, b_2, a_1, a_2. The definition is given by:

$$- Q(a(a_1, a_2), b(b_1, b_2)) = [-P(a(a_1, a_2), b(b_1, b_2)) + 1]/2.$$

The value of Q is always in the range [0,1] and it is equal to 0 when attributes are perfectly correlated. To calculate the distance $\rho(\alpha_1, \alpha_2)$ with respect to more than two attributes, we can use reducts in rough sets theory for that purpose [4].

Having defined the distance between tuples in the extended tinnitus database, we are ready to search for action rules showing the expected changes in the total score triggered by changes in patient's emotions. Such rules have been discovered by $Ac4FtMiner$ and they are listed below:

Rule 1: Attributes in the antecedent part of the rule are "Active", "Fatigued", "Worn-out". Attribute in the succedent part of the rule is "Total Score" from THI. Quantifiers defined as after and before state frequency must be greater than or equal to 4.00. The correlation index for active and worn-out is 0.3714, for active and fatigued is 0.4038 and for Worn-out and fatigued is 0.9366.

Action rule generated is as follows:

$$[Active(< 3; 4)) \to Active(< 2; 3))] \wedge [Fatigued(< 2; 3)) \to Fatigued(< 0; 1))] \wedge [Worn - out(< 2; 3)) \to Worn - out(< 0; 1))] \Rightarrow [ScT(< 18; 36 >) \to ScT(< 0; 16 >)]$$

The rule above states that if the emotional state of the patient shows improvement (lower the score, better it is), the total score of the patient also improves. Also, our research shows that the doctor can use a music recommender system to identify the right piece of music to be played in order to improve patient's emotional state and the same treat Tinnitus. The confidence of the rule is 0.36. For the antecedent part of the rule, distance between the attributes based on

Pearson correlation coefficient is 5.9821. Worn-out and fatigued are highly correlated and patient can be represented by just one of the emotions. The total score changes show that severity of tinnitus changes from mild to slight.

Rule 2: Attributes in the antecedent part of the rule are "Active", "Annoyed", "Worn-out". Attribute in the succedent part of the rule is "Total Score" from THI. Quantifiers defined as after and before state frequency must be greater than or equal to 4.00. The correlation index for Active and Worn-out is 0.3714, for Active and Annoyed is 0.6610 and for Worn-out and Annoyed is 0.4094.

Action rule generated is as follows:

$[Active(< 4; 5 >) \rightarrow Active(< 2; 3)))] \wedge [Annoyed(< 4; 5 >) \rightarrow Annoyed(< 2; 3 >)] \wedge [Worn - out(< 4; 5 >) \rightarrow Worn - out(< 0; 1))] \Rightarrow [ScT(< 58; 76 >) \rightarrow ScT(< 0; 16 >)]$

This rule also states that improvements in emotional state of the patient affect the total score in a positive way. Effect on life by tinnitus is greatly improved. Confidence of the rule is 0.45. For the left hand side of rule, distance between the attributes based on Pearson correlation coefficient is 19.94. The total score changes show that severity of tinnitus changes from severe to mild.

Rule 3: Attributes in the antecedent part of the rule are "Discouraged", "Energetic", "Lively". Attribute in the succedent part of the rule is "Total Score" from THI. The correlation index for Discouraged and Energetic is 0.6984, for Energetic and Lively is 0.8035 and for Discouraged and Lively is 0.8088.

Action rule generated is as follows:

$[Discouraged(< 4; 5 >) \rightarrow Discouraged(< 0; 1))] \wedge [Energetic(< 4; 5 >) \rightarrow Energetic(< 0; 1))] \wedge [Lively(< 4; 5 >) \rightarrow Lively(< 0; 1))] \Rightarrow [ScT(< 72; 108 >) \rightarrow ScT(< 0; 16 >)]$

This rule also states that improvements in emotional state of the patient affect the total score in a positive way. Effect on life by tinnitus is greatly improved. Confidence of the rule is 0.714. For the left hand side of rule, distance between the attributes based on Pearson correlation coefficient is 8.86. The total score changes show that severity of tinnitus changes from catastrophic to mild.

These three rules show correlations between the improvement in the emotional state of a patient and improvement in tinnitus treatment. From the extended tinnitus database, we also extracted action rules (not listed in the current paper) showing that larger improvement in patient's emotions yields larger improvement in the total score. Clearly, music invokes emotions in most of us. The same, music may be used as a tool to treat tinnitus patients. By identifying a music piece that can invoke possibly highest positive emotions listed in EIQ for a given patient, we should guarantee a nice speed up of his/her successful tinnitus treatment. Clearly, emotions invoked by music are very personalized. This is why we need to build personalized recommender systems for tinnitus treatment based on personalized systems for automatic indexing of music by emotions.

5   Conclusions and Acknowledgements

TRT is a complex treatment process, which generates a lot of data over time: some attributes have relatively stable values while others may be subject to change as the doctors are tuning the treatment parameters while symptoms of patients are altering. Understanding the relationships between and patterns among treatment factors helps to optimize the treatment process. Interesting action rules about the relationship among new emotional features and total score of the patient were revealed which show that improvement in emotional state brought by music brings significant changes in the total score from THI recorded over the course of the treatment. The database needs to be further extended to include stable attributes like characteristics pertaining to patients to investigate further the relationship between tinnitus treatment and patient emotional state. This will help to develop more specific rules and allow doctors to prescribe more personal treatment using music. The emotional indexing questionnaire can be used to better understand the emotional state of the patients.

This work was partially supported by the Research Center of PJIIT, supported by the Polish National Committee for Scientific Research (KBN).

References

1. Baguley, D.M.: What progress have we made with tinnitus. Acta Oto-Laryngologica 556, 4–8 (2006)
2. Greco, S., Matarazzo, B., Pappalardo, N., Slowiński, R.: Measuring expected effects of interventions based on decision rules. Journal of Experimental and Theoretical AI 17(1-2), 103–118 (2005)
3. Grekow, J., Raś, Z.W.: Emotion Based MIDI Files Retrieval System. In: Raś, Z.W., Wieczorkowska, A.A. (eds.) Advances in Music Information Retrieval. SCI, vol. 274, pp. 261–284. Springer, Heidelberg (2010)
4. Grzymala-Busse, J.: Managing Uncertainty in Expert Systems. Kluwer, Boston (1991)
5. He, Z., Xu, X., Deng, S., Ma, R.: Mining action rules from scratch. Expert Systems with Applications 29(3), 691–699 (2005)
6. Jastreboff, P.J., Gray, W.C., Gold, S.L.: Neurophysiological approach to tinnitus patients. American Journal of Otolaryngology 17, 236–240 (1995)
7. Jastreboff, P.J., Hazell, J.W.P.: Tinnitus Retraining Therapy - implementing the neurophysiological model. Cambridge University Press (2004)
8. Juslin, P.N., Sloboda, J.A. (eds.): Handbook of music and emotion, theory, research, applications. Oxford Univ. Press (2010)
9. Li, T., Ogihara, M.: Detecting emotion in music. In: Proceedings of the 4th International Conference on Music Information Retrieval (ISMIR 2003), Washington, DC (2003)
10. Meikle, M.B., et al.: The tinnitus functional index: development of a new clinical measure for chronic, intrusive tinnitus,
http://www.ncbi.nlm.nih.gov/pubmed/22156949
11. Nekvapil, V.: Data mining in the medical domain. Lambert Academic Publishing (2010)

12. Qiao, Y., Zhong, K., Wang, H.-A., Li, X.: Developing event-condition-action rules in real-time active database. In: Proceedings of the 2007 ACM Symposium on Applied Computing, pp. 511–516. ACM (2007)

13. Raś, Z.W., Dardzińska, A.: From data to classification rules and actions. International Journal of Intelligent Systems 26(6), 572–590 (2011)

14. Raś, Z.W., Wieczorkowska, A.: Automatic Indexing of Audio, Emotion Indexing Questionnaire, http://www.mir.uncc.edu/QuestionareConsent.aspx

15. Raś, Z.W., Wieczorkowska, A.A. (eds.): Advances in Music Information Retrieval. SCI, vol. 274. Springer, Heidelberg (2010)

16. Rauch, J., Šimůnek, M.: Action Rules and the GUHA Method: Preliminary Considerations and Results. In: Rauch, J., Raś, Z.W., Berka, P., Elomaa, T. (eds.) ISMIS 2009. LNCS (LNAI), vol. 5722, pp. 76–87. Springer, Heidelberg (2009)

17. Rauch, J.: Action4ft-Miner Module. Lisp-Miner Project, http://lispminer.vse.cz/

18. Thompson, P., Zhang, X., Jiang, W., Ras, Z.W., Jastreboff, P.: Mining tinnitus database for knowledge. In: Berka, P., Rauch, J., Zighed, D. (eds.) Data Mining and Medical Knowledge Management: Cases and Applications, pp. 293–306. IGI Global (2009)

19. Trohidis, K., Tsoumakas, G., Kalliris, G., Vlahavas, I.: Multi-label classification of music into emotions. In: Proceedings of International Symposium on Music Information Retrieval (ISMIR), pp. 325–330 (2008)

20. Wang, K., Jiang, Y., Tuzhilin, A.: Mining actionable patterns by role models. In: Proceedings of the 22nd Inter. Conf. on Data Engineering, April 3-7, pp. 16–26 (2006)

21. Wieczorkowska, A., Synak, P., Ras, Z.W.: Multi-Label Classification of Emotions in Music. In: Kłopotek, M.A., Wierzchon, S.T., Trojanowski, K. (eds.) Intelligent Information Processing and Web Mining. AISC, vol. 35, pp. 307–315. Springer, Heidelberg (2006)

22. Yang, Q., Chen, H.: Mining case for action recommendation. In: Proceedings of ICDM 2002, pp. 522–529. IEEE Computer Society (2002)

23. Zhang, X., Thompson, P., Raś, Z.W., Jastreboff, P.: Mining Tinnitus Data Based on Clustering and New Temporal Features. In: Biba, M., Xhafa, F. (eds.) Learning Structure and Schemas from Documents. SCI, vol. 375, pp. 227–245. Springer, Heidelberg (2011)

24. Zhang, X., Raś, Z.W., Jastreboff, P.J., Thompson, P.L.: From tinnitus data to action rules and tinnitus treatment. In: Proceedings of 2010 IEEE Conference on Granular Computing, Silicon Valley, CA, pp. 620–625. IEEE Computer Society (2010)

Adapting to Natural Rating Acquisition with Combined Active Learning Strategies

Mehdi Elahi[1], Francesco Ricci[1], and Neil Rubens[2]

[1] Free University of Bozen-Bolzano, Bozen-Bolzano, Italy
`mehdi.elahi@stud-inf.unibz.it, fricci@unibz.it`
`http://www.unibz.it`
[2] University of Electro-Communications, Tokyo, Japan
`neil@hrstc.org`
`http://www.uec.ac.jp`

Abstract. The accuracy of collaborative-filtering recommender systems largely depends on the quantity and quality of the ratings added to the system over time. Active learning (AL) aims to improve the quality of ratings by selectively finding and soliciting the most informative ratings. However previous AL techniques have been evaluated assuming a rather artificial scenario: where AL is the only source of rating acquisition. However, users do frequently rate items on their own, without being prompted by the AL algorithms (natural acquisition). In this paper we show that different AL strategies work better under different conditions, and adding naturally acquired ratings changes these conditions and may result in a decreased effectiveness for some of them. While we are unable to control the naturally occurring changes in conditions, we should adaptively select the AL strategies which are well suited for the conditions at hand. We show that choosing AL strategies adaptively outperforms any of the individual AL strategies.

Keywords: Recommender systems, active learning, combined-strategies, rating elicitation.

1 Introduction and State of the Art

In this paper we focus on collaborative filtering (CF) Recommender Systems (RS) [5] that recommend items to a user based on the collective ratings of other users. The CF rating prediction accuracy does depend on: the characteristics of the prediction algorithm, and the number, the distribution, and the quality of the ratings known by the system. In general, the more informative about the user's preferences the collected ratings are, the higher is the recommendation accuracy.

In previous research, briefly described here, it has been therefore stressed that it is important to keep acquiring new and useful ratings from the users by adopting Active Learning strategies [9]. Different techniques have been proposed, some being non-personalized such as choosing popular items or items with diverse ratings (higher entropy of ratings), and some personalized such as using a decision

L. Chen et al. (Eds.): ISMIS 2012, LNAI 7661, pp. 254–263, 2012.

tree to choose what to ask based on previous ratings of the users. These studies focussed especially on the cold start stage, i.e., where a new user is added to the system [10,8,11,4]. However, both the system and users do not remain in the cold start stage for long, as users start to interact with the system and the system starts to evolve. In realistic settings an active learning rating elicitation strategy can't be viewed as the only source, or tool, for collecting new ratings from the users. Users may freely and voluntary enter new ratings; yet all the previously mentioned elicitation strategies were evaluated in isolation from this natural ongoing system usage.

The need for evaluating the behavior of RS as it evolves is starting to receive its due attention. In particular, scholars are starting to evaluate system's performance (accuracy, diversity and robustness) from the temporal perspective where the users are rating items and the database is growing (although without considering the application of any active learning approaches) [1,6]. This is radically different from the typical evaluations where the rating dataset is decomposed into the training and test sets without considering the timestamps of the ratings. [1,6] have shown that in such scenarios the accuracy of the system might not improve even though ratings are added to the system.

Previous works have considered that ratings are acquired either actively during the cold start stage [10,8,11,4], or in a natural manner throughout the system's lifetime [1,6]. However, in practice, ratings may be acquired in both ways. In addition, recently [3] have shown that actively requesting and acquiring new ratings is useful for the overall system's performance at any stage of the system's evolution, and not just at the beginning (cold start). Motivated by the above, we propose a novel evaluation methodology, which we claim to be more realistic and indicative of the real performance of a rating elicitation strategy, i.e., combining AL with the natural acquisition of items and evaluating it throughout the evolution of RS. We show that especially in these settings, a single AL strategy cannot perform consistently well. To overcome this limitation, we propose active learning strategies that combine other AL strategies by (1) voting mechanism, or by (2) adaptively selecting seemingly best suited AL strategy. We believe that these results provide guidelines and conclusions that would help the deployment of AL rating elicitation in real RSs.

The rest of the paper is organized as follows. In section 2 we introduce the rating elicitation strategies that we have analyzed. In section 3 we present the simulation procedure that we designed to evaluate their effect on the system's recommendation performance (MAE and NDCG). The results of our experiments are shown in section 4, and in section 5 we summarize the outcome of this research.

2   Rating Elicitation Strategies

A rating dataset R is a $n \times m$ matrix of real values (ratings) with possible null entries. The variable r_{ui}, denotes the entry of the matrix in position (u, i), and contains the rating assigned by user u to item i, typically an integer between 1 and 5. r_{ui} could store a null value representing the fact that the system does not know the opinion of the user on that item.

A *rating elicitation strategy* S is a function $S(u, N, K, U_u) = L$ which returns a list of items $L = \{i_1, \ldots, i_M\}$ whose ratings should be elicited from the user u, where N is the maximum number of items that the strategy should return, K is the rating dataset of known ratings, i.e., the ratings (of all the users) that have been already acquired by the RS at a certain point. Finally, U_u is the set of items whose ratings have not yet been elicited from u, hence potentially interesting. The elicitation strategy enforces that $L \subset U_u$ and will not repeatedly ask a user to rate the same item; i.e. after the items in L are shown to a user they are removed from U_u.

Every elicitation strategy analyzes the dataset of known ratings K and scores the items in U_u. If the strategy can score at least N different items, then the N items with the highest score are returned. Otherwise a smaller number of items is returned. It is important to note that the user may have not experienced the items whose ratings are requested; in this case the system will not increase the number of known ratings. In practice, following a strategy may result in collecting a larger number of ratings, while following another one may results in fewer but more informative ratings. These two properties (rating quantity and quality) play a fundamental role in rating elicitation.

In a previous work we evaluated several individual strategies including: popularity, log(pop)∗entropy, binary-prediction, highest-predicted, lowest-predicted, highest-lowest-predicted, voting, and random [3]. In this paper, since the observed behaviors of some strategies are similar, and to make our presentation clearer, we consider and illustrate just three individual strategies, which are actually representative of some others. We have chosen log(popularity)∗entropy since it has been always indicated as an effective one [7], highest-predicted since it is a good representative for other prediction based strategies, and random. Moreover, we introduce here two novel strategies, voting and switching, which we call "combined", since they aggregate and combine the previously mentioned individual strategies (or any other selection of strategies).

*log(popularity) * entropy:* the score for the item i is computed by multiplying the logarithm of the popularity of i, i.e., the number of known ratings for i, with the entropy of the ratings for i in K. This strategy tries to combine the effect of the popularity score, which is discussed above, with the heuristics that favors items with more diverse ratings (larger entropy) which provide more useful (discriminative) information about the user's preferences [2,7].

Highest Predicted: based on the ratings in K, predictions are computed for all the items in U_u and the scores are set equal to these predicted values. We use Matrix Factorization with gradient descent optimization to generate rating predictions [5]. The idea behind the highest-predicted strategy is that the best recommendations could also be more likely to have been experienced by the user and obtained ratings could also reveal important information about user's preferences. Moreover, this is the default strategy for RSs, i.e., enabling the user to rate the recommendations.

Random: the score for an item is a random integer. This is a baseline strategy, used for comparison.

Combined with Voting: the score for the item i is the number of votes given by the committee of the previously mentioned strategies: *log(pop)*entropy, highest predicted,* and *random.* Each of these strategies produces its top 100 candidates for rating elicitation, and then the items appearing more often in these lists are selected. This strategy obviously depends on the selected voting strategies.

Combined with Switching: every time this strategy is used, a certain percentage (40% in our experiments) of the users (exploration group) are randomly selected for choosing the best performing individual strategy and applying it on the remaining users (60%). Each individual strategy is applied to an equal number of random users in the exploration group: it selects items to be rated by these users, and acquires their ratings if they are present in a set of hold out ratings X, which represents the ratings that users could give but they have not provided yet to the system. How X is generated is described in the next section. Moreover, based on the ratings in K, a factor model is trained and its MAE and NDCG for these newly acquired ratings are computed. We also compute the probability for an individual strategy to acquire the ratings for the selected items (estimated as the ratio of the number of acquired ratings over the number of items requested to be rated). Finally, the score of each individual strategy is calculated by multiplying this probability either by the rating prediction error (MAE) on the acquired ratings, if MAE is the target metric to minimize, or by (1 - NDCG) in the other case. The strategy with the highest score is then selected. Hence, the combined switching strategy is selecting the individual strategy that was able to acquire from the exploration group the largest number of ratings, for items for which the system prediction is currently most erroneous (either for MAE or NDCG). Conjecturing that these are the most informative items, and that the selected strategy will have the same behavior on the remaining users (60% in our experiments), the winner strategy is then applied for eliciting ratings from the remaining users.

3 Evaluation Approach

We designed a procedure to simulate the evolution of the RS's performance by mixing the ratings acquired by an active learning strategy (individual or combined) with the ratings entered naturally by the users without being explicitly requested, just as it happens in actual settings. To accomplish this goal, we have used the larger version of the Movielens dataset (1,000,000 ratings) for which we considered only the ratings of users that were active and rated movies for at least 8 weeks (2 months).

We have split the available data into three matrices K, X and T. We inserted into K, the set of known ratings, those acquired by Movielens in the weeks 1-4 (first month) of the system usage (14,195 ratings). Then we randomly split the remaining ratings by inserting 70% of them into X (251,241) and 30% into T (111,867). X represents the ratings that are available to be elicited, while T are ratings that are never elicited and are used for testing the system's performance. We perform a simulated iteration every week, namely each day of the week (starting from the second week) an active learning strategy requests each user, who already has some non-null ratings in K, to rate 40 items. If these ratings are present in X, they are added to K. This step is repeated for 7 days (1 week). Then, all the ratings in the Movielens dataset that according to the timestamps were acquired in that week are also added to K. Finally, the system is trained using the ratings in K and then tested on the ratings in T that users actually entered during the following week (according to the timestamps). This procedure is repeated for $I = 48$ weeks (1 year). In order to precisely describe the evaluation procedure, we use the following notation, where n is the week index:

K_n: is the set of ratings known by the system at the end of the week n. These are the ratings that have been acquired up to week n. They are used to train the prediction model, compute the active learning rating elicitation strategies for week $n + 1$, and test the system's performance using the ratings contained in the test set of the next week $n + 1$, T_{n+1}.

T_{n+1}: is the set of ratings timestamped during the week $n+1$ that are used as a test set to measure the system's performance after the ratings in the previous weeks have been added to K_n.

AL_n: is the set of ratings elicited by a particular elicitation strategy, and is added to the known set (K_n) at week n. We note that these are ratings that are present in X but not in T. This is required for assuring that the active learning strategies are not modifying the test set and that the system's performance, under the application of the strategies, is consistently tested on the same set of ratings.

X_n: is the set of ratings in X, timestamped in week n that are not in the test set T_n. These ratings, together with the ratings in T_n, are all of the ratings acquired in Movielens during the week n, and therefore are considered to have been naturally provided by the (simulated) users without being asked by the system (natural acquisition). We note that it may happen that an elicitation strategy has already acquired some of these ratings, i.e., the intersection of AL_n and X_n may be not empty. In this case, only those not yet actively acquired are added to K_n. The testing of an active learning strategy S now proceeds in the following way:

- System initialization: week 1 to 4 (1 month)
  1 The entire set of ratings is partitioned randomly into the two matrices X, T.
  2 The non-null ratings in X_1 and T_1 are added to $K_1 : K_1 = X_1 \cup T_1$
  3 U_u, the unclear set of user u is initialized to all the items i with a null value k_{ui} in K_1.
  4 The rating prediction model is trained on K_1, and MAE, Precision, and NDCG are measured on T_2.
- For all the weeks n starting from $n = 5$
  5 Initialize K_n with all the ratings in K_{n-1}.
  6 For each user u with at least 1 rating in K_{n-1}:

- Using strategy S a set of items $L = S(u, N, K_{n-1}, U_u)$ is computed.
- The set L_e is created, containing only the items in L that have a non-null rating in X. The ratings for the items in L_e are added to AL_n
- Remove from U_u the items in L: $U_u = U_u \setminus L$.

7 Add to K_n the ratings timestamped in week n and those elicited by S:
$K_n = AL_n \cup X_n \cup T_n$.

8 Train the factor model on K_n.

9 Compute MAE and NDCG on T_{n+1}.

4   Results

Figure 1 shows the evolution of the system MAE in the first 48 weeks under the application of the considered strategies. Here MAE is normalized with respect to the MAE of the baseline, i.e., when only the natural acquisition of the ratings is used: $(\frac{MAE_{Strategies}}{MAE_{Baseline}}) - 1$. Additionally, in figure 2 we plot the evolution of the MAE when the ratings are added only by the strategies themselves (i.e. at step 7 the ratings timestamped in week n are not added). Comparing these figures makes it clear that the natural acquisition of ratings can have a huge impact on the accuracy of the system. Without natural acquisition MAE fluctuates more, since the system's performance is only determined by the ratings acquired with active learning, which acquires fewer ratings. The traditional way of evaluating AL strategies in isolation is misleading, since ratings do get frequently added in a natural manner (besides being added by AL strategies), and do have a significant effect on the performance of AL strategies. Hence, evaluating active learning strategies in combination with the natural addition of ratings provides a more realistic and accurate evaluation.

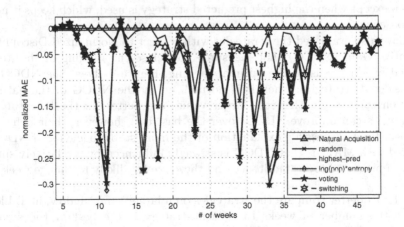

Fig. 1. Normalized system MAE under the effect of AL strategies and natural acquisition (48 weeks)

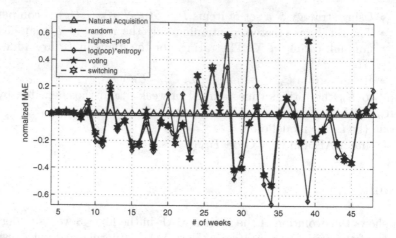

Fig. 2. Normalized system MAE under the effect of AL strategies without natural acquisition (48 weeks)

To better illustrate the MAE evolution, Figure 3 shows the system's performance, already presented in Figure 1, focusing only on the first 14 weeks (lower values are better). Here, it is clear that AL strategies can provide the potential benefit (most evident at week #11), but also that their efficiency may vary (e.g. week #11 vs. #13). It should be noted that there is a huge fluctuation of MAE, from week to week. This is caused by the fact that every week we test the system's performance on the next week's ratings. Hence, the difficulty of making good predictions and acquiring informative ratings may differ from week to week. The result shows that up to the seventh week, the performance of the strategies are similar. However, starting from week seven, the system MAE decreases except when the highest predicted strategy is used, which keeps it near to the baseline.

We have also evaluated the strategies with respect to Normalized Discounted Cumulative Gain (NDCG) that measures how close the ranking of the items predicted by the RS is to the optimal ranking. Figure 4 shows the NDCG of the considered strategies, when it is normalized by the NDCG of the natural acquisition baseline. In this case larger values are better. In the long run all the strategies can improve NDCG (over the baseline), however, their behavior varies over time. A possible explanation of the more stable behavior of NDCG, compared with MAE, is that NDCG is strongly influenced by a relatively small number of predicted top-rated items, and these are less likely to change week by week.

In order to better compare the strategies on a larger time window, in Table 1 we show the number of weeks that each strategy is the best or the second best among all the strategies. When the natural acquisition is considered, the three strategies that produce the lowest MAE are log(pop)*entropy, voting and switching. We can also observe that highest-predicted does not perform very

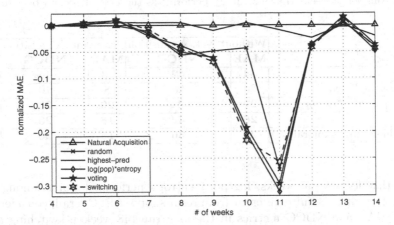

Fig. 3. Normalized system MAE under the effect of AL strategies and natural acquisition (14 weeks)

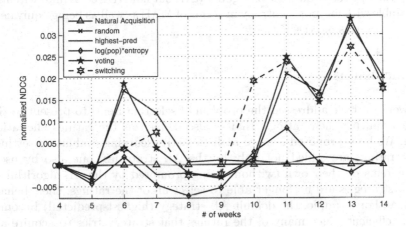

Fig. 4. Normalized system NDCG under the effect of AL strategies and natural acquisition (14 weeks)

differently from the baseline. The main reason is that this strategy is not acquiring additional ratings besides those already collected by the natural process, i.e., the user would rate these items on his own initiative. The other strategies are able to elicit more ratings, including those that the user will rate later on, i.e., in the successive weeks. When the natural acquisition is not considered, the best strategies are random, voting, switching and log(pop)*entropy. We stress again the different performance of the AL strategies when evaluated in isolation, and with the addition of naturally acquired ratings.

Considering NDCG, Table 1 shows that with natural acquisition the best strategies are voting, switching and log(pop)*entropy. However, without natural acquisition the best strategies are random, voting and switching. We should

Table 1. Number of weeks that a strategy performs as the best or second best (in 48 weeks)

Strategy	Without Nat. Acquisition		With Nat. Acquisition	
	MAE	NDCG	MAE	NDCG
Random	34	42	13	16
Highest predicted	20	28	8	10
log(popularity)*entropy	22	28	29	20
Voting	26	42	18	23
Switching (40% exploration)	22	36	15	20

mention that for several weeks, log(pop)*entropy performed worst among all the strategies which is quite different from results for MAE. Overall, considering both the MAE and NDCG metrics in the long run (48 weeks), switching and voting achieved remarkable results since both performed as good as the best strategies with respect to one of these metrics. For instance, as explained before, log(pop)*entropy can produce a very good MAE but not NDCG. While switching is comparable with log(pop)*entropy in terms of MAE as well as being equivalent to random in terms of NDCG.

5   Conclusions

In this work we have addressed the problem of selecting items to present to the users for acquiring their ratings; that is also defined as the ratings elicitation problem. We have proposed a more realistic active learning evaluation settings in which ratings are added not only by the AL strategies, but also by users that rate items on their own (without being prompted by the AL algorithms). Since AL strategies are no longer able to select all of the ratings, their behavior changes. For example, the default AL strategy (highest-predicted) becomes much less efficient, since many of the ratings that strategy tries to acquire are anyway added by the users. By evaluating the system's performance on a weekly basis, it becomes apparent that no single AL strategy performs consistently well due to changes in the rating set (caused by natural acquisition). We demonstrate that it is possible to adapt to the changes in the characteristics of the rating set, by proposing AL strategies that combine individual AL strategies, either by a voting mechanism, or by adaptively selecting them, from a pool of individual AL strategies, based on the estimation of how well each individual AL strategy is able to cope with the conditions at hand. For future work, we plan to fine-tune the switching strategy with a better method to minimize the exploration process. This can be done in a dynamic way so that in early weeks, the system concentrates more on exploration and later on exploitation. Moreover, we are implementing an online music recommender system where we will test the proposed methods, together with more recent strategies (e.g. [11,4]), in a live user experiment.

References

1. Burke, R.: Evaluating the dynamic properties of recommendation algorithms. In: Proceedings of the Fourth ACM Conference on Recommender Systems, RecSys 2010, pp. 225–228. ACM, New York (2010)
2. Carenini, G., Smith, J., Poole, D.: Towards more conversational and collaborative recommender systems. In: Proceedings of the 2003 International Conference on Intelligent User Interfaces, Miami, FL, USA, January 12-15, pp. 12–18 (2003)
3. Elahi, M., Repsys, V., Ricci, F.: Rating Elicitation Strategies for Collaborative Filtering. In: Huemer, C., Setzer, T. (eds.) EC-Web 2011. LNBIP, vol. 85, pp. 160–171. Springer, Heidelberg (2011)
4. Golbandi, N., Koren, Y., Lempel, R.: Adaptive bootstrapping of recommender systems using decision trees. In: Proceedings of the Forth International Conference on Web Search and Web Data Mining, WSDM 2011, Hong Kong, China, February 9-12, pp. 595–604 (2011)
5. Koren, Y., Bell, R.: Advances in collaborative filtering. In: Ricci, F., Rokach, L., Shapira, B., Kantor, P. (eds.) Recommender Systems Handbook, pp. 145–186. Springer (2011)
6. Lathia, N.K.: Evaluating Collaborative Filtering Over Time. PhD thesis, University College London (June 2010)
7. Rashid, A.M., Albert, I., Cosley, D., Lam, S.K., Mcnee, S.M., Konstan, J.A., Riedl, J.: Getting to know you: Learning new user preferences in recommender systems. In: Proceedings of the 2002 International Conference on Intelligent User Interfaces, IUI 2002, pp. 127–134. ACM Press (2002)
8. Rashid, A.M., Karypis, G., Riedl, J.: Learning preferences of new users in recommender systems: an information theoretic approach. SIGKDD Explor. Newsl. 10, 90–100 (2008)
9. Rubens, N., Kaplan, D., Sugiyama, M.: Active learning in recommender systems. In: Ricci, F., Rokach, L., Shapira, B., Kantor, P. (eds.) Recommender Systems Handbook, pp. 735–767. Springer (2011)
10. Schein, A.I., Popescul, A., Ungar, L.H., Pennock, D.M.: Methods and metrics for cold-start recommendations. In: SIGIR 2002: Proceedings of the 25th Annual International ACM SIGIR Conference on Research and Development in Information Retrieval, pp. 253–260. ACM, New York (2002)
11. Zhou, K., Yang, S.-H., Zha, H.: Functional matrix factorizations for cold-start recommendation. In: Proceeding of the 34th International ACM SIGIR Conference on Research and Development in Information Retrieval, SIGIR 2011, Beijing, China, July 25-29, pp. 315–324 (2011)

Intelligence in Interoperability with AIDA

Hugo Peixoto[1], Manuel Santos[3], António Abelha[2], and José Machado[2]

[1] Centro Hospitalar do Tâmega e Sousa E.P.E., Sistemas de Informação,
Lugar do Tapadinho, Penafiel, Portugal
hpeixoto@chts.min-saude.pt
[2] University of Minho, Computer Science and Technology Centre (CCTC),
Campus de Gualtar, Braga, Portugal
{abelha,jmac}@di.uminho.pt
[3] University of Minho, ALGORITMI Center,
Campus de Azúrem, Guimarães, Portugal
mfs@dsi.uminho.pt

Abstract. Healthcare systems have to be addressed in terms of a wide variety of heterogeneous, distributed and ubiquitous systems speaking different languages, integrating medical equipments and customized by different entities, which in turn were set by different people aiming at different goals. Demands of information within the healthcare sector range from clinically valuable patient-specific information to a variety of aggregation levels for follow-up and statistical and/or quantifiable reporting. The main goal is to gathering this information and present it in a readable way to physicians. In this work we show how to achieve interoperability in healthcare institutions using AIDA, an interoperability platform developed by researchers from the University of Minho and being used in some major Portuguese hospitals.

Keywords: Semantic Interoperability, Ambient Intelligence, Electronic Health Record.

1 Introduction

Healthcare systems have been for some years a very attractive domain for Computer Science (CS) researchers. Even more, such systems have great potential for information integration and automation, and this is an issue of study in which medicine and agent-based technologies and methodologies for problem solving may overlap. Furthermore, healthcare systems have to be addressed in terms of a wide variety of heterogeneous, distributed and ubiquitous systems speaking different languages, integrating medical equipment and customized by different entities, which in turn were set by different people aiming at different goals [1].

Demands of information within the healthcare sector range from clinically valuable patient-specific information to a variety of aggregation levels for follow-up and statistical and/or quantifiable reporting. Gathering this information and present it in a readable way to physicians it's an interesting task. This lead us to consider the solution to a particular problem, to be part of an integration process

L. Chen et al. (Eds.): ISMIS 2012, LNAI 7661, pp. 264–273, 2012.

of different sources of information, using rather different protocols, in terms of an Agency for the Integration, Diffusion and Archive (AIDA) of Medical Information, bringing to the healthcare arena new methodologies for problem solving and knowledge representation, computational models, technologies and tools, which will enable ambient intelligence or ubiquitous computing based practices at the healthcare facilities [2]. With access granted to clinical and historical databases, agent technology may provide answers to those who give assistance to patients in time and medical evidence [3].

Every day new applications are developed to assist physicians on their work and healthcare is turning into a science based on information and reputation. We want to find out how usability is maintained in this new landscape of social and ubiquitous computing, when applied to healthcare units, a domain that presents the strongest social focus, and in this way offering the biggest challenge for socially aware software systems design [4]. Nowadays, the exchange and share of clinical knowledge among medical information systems is an important feature to improve healthcare systems, quality of the diagnosis, but mainly, to improve quality in patient treatment.

Semantic Interoperability and Ambient intelligence are seen in this work as the key to solve this problem. Electronic Semantic Health Record (ESHR) will provide the base for this work. The guarantee of a homogeneous information system in a health care unity will produce enormous benefits to the institution. Reduction in diagnosis and appointments time since information about a patient is available at one time in the same place would give doctors more time to treat patients better. Less medical errors would be expected due to better quality of information [5].

2   Interoperability

Nowadays information technologies (ITs) in medicine and healthcare are facing a scenario in which a variety of healthcare providers have introduced ITs in their everyday workflows with a certain degree of independence. This independence may be the cause of difficulty in interoperability between information systems [5]. The overload of information systems within an healthcare facility may lead to problems in accessing the total information needed, since it is hard for a physician to access all information sources in an acceptable period of time. In the last decade, Health Information Systems (HIS) have gained great importance and have grown in quality and in quantity. With this information overload, it is necessary to infer what information is relevant to be registered in the EHR and decision support systems must allow for reasoning on incomplete, ambiguous and uncertain knowledge. Demands of information handling within the healthcare sector range from clinically valuable patient-specific information to a variety of aggregation levels for follow-up and statistical and/or quantifiable reporting.

Researchers in the field of Hospital Information Systems have focused special attention to the field of quality of information. A Health Unit is computationally represented by a heterogeneous set of applications that speak different languages and are customized by different customers. So a practical and effective communication platform between information systems is paramount taking into consideration the quality of information [5].

Each service has small database management systems where specific patient data are registered depending on pathologies or specific interests. This computational tissue generates development problems. However, these applications are used by people with good satisfaction despite they do not allow a transversal vision of the patient data along different services or specialties, they can not grow easily and sometimes they do not attend secure and confident procedures. Running applications in distributed environment is a huge problem when applications have not been developed to share knowledge and actions.

Information Systems capabilities are increasingly exposed and exploited, however, software developers and users must lead with new challenges. The EHR is already a topic widely discussed and explored that has brought innovations and advances every day. However the new requirements are to manage all information that is produced in health facilities to ensure quality and easy exchange between HI. It is necessary to use new emerging technologies.

Interoperability has been rather confined within the realms of IT and technology. Although IT plays a key role in making businesses interact seamlessly, such an information exchange infrastructure is meaningless if the other core aspects of business collaborations are not interoperable. Hence, the concept of Business Interoperability goes beyond IT, into organizational aspects of businesses, and includes the level of people-to-people interactions. Smoother workflows also mean that business processes originating in one organization can seamlessly flow into a collaborating partner organization without getting caught up in bureaucratic red tape. Systems for conflict resolution and Intellectual Property Management can further ensure Business Interoperability.

In the last years many projects have pursued the interoperability of EHR information systems. The different approaches have proposed solutions based on specific standards and technologies in order to satisfy the needs of a particular scenario, but no global interoperability frameworks have been provided so far. Some countries are already planning a unified medical language to ensure that the information is stored according to the same syntax and semantic. The Unified Medical Language System (UMLS) is a project of the US national Library of Medicine (NLM) and provides a conceptual framework for concept categorization. Information access is simple and effective, providing to users information with quality [5].

There are some research groups around the world working on semantic web, including the clinical area. Jentzsch and some colleagues developed a system to support the pharmaceutical area that allows the connection between drugs, treatments, cures and laboratory tests via semantic web. This system shows how to search for a Chinese medicinal herb that met the requirements for patient administration [6]. Jentzsch and his staff show how to build decision support systems in clinical activity. These technologies will be of great interest, requiring only some adaptation in HIS and hospital software providers [6].

2.1   Semantic Interoperability

The information to be transferred must be standardized and normalized in order avoid different structures and misinterpretations. We must also take into account the data semantics, so information can be understood by different systems. In addition, the use of standards ensures the best communication between health professionals and interoperability between systems, allowing some automation in the hospital recording. The standards used in EHR are divided into three different purposes:

- standards for representing clinical information;
- communication standards; and
- image standards.

International Classification of Diseases, Ninth Revision, Clinical Modification (ICD-9-CM), Systematized Nomenclature of Medicine-Clinical Terminology (SNOMED-CT) and International Classification for Nursing Practice (ICNP) are standards for classification of diseases and therapeutic clinics, where each therapy or disease is associated with a code recognized anywhere in the world. The use of these standards ensures that the EHR can be readable by any clinician in the world, allowing machines to interpret symptoms and assisting the clinicians in making a diagnosis and treatment plan decision [7]. As communication standards, the AIDA-PCE adopts the Health Level Seven (HL7) as a protocol for exchanging messages, and web architectures and service-oriented architectures (SOA).

There are a high number of benefits of semantic web but there are also some disadvantages. One negative factor is the complexity of implementation and the specific domain of the medical field.

3   Ambient Intelligence

Ambient intelligence is related with an atmosphere where rational and emotional intelligence is omnipresent [2]. In an ambient intelligent environment, people are surrounded with networks of embedded intelligent devices to gather and diffuse information around physical places, forming a ubiquitous network around an integrated global middleware accepting specific requests and data from heterogeneous sources, and providing ubiquitous information, communication and services. Intelligent devices are available whenever needed, enabled by simple or effortless interactions, attuned to senses, adaptive to users and contexts, and acting autonomously. High quality information and content may therefore be available to any user, anywhere, at any time, and on any device. Users are aware of their presence and context and digital environments are sensitive, adaptive, and responsive to needs, habits, gestures and emotions [2]. In virtual health care environments, they can not be separated from medical informatics, biomedical informatics or bio-informatics, aggregating electronic health records, decision support, telemedicine, knowledge representation and reasoning, knowledge discovery and computational biology. Radiological films, pathology slides and laboratory reports can be viewed in remote places. Remote robotics is used in surgery and telemedicine is becoming popular. However applications are used for discrete clinical and medical activities in specific areas and services, in particular diagnostics and pathologies.

Ambient Intelligence benefited from an exponential growth of Internet use on the last few years. New rapid web advancements are emerging, transferring technology benefits sometimes without a solid theoretic underpinning [2]. Although web browsers support many features that facilitate the development of user-friendly applications and allow users to run application anywhere without installing flat software packages in order to run remote applications. Storage and information access over the web encourages the information and knowledge re-use and the offer of global information and resources. The vitality of a web-based system lies in its integration potential, in supporting communities of virtual entities and in the gathering, organization and diffusion of information. Operating on the web means the use of documents or programs that contain images, audios, videos and interactive tools in addition to text. Scripting languages are used to build high level programs improving distribution, as well as information and knowledge sharing, increasing quality software and reducing costs [2].

4   Implementation

4.1   AIDA Framework

To build systems for real healthcare environments, the infrastructure must meet a range of basic requirements with respect to security, reliability and scaling. With access granted to Clinical and Historical Databases, agent technology may provide answers to those who give assistance to patients with a maximum of quality and medical evidence [8]. Communications are sometimes limited by old infrastructures and new projects collide with financial restrictions and bureaucratic delays. The homogeneity of clinical, medical and administrative systems is not possible due to financial and technical restrictions, as well as functional needs. The solution is to integrate, diffuse and archive this information under a dynamic framework, in order to share this knowledge with every information system that needs it. Indeed, to build systems for real healthcare environments, the infrastructure must meet a range of basic requirements with respect to security, reliability and scaling. With access granted to Clinical and Historical Databases, agent technology may provide answers to those who give assistance to patients with a maximum of quality and medical evidence [9]. Figure 1 shows the schematic representation of AIDA framework. In this schema, it is possible to understand the workflow of information, as well as integration and interoperability.

AIDA is an agency that provides intelligent electronic workers, here called pro-active agents and in charge of tasks such as communicating with the heterogeneous systems, sends and receives information (e.g., medical or clinical reports, images, collections of data, prescriptions), managing and saving the information and answering to requests, with the necessary resources to their correct and an on time accomplishment [9].

AIDA also supports web-based services to facilitate the direct access to the information and communication facilities set by th1ird parties, i.e., AIDA construction follows the acceptance of simplicity, the conference of the achievement of common goals and the addressing of responsibilities. The main goal is to integrate, diffuse and archive large sets of information from heterogeneous sources (i.e., departments, services, units, computers, medical equipments); AIDA also provides tools in order to implement communication with human beings based on web-based services. Under these presuppositions, a healthcare information system (HIS) will be addressed in terms of Figure 1:

- Administrative Information System (AIS), which intends to represent, manage and archive the administrative information during the episode. Being an episode a collection of all the operations assigned to the patient since the beginning of the treatment until the end;

- The Medical Support Information System (MIS), which intends to represent, manage and archive the clinical information during the episode;

- The Nursing Support Information System (NIS), which intends to represent, manage and archive the nursing information during the episode; and

- The Electronic Medical Record Information System (EMR); and

- The Information Systems (DIS) of all the departments or services, in particular of the laboratories (Labs), Radiological Information System (RIS) and Medical Imaging (PACS - Picture Archive and Communication System), which deals with images in a standard format, the DICOM one.

The architecture presented was envisaged to support medical applications in terms of AIDA and EHR, a form of web spider of an intelligent information processing system, its major subsystems, their functional roles, and the ow of information and control among them, with adjustable autonomy.

Healthcare staff acquires this information and its value is automatically stored and distributed to where it is needed. Every document created within a specialized service respect this rules, making different and individualized departments closer. The coding and ordering features are very useful to link different data to one specific problem, as coded data is much easier to access and it is recommended for decision support using Artificial Intelligence. The electronic ordering embedded in EHR can be used not only to obtain medical equipment or pharmacological prescriptions, but also for acquiring laboratory and imaging studies outside the service where it is used. Furthermore, it may enable the centralization of exam display, allowing different services to share results concerning the same patient, diminishing costs on unnecessary exams, and above all, improving the quality of service being provided [9].

There are also different access permissions when dealing with medical data. Although it can only be viewed by the authorized personnel from any terminal

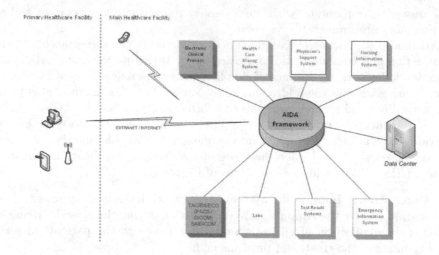

Fig. 1. AIDA- Agency for Integration Diffusion and Archive

inside the healthcare facility or even on its own laptop or PDA, the access must be flexible in order to enable the professionals to access it when needed. In other words, the access to the medical information of the patient is as important in terms of privacy as in terms of significance for medical situations. On the other hand, interfaces must be intuitive and easy to use. Messaging enables one to create, send and retrieve messages online. It may be very useful for handling data, images or even file exchange. Encryption and the right protocols of trading are also paramount. Messaging systems are extremely important not only for the internal workflow in a healthcare institution, but as well as an essential component for the development of group work, namely in the area of diagnostic that is supported by decision support systems.

4.2   Electronic Semantic Health Record

Adoption of Electronic Health Record (EHR) is well known in several countries all over the world. Canada, Australia, England and United States have already started their own way to achieve an infrastructure for national Health Information Systems [10]. All of these projects share the main elements and focus on the same important subject, interoperability and integration of HIS in spite of interoperability between healthcare providers being a hard task [8]. Unfortunately, information emerges from an assortment of sources, from informatics applications, medical equipments and physicians' knowledge introduced in the Electronic Health Record. Decision support systems are enhanced by quality of information and that can only be achieved with good collection of all the data from the patient. With this information overload, it is necessary to infer what information is relevant to be registered in the EHR and decision support systems must allow for reasoning on incomplete, ambiguous and uncertain knowledge.

Demands of information handling within the healthcare sector range from clinically valuable patient-specific information to a variety of aggregation levels for follow-up and statistical and/or quantifiable reporting.

EHR is a repository of information concerning an individual in an electronic format. It is stored and transmitted securely and may be accessed by multiple users [11]. The main objective is to ensure ubiquity, i.e. information is accessible at anytime and anywhere. The lack of integration between the different HIS is not only an obstacle for a more effective clinical practice, but it may lead to a suboptimal care for the patient. Although, this is an accepted definition Hayrinen proved that different researches describe in distinct ways the same thing, and EHR definition is one example [12].

Hayrinen published, in the International Journal of Medical Informatics, an exploratory approach to review the impact of use of the HER as well all the definitions associated with EHR. Hayrinen tries to understand the ongoing studies on several countries and evaluate the possibility of current technologies and underlying architectures and on exploring the health care registers as a source for evidence-based medicine. Automated search was used by the author in several known databases such as PubMed, Cinalh and Cochrane to find recent published works on the area.

In Europe various organizational, institutional, governmental and private initiatives are in course, having the some common main proposes: the standardization, the definition of functional models, minimal data sets and interoperability. EHR depends on three areas:

- Terminology;
- Structure; and
- Interoperability in Communication [13].

Semantic Web has been the subject of much debate and several studies in recent years by the scientific community and the main issue is whether he can fill some gaps in the information systems of health units and perhaps bring an additional quality. W3C consortium is one of the world leading task forces where several research teams gather their work and present it to scientific community in seek for approval and newer approaches for improvements [14].

Several papers come out from this consortium presenting state of the work and tasks executed between publications. Paper derive from explanatory to exploratory and descriptive but the main investigation query and theme is how to use semantic as a mean to achieve full interoperability between systems under the main theme of healthcare.

Jentzsch and colleagues presents the use of semantic as linked data to drugs systems and enabling integration with other web based systems for studies and result retrieving. Their main efforts go to provide the correct drug for the treatment of the patient and study whether their system meets the desired requirements. For the big test for their system they apply it to the search of Chinese herbs that may help in certain diseases. The final results prove that their system is able to provide extended studies for retrieving the correct herb, such as gene study of the patient, side effects expected and how active ingredients can interfere with patient health [6]. As well as Jentzsch, Clark followed a case study for presenting his work in the Semantic Web in Health Care and Life Sciences Interest Group (HCLS) and the goal is to develop cures for highly complex diseases, such as neurodegenerative disorders, even thought it requires extensive interdisciplinary collaboration and exchange of biomedical information in context. This research team presents the process of integration of semantic ontology in SWAN Alzheimer Knowledge Base. The description provides the user knowledge of ontology that can improve search and connection between systems for extraction of information in easier queries. This project is fully integrated in the Massachusetts General Hospital and provides better access to information relevant for better diagnosis and scientific studies of Alzheimer disease [15].

These studies prove that semantic can promote better results for decision support systems since quality of information is preserved during all processes and correct data is accessed when certain query is performed. Making usage of such technology and combining it with AIDA work methodology an Electronic Semantic Health Record improves quality of information accessed inside the institution. Making a semantic approach to current health records is possible to enhance communication between health care facilities and information can be shared without any lost in quality since queries are most accurate and specific. Semantic enhances linked data and provides the tools to compare data that is most of the time separate but that can together provide accurate information and most important quality information for users. Our main goal is to take semantic advantages and pull out graphical interactions from disperse data.

5   Conclusions

This paper presents an innovative intelligent framework responsible for interoperability in healthcare units. The homogeneity of clinical, medical and administrative systems was not possible due to financial and technical restrictions, as well as functional needs. The solution was to integrate, diffuse and archive this information under a dynamic framework, in order to share this knowledge with every information system that needs it.

AIDA platform has, until now, demonstrate to be an useful tool in order to interoperate in a healthcare facility, being used in several portuguese hospitals in Elvas, Portalegre, Guimarães, Pemafiel, Amarante and Oporto. Starting from AIDA, projects in the area of Business Intelligence are now at an ending point. AIDA has shown to be very useful in the process of extracting, loading and transforming info rmation in order to extract knowledge.

Acknowledgments. We are thankful to the Centro Hospitalar do Tâmega e Sousa (CHTS), in Penafiel, Portugal, for their help in implementing and testing of the above referred as AIDA, which is now being largely used in their premises. This work is funded by National Funds through the FCT - Portuguese Foundation for Science and Technology within project PEst-OE/EEI/UI0752/2011.

References

1. Machado, J., Abelha, A., Novais, P., Neves, J., Neves, J.: Quality of service in healthcare units. Int. J. Computer Aided Engineering and Technology 2(4) (2010)
2. Machado, J., Abelha, A., Neves, J., Santos, M.: Ambient intelligence in medicine. In: Proceedings of the IEEE-Biocas, Biomedical Circuits and Systems Conference, Healthcare Technology. Imperial College, London (2006)
3. Maes, P.: Modeling adaptive autonomous agents. Artificial Intelligence Magazine
4. Marreiros, G., Santos, R., Ramos, C., Neves, J., Bulas-Cruz, J.: ABS4GD: a multi-agent system that simulates group decision processes considering emotional and argumentative aspects. AAAI Spring Symposium Series. Stanford University (2008)
5. Peixoto, H., Machado, J., Neves, J., Abelha, A.: Semantic Interoperability and Health Records. In: Takeda, H. (ed.) E-Health 2010. IFIP AICT, vol. 335, pp. 236–237. Springer, Heidelberg (2010)
6. Jentzsch, A., Zhao, J., Hassanzadeh, O., Cheung, K., Samwald, M., Andersson, B.: Linking Open Drug Data. iTriplification Challenge (2009)
7. Duarte, J., Portela, C.F., Abelha, A., Machado, J., Santos, M.F.: Electronic Health Record in Dermatology Service. In: Cruz-Cunha, M.M., Varajão, J., Powell, P., Martinho, R. (eds.) CENTERIS 2011, Part III. CCIS, vol. 221, pp. 156–164. Springer, Heidelberg (2011)
8. Rigor, H., Machado, J., Abelha, A., Neves, J., Alberto, C.: A Webbased system to reduce the nosocomial infection impact in healthcare units. In: Webist, Madeira (2008)
9. Abelha, A., Machado, J., Santos, M., Allegro, S., Rua, F., Paiva, M., Neves, J.: Agency for Integration, Diffusion and Archive of Medical Information. In: Proceedings of the Third IASTED International Conference - Artificial Intelligence and Applications, Benalmadena, Spain (2002)
10. Sujansky, W.: Heterogeneous Database Integration in Biomedicine. Journal of Biomedical Informatics (2001)
11. Eichelberg, M., Aden, T., Riesmeier, J., Dogac, A., Laleci, G.: A survey and Analysis of Electronic Healthcare Record Standards. ACM Computing Surveys V(N 20YY), 1–47 (2005)
12. Häyrinen, K., Saranto, K., Nykänen, P.: Defenition, structure, content, use and impacts of electronic health records: A review of research literature. International Journal of Medical Informatics 77, 291–304 (2008)
13. Neves, J., Santos, M., Machado, J., Abelha, A., Allegro, S., Salazar, M.: Electronic Health Records and Decision Support - Local and Global Perspectives. WSEAS Transactions on Biology and Biomedicine 5(8), 189–198 (2008) ISSN 1109-9518
14. W3C - World Wide Web Consortium, http://www.w3C.org
15. Clark, T., Ocana, M.: Semantic Web Applications in Neuromedicine (SWAN) Ontology (2009), http://www.w3.org/TR/hcls-swan/

An Intelligent Patient Monitoring System

Rui Rodrigues[1], Pedro Gonçalves[1], Miguel Miranda[1], Carlos Portela[2],
Manuel Santos[2], José Neves[1], António Abelha[1], and José Machado[1]

[1] University of Minho, Computer Science and Technology Centre (CCTC),
Campus de Gualtar, Braga, Portugal
a47059@alunos.uminho.pt,
{pgoncalves,miranda,jneves,abelha,jmac}@di.uminho.pt
[2] University of Minho, ALGORITMI Center,
Campus de Azurem, Guimarães, Portugal
{cfp,mfs}@dsi.uminho.pt

Abstract. Intensive Care Units (ICUs) are a good environment for the
application of intelligent systems in the healthcare arena, due to its critical environment that requires diagnose, monitoring and treatment of
patients with serious illnesses. An intelligent decision support system
- *INTCare*, was developed and tested in CHP (Centro Hospitalar do
Porto), a hospital in Oporto, Portugal. The need to detect the presence or absence of the patient in bed, in order to stop the collection
of redundant data concerning about the patient vital status led to the
development of an RFID localisation and monitoring system - *PaLMS*,
able to uniquely and unambiguously identify a patient and perceive its
presence in bed in an ubiquitous manner, making the process of data collection and alert event more accurate. An intelligent multi-agent system
for integration of PaLMS in the hospital's platform for interoperability
(AIDA) was also developed, using the characteristics of intelligent agents
for the communication process between the RFID equipment, the INT-
Care module and the Patient Management System (PMS), using the HL7
standard embedded in agent behaviours.

Keywords: Medical Informatics, Patient Monitoring System, Intensive
Care Unit, Ambient intelligence, RFID; Multi-Agent System, HL7.

1 Introduction

There is a growing demand for the help of technology in the healthcare industry
to aid medical professionals in tasks such as patient identification and information storage, patient monitoring of vital signs, amongst others. Naturally, to
respond to this needs, new trends in technology become target of research in
hospital scenarios, and Ambient Intelligence (AmI) have been gaining a prepon-
derant role in the last years in such ambients. CHP (Centro Hospitalar do Porto),
a hospital in Oporto, Portugal, developed an intelligent decision support system
called INTCare [18] [22], aiming the real-time monitoring of patients, predicting
the dysfunction or failure of six organic systems within a short period of time,

L. Chen et al. (Eds.): ISMIS 2012, LNAI 7661, pp. 274–283, 2012.

and the patient outcome in order to help doctors deciding on the better treatments or procedures for each patient. This system also alerts the medical staff when vital signs of the patients are out of the normal range, starting an alert process. The Intensive Care Unit (ICU) become the unit chosen to implement such intelligent system due to its critical environment that require diagnose, monitoring and treatment of patients with serious illnesses. [18].

To complement the pervasive and intelligent system INTCare, a process of automatic identification of the patient is required. Such system must identify the patient in an unique and unambiguously manner, and also be able to admit, discharge and transfer a patient in a intelligent and pervasive way, sparing time from the medical staff in this process, allowing them to deal with other more important tasks.

PaLMS - a Patient Localisation and Monitoring System is a project aiming the simulation of a Radio-Frequency Identification equipment in the CHP's ICU unit, able to: (1) identify and admit the patient in the unit; (2) monitor his presence in the bed, prevent redundant data acquisition in situations when the patient isn't present, and prevent wrong alert situations by the INTCare system; (3) discharging or transferring the patient in a pervasive form, simplifying the process. Also this RFID-based identifying and monitoring system must be able to communicate with other hospital information services, thus adopt an interoperable form of communication. Multi-Agent Systems, the HL7 protocol to achieve interoperability, and RFID events, enabling ambient intelligence, will be the key aspects to achieve the goals.

2 Background

2.1 Ambient Intelligence

AmI is considered to be a new paradigm that supports the idea of enriching an environment with technology, mainly sensors and network-connected devices, building a system able to take decisions that benefit the users, based on real-time information gathered and historical data accumulated. In AmI, people are empowered through a digital environment that is aware of their presence and context, and is sensitive, adaptive, and responsive to their needs, habits, gestures and emotions. [3] [11].

Most of the times the concept of AmI is referred associated with other terms. Transparency, ubiquitous computing and pervasive computing, some already mentioned, are good examples. All of them are aligned towards the concept of the *disappearing computer*, described by Weiser [23], in which is pointed out that *"The most profound technologies are those that disappear. They weave themselves into the fabric of everyday life until they are indistinguishable from it"*.

This concept, associated with pervasive and ubiquitous computing, are concepts evolving in a plethora of applications in healthcare [17] [22]. The applications of AmI in hospitals can vary from enhancing safety for patients and professionals by following the evolution of patients after surgical intervention, monitor the patients and outpatients movements in hospital, tracking equipments and medical staff, and also for prevention, healthcare maintenance and checkups, benefitting the hospital management and quality of service provided [21] [15].

2.2 Sensors and Monitoring Systems

Many of the existing and emerging wireless networks such as cellular-oriented (2G/3G/4G), wireless LANs, satellites, and short range technologies like sensors, RFID, Bluetooth, Zigbee, and personal area networks can support the requirements for pervasive healthcare services, namely: comprehensive coverage; reliable access and transmission of medical information; location management; support for patient mobility [5] [4].

RFID (Radio-Frequency Identification) is an emerging technology in healthcare that can facilitate automating and streamlining safe and accurate patient identification, tracking, and processing important health related information in health care sector, therefore a contributing technology to the implementation of ambient intelligence scenarios in hospitals [25] [7].

RFID technology mainly consists of a transponder (tag) which contains electronically stored information, usually an EPC code, and is able to be read from up to several meters, an antenna reader and an information system.

2.3 HL7

The growing need for interoperability lead to the development of several health-related standards. *HL7* - standing for Health Level Seven, is an ANSI-accredited standard dedicated to providing a comprehensive framework and related standards for the exchange, integration, sharing, and retrieval of electronic health information that supports clinical practice and the management, delivery and evaluation of health services [1] [16]. Seven represents the seventh layer of the Open System Interconnection (OSI) communication model. The HL7 standard is not bound to this architecture, but it is the most widely used in healthcare interoperability [14].

2.4 Multi-Agent Systems - MAS

Agents and the *Agent-Oriented Programming* are concepts correlated to the field of artificial intelligence, and their importance is growing in the healthcare environment, namely in the quest for interoperability. Agents are commonly considered as a computing artefact used in software or hardware devices, capacitated with the properties of: autonomy - acting independently from the action of humans; reactivity - such entities are situated in an environment that can perceive through sensors and act in reaction to stimuli (e.g., revising their beliefs according to or in reaction to new inputs; pro-activity - exhibiting intelligent problem solving capabilities, for example, towards the achievement of short or long time goals; social-behaviour - agents are aware of one another and are able to interact between them [13]. MAS consist on a set of agents that communicate with each other and work together towards common goals, with a degree of reactivity and / or reasoning [24].

2.5   INTCare and AIDA

INTCare, an intelligent decision support system for intensive medicine, was developed by both researchers from the University of Minho and CHP, implemented in the hospital's ICU. It makes use of intelligent agents capable of autonomous actions to meet its goals [9] [20]. This system predicts in real-time organ failure and mortality assessment and, according to these predictions, it suggests therapeutic treatments. The system includes some features such as [20]: online learning; real-time monitoring; adaptability data mining models; decision models; optimisation; intelligent agents; accuracy; safety; pervasiveness; privacy; secure access from exterior; user policy.

AIDA stands for Agency for Integration, Diffusion and Archive of Medical Information. It was created by both researchers from University of Minho and CHP, although it is already fully implemented in several major portuguese healthcare institutions. AIDA can be described an agency that provides intelligent electronic workers (agents) that present a pro-active behaviour, and are in charge of tasks such as: communications among the sub-systems that make the whole one, sending and receiving information (e.g. medical or clinical reports, images, collections of data, prescriptions), managing and saving the information and answering to information requests, in time. The main goal is to integrate, diffuse and archive large sets of information from heterogeneous sources (departments, services, units, computers, medical equipments) [2].

3   HL7 in Multi-Agent Systems

PaLMS is an event-based monitoring system developed using HL7 standard messages embedded in a multi-agent programming ambient. In an environment where the demand for middleware both for production and legacy systems are constant, the agent paradigm demonstrates an intuitive advantage in organisational development in terms of creation of such services, and are considered to have great potential towards interoperability in healthcare information systems [8].

To avoid the limitation of the usual of information systems found in healthcare institutions, in which equipments usually either communicate through the usage of standards in a loosely-coupled manner, i.e. directly with an information system (Radiological Information System, Cardiological Information System, ...) or with a proprietary system which can in its term be compatible or not with other information systems, the adoption of HL7 standard in all the communication process represents an important step in extending the interoperability for the development of health information exchange. Also the agent paradigm offers the hospital information services, besides its characteristics of modularity, scalability and adaptability, the empowerment of being able to solve different problems through intelligent agents, hard to overcome before [11] [12] [10].

Besides the fact that the HL7 standard is completely distinct from agent communication standards, it is possible to implement an HL7 service under an agent paradigm. This brings the advantage of using the agent system's vast interoperability capability, able to be embedded with the most particular behaviours. These behaviours can become increasingly effective if they use machine learning and other artificial intelligence techniques in order to adapt to the existing environment and being able to prevent errors and correct the flow of information and extraction of knowledge within the institution.

Among the possible tools for analysis and development of PaLMS, the *WADE* platform (Workflows and Agents Development Environment) was the chosen one. WADE is a domain independent platform built on top of the JADE framework (Java Agent DEvelopment Framework), developed using the cross-platform technology Java, which allows a strong connection among two similar computational paradigms, such as The Agent and The Object Oriented Programming ones, as well as the possibility to easily integrate the rich Java libraries into the agent's behaviours.

4   RFID Event-Simulation

The project aims the simulation of a full RFID system installed in the ICU room of CHP. Ideally, at least one RFID antenna should be placed in each bed in order to detect the patient presence in bed, or his absence. This number of RFID antennas will be optimised, studying the antenna's range in the unit room. At the admission stage, when a patient enters the ICU unit for healthcare services, a bracelet with a RFID tag is placed in the patient arm or leg. The bracelet's tag contains its own EPC code, similar to the bar-code, which is unique and distinguishable from any other EPC code. A bed monitoring system is required in order to associate each bed to the RFID's antennas placed in the unit. This bed monitoring system communicates with each bed, receiving information, and communicates with the hospital's AIDA platform to receive and send a number of events, concerning the patient's status.

The hospital's PMS communicates with the ICU's service through HL7 messages, which will be handled by the PaLMS platform. This platform is inserted in the AIDA, as well as the PMS. First, we enumerate all the type of events concerning the patient "status" in the hospital.

HL7 is written from the assumption that a real-time event in a clinical context creates the need for data to flow among systems. This real-world event is called a *trigger event*. There are many trigger events available [6], and some take special importance in this project. Each event created and described further on in this chapter, are associated with event triggers, more specifically, some trigger events inserted in the "ADT" type of message.

We have developed six type of events, represented in Figure 1, and described further:

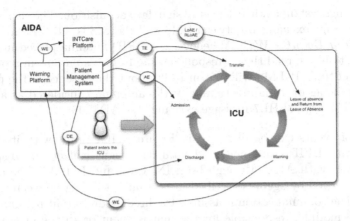

Fig. 1. Simulation of the flow of events with HL7 messages, enabling the communication process between heterogeneous systems among the different stages of the patient inside the ICU

1. **Admission Event - AE** - PMS sends an HL7 message to the ICU informing about the patient admission in the ICU, requiring an EPC tag to associate. The medical staff associate a bracelet, with its EPC code, to the admission request, and associate to the patient's bracelet an unoccupied bed. The antenna placed in the bed starts the reading process, periodically, verifying the patient's presence. The HL7 trigger-event ADT_A01 represents the "admit patient" message, used for admitted patients only. This is the type of HL7 message received and handled in the ICU by the PaLMS platform.
2. **Discharge Event - DE** - At the time the patient leaves the unit and the hospital itself, a discharge event is created by the hospital's PMS system. HL7's ADT_A03 message represents the "discharge of patients", associated to DE. PaLMS receives this message and ends the antenna's reading process as well as the patient's vital data acquisition by the INTCare platform.
3. **Transfer Event - TE** - An HL7 message is sent by the PMS informing the patient's transfer to another unit. PaLMS receives this message, sent as a ADT_A02 trigger-event, and acts similarly to the DE, stopping the data collection of INTCare and the reading process from the RFID equipment.
4. **Leave of Absence Event - LoAE** - A patient may have to get out of bed, for instance, to do exams, without leaving the unit permanently. This is considered as a leave of absence, with the corresponding HL7 message - ADT_A21, sent by PMS to the ICU informing the temporarily absence of the patient. The RFID stops the reading process, until the patient returns, as well as the INTCare monitoring system stops the patient's data acquisition. This way, the values of data signs acquired by the INTCare (which gathers data uninterruptedly) get out of range, but no warning signals are sent to medical staff since there's no one in the bed.
5. **Return from Leave of Absence Event - RLoAE** - At the time the patient returns, the antenna starts the reading process again and INTCare

monitoring system restarts the vital sign data acquisition. The correspondent HL7 message exchanged between PMS and PaLMS is ADT_A22.

6. **Warning Event - WE** - Whenever the RFID antenna placed in a certain bed isn't able to read the correspondent tag from it's patient in a predefined period of time, PaLMS will inform PMS about the absence of the patient, a misplaced tag or a malfunctioning of the antenna, considered as a warning event. The type of HL7 message exchanged is ADT_08.

This flow of events can be observed in Figure 1. ICU unit, with its ten beds equipped with RFID equipment and vital data signs collect and display systems, interact with AIDA through both INTCare and PaLMS. These interactions are achieved by agents described in Section 5, sending between them HL7 messages. The communication is assured by three modules that integrate AIDA: the PaLMS module, responsible for the management of all the events relating PaLMS system (AE, RE, etc.); the INTCare Module, responsible for the vital signs data acquisition and processing; the alert module, responsible for the alert situations, (concerning the vital signs of the patient which became, for instance, out of the normal range, or the RFID monitoring system that detected the absence of the bracelet from a certain patient).

5   Agent Architecture

Figure 2 represents the agent communication process after the implementation of PaLMS in the perspective of the modules that incorporates the AIDA platform. Agents in this project are responsible for all the exchange of information between AIDA's modules and the ICU hardware equipment. Therefore, several agents were created.

First of all, there is the need for an agent to act as an HL7 server, since the HL7 messages are sent and received typically by a client-server configuration. This agent, the *Patient Information Server Agent*, receives information about patients, such as admissions, discharges, transfers, leave of absences and the correspondent return from leave of absences. This information is sent by the AIDA platform, trough its Patient Management System[1]. At this stage, the server agent communicates the action to other two agents. The *PaLMS HL7 Message Server Agent* and the *Patient Warning Handler Agent*. The first one is an agent which also acts as a server for HL7 messages, receives the events transmitted, and automatically informs the RFID equipment to the action required. If the event is an AE, the medical staff in the ICU will receive the tag to associate to the patient and will proceed to the identification process. If the event is an DE, TE or LoAE, this agent will inform the RFID system to stop the reading process. in case of a RLoAE, the RFID is informed that the reading process shall start again. Also the *PaLMS HL7 Server Message* agent deals with the RFID data obtained in the bed. When a patient is not present, an HL7 event will be

---

[1] In Portugal, the patient management system adopted and implemented in almost all of healthcare institutions is SONHO [19].

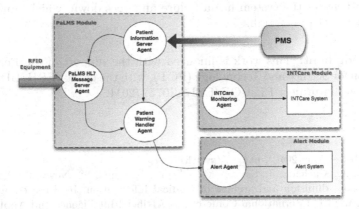

Fig. 2. Description of the communication of the three modules inside AIDA, enabled by a series of agents responsible for incoming messages and tasks according to them.

generated by the RFID equipment. This require a software monitor for the RFID equipment, able to create HL7 messages, which can be developed in the future, or can be created by the software of the equipment's developer. In order to stop the warning signals caused by vital signs out of range, the system must also communicate with the second agent mentioned. The *Patient Warning Handler Agent* will receive inputs from both *PaLMS HL7 Message Server Agent* and *Patient Information Server Agent*, because, for instance, a DE must stop the data storage and warning signals and this has to be informed to INTCare Module and to the Alert Module. Inherently, the *Patient Warning Handler Agent* has to communicate to the two modules associated to the INTCare system and the alert system.

6   Conclusions and Future Work

As this system aims for the detection of the patient in the bed, making the data collection of the patient dependent from this information acquired by an RFID system through antennas and bracelet tags, we can say that the PaLMS system ensures this kind of monitoring. The MAS, with its components fully capable of communicating between them and with their tasks well defined, ensures that the data storage of the patients' vital signs stop when the patient isn't in bed, and consequently the alert process so that the medical staff is warned about any error situation is cancelled when its not necessary.

Also, this system open the perspective of patient monitoring in other hospital units. RFID's potentiality can assure the tracking of patients, medical staff, equipments, person flow inside units, whether they are medical staff or not. It can be an interesting perspective for a future implementation in the hospital. Also, the development of middleware for the RFID equipment is a possibility, with an interface to medical staff interact directly. Persistence of data exchanged between agents would be achieved by inserting it into databases, preventing data loss whenever the system has a failure and goes down, which could be an important aspect to research.

Acknowledgment. This work is financed with the support of the Portuguese Foundation for Science and Technology (FCT), with the grant SFRH/BD/70549/ 2010 and within project PEst-OE/EEI/UI0752/2011.

References

1. Hl7 website (May 2012), http://www.hl7.org/
2. Abelha, A., Machado, J., Santos, M., Allegro, S., Paiva, M., Neves, J.: Agency for integration, diffusion and archive of medical information. In: Proceedings of the Third IASTED International Conference - Artificial Intelligence and Applications, Benalmadena, Spain (2003)
3. Augusto, J., Mccullagh, P.: Ambient intelligence: Concepts and applications. Computer Science and Information Systems 4(1), 1–27 (2007)
4. Benini, L., Farella, E., Guiducci, C.: Wireless sensor networks: Enabling technology for ambient intelligence. Microelectronics Journal 37(12), 1639–1649 (2006)
5. Dağtaş, S., Pekhteryev, G., Şahinoğlu, Z., Çam, H., Challa, N.: Real-time and secure wireless health monitoring. International Journal of Telemedicine and Applications 2008(135808) (2008)
6. Dolin, R.H., Alschuler, L., Beebe, C., Biron, P.V., Boyer, S.L., Essin, D., Kimber, E., Lincoln, T., Mattison, J.E.: The HL7 Clinical Document Architecture. Journal of the American Medical Informatics Association 8(6), 552–569 (2001)
7. Fisher, J.A., Monahan, T.: Tracking the social dimensions of rfid systems in hospitals. International Journal of Medical Informatics 77(3), 176–183 (2008)
8. Isern, D., Sánchez, D., Moreno, A.: Agents applied in health care: A review. International Journal of Medical Informatics 79(3), 145–166 (2010)
9. Jennings, N.R.: On agent-based software engineering. Artificial Intelligence 117(2), 277–296 (2000)
10. Lopez, D.M., Blobel, B.G.: A development framework for semantically interoperable health information systems. International Journal of Medical Informatics 78(2), 83–103 (2009)
11. Machado, J., Abelha, A., Neves, J., Santos, M.: Ambient intelligence in medicine. In: IEEE 2006 Biomedical Circuits and Systems Conference Healthcare Technology, BioCAS 2006, pp. 94–97 (2006)
12. Machado, J., Alves, V., Abelha, A., Neves, J.: Ambient intelligence via multiagent systems in the medical arena. Engineering Intelligent Systems for Electrical Engineering and Communications 15(3), 151–157 (2007)
13. Machado, J., Miranda, M., Abelha, A., Neves, J., Neves, J.: Modeling Medical Ethics through Intelligent Agents. In: Godart, C., Gronau, N., Sharma, S., Canals, G. (eds.) I3E 2009. IFIP AICT, vol. 305, pp. 112–122. Springer, Heidelberg (2009)
14. Miranda, M., Pontes, G., Gonçalves, P., Peixoto, H., Santos, M., Abelha, A., Machado, J.: Modelling Intelligent Behaviours in Multi-agent Based HL7 Services. In: Lee, R. (ed.) Computer and Information Science 2010. SCI, vol. 317, pp. 95–106. Springer, Heidelberg (2010)
15. Miranda, M., Abelha, A., Santos, M., Machado, J., Neves, J.: A Group Decision Support System for Staging of Cancer. In: Weerasinghe, D. (ed.) eHealth 2008. LNCIST, vol. 1, pp. 114–121. Springer, Heidelberg (2009)

16. Ohe, K., Kaihara, S.: Implementation of hl7 to client-server hospital information system (his) in the university of tokyo hospital. Journal of Medical Systems 20, 197–205 (1996)
17. Orwat, C., Graefe, A., Faulwasser, T.: Towards pervasive computing in health care - a literature review. BMC Med. Inform. Decis. Mak. 8, 26+ (2008)
18. Portela, C.F., Santos, M.F., Silva, Á., Machado, J., Abelha, A.: Enabling a Pervasive Approach for Intelligent Decision Support in Critical Health Care. In: Cruz-Cunha, M.M., Varajão, J., Powell, P., Martinho, R. (eds.) CENTERIS 2011, Part III. CCIS, vol. 221, pp. 233–243. Springer, Heidelberg (2011)
19. Ribeiro, J.V., Geirinhas, P.M.: Icnp in sonho. In: Mortensen, R.A. (ed.) ICNP in Europe: Telenurse. Technology and Informatics, pp. 131–136. IOS Press (1997)
20. Santos, M.F., Portela, F., Vilas-Boas, M., Machado, J., Abelha, A., Neves, J., Silva, A., Rua, F.: Information architecture for intelligent decision support in intensive medicine. W. Trans. on Comp. 8(5), 810–819 (2009)
21. Varshney, U.: Pervasive healthcare and wireless health monitoring. Mobile Networks and Applications 12, 113–127 (2007)
22. Vilas-Boas, M., Gago, P., Portela, F., Rua, F., Silva, Á., Santos, M.F.: Distributed and real time data mining in the intensive care unit. In: 19th European Conference on Artificial Intelligence - ECAI Lisbon, pp. 51–55 (2010)
23. Weiser, M.: The computer for the twenty-first century. Scientific American 265(3), 94–104 (1991)
24. Weiss, G.: Multiagent Systems: A Modern Approach to Distributed Artificial Intelligence. The MIT Press (July 2000)
25. Wicks, A.M., Visich, J.K., Suhong, L.: Radio frequency identification applications in hospital environments. Hospital Topics 84(3), 3–8 (2006)

Corpus Construction for Extracting
Disease-Gene Relations

Hong-Woo Chun, Sa-Kwang Song, Sung-Pil Choi*, and Hanmin Jung

Korea Institute of Science and Technology Information (KISTI)
245 Daehak-ro, Yuseong-gu, Daejeon, South Korea
{hw.chun,esmallj,spchoi,jhm}@kisti.re.kr

Abstract. Many corpus-based statistical methods have been used to tackle issues of extracting disease-gene relations (DGRs) from literature. There are two limitations in the corpus-based approach: One is that available corpora for training a system are not enough and the other is that previous most research have not deal with various types of DGRs but a binary relation. In other words, analysis of presence of relation itself has been a common issue. However, the binary relation is not enough to explain DGR in practice. One solution is to construct a corpus that can analyze various types of relations between diseases and their related genes.

This article describes a corpus construction process with respect to the DGRs. Eleven topics of relations were defined by biologists. Four annotators participated in the corpus annotation task and their inter-annotator agreement was calculated to show reliability for the annotation results.

The gold standard data in the proposed approach can be used to enhance the performance of many research. Examples include recognition of gene and disease names and extraction of fine-grained DGRs. The corpus will be released through the GENIA project home page.

1 Introduction

Many natural language processing (NLP) techniques are used to automatically extract and analyze meaningful information from literature. In biomedical domain, NLP techniques-based named entity recognition (NER) [1] and relation extraction (RE) [2] are also encouraging issues. One of the most practical research topics deals with specific diseases and their relevant genes or proteins. Many related research focused on only presence of relations between diseases and genes or proteins in a given sentence. In other words, they focused on analyzing not deep semantic meaning of relations but surface expression of them. To extract deep semantic meaning of disease-gene relations (DGRs) automatically, a corpus-based statistical approach is a popular approach, but this approach needs gold-standard data.

---

* Corresponding author.

L. Chen et al. (Eds.): ISMIS 2012, LNAI 7661, pp. 284–292, 2012.

This article describes the process of construction of a gold-standard data as for the relations between diseases and their relevant genes. We focus on not only the cases of binary relations, but also the case of fine-grained relations such as causal factor (etiology), marker of diseases (clinical marker). The binary relation includes all kinds of fine-grained relations and each of the fine-grained relations does not exclusive: a disease-gene pair can be included more than one fine-grained relation. In this article, we call these fine-grained relation *topic-classified relation*, and eleven topics were defined by biologists and medical doctors. Considering popular diseases as research topics and a number of published papers, we decided *prostate cancer* as a disease in the proposed approach. Example 1 includes a relation between prostate cancer (prostate adenocarcinoma) and gene (PSA) and PSA is a clinical marker of *prostate adenocarcinoma*. Small G and D with a square bracket present a gene and a disease names, respectively. This disease and gene name expression style is used for all example sentences in this article.

Ex. 1. The clinical utility of $[PSA]_G$ and PAP for early detection of $[prostate$ $adenocarcinoma]_D$, however, requires distinction between prostate adenocarcinoma and prostate nodular hyperplasia.

Such corpus can be used to develop an automatic NE recognition and DGR extraction systems. Such systems are receiving increased attention, particularly from medical doctors, pharmacists and biologists, as they have the potential of reducing the burden on researchers to explore the extensive pool of literature. Moreover, they can receive some help to give a diagnosis of disease and develop medicines if DGR information is provided.

In most of the previous articles only algorithms for information extraction have been described, but the process of corpus construction has not mentioned. However, not only algorithms but also a right corpus with respect to the corresponding purpose play a role in extracting information from literature.

2 Related Work

Because the tasks of corpus construction are highly dependent upon target applications, DGR extraction research using the corpus-based statistical methods are introduced in this section.

Chen *et al.* proposed a method for collecting Alzheimer's disease-related proteins [3]. There were 65 seed genes collected from the OMIM database and mapped to 70 Alzheimer's disease-related proteins in HUGO and SwissProt databases. They then show 765 ranked proteins which were collected using protein relations of the online predicted human relation database (OPHID). Rosario *et al.* classified seven semantic relations between entities disease and treatment using several machine learning techniques, including *hidden Markov models* and textitneural networks [4]. The relations were *cure, only disease, only treatment, prevent, vague, side effect,* and textitno cure. Some of these semantic relations described binary relations between diseases and treatments. Using 3,662 labeled

Table 1. Correlation between cancers and genes: Results of Chen *et al.*

(%)	astrocytoma	bladder cancer	brain tumor	breast cancer	carcinoid tumor	cervical cancer
CDK4	34.29	0.0	33.09	0.0	0.0	0.0
EGFR	38.57	0.0	34.41	33.72	0.0	0.0
CCND3	30.86	0.0	30.44	0.0	0.0	0.0
SIL	0.0	0.0	0.0	0.0	0.0	31.58
JUN	0.0	31.20	0.0	30.22	0.0	0.0
⋮	⋮	⋮	⋮	⋮	⋮	⋮

sentences in MEDLINE abstracts and dynamic hidden Markov models, the authors achieved an F-measure of 71%.

Chen *et al.* focused on the cause of cancer and calculated relevance scores between cancers and genes using co-occurring frequency of cancer and gene names in MEDLINE abstracts. Table 1 shows the relevance scores between cancers and genes they provided in the paper. They dealt with 59 cancers, 3,270 chromosome regions, and 18,692 genes [5].

Although above related work might have consumed time and human efforts to build the corpora, they have focused on not the corpus construction but introduction of results that were found using the corpus.

3   Construction of Gold Standard Data

3.1   Goals and Method

We aim to construct a corpus for extracting relations between disease names and their relevant gene names from the biomedical literature. When a DGR extraction system is developed using the corpus, the contribution can be mentioned as follows:

- We classified disease-gene relations based on various topics to enable human genetics experts and oncologists to use the results of the proposed approach.
- We identified disease and gene names with ID tags that are used in six publicly available biological databases to determine the utility of the proposed approach. Numerous previous studies have identified entities or relations that are not grounded in any explicit external model of the world, but rather simply point to substrings in the input text. Such outputs are of intrinsically limited value [6]. Considering scalability of the results, links to other databases are valuable.

Figure 1 is architecture of the corpus construction. Our system first collects sentences that contain at least one pair of disease and gene names, using the dictionary-based longest matching technique. Four experts were participated in the annotation task. To show the reliability for the annotation results (gold standard data), inter-annotator agreement rate was calculated. Finally, topic-classified relations with their evidential contexts were collected.

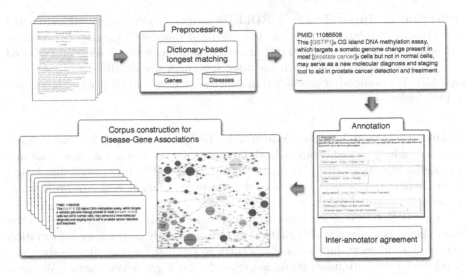

Fig. 1. Architecture of Corpus Construction

Table 2. Selected unique identifiers of semantic types (TUIs)

T019	Congenital abnormality
T020	Acquired abnormality
T033	Finding
T037	Injury or poisoning
T046	Pathologic function
T047	Disease or syndrome
T048	Mental or behavioral dysfunction
T049	Cell or molecular dysfunction
T050	Experimental model of disease
T184	Sign or symptom
T190	Anatomical abnormality
T191	Neoplastic process

3.2   Corpus Construction

To determine the utility of the results of the proposed approach, we applied a dictionary matching method to recognize disease and gene names. The dictionary matching method used disease and gene dictionaries. Thus, all disease and gene names in the results have ID tags that are used in the publicly available biological databases. When a disease-gene pairs sentence contained more than one disease name and more than one gene name, the system made sufficient copies of the sentence to accommodate all possible disease-gene name pairs. We call these copies *disease-gene pair instance sentences. Instance sentence* presents the disease-gene pair instance sentence in this article.

Collection of Input Data. MEDLINE abstracts were collected using 248 prostate cancer-related names selected from our disease dictionary. From these abstracts, we generated 2,503,037 instance sentences using dictionary-based longest matching technique. All instance sentences are candidates for analyzing DGR. We randomly chose 3,900 instance sentences and they were annotated by four biologists.

Dictionary-Based Longest Matching. To link each output disease or gene name to publicly available biomedical databases, we created human gene and disease dictionaries by merging the entries of numerous public biomedical databases. These dictionaries provide gene- and disease-related names and cross-references between the original databases.

A unique *EntrezGene* identifier for genetic loci is assigned to each entry in the human gene dictionary, which enabled us to consistently merge gene information contained in different databases. Each entry in the merged gene dictionary holds all relevant literature information associated with a given gene name. We used five public databases to build the gene dictionary: *HUGO, EntrezGene, SwissProt, RefSeq, and DDBJ.*Each entry in the merged gene dictionary consists of five attributes: gene name, gene symbol, gene product, chromosomal band, and PubMed ID tags. The current version of the gene dictionary contains a total of 34,959 entries with 19,815 HUGO-approved gene symbols, 19,788 HUGO-approved gene names, and 29,470 gene products. Note that there are numerous alias gene symbols and gene names in these entries.

We used the Unified Medical Language System (UMLS) to collect disease-related vocabulary. We selected 12 unique identifiers of semantic types (TUIs) that correspond to disease names, abnormal phenomena, or symptoms (Table 2). From these TUIs, we extracted 431,429 unique identifiers for strings (SUIs) and stored them as a disease-related lexicon.

3.3   Types of Annotation and Definition of a Relation and Topics

The types of annotation in our corpus are the following:

- Disease and gene names:
  Biologists annotated whether recognized disease and gene names by the dictionary matching method were *correct*.
- Relations between disease and gene names:
  Biologists annotated whether a binary relation *existed* between disease and gene names.
- Topic-classified relations:
  Biologists annotated topic-classified relations based on eleven topics.

For the annotation of binary relations between gene and prostate cancer names, the biologists considered three aspects. In other words, the annotator judged a co-occurrence as *correct* if any of the following three types of relations had been described in the co-occurrence.

Table 3. Inter-agreement rate

	F1(%)	Number of positives in 300 instances			
		Annotator1	Annotator2	Annotator3	Annotator4
Gene	96.88	238	247	246	257
Disease	99.75	300	299	300	298
Relation	94.72	229	240	232	250
Etiology	87.50	4	4	4	4
Clinical marker	69.53	88	92	86	88
Study description	60.47	69	87	76	59
Genetic Variation	80.62	31	28	33	14
Gene expression	70.85	74	78	93	91
Pharmacology	63.23	26	27	31	19
Functions of gene products	59.84	42	49	39	42
Sub-cellular localization	50.17	12	17	13	6
Risk factor	58.39	16	13	19	12
Tumor biology	48.63	44	54	47	37
Negation	49.65	13	11	8	7

1. Pathophysiology, mechanisms of prostate cancer, including etiology, causes of prostate cancer.
2. Therapeutic significance of genes or gene products; specifically, classification of genes or gene products based on their current therapeutic use and their potential as therapeutic targets.
3. Use of genes and gene products as markers for prostate cancer risk, diagnosis, and prognosis.

In addition to the binary relation, we classified disease and gene names and their relations based on the following eleven topics: *Etiology, Clinical marker, Study description (method), Genetic variation, Gene expression, Pharmacology, Functions of gene products, Sub-cellular localization, Risk factor, Tumor biology, and Negation.* All topics are not mutually exclusive, so certain instance sentence can be classified by the multiple topics.

3.4   Inter-agreement Rate for Annotation

Instance sentences were annotated by four biologists, based on their domain knowledge and the annotation policy. Although all annotators are experts and have the same annotation policy, they have many different opinions. Table 3 explains the inter-agreement rates. For calculating the inter-agreement rates, 300 instance sentences were randomly selected and were annotated for disease and gene name recognition, relation extraction, and extraction of eleven topics. We assumed that one annotator's opinion is correct, and calculated precision, recall, and F1-measure for the remained annotators' opinions. We conducted the calculations four times and averaged the performances of them. For the disease and gene name recognition and relation extraction, there are a lot of positives

Table 4. Number of positives in the annotated corpus

	Number of positives (Distribution(%))
Relation	3,167 (81.2)
Etiology	52 (1.3)
Clinical marker	1,231 (31.6)
Study description	1,040 (26.7)
Genetic Variation	277 (7.1)
Gene expression	1,060 (27.2)
Pharmacology	352 (9.0)
Functions of gene products	561 (14.4)
Sub-cellular localization	158 (4.1)
Risk factor	183 (4.7)
Tumor biology	605 (15.5)
Negation	145 (3.7)

Table 5. Analysis of prostate cancer and gene name pairs

	# of pairs in only POS	# of pairs in only NEG	# of pairs in POS/NEG
Relation	967	271	108
Etiology	8	124	214
Clinical marker	92	90	164
Study description	260	576	239
Genetic Variation	33	113	200
Gene expression	81	63	202
Pharmacology	31	106	209
Functions of gene products	47	86	213
Sub-cellular localization	9	117	220
Risk factor	8	119	219
Tumor biology	45	75	226
Negation	37	0	343

Notes) POS: positive instance sentence,
NEG: negative instance sentence,
POS/NEG: both positive and negative instance sentence.

and they achieved comparatively high inter-agreement rates. However, the inter-agreement rates for extraction of topic-classified relations are relatively low, and F1-measure was not a proper measure for the topics that have few positives.

3.5   Data Analysis

The number of annotated instance sentences is 3,900. The instance sentences are from 2,471 abstracts, so there are some of the instance sentences are from the same abstract. The annotated instance sentences contain 3,167 (81.2%) positive relations. Table 4 show the number of positives for all topics in annotated instance sentences. The percentages in tables are the distribution of positives of each topic for all instance sentences.

We analyzed our data based on pairs. Table 5 shows the number of pairs based on given conditions. The number of all disease-gene pairs is 1,346. There are many pairs appeared in both positive and negative instance sentences. It means the binary relations and topic-classified relations were annotated not depending on only two entity names and contexts provide important clues to classify the binary relation and topic-classified relations.

4   Conclusion

Extracting disease-gene relations from literature is an encouraging research topic in the biomedical domain. To extract various topic-classified relations and apply a corpus-based statistical approach, a robustly constructed gold-standard data is necessary. All processes for corpus construction are described, and quality and reliability are shown. Not only F-measure but also Kappa score that has been shown as a proper measure for the inter-annotator agreement will be calculated in the near future.

Prostate cancer and its relevant genes with respect to eleven topics are focused on the corpus. They were defined by biologists and medical doctors based on popular research topics, number of published papers, and practical usage. Other corpus with respect to gastric cancer is under construction. A preliminary research using the corpus has been done. Recognition of prostate cancer and gene name, and extraction of relations are those research and have shown encouraging results [7]. The corpus will be released through the GENIA project home page.

References

1. Song, S.-K., Choi, Y.-S., Chun, H.-W., Jeong, C.-H., Choi, S.-P., Sung, W.-K.: Multi-words Terminology Recognition Using Web Search. In: Kim, T.-H., Gelogo, Y. (eds.) UNESST 2011. CCIS, vol. 264, pp. 233–238. Springer, Heidelberg (2011)
2. Chun, H.W., Jeong, C.H., Song, S.K., Choi, Y.S., Choi, S.P., Sung, W.K.: Composite Kernel-based Relation Extraction using Predicate-Argument Structure. In: Kim, T.-H., Adeli, H., Ma, J., Fang, W.-C., Kang, B.-H., Park, B., Sandnes, F.E., Lee, K.C. (eds.) UNESST 2011. CCIS, vol. 264, pp. 269–273. Springer, Heidelberg (2011)
3. Chen, J.Y., Shen, C., Sivachenko, A.Y.: Mining Alzheimer disease relevant proteins from integrated protein interactome data. In: The Pacific Symposium on Biocomputing (PSB), pp. 367–378 (2006)
4. Rosario, B., Hearst, M.A.: Classifying Semantic Relations in Bioscience Texts. In: Proc. of the Annual Meeting of the relation of Computational Linguistics (ACL), pp. 431–438 (2004)

5. Chen, S., Wen, K.: An integrated system for cancer-related genes mining from biomedical literatures. International Journal of Computer Science and Applications 3(1), 26–39 (2006)
6. Chun, H.W., Tsuruoka, Y., Kim, J.D., Shiba, R., Nagata, N., Hishiki, T., Tsujii, T.: Extraction of gene-disease relations from MEDLINE using domain dictionaries and machine learning. In: The Pacific Symposium on Biocomputing (PSB), pp. 133–154 (2006)
7. Chun, H.W., Tsuruoka, Y., Kim, J.D., Shiba, R., Nagata, N., Hishiki, T., Tsujii, T.: Automatic recognition of topic-classified relations between prostate cancer and genes using MEDLINE abstracts. BMC Bioinformatics 7, S4 (2006)

On Supporting Weapon System Information Analysis with Ontology Model and Text Mining

Jung-Whoan Choi[1], Seungwoo Lee[2,*], Dongmin Seo[2], Sa-Kwang Song[2], Hanmin Jung[2], Sang Hwan Lee[2], and Pyung Kim[3]

[1] Defense Agency for Technology and Quality,
Cheongnyang P.O. Box 276, Seoul, 130-650, Korea
[2] Korea Institute of Science and Technology Information,
245 Daehangno, Yuseong-gu, Daejeon, 305-806, Korea
[3] Jeonju National University of Education,
50 Seohak-ro, Wansan-gu, Jeonju, 560-757, Korea
c0802@hanafos.com,
{swlee,dmseo,esmallj,jhm,sanglee}@kisti.re.kr, pyung@jnue.kr

Abstract. Issues and changes on weapon systems of a country is very critical to its adjacent countries and therefore each country needs to collect, analyze, and monitor them. Information analysis in defense domain requires very high accuracy due to the characteristics of the domain in view points of national security. However, it is impossible to manually analyze defense-related data with limited manpower because such data is huge and also updated and added frequently. In this paper, we present an implementation of a defense-domain information analysis supporting system, especially focusing on information analysis of weapon systems. It categorizes all weapon systems, extracts their specification data, descriptions, development, and status, and furthermore provides their trends based on ontology modeling and text mining.

Keywords: defense information analysis, weapon system, ontology, text mining.

1   Introduction

The growth of the Internet has enabled to publish and share huge amount of data ranging from academic to social or even private data through Web. Defense-related data is no exception. There are many Web sites or databases providing defense and military information[1]. They deal with issues and changes on weapon systems of each country as well as world-wide political issues. Weapon system usually includes supporting equipment of a weapon as well as the actual weapon hardware itself. An issue or change on weapon systems of a country may be very critical to its adjacent countries because it could threaten their national defense and security. There is also interoperability between offensive and defensive weapons. For example, if a new

---

[*] Corresponding author.
[1] See IHS Jane's Defense (http://www.janes.com) and Yongwon Yu's Military World (http://bemil.chosun.com), for examples.

L. Chen et al. (Eds.): ISMIS 2012, LNAI 7661, pp. 293–300, 2012.

offensive weapon is developed or imported by a country then its adjacent countries also need to develop or import a new defensive weapon in response to that offensive weapon, and vice versa. Therefore, a country need to collect, analyze and monitor all issues and changes on weapon systems, especially, in order to make a decision on which and how many weapon systems it should develop or import.

Unlike general information analysis, defense-domain information analysis requires very high accuracy due to the characteristics of the domain in the point of national security view. It also needs highly domain-specific knowledge such as terminologies and entity-relationships on weapon systems. However, it is impossible to manually analyze defense-related data – usually unstructured or semi-structured documents – with limited manpower because such data is huge, and also updated and added frequently. In addition, manual work by experts does not guarantee the reliability of the analyzed information because it can cover only small part of data while it requires very high costs. In order to resolve this issue, Choi and Lim [1] proposed and designed a virtual customized information analysis system which utilizes natural language processing, text mining and data mining as base technologies and reduces manual work of experts. In this paper, we introduce the first implementation of such a defense-domain information analysis system, especially focusing on information analysis of weapon systems. We first model weapon system information as an ontology and then constructs ontology instances from defense-domain literatures using text mining. Ontology is especially used for standardizing specification data and representing their hierarchy, and text mining is used for extracting specification data (i.e., attributes and values), descriptions, development, and status of weapon systems from text. The system finally serves classification of all weapon systems with the extracted information and analyzes their trends.

The paper is organized as follows. Section 2 presents some previous works related to military information analysis. Section 3 identifies the requirements and describes ontology model and text mining applied for analyzing information on weapon systems and Section 4 explains the analysis services provided at current implementation. The conclusion is given in Section 5.

2   Related Work

There were several works dealing with defense and military information in the quite different points of view. First, Grossman et al. [2] and Geiselman and Samet [3] explored military intelligence. Grossman et al. [2] observed team-based distributed work and investigated the contrasts between weak and strong information analysis and synthesis as a type of cognitive work. Geiselman and Samet [3] summarized tactical intelligence data to produce useful and effective intelligence-message summaries.

Second, Kim et al. [4] performed simulation and analysis of weapon system itself. They identified requirements for analyzing weapon systems and designed architecture of a simulation engine.

Third, some other studies explored management and analysis of defense and military information. Ceruti [5] explored challenges with which information system professionals face on managing data and knowledge in the Department of Defense (DoD), and suggested that broad technological trends might enable the DoD to meet the present and future information-management challenges. Wang [6] designed a data classification algorithm for military information analysis system. Choi and Lim [1] designed a customized virtual defense information analysis system and described its core technologies, which could be powered by a terminology dictionary of defense science and technology [7].

There were also many works on management and analysis of information from the other domains. A legal domain navigator was designed and implemented using Semantic Web and text mining technologies with a thesaurus to identify and classify each fragment of legal documents as well as links between the fragments [8]. It made the information more visible, accessible and presentable to users. Semantic Web technologies were much more actively applied to an automated technology intelligence system, InSciTe [9]. The system utilized ontology modeling, semantic repository, inference, and verification to provide statistics, trends and competitive analysis between agents from large amount of technical documents. Much deeper analysis such as future forecasting on technology growth was tried [10] in science and information technology area.

This paper introduces an implementation of a defense-domain information analysis system using Semantic Web and text mining technologies. Focusing on weapon systems, we extract and analyze their information from reference and news documents in defense and military domain by introducing an ontology model, which makes the analyzed information more visible, accessible and flexibly presentable to users.

3 Modeling and Extraction of Weapon System Information

3.1 Requirement Analysis

Weapon system is one of the most important targets for defense information analysis. So, we focused on analysis of weapon systems at the first implementation of defense information analysis system. The analysis system aims to improve accessibility to information of weapon systems and efficiency of analysis work of human experts on overseas defense information, especially including weapon systems and products. In more detail, we identified following requirements for defense information analysis from interviews with three human experts:

- The analysis system should enable users to search weapon systems with their categories, specification data, and product names and to link flexibly with each other and related documents.
- The analysis system should also enable users to search and link information on development, applicability, and necessity of each weapon system.
- User's work procedure is usually as follows: a user first searches specific weapon products by specifying their categories and specification data, and then identifies

other specification data of the products, and finally analyzes trends related to the products.

3.2   Definition of Weapon System Information

In order to satisfy these requirements, we first identified and defined information of weapon systems and then modeled an ontology representing such information well, which is described in the next subsection. The information of weapon systems includes classification scheme with hierarchy, types, specification data, variants, development history and current status of each weapon system. It also includes the name of country and organization which developed each weapon.

The classification scheme has hierarchical categories with maximum of seven levels and four categories at the first levels such as '*Air*', '*Land*', '*Sea*', and '*System*'. Each category can be related to more than one type of weapon products. That is, the type of weapon products forms additional hierarchy in the classification scheme of weapon systems. Specification data may also have hierarchy. For example, '*aiming speed*' has '*elevation*' and '*traverse*' as sub-specification data. In fact, our data has maximum of 4 levels of specification.

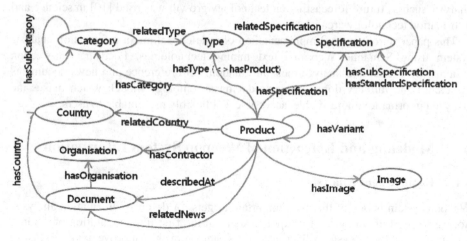

Fig. 1. Ontology model of weapon system information

3.3   Ontology Modeling

Fig. 1 describes our ontology model which represents the information of weapon systems. This figure shows only classes and object properties between them for simplification.

The class '*Category*' and its property '*hasSubCateogry*' represent the classification scheme of weapon systems with hierarchy. Similarly, the class '*Specification*' and its property '*hasSubSpecification*' represent specification data of weapon products with hierarchy. Specification data is usually modeled as properties, not classes, but it is not the case in weapon systems, because weapon systems require hierarchical specification data. For example, *speed* should be modeled as upper specification of *landing speed* and *takeoff speed*. So, we had to model specification data as a class, not a property. Each *'Product'* can be classified by both of the product type *'Type'* and weapon system hierarchy *'Category'* and may have variant products (represented by the property *'hasVariant'*) and images (represented by the class *'Image'* and the property *'hasImage'*). The class *'Country'* and *'Organization'* represent a country and an organization, respectively, developed a weapon product. The class *'Document'* represents a source document describing a weapon product.

3.4   Extraction of Weapon System Information

Based on the definition of weapon system information and the ontology model described in the previous subsections, we extracted weapon system information which includes classification hierarchy, type, specification data, description, development history, current status, and variants of a weapon product. Our defense-related documents are structured in XML and most of such information except specification data is itemized by XML tags in the documents and therefore can be extracted by parsing XML tags (including table tags). We utilized the terminology dictionary for defense science and technology [7] to identify and extract weapon names and specification data.

Although more complex and useful information of weapon systems is represented in plain text in the documents, we restricted the target of information to extract in current implementation into structured and semi-structured text units, not parsing of plain text, to guarantee the quality of extracted information. Unstructured data will be attacked at the next implementation.

4   Weapon System Analysis Services

Using the ontology model and extracted information of weapon systems, we designed and implemented the weapon system analysis services which include description and development history analysis, classification and type analysis, specification data analysis, and news tracking analysis of a weapon product.

The description and development history about a concerned weapon product are provided with several textual statements which also describe some specification data of the product. The service also provides the current development and disposition status and variants of a weapon product.

Fig. 2 shows classification hierarchy and type, with an image, to which a concerned weapon product *'KS-19'* belongs when a user searches this product. It also says that this product was developed at *'Russian state factories'* in *'Russian Federation'*. The links in classification, type, country and organization can lead to further search for products connected to the links.

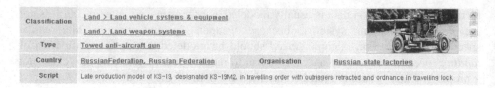

Classification	Land > Land vehicle systems & equipment
	Land > Land weapon systems
Type	Towed anti-aircraft gun
Country	RussianFederation, Russian Federation	Organisation	Russian state factories
Script	Late production model of KS-19, designated KS-19M2, in travelling order with outriggers retracted and ordnance in travelling lock

Fig. 2. Example of classification and type analysis on '*KS-19*'

Specification

crew				15
effective range				
	horizontal			21,000 m
	vertical			13,700 m
		proximity fuzed		15,000 m
		time fuzed		12,700 m
elevation-traverse				
	speed			

[KS-19]에 대한 검색결과 총 4건 [전체]

no	Statement
1	each end of the carriage and one either side on outriggers. The KS-19...
2	is intended primarily for the KS-19 anti-aircraft gun and always uses...
3	59; 100 mm BS-3 field gun; 100 mm KS-19 anti-aircraft gun; NORINCO 100...
4	59); 100 mm BS-3 field gun; 100 mm KS-19 anti-aircraft gun; NORINCO 10...

Fig. 3. Example of specification data analysis on '*KS-19*'

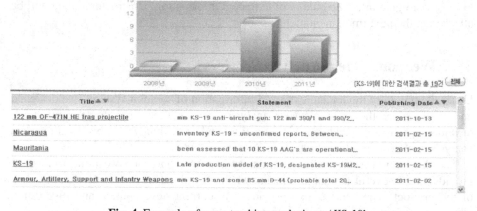

[KS-19]에 대한 검색결과 총 19건 [전체]

Title ▲▼	Statement	Publishing Date ▲▼
122 mm OF-471N HE frag projectile	mm KS-19 anti-aircraft gun; 122 mm 390/1 and 390/2...	2011-10-13
Nicaragua	Inventory KS-19 - unconfirmed reports. Between...	2011-02-15
Mauritania	been assessed that 10 KS-19 AAG's are operational...	2011-02-15
KS-19	Late production model of KS-19, designated KS-19M2...	2011-02-15
Armour, Artillery, Support and Infantry Weapons	mm KS-19 and some 85 mm D-44 (probable total 20...	2011-02-02

Fig. 4. Example of news tracking analysis on '*KS-19*'

Fig. 3 shows specification data analysis of a concerned weapon product '*KS-19*'. It shows hierarchical specification with values and also provides links to corresponding documents where the specification data occurs. Those specification data were extracted fully from semi-structured data and experimentally from the limited number of unstructured documents describing two specific weapon systems.

Fig. 4 shows a snapshot of news tracking analysis on a concerned weapon product '*KS-19*'. The upper graph of Fig. 4 provides trends by year on that weapon and says that the weapon has been emerged suddenly from 2010. The lower list provides contextual statements and links to documents related to that weapon. So, users can track the documents related to that weapon by date.

This weapon system analysis service aims to provide human experts with easy access to overseas defense information and to help them to efficiently analyze the information of weapon systems and products.

5   Conclusion

Information analysis in defense domain requires very high accuracy due to the characteristics of the domain. Human experts first search specific weapon products by specifying their categories and specification data then identify other specification data of the products, and finally analyze trends related to the products. In order to help them to accurately access and analyze the overseas defense information, especially including weapon systems and products, we designed and implemented an ontology-based weapon system analysis service which includes classification hierarchy, type, specification data, description, development history, current status, variants, and trends of a weapon product. We designed an ontology model appropriate for representing weapon system information and extracted and analyzed the information from defense-domain documents by using simple text mining. The implemented service made the analyzed weapon system information more visible, accessible, and presentable to users.

The current implementation used only semi-structured data and specification data of only limited number of products was experimentally mined from plain text. For further study, we plan to attack unstructured data. We also plan to intensify navigational and forecasting capability of the analysis service by applying multi-faceted navigation and prediction model introduced in [8] and [10].

References

1. Choi, J.-W., Lim, C.-O.: Customized Information Analysis System Using National Defense News Data. Korea Contents Association Journal 10(12), 457–465 (2010) (in Korean)
2. Grossman, J.B., Woods, D.D., Patterson, E.S.: Supporting the Cognitive Work of Information Analysis and Synthesis: A Study of the Military Intelligence Domain. In: The Human Factors and Ergonomics Society Annual Meeting, pp. 348–352 (2007)
3. Geiselman, R.E., Samet, M.G.: Summarizing Military Information: An Application of Schema Theory. Journal of the Human Factors and Ergonomics Society 22(6), 693–705 (1980)
4. Kim, T.S., Chang, H.J., Lee, J.M., Lee, K.S.: A Modeling & Simulation Engine for Analyzing Weapons Effectiveness: Architecture. Korea Society for Simulation Journal 19(2), 51–62 (2010) (in Korean)

5. Ceruti, M.G.: Data Management Challenges and Development for Military Information Systems. IEEE Transactions on Knowledge and Data Engineering 15(5), 1059–1068 (2003)
6. Wang, Z.-J.: A Data Classifying Algorithm in Military Information Analysis System. Command Control & Simulation (April 2011)
7. Choi, J.-W., Choi, S., Kim, L., Park, Y., Jeong, J., An, H., Jung, H., Kim, P.: Application and Process Standardization of Terminology Dictionary for Defense Science and Technology. Korea Contents Association Journal 11(8), 247–259 (2011) (in Korean)
8. Lee, S., Kim, P., Seo, D., Kim, J., Lee, J., Jung, H., Dirschl, C.: Multi-faceted Navigation of Legal Documents. In: IEEE International Conferences on Internet of Things, and Cyber, Physical and Social Computing, pp. 537–540 (2011)
9. Lee, S., Lee, M., Jung, H., Kim, P., Seo, D., Kim, T.H., Lee, J., Sung, W.-K.: Using Semantic Web Technologies for Technology Intelligence Services. In: Zhong, N., Callaghan, V., Ghorbani, A.A., Hu, B. (eds.) AMT 2011. LNCS, vol. 6890, pp. 333–344. Springer, Heidelberg (2011)
10. Kim, J., Lee, S., Lee, J., Lee, M., Jung, H.: Design of TOD Model for Information Analysis and Future Prediction. In: Kim, T.-H., Adeli, H., Ma, J., Fang, W.-C., Kang, B.-H., Park, B., Sandnes, F.E., Lee, K.C. (eds.) UNESST 2011. CCIS, vol. 264, pp. 301–305. Springer, Heidelberg (2011)

Some Experimental Results Relevant to the Optimization of Configuration and Planning Problems

Paul Pitiot[1,2], Michel Aldanondo[1], Elise Vareilles[1], Linda Zhang[3],
and Thierry Coudert[4]

[1] Toulouse University - Mines Albi, CGI Lab, Albi, France
{paul.pitiot,michel.aldanondo,elise.vareilles}@mines-albi.fr
[2] 3IL-CCI, Rodez France
p.pitiot@aveyron.cci.fr
[3] IESEG School of Management – LEM CNRS, Lille, France
l.zhang@ieseg.fr
[4] Toulouse University - ENI Tarbes, LGP Lab. Tarbes, France
thierry.coudert@enit.fr

Abstract. This communication deals with mass customization and the association of the product configuration task with the planning of its production process while trying to minimize cost and cycle time. We consider a two steps approach that first permit to interactively (with the customer) achieve a first product configuration and first process plan (thanks to non-negotiable requirements) and then optimize both of them (with remaining negotiable requirements). This communication concerns the second optimization step. Our goal is to evaluate a recent evolutionary algorithm (EA). As both problems are considered as constraints satisfaction problems, the optimization problem is constrained. Therefore the considered EA was selected and adapted to fit the problem. The experimentations will compare the EA with a conventional branch and bound according to the problem size and the density of constraints. The hypervolume metric is used for comparison.

Keywords: mass customization, aiding configuration, aiding planning, constraint satisfaction problem, evolutionary algorithm.

1 Introduction

This paper deals with mass customization and more accurately with aiding the two activities, product configuration and production planning, achieved in a concurrent way. According to the preferences of each customer, the customer requirements (concerning either the product or its production) can be either non-negotiable or negotiable. This situation allows considering a two-step process that aims to associate the two conflicting expectations, interactivity and optimality. The first interactive step, that sequentially processes each non-negotiable requirement, corresponds with a first configuration and planning process that reduces the solution space. This process is present in many commercial web sites using configuration techniques like automotive

L. Chen et al. (Eds.): ISMIS 2012, LNAI 7661, pp. 301–310, 2012.

industry for example. Then, a second process optimizes the solution with respect to the remaining negotiable requirements. As the solution space can quickly become very large, the optimization problem can become hard. Thus, this behavior is not frequent in commercial web sites. Meanwhile some scientific works have been published on this subject (see for example [1] or [2]) and the focus of this article is on the optimization problem. In some previous conferences we proposed an interesting adapted evolutionary algorithm for this problem [3]. However, the presentation was rather descriptive and experimentations were not significant. Therefore, the goal of this paper is to compare this algorithm with a classical branch and bound. This initial section introduces the problem and the organization of the paper.

1.1   Concurrent Configuration and Planning Processes as a CSP

Deriving the definition of a specific or customized product (through a set of properties, sub-assemblies or bill of materials, etc...) from a generic product or a product family, while taking into account specific customer requirements, can define product configuration [4]. In a similar way, deriving a specific production plan (operations, resources to be used, etc...) from some kind of generic process plan while respecting product characteristics and customer requirements, can define production planning [5]. As many configuration and planning studies (see for example [6] or [5]) have shown that each problem could be successfully considered as a constraint satisfaction problem (CSP), we have proposed to associate them in a single CSP in order to process them concurrently.

This concurrent process and the supporting constraint framework present three main interests. Firstly, they allow considering constraint that links configuration and planning in both directions (for example: a luxury product finish requires additional manufacturing time or a given assembly duration forbids the use of a particular kind of component). Secondly they allow processing in any order product and planning requirements, and therefore avoid the traditional sequence: configure product then plan its production [7]. Thirdly, CSP fit very well on one side, interactive process thanks to constraint filtering techniques, and on the other side, optimization thanks to various problem-solving techniques. However, we assume infinite capacity planning and consider that production is launched according to each customer order and production capacity is adapted accordingly.

In order to illustrate the addressed problem we consider a very simple example dealing with the configuration and planning of a small plane. The constraint model is shown in figure 1. The plane is defined by two product variables: number of seats (Seats, possible values 4 or 6) and flight range (Range, possible values 600 or 900 kms). A constraint Cc1 forbids a plane with 4 seats and a range of 600 kms. The production process contains two operations: sourcing and assembling. (noted Sourc and Assem). Each operation is described by two process variables: resource and duration: for sourcing, the resource (R-Sourc, possible resources "Fast-S" and "Slow-S") and duration (D-Sourc, possible values 2, 3, 4, 6 weeks), for assembling, the resource (R-Assem, possible resources "Quic-A" and "Norm-A") and duration (D-Assem, possible values 4, 5, 6, 7 weeks). Two constraints linking product and process variables modulate configuration and planning possibilities: one linking seats with sourcing, Cp1 (Seat, R-Sourc, D-Sourc), and a second one linking range with the assembling, Cp2 (Range, R-Assem, D-Assem). The allowed combinations of each constraint are shown in the 3 tables of figure 1. Without taking constraints into account, this model shows a combinatory of 4 for the product (2x2) and 64 for the production process (2x4) x (2x4) providing a combinatory of 256 (4 x 64) for the whole problem. Considering constraints lead to 12 solutions for both product and production process.

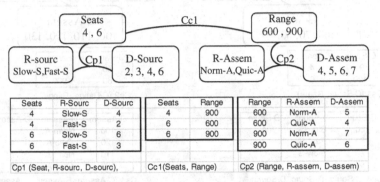

Seats	R-Sourc	D-Sourc		Seats	Range		Range	R-Assem	D-Assem
4	Slow-S	4		4	900		600	Norm-A	5
4	Fast-S	2		6	600		600	Quic-A	4
6	Slow-S	6		6	900		900	Norm-A	7
6	Fast-S	3					900	Quic-A	6

Cp1 (Seat, R-sourc, D-sourc),	Cc1(Seats, Range)	Cp2 (Range, R-assem, D-assem)

Fig. 1. Example of a concurrent configuration and planning CSP model

1.2   Optimizing Configuration and Planning Concurrently

Given previous problem, various criteria can characterize a solution: on the product configuration side, performance and product cost, and on the production planning side, cycle time and process cost. In this paper we only consider cycle time and cost. The cycle time matches the ending date of the last production operation of the configured product. Cost is the sum of the product cost and process cost. We are consequently dealing with a multi-criteria optimization problem. As these criteria are in conflict, it is better for decision aiding to offer the customer a set of possible compromises in the form of Pareto Front.

In order to complete our example, we add a cost and cycle time criteria as represented in figure 2. For cost, each product variable and each process operation is associated with a cost parameter and a relevant cost constraint: (C-Seats, Cs1), (C-Range, Cs2), (C-Sourc, Cs3) and (C-Assem, cs4) detailed in the tables of figure 2. The total cost is obtained with a numerical constraint and the cycle time, sum of the two operation durations, is also obtained with a numerical constraint as follow:

Total cost = C-Seats + C-Range + C-Sourc + C-Assem.
Cycle time = D-Sourc + D-Assem

The twelve previous solutions are shown on the figure 3 with the Pareto front gathering the optimal ones. In this figure, all solutions are present. When non-negotiable requirements are processed during interactive configuration and planning, some of these solutions will be removed. Once all these requirements are processed, the identification of the Pareto front can be launched in order to propose the customer a set of optimal solutions.

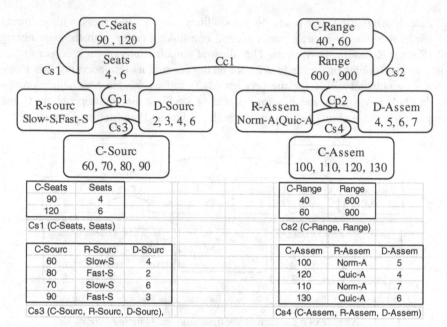

C-Seats	Seats
90	4
120	6

Cs1 (C-Seats, Seats)

C-Range	Range
40	600
60	900

Cs2 (C-Range, Range)

C-Sourc	R-Sourc	D-Sourc
60	Slow-S	4
80	Fast-S	2
70	Slow-S	6
90	Fast-S	3

Cs3 (C-Sourc, R-Sourc, D-Sourc),

C-Assem	R-Assem	D-Assem
100	Norm-A	5
120	Quic-A	4
110	Norm-A	7
130	Quic-A	6

Cs4 (C-Assem, R-Assem, D-Assem)

Fig. 2. Concurrent configuration and planning model to optimize

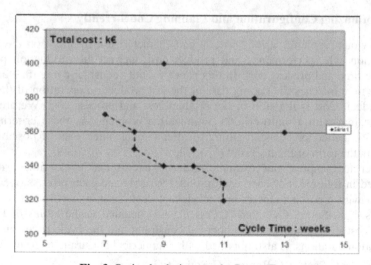

Fig. 3. Optimal solutions on the Pareto Front

A strong specificity of this kind of problems is that the solution space is large. It is reported in [8] that a configuration solution space of more than $1.4*10^{12}$ is required for a car configuration problem. When planning is added, the combinatorial structure can become huge. Specificity lies in the fact that the shape of the solution space is not continuous and in most cases shows many singularities. Furthermore, the multi-criteria aspect and the need for Pareto optimal results are also strong problem expectations. These points explain why most of the articles published on this subject (as for example [9]) consider genetic or evolutionary approaches to deal with this problem. However classic evolutionary algorithms have to be adapted in order to take into account the constraints of the problem as explained in [10]. Among these adaptations, the one we have proposed in [3] is an evolutionary algorithm with a specific constrained evolutionary operators and our goal is to compare it with a classical branch and bound approach.

In the following section we characterize the optimization problem and briefly recall the optimization techniques. Then experimentation results are presented and discussed in the last section.

2   Optimization Problem and Optimization Techniques

2.1   Optimization Problem

The problem of figure 2 is generalized as the one shown in figure 4. The optimization problem is defined by the quadruplet $<V, D, C, f>$ where V is the set of decision variables, D the set of domains linked to the variables of V, C the set of constraints on variables of V and f the multi-valuated fitness function. Here, the aim is to minimize both cost and cycle time. The set V gathers: the product descriptive variables and the resource variables. The set C gathers constraints (Cc and Cp). Cost variables and operation durations are deduced from the variables of the set V thanks to the remaining constraints.

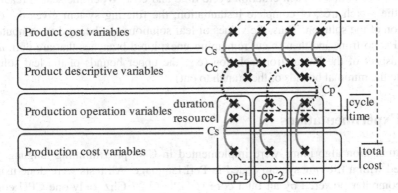

Fig. 4. Constrained optimization problem

Experimentations will consider different problem sizes: different numbers of product variables, different number of production operations and different number of possible values for these variables. Different constraint densities (percentage of excluded combinations of values) will be also considered.

2.2   Optimization Techniques

The proposed evolutionary algorithm is based on SPEA2 [11] with an added constraints filtering process that avoids infeasible individuals (or solutions) in the archive. This provides the six steps following approach:

1. Initialization of individual set that respect the constraints (thanks to filtering),
2. Fitness assignment (balance of Pareto dominance and solution density)
3. Individuals selection and archive update
4. Stopping criterion test
5. Individuals selection for crossover and mutation operators (binary tournaments)
6. Individuals crossover and mutation that respect constraints (thanks to filtering)
7. Return to step 2.

For initialization, crossover and mutation operators, each time an individual is created or modified, every gene (decision variable of V) is randomly instantiated into its current domain. To avoid the generation of unfeasible individuals, the domain of every remaining gene is updated by constraint filtering. As filtering is not full proof, inconsistent individuals can be generated. In this case a limited backtrack process is launched to solve the problem. For full details please see [3].

The key idea of the Branch and Bound algorithm is to explore a search tree but using a cutting procedure that stops exploration of a branch when a better branch has already been found. The first tool is a splitting procedure that corresponds to the selection of one variable of the problem and to the instantiation of this variable with each possible value. The second tool is a node-bound evaluation procedure. The filtering process is used to achieve this task with a partial instantiation and is able to evaluate if the partial instantiation is consistent with the constraints of the problem, and, if this is the case, to provide the lower bound of each criterion cycle time and cost. When the search reaches a leaf of the search tree, or complete instantiation, the filtering system gives the exact evaluation of the solution. Thus, the values of leaf solutions can be used to compute the current Pareto front and then to cut remaining unexplored branches that are dominated by any aspect of the Pareto front solution (e.g. the upper bounds of the leaf solution dominate the minimal bounds of the branch to cut).

3   Experimentations

The optimization algorithms were implemented in C++ programming language and interacted with a filtering system coded in Perl language. All tests were done using a laptop computer powered by an Intel core i5 CPU (2.27 Ghz, only one CPU core is used) and using 2.8 GB of RAM. These tests compared the behavior of our constrained EA algorithm with the exact branch-and-bound algorithm.

3.1   First Experimentation: Problem Size and Constraint Densities

An initial first model, named "full model" is considered. It can be consulted and interactively used at http://cofiade.enstimac.fr/cgi-bin/cofiade.pl select model 'Aircraft-CSP-EA-10'. It gathers five product variables with a domain size between 4 and 6, six production operations with a number of possible resources between 3 and 25. Without constraints consideration, the solution space of the product model is 5,184, and the planning model is 96,000. The size of the global problem model is 497,664,000. A second model, named "small model", has been derived from the previous one with the suppression of a high combinatory task and a reduction of one domain size. This reduces the planning problem size to 12,000 and global model 6,220,800.

In order to evaluate the impact of constraints density, two versions of each model were produced: one with a "weak density" of constraints (20% of possible combinations are excluded in each constraint Cc and Cp) and the other with a "high density" of constraints (50% excluded). These values are frequently met in industrial configuration situations. This provides four models characteristics in table 1.

Solution quantity	Without constraints	Low density	High density
Small model	6 220 800	595 000	153 000
Full model	497 664 000	47 600 000	12 288 000

Table 1. Problems characteristics

For the small models, evolutionary settings are tuned to: population size: 50; archive size: 40; P_{mut}: 0.4; P_{cross}: 0.8. The ending criterion used is a time limit of half an hour. For the full models, we adapt settings for a wider search: population size: 150; archive size: 100; P_{mut}: 0.4; P_{cross}: 0.8. The ending criterion used is the time required by the BB algorithm. In order to analyze the two optimization approaches, we compare the hypervolume evolution during optimization process. Hypervolume metric has been defined in [12]. It measures the hypervolume of space dominated by a set of solutions and is illustrated in Figure 5.

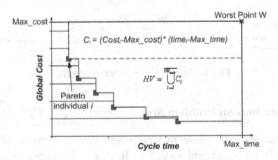

Fig. 5. Hypervolume linked to a Pareto front

Results are presented in figure 6 where EA curves are average results for 30 executions. Both algorithms start with a lapse of time where performance is null. For the BB algorithm, this corresponds to the time needed to reach a first leaf on the search tree, while for the EA; it corresponds to the time consumed to constitute the initial population.

For the small models (first two curves), the BB algorithm reaches the optimal Pareto front much faster compared with EA performance. On the other hand, when the problem size increases, the EA is logically better than the BB algorithm on the full model. For example, on the low-constrained model, the BB algorithm took 20 times longer to reach a good set of solutions (less than 0.5% of the optimal hypervolume).

The impact of constraints density could also be discussed. As it can be seen, the BB algorithm performance is improved when the density of constraints is high. Indeed the filtering allows more branches to be cut on the search tree, in such way that the algorithm reaches leaf solutions and, consequently, optimal solutions more quickly. The EA performance moves in the opposite way. The more the model is constrained, the more the random crossover operation will have to backtrack to find feasible solutions, and thus the time needed by the algorithm will be consequent.

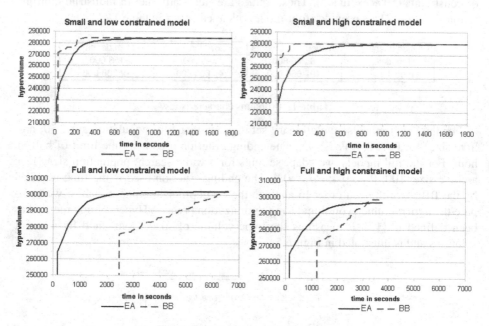

Fig. 6. First experimentation results

3.2   Experimentations on Problem Size

In order to try to identify the problem size where EA is more suitable than BB, we have modified the low constrained model as follows. We consider now a model gathering six product variables and six production operations with three possible values for each, and sequentially add either a product variable and or an operation. The range of study is between 12 and 16 decision variables with three possible values for each. Relevant solution spaces without constraint vary between $1.6*10^6$ and $43*10^6$.

The results are shown in the left part of figure 7. The vertical axis corresponds with the computation time and the horizontal one with the number of decision variables. For BB curves, it shows the time to reach the optimal solution. For the EA curve it shows the time required for nine EA runs over ten to reach the optimal solution. Order of magnitude are close for both around 13 or 14 variables corresponding with a solution space around $2*10^6$ to $5*10^6$ comparable with our previous small model size.

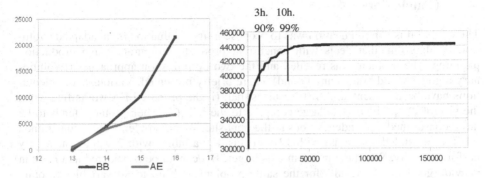

Fig. 7. Experimentation results dealing with problem size

As we already mentioned, industrial models are frequently larger than that. We therefore try our EA approach with a low constrained model with 30 variables and a solution space around 10^{16}. The stopping criterion is "2 hours without improvement". The right part of figure 7 shows that the optimization process has stopped after 48 hours (172800 seconds). It can be noticed that 90% of the final score was obtained after 3 hours and 99% in 10 hours. This allows underlining the good performance of our approach when facing large low constrained problem.

Finally we also try to break optimization in two steps. The idea is: (i) compute quickly a low quality Pareto, (ii) select the area that interest the customer (iii) compute a Pareto on the restricted area. The restricted area is obtained by constraining the two criteria total cost and cycle time (or interesting area) and filtering these reductions on the whole problem. The search space is greatly reduced and the second optimization much faster. This is shown in figure 8 where the left part shows the single step process with 10 and 60 minutes Pareto and the right part shows the restricted area with the two previous curve and the one corresponding with a 10 minutes Pareto launched on the restricted area. It shows that the sequence of two optimization steps of 10 minutes provide a result almost equivalent (only 3 points are missing) to a 60 minutes optimization process.

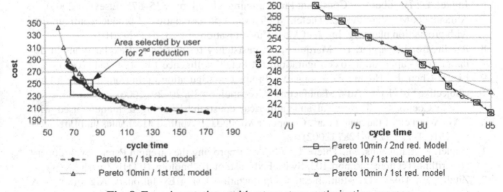

Fig. 8. Experimentations with a two steps optimization process

4   Conclusions

The goal of this communication was to propose a first evaluation of an adapted evolutionary algorithm that deals with concurrent product configuration and production planning. The problem was recalled and the two optimization approaches (Evolutionary algorithm and branch and bound) where briefly presented. Various experimentations have been presented. A first result is that: (i) the proposed EA works fine when the size of the problem gets large compare to the BB, (ii) when problem tends to be more constrained the tendency goes to the opposite. When problem is low constrained (90% of excluded solutions) with 13-14 decision variables with 3 values each, they perform equally. When the problem gets larger, BB cannot be considered and EA can provide good quality results for the same problem with up to 30 variables (around 10^{16} solutions - 90% unfeasible). Finally some ideas about a two steps optimization process have shown that the proposed approach is quite promising for large problems. These are first experimentation results and we are now working on comparing our proposed EA with some penalty function approaches.

References

1. Hong, G., Hu, L., Xue, D., Tu, Y., Xiong, L.: Identification of the optimal product configuration and parameters based on individual customer requirements on performance and costs in one-of-a-kind production. Int. J. of Production Research 46(12), 3297–3326 (2008)
2. Aldanondo, M., Vareilles, E.: Configuration for mass customization: how to extend product configuration towards requirements and process configuration. Journal of Intelligent Manufacturing 19(5), 521A–535A (2008)
3. Pitiot, P., Aldanondo, M., Djefel, M., Vareilles, E., Gaborit, P., Coudert, T.: Using constraints filtering and evolutionary algorithms for interactive configuration and planning. In: IEEM 2010, Macao China, pp. 1921–1925. IEEE Press (2010)
4. Mittal, S., Frayman, F.: Towards a generic model of configuration tasks. In: Proc. of IJCAI, pp. 1395–1401 (1989)
5. Barták, R., Salido, M., Rossi, F.: Constraint satisfaction techniques in planning and scheduling. Journal of Intelligent Manufacturing 21(1), 5–15 (2010)
6. Junker, U.: Handbook of Constraint Programming, ch. 24, pp. 835–875. Elsevier (2006)
7. Aldanondo, M., Vareilles, E., Djefel, M.: Towards an association of product configuration with production planning. Int. J. of Mass Customisation 3(4), 316–332 (2010)
8. Amilhastre, J., Fargier, H., Marquis, P.: Consistency restoration and explanations in dynamic csps - application to configuration. Artificial Intelligence 135, 199–234 (2002)
9. Li, L., Chen, L., Huang, Z., Zhong, Y.: Product configuration optimization using a multiobjective GA. I.J. of Adv. Manufacturing Technology 30, 20–29 (2006)
10. Coello Coello, C.: Theoretical and numerical constraint-handling techniques used with EAs: A survey of the state of art. Computer Methods in Applied Mechanics and Engineering 191(11-12), 1245–1287 (2002)
11. Zitzler, E., Laumanns, M., Thiele, L.: SPEA2: Improving the Strength Pareto Evolutionary Algorithm, Technical Report 103, Swiss Fed. Inst. of Technology (ETH), Zurich (2001)
12. Zitzler, E., Thiele, L.: Multiobjective Optimization Using Evolutionary Algorithms - A Comparative Case Study. In: Eiben, A.E., Bäck, T., Schoenauer, M., Schwefel, H.-P. (eds.) PPSN 1998. LNCS, vol. 1498, pp. 292–301. Springer, Heidelberg (1998)

Resolving Anomalies in Configuration Knowledge Bases

Alexander Felfernig, Florian Reinfrank, and Gerald Ninaus

Institute for Software Technology
Graz University of Technology
Inffeldgasse 16b, 8020 Graz, Austria
{afelfern,freinfra,gninaus}@ist.tugraz.at

Abstract. Configuration technologies are well established in different product domains such as financial services, cars, and railway interlocking stations. In many cases the underlying configuration knowledge bases are large and complex have a high change frequency. In the context of configuration knowledge base development and maintenance, different types of knowledge base anomalies emerge, for example, inconsistencies and redundancies. In this paper we provide an overview of techniques and algorithms which can help knowledge engineers and domain experts to tackle the challenges of anomaly detection and elimination. Furthermore, we show the integration of the presented approaches in the ICONE configuration knowledge base development and maintenance environment.

Keywords: Configuration Knowledge Engineering, Anomaly Detection, Evaluation.

1 Introduction

Knowledge bases (KB) describe a part of the real world (e.g., the set of valid product configurations). The implementation of a KB is typically done within the scope of a cooperation between domain experts and knowledge engineers [1, 2]. configuration knowledge bases (CKB) and can be represented, for example, as a constraint satisfaction problem (CSP [3]).

Configuration knowledge bases (CKB) represent the complete set of valid product configurations. In the real world the set of valid product configurations changes over time and the CKB must be updated. Humans in general and knowledge engineers in particular tend to keep efforts related to knowledge acquisition and maintenance as low as possible. Due to cognitive limitations [4, 5, 6] anomalies such as inconsistencies and redundancies are inserted into CKBs. The task of CKB maintenance gets even more complicated, if more than one knowledge engineer has to develop and maintain the knowledge base.

In this paper we pick up the problem of anomalies in configuration knowledge bases and make two research contributions: first, we provide an overview of the state of the art in anomaly detection: we present different approaches for the detection of inconsistent / redundant constraints and show, which conditions influence the performance of the corresponding algorithms. Second, we show how anomalies can be detected and presented to knowledge engineers in a configuration knowledge base maintenance and development environment (ICONE).

L. Chen et al. (Eds.): ISMIS 2012, LNAI 7661, pp. 311–320, 2012.

The remainder of this paper is organized as follows: Section 2 introduces a working example. Section 3 provides an overview of the state of the art in anomaly detection. Section 4 compares different types of anomaly detection algorithms. Section 5 provides an overview of the ICONE knowledge engineering environment which includes the presented anomaly detection mechanisms. Finally, in Section 6 we conclude this paper with a summary and a discussion of relevant issues for future research.

2   Example Configuration Knowledge Base

Figure 1 shows a bike configuration model - this model will be used as a working example throughout our paper. The notation of this model is based on the feature model notation introduced by Benavides et al. [7].[1]

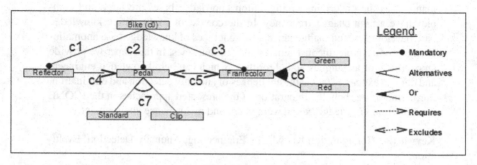

Fig. 1. Feature Model of the bike configuration example

The following configuration knowledge base (CKB) reflects a constraint-based (CSP-based [3]) representation of the configuration model depicted in Figure 1.

$c_0 : Bike = true$;
$c_1 : Bike = true \leftrightarrow Reflector = true$;
$c_2 : Bike = true \leftrightarrow Pedal = true$;
$c_3 : Bike = true \leftrightarrow Framecolor = true$;
$c_4 : Reflector = true \rightarrow Pedal = true$;
$c_5 : \neg(Pedal = true \wedge Framecolor = true)$;
$c_6 : Framecolor = true \leftrightarrow (Green = true \vee Red = true)$;
$c_7 : (Standard = true \leftrightarrow (Clip = false \wedge Pedal = true)) \wedge (Clip = true \leftrightarrow (Standard = false \wedge Pedal = true))$;

The CKB contains several types of anomalies. In the following we will discuss these types of anomalies and show how to automatically detect the anomalies in a CKB.

3   Anomalies in Configuration Knowledge Bases

Anomalies are patterns in data that do not conform to a well defined notion of normal behavior [8]. In the following we describe two different types of anomalies in a configuration knowledge base (CKB): *inconsistencies* and *redundancies*.

---

[1] For a more detailed discussion of these modeling concepts we refer the reader to [7].

Inconsistencies

Inconsistencies are probably the most discussed type of anomaly. A CKB is inconsistent if there does not exist a consistent and complete solution (i.e., a valid instance) to the constraint satisfaction problem (CSP) defined by the CKB. In such situations we have to identify at least one conflict set which is a set of constraints of the CKB.

If an inconsistency exists in the CKB we are able to identify one or more conflict sets $CS \subseteq CKB$ [9] which do not allow the determination of a solution. In our working example the constraint set $CS = \{c_0, c_2, c_3, c_5\}$ represents a conflict set since a constraint solver is not able to determine a solution for CS. A conflict set CS is minimal if there does not exist a conflict set CS' with $CS' \subset CS$. The standard algorithm for detecting a minimal conflict set is QuickXPlain [10]. QuickXPlain is based on the idea of divide-and-conquer where the basic strategy of the algorithm is to filter out constraints - which are not part of a minimal conflict - as soon as possible (see Algorithms 1 and 2).

The following algorithms have two sets as parameters which are representing the CKB: the set B is a set of constraints which can not be part of a conflict set because, for example, they have already been tested by the knowledge engineer. C is a set of constraints which can contain a conflict set because, for example, errors during the last CKB update are inserted. The union of both sets $C \cup B$ represents the CKB.

Algorithm 1. QUICKXPLAIN(B, C):Δ

   $\{B$: set of not diagnosable constraints$\}$
   $\{C$: set of diagnosed constraints$\}$
   if $isEmpty(C)$ *or* $consistent(B \cup C)$ **then**
     return $\emptyset$;
   else
     return $QuickXPlain'(B, B, C)$;
   end if

If $B \cup C$ is inconsistent and $C \neq \emptyset$ the preconditions of applying QuickXPlain are fulfilled and Algorithm 2 will detect a minimal conflict. This algorithm divides C into C_1 and C_2 and checks, whether $B \cup C_1$ is inconsistent. If it is inconsistent, an empty set will be returned, otherwise C_1 will be divided and tested again. If $singleton(C)$ is true the constraint $c \in C$ will be part of the minimal conflict set and inserted into Δ_1. Δ_2 receives a constraint set which also will be a part of the minimal conflict and can be $\emptyset$ or $\Delta_2 \subseteq C_1$. This set All constraints in Δ_1 and Δ_2 are part of the minimal conflict set and returned by the algorithm. $\Delta_1 \cup \Delta_2 = \emptyset$ means that all conflict sets are in B and the CKB can not be consistent without removing constraints from B.

Where minimal conflict sets represent minimal sets of constraints which do not allow the calculation of a solution, diagnoses are minimal sets of constraints which have to be deleted from CKB such that a solution can be identified for the remaining set of constraints. More formally, a diagnosis $\Delta \subseteq CKB$ is a set of constraints such that $CKB - \Delta$ is consistent, i.e., there exist at least one solution for $CKB - \Delta$. A diagnosis Δ is minimal if there does not exist a diagnosis Δ' with $\Delta' \subset \Delta$. In the example CKB described in Section 2 the constraint c_5 is a diagnosis because $CKB - \{c_5\}$ is consistent. In contrast to QuickXPlain the FastDiag algorithm [11] (see Algorithms 3 and 4) calculates one minimal diagnosis $\Delta \subset CKB$ s.t. $CKB - \Delta$ is consistent. Similar

---

Algorithm 2. QUICKXPLAIN'$(B, \Delta, C = \{C_1, ..., c_r\}):\Delta$

$\{B$: Set of diagnosed constraints which are not part of a diagnoses$\}$
$\{\Delta$: Set of diagnosed constraints which are part of a diagnoses$\}$
$\{C$: Set of constraints which will be diagnosed$\}$
if $\Delta \neq \emptyset$ *and inconsistent*(B) **then**
    return $\emptyset$;
end if
if *singleton*(C) **then**
    return C;
end if
$k \leftarrow \lceil \frac{r}{2} \rceil$;
$C_1 \leftarrow \{c_1, ..., c_k\}$;
$C_2 \leftarrow \{c_{k+1}, ..., c_r\}$;
$\Delta_1 \leftarrow QuickXPlain'(B \cup C_1, C_1, C_2)$;
$\Delta_2 \leftarrow QuickXPlain'(B \cup \Delta_1, \Delta_1, C_1)$;
$return(\Delta_1 \cup \Delta_2)$;

---

to QuickXPlain [10], FastDiag [11] exploits a divide-and-conquer strategy. However the focus is different: it determines a minimal diagnosis as opposed to a minimal conflict set determined by QuickXPlain.

Similar to QuickXPlain Algorithm 3 receives a set C which contains all diagnosable constraints. Contrary to Junker's QuickXPlain, CKB contains all constraints. If C is an empty set, FastDiag has no diagnosable set and the algorithm skips all further steps. It also stops, if the set $CKB - C$ is inconsistent, because this set (defined as B in Quick-XPlain) contains inconsistencies, but it will not be diagnosed. If both preconditions are fullfilled, Algorithm 4 will diagnose C.

Algorithm 3. FASTDIAG$(C, CKB):\Delta$

$\{C$: Set of constraints which will be diagnosed$\}$
$\{CKB$: inconsistent configuration knowledge base including all constraints$\}$
if $C = \emptyset \vee inconsistent(CKB - C)$ **then**
    return $\emptyset$;
else
    return $DIAG(\emptyset, C, CKB)$
end if

First of all, DIAG checks whether CKB is consistent. If it is consistent, each subset of CKB is also consistent and no constraint in CKB will be a part of the diagnosis. C will be divided into two subsets C_1 and C_2. Each subset will be removed from CKB separately and within a recursion, which means that the subsets will be further divided, if an inconsistency is still given. If $CKB - C_1$ is consistent, we can say that C_2 is consistent and an empty set will be returned. If it is inconsistent at least one constraint in C_1 must be part of the diagnosis and therefore C_1 will be divided and tested again unless $|C| = 1$. In this case DIAG returns this constraint as a part of the diagnosis. The algorithm returns a set $\Delta_1 \cup \Delta_2$ of constraints which represent a minimal diagnosis.

Algorithm 4. DIAG(Δ, $C = \{C_1, ..., c_r\}$, CKB):Δ

$\{\Delta$: Set of diagnosed constraints which are part of a minimal diagnoses$\}$
$\{C$: Set of constraints which will be diagnosed$\}$
$\{CKB$: Set of diagnosed constraints which are not part of a diagnoses$\}$
if $\Delta \neq \emptyset$ *and consistent*(CKB) **then**
    return $\emptyset$;
end if
if *singleton*(C) **then**
    return C;
end if
$k \leftarrow \lceil \frac{r}{2} \rceil$;
$C_1 \leftarrow \{c_1, ..., c_k\}$;
$C_2 \leftarrow \{c_{k+1}, ..., c_r\}$;
$\Delta_1 \leftarrow DIAG(C_2, C_1, CKB - C_2)$;
$\Delta_2 \leftarrow DIAG(\Delta_1, C_2, CKB - \Delta_1)$;
$return(\Delta_1 \cup \Delta_2)$;

Redundancies

Redundancies are anomalies which do not influence the behavior of, i.e., do not change the semantics of the CKB. A constraint c_a is redundant, if the deletion of the constraint will not influence the behavior of the CKB, more formally described as $CKB - \{c_a\} \models c_a$. A constraint c_a is said to be non-redundant if the negation of CKB (i.e. $\overline{CKB}$) is consistent with $CKB - \{c_a\}$. The redundancy detection algorithms can be applied only if the CKB is consistent and no inconsistencies are in CKB.

The first approach to the detection of redundancies has been proposed by Piette [12]. The approach is the following: a CKB with it's negotiation must be inconsistent, formally described as $CKB \cup \overline{CKB} = \emptyset$. By removing each constraint c_a separately from CKB the algorithm checks, whether the result of $CKB - \{c_a\} \cup \overline{CKB}$ is still inconsistent. If this is the case, then the constraint is redundant and can be removed.

Algorithm 5. SEQUENTIAL(CKB): Δ

$\{CKB$: configuration knowledge base$\}$
$\{\overline{CKB}$: the complement of $CKB\}$
$\{\Delta$: set of redundant constraints$\}$
$CKB_t \leftarrow CKB$;
for all c_i in CKB_t **do**
    if *isInconsistent*$(CKB_t - c_i \cup \overline{CKB})$ **then**
        $CKB_t \leftarrow CKB_t - \{c_i\}$;
    end if
end for
$\Delta \leftarrow CKB - CKB_t$;
return Δ;

After each constraint has been checked separately, CKB_t is a non-redundant constraint set (minimal core) which means, that $\Delta = CKB - CKB_t$ is a set of redundant constraints in CKB and these are returned.

An alternative approach (CoreDiag) has been proposed by Felfernig et al. [13]. Instead of a linear approach they adapt the QuickXPlain algorithm. The divide-and-conquer approach of this algorithm checks whether removing a set of constraints C leads to an inconsistency, formally described as $CKB - C \cup \overline{CKB} = \emptyset$. If it is not inconsistent, C must be further divided and tested again. Similar to SEQUENTIAL the CoreDiag algorithm also has CKB as input.

Algorithm 6. CoreDiag (CKB): Δ

$\{CKB = \{c_1, c_2, ..., c_n\}\}$
$\{\overline{CKB}:$ the complement of $CKB\}$
$\{\Delta:$ set of redundant constraints$\}$
$\overline{CKB} \leftarrow \{\neg c_1 \vee \neg c_2 \vee ... \vee \neg c_n\};$
$return(CKB - \text{CoreD}(\overline{CKB}, \overline{CKB}, CKB));$

CoreD (see Algorithm 7) checks, if $B \subseteq CKB$ is inconsistent. An inconsistency of $B \cup \overline{CKB}$ means that the subset is not redundant and no constraint of B will be a part of Δ. $singleton(C)$ is true means that this constraint is part of the diagnosis and will be returned. Otherwise the constraint set C will be further divided and the subsets will be checked recursively.

Algorithm 7. CoreD($B, \Delta, C = \{c_1, c_2, ..., c_r\}$): Δ

$\{B:$ Set of diagnosed constraints which are not part of a diagnoses$\}$
$\{\Delta:$ Set of diagnosed constraints which are part of a diagnoses$\}$
$\{C:$ set of constraints to be checked for redundancy$\}$
if $\Delta \neq \emptyset$ *and* $inconsistent(B)$ **then**
    $return\ \emptyset;$
end if
if $singleton(C)$ **then**
    $return(C);$
end if
$k \leftarrow \lceil \frac{r}{2} \rceil;$
$C_1 \leftarrow \{c_1, c_2, ..., c_k\};$
$C_2 \leftarrow \{c_{k+1}, c_{k+2}, ..., c_r\};$
$\Delta_1 \leftarrow \text{CoreD}(B \cup C_2, C_2, C_1);$
$\Delta_2 \leftarrow \text{CoreD}(B \cup \Delta_1, \Delta_1, C_2);$
$return(\Delta_1 \cup \Delta_2);$

Which algorithm should be used and which preconditions influence the selection of the algorithm will be described in an empirical study in the next Section.

4 Comparing Anomaly Detection Algorithms

For the comparison we have generated 16 different configuration knowledge bases containing a random number of variables ($0 < |V| < 50$), constraints ($0 < |C| < 50$), and anomalies (between 20% and 50%). Anomalies are generated randomly by a random generator. The generation is done by using the Betty feature model generator[2]. Each

---

[2] http://www.isa.us.es/betty/welcome

CKB has been calculated 20 times in which the order of the constraints has changed randomly on a CPU @ 2.4Ghz x2 and 24GB RAM.

Detection of Inconsistencies

The detection of inconsistencies leads to a set of diagnoses. We used three different set-ups for the performance analysis, differing in the number of diagnoses which are calculated. Figure 2 presents results for calculating 1, 5, and all diagnoses. For the calculation of diagnoses with the QuickXPlain algorithm (Algorithm 1 and 2) and for getting more than one diagnosis from FastDiag (Algorithm 3 and 4) we have used Reiter's HS-tree [14]. With this tree it is possible to calculate all conflicts and diagnoses for a given CKB by transferring each constraint from the result of QuickXPlain / FastDiag separately from the set of diagnoseable constraints C to the set of not diagnoseable constraints (B in QuickXPlain respectively AC in FastDiag). For a more detailed description of the HS-tree we refer the reader to [14].

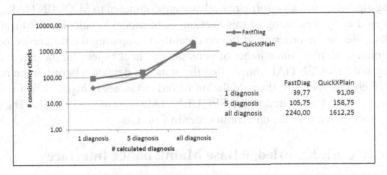

Fig. 2. Number of consistency checks for calculating diagnoses

Results show that the FastDiag approach has an overall advantage compared to the QuickXPlain algorithm, if one diagnosis has to be calculated. The reason for that is, that the calculation of a diagnosis in the HS-tree, when using the FastDiag approach, is done when the first node in the HS-tree is calculated. On the other hand the conflict set approach must expand a path from the root to a leaf before having a diagnosis (see Section 3). The advantage is still given, if 5 diagnoses have to be calculated, but the advantage of FastDiag measured by the number of calls is decreasing.

Finally, if all diagnoses have to be calculated, QuickXPlain performs better, because the reuse of previously calculated conflict sets within the HS-tree increases and the number of expanded nodes in the HS-tree is lower.

Detection of Redundancies

We received the following evaluation results for the calculation of redundancies: if the number of redundant constraints r in relation to the total number of constraints n is high, the CoreDiag algorithm performs better and has an approx. 40% runtime advantage.

The SEQUENTIAL approach performs better, if the redundancy rate is lower than 50% and looses the performance advantage, if the CKB contains between 50% and 75%

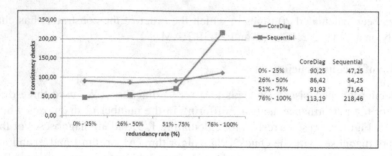

Fig. 3. Number of consistency checks of redundancy detection algorithms

redundant constraints. This confirms the complexity of CoreDiag ($2r \times log2(\frac{n}{r}) + 2r$) and SEQUENTIAL ($n$) where r is the number of redundant constraints and n is the number of all constraints in CKB. Beginning with a redundancy rate $\frac{r}{n} > 0.6$ the CoreDiag algorithm has a performance advantage compared to SEQUENTIAL.

As argued at the beginning of this section, each model is calculated 20 times and the CKBs differ in the ordering of the constraints. Comparing the difference between the maximum and minimum number of consistency checks spent for the calculation of CoreDiag and SEQUENTIAL shows, that the standard deviation between the minimal number of checks spent for the calculation of redundancies is higher when using the CoreDiag (38.67%) compared to the SEQUENTIAL algorithm (19.34%). The reason for that is the influence of the constraint order in CoreDiag.

5   The ICONE Knowledge Base Maintenance Interface

In this section we describe, how we can support knowledge engineers when maintaining a CKB. As mentioned before, changes in valid configurations and distributed knowledge maintenance scenarios lead to anomalies in the CKB. For such scenarios maintenance tools are used which should help the knowledge engineer to develop a CKB without anomalies [15]. The ICONE interface supports the knowledge engineer when maintaining the knowledge base.[3]

The application allows knowledge engineers to define variables and constraints. If anomalies occur in the knowledge base, the refactoring web page shown in Figure 4 supports the knowledge engineer when dealing with the anomalies:

- Anomaly detection: the system automatically detects anomalies (see Section 3). Based on characteristics such as the number of detected redundancies in previous diagnoses, ICONE selects the algorithm with the probably best runtime performance automatically based on the runtime experience of previous anomaly detections.
- For having a higher acceptance from the knowledge engineer, the system explains the anomalies. For example a diagnosis can be explained in terms of the conflicts that are resolved by applying the diagnoses.

---

[3] ICONE is an acronym for 'Intelligent Assistance for Configuration Knowledge Base Development and Maintenance'.

– Recommendations: the right part of the web page shows recommendations for the knowledge engineer. Especially, if many knowledge engineers are working together, the recommendations can help to keep knowledge engineers up to date.

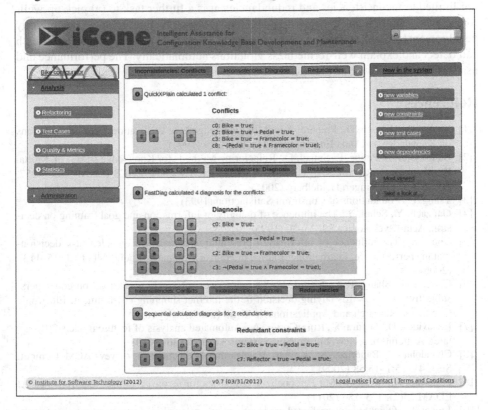

Fig. 4. ICONE anomaly presentation interface. The left part represents the navigation. On the right side, ICONE shows recommendations for the optimization of the knowledge base. In the middle of the web page, ICONE shows all anomalies which are detected. The top list shows all calculated *diagnoses* for inconsistencies. Part 2 lists all detected *conflict sets*. The third part informs the knowledge engineer about the detected *redundancies*. All of the anomalies can be enabled / disabled and the constraints can be deleted. Mailing anomalies, starting a forum, and getting explanations from ICONE for each anomaly is also supported by the ICONE interface.

6   Conclusion and Further Research

This paper gives an overview of the state-of-the art in the detection of anomalies (inconsistencies and redundancies) in configuration knowledge bases. We show how we can detect anomalies in the configuration knowledge base automatically. Our empirical study has shown that the applicability of the diagnosis algorithms differs depending on different factors such as the number of redundant constraints and the number of calculated diagnoses. The number of constraints and variables do not have an impact on the ranking of the algorithms.

The ICONE interface is an approach to support knowledge engineers when maintaining a CKB. This tool shows all anomalies which are discussed in this paper, explains them, and allows the user to resolve the anomaly with one click. Since anomalies are not limited to inconsistencies and redundancies, it is a further task to (a) pick up well-formedness violations from other domains like software feature models [7], ontology development [16] or knowledge management [17] and (b) to develop algorithms which can detect and explain well-formedness violations automatically. The performance and usability of the tool will be part of our future research.

References

[1] Zhang, D., Nguyen, D.: Prepare: A tool for knowledge base verification. IEEE Transactions on Knowledge and Data Engineering 6(6), 983–989 (1994)
[2] Baumeister, J., Puppe, F., Seipel, D.: Refactoring Methods for Knowledge Bases. In: Motta, E., Shadbolt, N.R., Stutt, A., Gibbins, N. (eds.) EKAW 2004. LNCS (LNAI), vol. 3257, pp. 157–171. Springer, Heidelberg (2004)
[3] Tsang, E.: Foundations of Constraint Satisfaction (1993)
[4] Ganzach, Y., Schul, Y.: The influence of quantity of information and goal framing on decision. Acta Psychologica 89, 23–36 (1995)
[5] Cheri, S.: The influence of information presentation formats on complex task decision-making performance. International Journal of Human-Computer Studies 64(11), 1115–1131 (2006)
[6] Chen, Y.C., Shang, R.A., Kao, C.Y.: The effects of information overload on consumers' subjective state towards buying decision in the internet shopping environment. Electronic Commerce Research and Applications 8, 48–58 (2009)
[7] Benavides, D., Segura, S., Ruiz-Cortés, A.: Automated analysis of feature models 20 years later: A literature review. Information Systems 35, 615–636 (2010)
[8] Chandola, V., Banerjee, A., Kumar, V.: Anomaly detection: A survey. ACM Comput. Surv. 41, 15:1–15:58 (2009)
[9] Felfernig, A., Schubert, M.: Personalized diagnoses for inconsistent user requirements. AI EDAM 25(2), 175–183 (2011)
[10] Junker, U.: Quickxplain: preferred explanations and relaxations for over-constrained problems. In: Proceedings of the 19th National Conference on Artifical Intelligence, AAAI 2004, pp. 167–172. AAAI Press (2004)
[11] Felfernig, A., Schubert, M., Zehentner, C.: An efficient diagnosis algorithm for inconsistent constraint sets. AI EDAM 26(1), 53–62 (2012)
[12] Piette, C.: Let the solver deal with redundancy. In: Proceedings of the 2008 20th IEEE International Conference on Tools with Artificial Intelligence, vol. 01, pp. 67–73. IEEE Computer Society, Washington, DC (2008)
[13] Felfernig, A., Zehentner, C., Blazek, P.: Corediag: Eliminating redundancy in constraint sets. In: Sachenbacher, M., Dressler, O., Hofbaur, M. (eds.) DX 2011. 22nd International Workshop on Principles of Diagnosis, Murnau, Germany, pp. 219–224 (2010)
[14] Reiter: A theory of diagnosis from first principles. Artificial Intelligence 32(1), 57–95 (1987)
[15] Felfernig, A., Blazek, P., Reinfrank, F., Ninaus, G.: User interfaces for configuration environments (to appear, 2012)
[16] Baumeister, J., Seipel, D.: Anomalies in ontologies with rules. Web Semantics: Science, Services and Agents on the World Wide Web 8, 55–68 (2010)
[17] Rech, J., Feldmann, R.L., Ras, E., Jedlitschka, A., Decker, B., Jennex, M.E.: Number 17 in premier reference source. In: Knowledge Patterns and Knowledge Refactorings for Increasing the Quality of Knowledge, pp. 281–328. IGI (2008)

Configuration Repair via Flow Networks

Ingo Feinerer[1], Gerhard Niederbrucker[2], Gernot Salzer[1], and Tanja Sisel[1]

[1] Technische Universität Wien, Vienna, Austria
{Ingo.Feinerer,Gernot.Salzer,Tanja.Sisel}@tuwien.ac.at
[2] Universität Wien, Vienna, Austria
Gerhard.Niederbrucker@univie.ac.at

Abstract. Reconfiguration and configuration repair are central tasks when designing and maintaining long-lived systems. Specifications evolve over time and modifying existing configurations in the hardware domain is prohibitively expensive. Consequently there is a demand for efficient methods to repair and change existing configurations where the number of components and a given layout of links between them is given. We present an efficient approach using network flow algorithms for finding and optimising links between a topology of components. This provides a natural formalism for modelling reconfiguration tasks and repairing configurations.

Keywords: formal methods, configuration repair, class diagrams, unified modeling language (UML), integer linear programming (ILP), flow networks.

1 Introduction

Suppose a company has a network of computers connected by switches. Each switch can be connected to at most twenty computers and each computer has to be connected to a switch. Now the company grows and adds more computers (and at some point also additional switches). To protect the infrastructure against failures the company furthermore plans to connect each computer to two or three switches instead of just one. Naturally this change should be accomplished with as little rewiring as possible while keeping the number of switches minimal for cost reasons. For such changing requirements a tool that determines the minimally required switches and also computes a conservative rewiring would be very helpful, i. e., reconfiguration and configuration repair is of central importance.

A configuration in our sense is a collection of objects partially related to each other. Usually they correspond to particular physical entities like computer components or car parts, but in principle may also be software components or data items during program execution. Configurations represent individual multi-part systems, like your computer or car, or some work of art composed of computer and car components arranged according to aesthetic instead of functional criteria. A configuration (task) is hereby "... characterised through a set

L. Chen et al. (Eds.): ISMIS 2012, LNAI 7661, pp. 321–330, 2012.

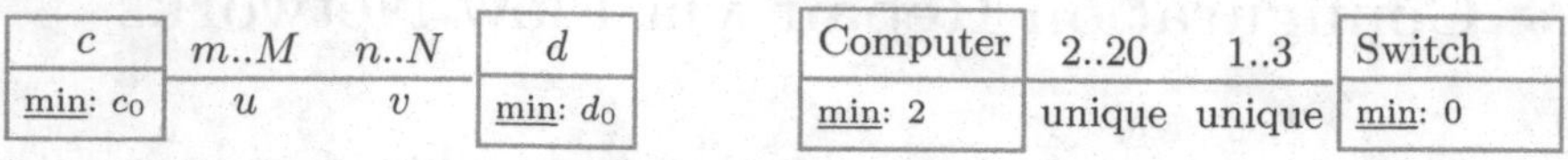

c	$m..M$	$n..N$	d
min: c_0	u	v	min: d_0

Computer	2..20	1..3	Switch
min: 2	unique	unique	min: 0

(a) $c_0, d_0 \geq 0$ specify the minimal number of objects in classes c and d. The multiplicities $m..M$ and $n..N$ constrain the number of objects ($u = v =$ unique) or links ($u = v =$ non-unique).

(b) A configuration satisfying this specification contains at least two computers, every computer is connected to 1 to 3 switches, and every switch is connected to at least two and at most 20 computers.

Fig. 1. Binary association: naming conventions and an example

of components, a description of their properties, namely attributes and possible attribute values, connection points (ports), and constraints on legal configurations." [7, p. 450]. A specification defines general properties of a type (class) and relations between them (association). Objects belong to classes (types) from which they inherit their properties. A specification fixes the conditions under which objects of a certain class are operable.

Criteria for the solution of a configuration task are the computation of the components to be included in the solution, type information on each of these components, the way the components are connected, and values for the attributes of each component [11]. For reconfiguration the task is "Given an existing product individual, a set of requirements and a reconfiguration model, provide a modified product individual fulfilling the requirements and the required changes, both correct with respect to the reconfiguration model." [12]

In repair scenarios the number of components and links, and the link layout and topology is typically part of the input. In order to obtain a configuration that minimises the costs caused by newly introduced components, links, or rearrangement of existing links, the main problem we address in this paper is the following one.

MINIMAL REPAIR
Input: A specification $\mathcal{S}$, a configuration $\mathcal{C}$ (in general not satisfying $\mathcal{S}$), an ordering $<$ on instances, and a notion of similarity of instances.
Output: A $<$-minimal configuration $\mathcal{C}'$ satisfying $\mathcal{S}$ that is similar to $\mathcal{C}$.

We propose an approach based on network flows which allows us to model link topologies in a natural way. It provides fine-grained control over costs for individual links and provides the foundation for efficient algorithms which can be used for configuration repair and reconfiguration.

2   Specifying Configurations with UML

In this section we describe the fragment of UML class diagrams that we need in this paper.

A *specification* $\langle \mathcal{C}, \mathcal{A} \rangle$ consists of a set of *classes* $\mathcal{C}$ and a set of *associations* $\mathcal{A}$. Classes and associations are represented by unique symbols. W. l. o. g. we consider only single associations in this paper. This works as the problems under investigation are compositional; the overall result can be constructed by taking the union of the results for the individual participating associations.

Each association a is characterised by two classes c and d, two intervals $m..M$ and $n..N$ (called multiplicities), where $0 \leq m \leq M$ and $0 \leq n \leq N$, and attributes u and v that may take the value *unique* or *non-unique* (see Fig. 1(a)). In this paper we only deal with symmetric associations, i. e., $u = v$. For a discussion of other combinations see [4].

A *configuration* is a pair $\langle \mathcal{O}, \mathcal{L} \rangle$, where $\mathcal{O}$ is a set of objects and $\mathcal{L}$ is a set of *links*. In our context $\mathcal{L}$ is a *multiset* of pairs of objects, i. e., it may contain duplicate pairs.

A configuration $\langle \mathcal{O}, \mathcal{L} \rangle$ is an *instance* of a specification $\langle \mathcal{C}, \mathcal{A} \rangle$ if every object $o \in \mathcal{O}$ is an instance of exactly one class $c \in \mathcal{C}$, every link $l \in \mathcal{L}$ is an instance of exactly one association $a \in \mathcal{A}$, and if every link l is well-typed, i. e., the linked objects have matching classes as defined by a.

A configuration $\langle \mathcal{O}, \mathcal{L} \rangle$ *satisfies* a specification $\langle \mathcal{C}, \mathcal{A} \rangle$ if all class lower bounds are respected, i. e., $|c| \geq c_0$ (where $|c|$ denotes the number of objects of class c in the configuration) for all $c \in \mathcal{C}$, and if the links $\mathcal{L}$ respect the multiplicities. The latter means that each object of a class c is linked to at least n and at most N objects of class d in the case of $v =$ unique, or that it participates in at least n and at most N links to some d-object in the case of $v =$ non-unique. Note that UML adheres to a look-across semantics: In Fig. 1(a), the attribute u and the interval $m..M$ control the links for each d-object, while v and $n..N$ restrict the d-objects and links to them for a single c-object. See Fig. 1(b) for an example.

The first problems to solve when dealing with such specifications and configurations are to determine whether a given configuration is a satisfying instance of a given specification (model checking) and whether a given specification admits at least one satisfying instance (satisfiability checking), and to compute optimal satisfying instances (minimal model computation). An efficient way to solve these problems is to use integer linear programming.

3 Computing Configurations Using ILP

Class diagrams can be translated to a system of inequalities forming an ILP program [5, 10]. A binary association as shown in Fig. 1(a) corresponds to the inequalities in Fig. 2(a), where x and y are variables for the cardinalities of classes c and d. Inequalities (1) model the multiplicity constraints, while the inequalities (2) enforce the lower bounds on the classes. The inequalities (3) are only necessary in the case of $u = v =$ *unique*. The number of links, ℓ, can be bounded by inequalities (4). By solving such an ILP program we can efficiently check the satisfiability of the specification and obtain at the same time the number of objects and links required for each class and association. The translation is correct and complete: For every valid configuration there exists a corresponding solution of the inequalities, and for every solution of the latter we can construct a valid configuration. Since the ILP solutions are closed under linear combinations and the minimum operator we can compute the minimal

$$Nx \geq my \qquad My \geq nx \qquad \qquad (1)$$
$$x \geq c_0 \qquad \quad y \geq d_0 \qquad \qquad (2)$$
$$xy \geq my \qquad xy \geq nx \qquad \qquad (3)$$
$$\max(nx, my) \leq \ell \leq \min(Nx, My) \quad (4)$$

(a) General case

$$3x \geq 2y \qquad 20y \geq 1x$$
$$x \geq 2 \qquad \quad y \geq 0$$
$$xy \geq 2y \qquad xy \geq 1x$$
$$\max(1x, 2y) \leq \ell \leq \min(3x, 20y)$$

(b) Example

Fig. 2. ILP program corresponding to Fig. 1

model for the specification as well. The approach can be also generalised to multiary associations and other combinations of uniqueness attributes [4,5].

As an example, the specification in Fig. 1(b) corresponds to the inequalities in Fig. 2(b). The minimal solution is $x = 2$, $y = 1$ and $\ell = 2$, meaning that we need at least two computers, one switch and two links.

The ILP program above tells us whether a specification admits configurations, and if so, how many objects and links we need for a minimal configuration. By distributing the required number of links uniformly among the objects we obtain one particular configuration. In general, however, we want to control the placement of links, as the links may correspond to cables or other entities whose installation incurs costs. In other words, among the configurations that are minimal regarding the number of objects and links we prefer those with minimal cost. Moreover, configurations have to be extended or to be adapted to a changed specification over time. In this case it is preferable to keep as many components and links from the old configuration as possible to keep the adaption costs minimal, even if the extended configuration is no longer minimal when considering the number of objects and links. In the next sections we show how such extension and repair problems can be solved efficiently.

4 Flow Problems

Flow networks model transportation problems, where goods flow between nodes under capacity and cost constraints. Typical problems are to maximise the flow through the network or to minimise the cost. In this section we give some basic definitions and introduce the minimal cost flow problem. For a comprehensive treatment of flow networks see [1].

A *flow network* is a directed graph (V, E), where $E \subseteq V \times V$, with a demand *demand*: $V \mapsto \mathbb{Z}$ for every node, bounds *low*: $E \mapsto \mathbb{N}$ and *high*: $E \mapsto \mathbb{N}$ on the flow over the edges, and the cost *cost*: $E \mapsto \mathbb{Z}$ for transporting a unit of flow along an edge.[1] A node v with $demand(v) > 0$ or $demand(v) < 0$ is called a sink or a source, respectively. A *flow* is a function $f: E \mapsto \mathbb{N}$; it is *feasible* if it satisfies the constraints

balance: $\displaystyle \sum_{(u,v) \in E} f((u,v)) - \sum_{(v,w) \in E} f((v,w)) = demand(v)$ for all $v \in V$

flow limits: $low(e) \leq f(e) \leq high(e)$ for all $e \in E$.

---

[1] $\mathbb{Z}$ denotes the set of integers, $\mathbb{N}$ the set of positive integers including zero.

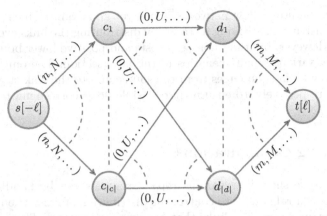

Fig. 3. Flow network corresponding to an instance of the specification in Fig. 1(a) with $|c|$ c-objects, $|d|$ d-objects and ℓ links. There is an edge from s to every c_i, from each c_i to every d_j, and from each d_j to t. Each edge e is labelled by $(low(e), high(e), cost(e))$, with the cost left unspecified here. Numbers in square brackets give the demand of nodes if different from zero. The variable U is set to 1 for an association with $u = v = $ unique, and to $\min(M, N, \ell)$ or any greater number in the case of $u = v = $ non-unique.

The first set of constraints specifies that the flow entering a node minus the one leaving it has to equal the demand at that node. The second set requires that the flow along each edge is within the given bounds.

MinCostFlow

Input: A flow network with functions *demand*, *low*, *high*, and *cost*.
Output: A feasible flow f such that $cost(f) = \sum_{e \in E} cost(e) f(e)$ is minimal.

Minimal cost flows can be computed in polynomial time [1]. One approach that fits our framework particularly well is to map the problem to integer linear programming. This way we can use the same tools for solving the inequalities of the last section and for computing minimal flows.

Now suppose we are given a specification as depicted in Fig. 1(a), and we have determined that we need $|c|$ and $|d|$ objects of classes c and d, respectively, with ℓ links in total. We construct the flow network given in Fig. 3.

Proposition 1. *Given an instance of a specification and a flow f in the corresponding network such that there are exactly $f((c_i, d_j))$ links between the objects c_i and d_j, then the instance satisfies the specification if and only if the flow is feasible, for arbitrary flow costs along the edges.*

Hence, a satisfying instance for a specification can be constructed as follows.

1. Translate the specification to inequalities.
2. Solve this ILP problem to determine the minimal number of objects necessary for a satisfying instance. Fix the number of links between the given bounds (e. g., choose the smallest number).
3. Solve the netflow problem to arrange the links in an admissible way.

At this point the use of flow networks is not yet an advance, since it is always possible to construct a satisfying instance by distributing the links evenly among the objects. However, by using non-zero costs and positive lower bounds we are able to model various repair scenarios in the next section. The only constraint is that the flow bounds on edges from the source and to the sink stay at (n, N) and (m, M), respectively, to ensure that feasible flows correspond to satisfying instances.

5   Repairing Configurations

In this section we show how various repair scenarios can be handled by flow networks. The central issue is to control the distribution of links when modifying configurations, e. g. retaining links that are costly to remove. The technique to achieve this is assigning appropriate costs to the edges in the flow network. We first discuss how costs influence link distribution and then apply the technique to repair scenarios.

5.1   Assigning Costs

Suppose that setting up links between objects incurs costs. If they are the same for every link, we may choose a uniform distribution. The total costs are proportional to the number of links: the fewer links we need, the cheaper the configuration gets. Taking the smallest possible number of links as required by other constraints (like multiplicities) is optimal. Once the number of links is fixed, the total cost is not affected by the arrangement of links.

If the cost varies with the objects, a uniform link distribution may be suboptimal. We obtain an optimal link distribution by first setting $cost((c_i, d_j))$ appropriately for all i and j in Fig. 3 and then solving MINCOSTFLOW. As an example, consider a specification with non-unique multiplicities 1..2 and an instance with two objects per class and a total of four links. Suppose that neighbouring objects (with the same index) may be linked with cost 1, while other links have a cost of 2 (Fig. 4). A uniform distribution of links (connecting every c-object with every d-objects) results in a total cost of 6, while the minimal cost of 4 is obtained by double links between neighbours (i. e., a flow of 2 units from c_i to d_i for $i = 1, 2$). Special care has to be taken when minimising the total cost while varying the number of links, ℓ, between the bounds computed by the inequalities. In the presence of negative costs more links may result in a cheaper configuration. This may even be the case when all costs are positive (more-for-less paradox [3]). (Negative costs can be used e. g. to model situations where removing existing links costs money, while using them saves costs.)

Now consider an even more diverse setting, where links between the same pair of objects may have different costs, or where the cost of a link increases with the number of links already connected to the object. Such constraints can be modelled by convex cost functions or by introducing additional network layers.

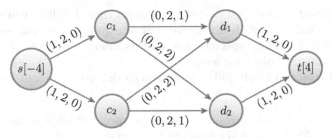

Fig. 4. The minimal cost of 4 is obtained for the flow $f((c_1, d_1)) = f((c_2, d_2)) = 2$ and $f((c_1, d_2)) = f((c_2, d_1)) = 0$

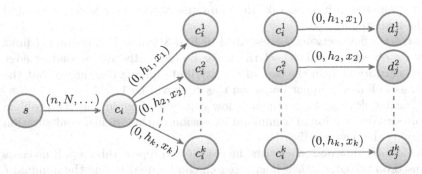

(a) Several link priorities per object   (b) Several link priorities per object pair

Fig. 5. Modeling situation where links to an object are associated with different costs

Figure 5 shows how to model links with different priorities. Each object is represented by $k + 1$ nodes in case the links fall into k priority groups; w.l.o.g. we assume $x_1 \leq x_2 < \cdots \leq x_k$. In Fig. 5(a) the d-nodes are connected to the intermediate nodes $c_i^1, \ldots, c_i^k$ instead of c_i directly; Figure 5(b) distinguishes k different classes of links between particular objects c_i and d_j. To obtain a flow of minimal cost the first h_1 units of flow have cost x_1, the next h_2 units have cost x_2, and so on. The flow via node c_i is bounded by $h_1 + \cdots + h_k$ as well as by the multiplicity $n..N$.

5.2   Configuration Completion

Configuration completion is the problem of making a partial configuration a proper instance of a specification by adding objects and links.

CONFIGURATIONCOMPLETION
Input: a specification $\mathcal{S}$ and a configuration $\mathcal{C} = (\mathcal{O}, \mathcal{L})$
Output: a (minimal) configuration $\mathcal{C}' = (\mathcal{O} \cup \mathcal{O}', \mathcal{L} \cup \mathcal{L}')$ that satisfies $\mathcal{S}$

Such problems arise when the specification changes, e.g. by requiring that more computers have to participate in the network. The aim here is to maintain the current configuration and just to extend it such that it fulfils again the specification. It is usually not desirable to view the problem as the search for a minimal model of the new specification, as this might require to rearrange the old components and links completely.

In our framework of ILP and flow networks this problem can be solved as follows.

1. Solve the inequalities corresponding to the specification S to determine the number of objects and links required by a minimal model.
2. If more objects are needed than available in $\mathcal{O}$, add an appropriate number, giving $\mathcal{O}'$.
3. For the number of links, ℓ, take the minimum of the lower bound computed in the first step and $|\mathcal{L}|$.
4. Construct a flow network as described in Fig. 3 and 5. If $\mathcal{L}$ contains l links between objects c_i and d_j, set the lower bound of the corresponding edge in the network to l. In the case of the attribute unique this means that the lower as well as the upper bound on this edge will be 1.
5. If a feasible flow exists, a minimal flow can be computed and the problem is solved: We have found a minimal extension of the original configuration satisfying the new specification.
   If the feasible flow does not exists, increase ℓ and repeat this step. If no costs are used on the edges, then binary search can be used to find the minimal ℓ leading to a solvable flow problem.
   If there is no feasible flow for any value ℓ up to the upper bound computed by ILP, then there is no extension that contains the original configuration.

If the above procedure fails to find an extension, we can increase the number of objects beyond the minimum computed by the ILP. In our example of the computer network this could mean that we buy a new switch in order to avoid re-cabling the existing network. If the cost of new components exceeds the cost of changing links, however, it is preferable to weaken the constraint that the original configuration has to be maintained by all means. This leads us to the problem of configuration repair that can also be addressed in our framework.

5.3   Configuration Repair

Configuration repair is the problem of modifying a configuration such that it becomes a proper instance of the specification. It includes extension, but as an additional degree of freedom existing links may be removed if they stand in the way.

CONFIGURATIONREPAIR
Input: a specification S and a configuration $\mathcal{C} = (\mathcal{O}, \mathcal{L})$
Output: a configuration $\mathcal{C}' = (\mathcal{O} \cup \mathcal{O}', \mathcal{L}')$ that satisfies S such that $\mathcal{L} \cap \mathcal{L}'$ is maximal.

To solve this problem we proceed as follows.

1.–3. See Sect. 5.2 on configuration completion.
4. Construct a flow network as described in Fig. 3 and 5. In the unique case, assign cost -1 to edges corresponding to links in $\mathcal{L}$, and cost 1 to all other edges. In the non-unique case the situation is more complex, as some but not all of the possible links between two objects may exist, i. e., in general we have two link priorities. Therefore, if $\mathcal{L}$ contains l links between objects c_i and d_j, introduce the nodes c_i, c_i^1, c_i^2, d_j, d_j^1, and d_j^2 (Fig. 5(b)). The edge between c_i^1 and d_i^1 is labelled with $(0, l, -1)$, the edge between c_i^2 and d_i^2 with $(0, \max(U - l, 0), 1)$.
5. A feasible flow always exists (Proposition 1). Hence the MinCostFlow problem can be solved, resulting in a configuration that resembles the original one but satisfies the specification.

Strictly speaking, this procedure does not solve ConfigurationRepair literally, as there is no guarantee that the number of links shared with the original is maximal. We can approach the solution, however, by increasing ℓ within the bounds computed by ILP, and we can increase the penalty of introducing new links by decreasing costs for existing links and increasing costs for new ones.

6 Related Work

The tasks of repairing existing configurations and the reconfiguration of established (legacy) systems have already a long tradition in knowledge-based configuration. Stumptner and Wotawa [14] use a diagnosis approach focusing on model-based reconfiguration. The two major factors identified for reconfiguration of an incorrect configuration are altered requirements and legacy systems. We are faced with the same set of requirements however we differ by using netflow algorithms instead of logics with a corresponding model semantics. Männistö et al. [12] further elaborate a conceptual model for reconfiguration and formalise the reconfiguration process as a sequence of reconfiguration operations to fulfil new requirements which are expressed as a set of conditions. Using UML in the configuration domain is established as well [4, 7]; Felferning et al. [8] show how the UML can be employed for modeling configuration knowledge bases and how old configurations can be reconfigured after diagnosis. Friedrich et al. [9] present an approach for reconfiguration based on model generation utilising answer set programming. Aschinger et al. [2] have recently introduced LoCo, a logic-based high-level representation language for expressing configuration problems. In this formalism the knowledge engineer specifies the possible number of connections between two component kinds, i. e., the number of links between objects of each class in our terminology. Their approach can then infer finite bounds on the number of components, and is therefore complementary to our achievements.

7 Conclusion

In this paper we continue our research programme that aims at efficient (i. e., polynomial) algorithms for product configuration. One distinctive feature of our programme is the extensive use of numeric methods like ILP and netflow algorithms. Framing configuration tasks as problems in symbolic logic frequently leads to algorithms of high complexity – a complexity that often is not inherent in the original problem but is introduced by using (too) expressive languages. In the past we showed how to check efficiently the satisfiability of certain specifications and to compute minimal configurations [5], and how to detect particular specification errors based on an analysis of the multiplicities [6]. In this paper we address the problem of reconfiguration, which is a central challenge when dealing with long-lived component systems. Our next steps will be the implementation of the algorithms proposed here into our prototype [13] and an investigation, how the choice of costs in the flow network influences reconfiguration. An important open issue is how to proceed when repair fails; our aim is to obtain information about bottlenecks from the failed reconfiguration attempts.

References

1. Ahuja, R.K., Magnanti, T.L., Orlin, J.B.: Network flows – theory, algorithms and applications. Prentice Hall (1993)
2. Aschinger, M., Drescher, C., Gottlob, G.: Introducing LoCo, a logic for configuration problems. In: Proceedings of LoCoCo 2011, Perugia, Italy (2011)
3. Charnes, A., Klingman, D.: The more for less paradox in the distribution model. Cahiers du Centre d'Etudes de Recherche Operationelle 13(1), 11–22 (1971)
4. Falkner, A., Feinerer, I., Salzer, G., Schenner, G.: Computing product configurations via UML and integer linear programming. Int. J. Mass Cust. 3(4) (2010)
5. Feinerer, I., Salzer, G.: Consistency and minimality of UML class specifications with multiplicities and uniqueness constraints. In: Proceedings of TASE 2007, pp. 411–420. IEEE Computer Society Press (2007)
6. Feinerer, I., Salzer, G., Sisel, T.: Reducing Multiplicities in Class Diagrams. In: Whittle, J., Clark, T., Kühne, T. (eds.) MODELS 2011. LNCS, vol. 6981, pp. 379–393. Springer, Heidelberg (2011)
7. Felfernig, A., Friedrich, G., Jannach, D.: UML as domain specific language for the construction of knowledge-based configuration systems. International Journal of Software Engineering and Knowledge Engineering 10(4), 449–469 (2000)
8. Felfernig, A., Friedrich, G., Jannach, D.: Conceptual modeling for configuration of mass-customizable products. AI in Engineering 15(2), 165–176 (2001)
9. Friedrich, G., Ryabokon, A., Falkner, A.A., Haselböck, A., Schenner, G., Schreiner, H.: (Re)configuration based on model generation. In: Drescher, C., Lynce, I., Treinen, R. (eds.) Proceedings of LoCoCo 2011. EPTCS, vol. 65, pp. 26–35 (2011)
10. Lenzerini, M., Nobili, P.: On the satisfiability of dependency constraints in entity-relationship schemata. Information Systems 15(4), 453–461 (1990)
11. Mailharro, D.: A classification and constraint-based framework for configuration. AI EDAM 12(4), 383–397 (1998)
12. Männistö, T., Soininen, T., Tiihonen, J., Sulonen, R.: Framework and conceptual model for reconfiguration. Tech. rep., AAAI Conf. Workshop. AAAI Press (1999)
13. Niederbrucker, G., Sisel, T.: Clews Website (2011), http://www.logic.at/clews
14. Stumptner, M., Wotawa, F.: Model-based reconfiguration. In: Proceedings Artificial Intelligence in Design, pp. 45–64. Kluwer Academic Publishers (1998)

Analyzing the Accuracy of Calculations
When Scoping Product Configuration Projects

Martin Bonev[*] and Lars Hvam

Department of Management Engineering, Technical University of Denmark,
Kgs. Lyngby, Denmark
{mbon,lahv}@dtu.dk

Abstract. Product configurators have increasingly been applied in industrial environments. With their help, companies providing customized products have managed to redesign their specification processes and to better handle the growing product variety. But despite the promising benefits, conducting configuration projects is still challenging. Assuming that configurators would naturally solve the existing flaws, both, researchers and professionals typically neglect the need for a making a precise scope for their implementation. Based on this theoretical and practical concern, the present study provides a detailed framework on how the highest potential and eventually the most benefits from using configuration systems can be identified. In particular, this paper investigates how the less explored domain of varying gross margins and calculations reveal a considerable potential for improvement by means of configuration.

Keywords: Product configuration, Configurator development, Sales configuration, Price calculation, Gross margin.

1   Introduction

Implementing mass customization strategies has helped organizations to meet this "customization-responsiveness squeeze" [8] and whilst to produce and to offer customized products with a reasonable high quality at nearly mass production prices [25]. To this end, industrial companies have been challenged to reorganize their way of doing business [3] in multiple dimensions. At the same time, a constantly growing product variety has lead to an increasing complexity of products and processes and thus to the need to better coordinate the way product specifications are performed [8].

The development progress of IT systems has enabled engineering-oriented companies to increasingly implement software-based expert systems, such as configuration systems, to support the product specifications of complex products [1]. Eventually, a growing number of successfully implemented configuration projects have been studied, where organizations have significantly improved their operational performance [7]. The thereby achieved positive effects typically address a series of implications, such as reduced lead times, fewer specification errors and better knowledge sharing and knowledge representation [1, 6, 7-10]. But while there are doubtlessly several advantages of using

---

[*] Corresponding author.

L. Chen et al. (Eds.): ISMIS 2012, LNAI 7661, pp. 331–342, 2012.

configuration systems in an engineering-oriented environment, when starting configuration projects, several risks need to be considered:

1. Since performing configuration projects is a rather complicated task [1, 20], it is difficult to anticipate the accruing development and implementation costs beforehand. If for instance a project turns out to be more costly than initially expected, the risk of failure would be relatively high, as the management board might no longer willing to support the investment.
2. Implementing a configuration system usually affects the internal workflow of an entire firm, starting from the sales to the production department. Reorganizing established workflows would then typically demand significant changes in the business process of organizations, where configuration systems have to be widely accepted and used. If the resistance of change thereby outranges the promised benefits, the configuration project is very likely to fail [1].

In order to keep the risk of abandoning a planned or even initiated project down, companies should focus on identifying the highest potential and eventually the most benefits from using configuration systems. After all, defining the scope is crucial for the success of the project, as an effective (doing the right things) and efficient (doing things right) approach for product configurations is becoming a highly relevant way on coping with rising complexity, whether organizational, product or process related. This paper therefore deals with the question on how to define a suitable scope for implementing configuration systems, which reveals the highest benefits for project implementation and the least risks for project failure. To answer this question, after introducing the research methods (section 2), the first part of this paper (section 3) provides a brief overview on existing approaches for the development and implementation of configuration systems, based on which a new framework is introduced. The second part (section4) refers of an industry case, where the developed ideas are directly tested for relevance and their assumptions are verified. The achieved results (section 5) are then analyzed with the aim to reflect on the previously developed hypothesis. A final conclusion is drawn in section 6, where the most important findings are summarized.

2   Research Methodology

The research methodology applied this paper is following an action research approach, where the researcher is actively involved in a transformation process on a real case and is thereby achieving scientific contributions [30]. This type of methodology requires separating the development procedure of the performed application (i.e. the industrial project) from the development methodology (i.e. the scientific contribution). Based on a foregoing literature study, the created ideas are applied on a collaborating partner, an Engineer-To-Order (ETO) manufacturer in the Danish precast construction industry. Since the construction business is a very complex environment, where only little IT tools have yet been widely applied [31], the industry is a particularly interesting research field for developing and testing out new ideas.

By including the cooperation of the industry case in the development process, the authors believe that more stable results can be implemented at a faster pace. To ensure the rigor of the data collection, first, qualitative methods (e.g. unstructured or semi-structured interviews, workshops and discussions, notes and observations) are used and help to achieve the required knowledge background. Inspired by the machinery industry, product and process modeling tools are hereby applied to qualify the operational performance and product complexity. Then, quantitative data is collected and analyzed to obtain triangulation of the gained insight.

3   Literature Review

3.1   Effects from Product Configuration

The literature dealing with the development and implementation of configuration systems suggests a number of ways on carrying out configuration projects in a systematic way [10, 26]. The majority of the studies is thereby focusing on defining the right development and implementation procedure, while only few of them investigate possible strategies for developing product configurators [1]. Either way, once projects have been initiated, a well defined framework for developing configuration systems obviously helps project leaders and domain experts to follow predefined phases, to employ best practices, established tools and suitable modeling techniques.

Table 1. Benefits from using product configuration systems

Potential Benefits	Author											
	Ardissono [6]	Blecker [21, 27]	Forza and Salvador [7-10]	Haug [1, 11-12]	Helo [18]	Hvam [13-14, 23, 26]	Song [15]	Tenhiälä [16]	Trentin [17, 22]	Tiihonen [24]	Tseng [28]	Yang [19]
Shorter Lead Times	x	x	x	x		x		x	x	x	x	x
Improved Quality of Product Specifications	x	x	x	x	x	x		x	x	x		
Better knowledge preservation			x	x		x			x		x	x
Fewer recources for product specification	x	x		x	x	x			x	x	x	x
Less routine work during specification process		x		x	x	x			x	x		
Less time for training new employees				x								
Improved delivery calculation	x		x	x		x		x				
Improved handling of product variety	x	x	x	x	x	x	x			x	x	
Improved order acquisition		x	x	x	x	x				x	x	
Less quotation to order deviation				x		x						
Fewer recources for quotation process		x	x	x	x	x						
Reduced complexity in the specification process	x	x	x	x			x			x		x
Better product quality			x					x	x			
Better adopting new products and processes						x			x			

To give reasons and justification for conducting configuration projects, academia usually limits to proving the benefits from using already successfully implemented systems [1, 20]. To this end, apart from a number of well described case studies [11-14], more recently extensive surveys have been conducted [17, 22]. Table 1 above lists a sample of the research dealing with mapping the benefits for engineering oriented companies when using configuration systems to support their business. As illustrated, the studies propose a series of benefits which companies potentially gain from using configuration systems. In most of the cases, they are directly related to the operational performance of organizations, which in an operations management domain concerns const efficiency, quality and delivery [32].

However, little attention has yet been paid on how to efficiently meet the wide-raging challenges that need to be overcome when initially considering the implementation of configuration systems. To confirm the improved performance quantitatively, researchers mainly analyze the lead time performance and the quality of the specification process. While for the first aspect, management tools such as a Gap Analysis or Value Stream Mapping (VSM) have been suggested [12, 14], the latter aspect has been less examined [22]. A reason for that can be that in general quality can be defined in several ways [26]. Crosby (1980) for example approaches the term from four viewpoints, as he addresses the conformance to requirements, prevention, performance with no defects, and the price for non conformance [33].

Considering this multidimensional perspective to quality, it is eventually much easier to measure the lead time performance of an organization, than the quality with which specifications are done. Thus apart from counting the defects (errors) of companies' specifications [26], additional analytical methods have to be employed to assure reliable statements about their quality. This implication has even gradually been reinforced by today's business environment, where firms which pursue mass customization strategies struggle with an increase of product, process and organizational complexity.

3.2   Quantifying the Accuracy in Cost Calculation

When investigating the economical perspective on how successful companies deliver their custom tailored products and services, it is useful to study the parameters that asses this performance. From a financial perspective, the so called Key Performance Indicators (KPIs) aim to summarize the ultimate results of a business, which may consist of a revenue ratio, gross margin deviation, Earning Before Interests and Taxes (EBIT) and profitability [35-36]. Experiences from collaborations with companies making complex customized products show that the majority encounter significant gross margin deviations [34]. Figure 1 below shows one of the industrial examples, a manufacturer providing customized building equipment, where the actual gross margins (GMs) of completed projects vary between -60% to +50%. The achieved GMs of individual projects have been sorted according to their success, assuming that projects with higher GM would be regarded as more successful.

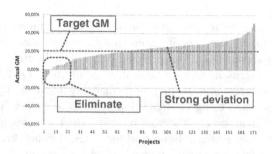

Fig. 1. Gross margin deviation for projects (adopted from [34])

Even though the manufacturer has estimated a 20% margin for calculating all his quotations, the post calculation reveals a very different picture of the obtained GM. No doubt, there might be many reasons why companies are experiencing such a significant variation. But assuming that a relatively fixed GM (20 ± 5%) is pre-estimated, in general, it can be concluded that unexpected variations on actual GMs result from poorly made pre-estimations on costs for making specifications, manufacturing and for providing services. As indicated in Figure 1, at this point we argue that more accurate pre-calculations help companies to decide on their product portfolio and accordingly to evaluate beforehand which projects are profitable. Better performed estimations would thereby help to improve the quality of specifications and products by means of an improved conformance of the requirements [22]. To fill this gap and to come a step closer to our initial question on how to ensure a successful planning and implementation of configuration systems, we propose the following hypothesis:

Hypothesis 1: Investigating the deviation between GMs and pre and cost calculations is positively related to the resulting potential benefits from implementing configuration systems.

3.3   Introducing the Framework

To clarify the hypothesis, we introduce a framework for making the right decisions when investigating the most suitable scope for implanting configuration systems. The framework is based on the procedure for the development and implementation of configuration projects introduced by Hvam et al. (2008) [26].

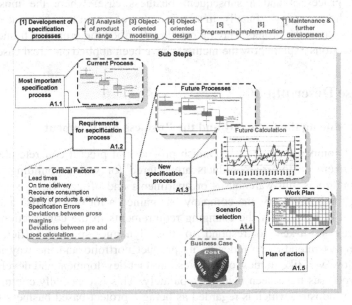

Fig. 2. Development of specification processes

By following the lifecycle of a configuration project, the procedure suggests conducting projects in 7 major phases, starting from the panning phase (development of specification processes) first. The authors argue that at the beginning, engineering companies should investigate the way their custom tailored products and services are specified (order fulfillment) and how the communication to the customer (order acquisition) is organized. Analyzing the specification process would allow firms to draw conclusions on their current operational performance and to uncover vulnerability. The objective of this phase is to develop a better performing future specification process, which is supported by a configuration system.

As illustrated in Figure 2 above, the authors describe this first phase in 5 sub steps, in which well established modeling techniques and analyzing methods are used. In this paper, we draw our attention in particular on the less examined approaches in literature (marked in red). For a more detailed description of the entire steps, we recommend Hvam et al. (2008) [26]. Once the current specification activities have been mapped (step 1), in the next step, the requirements for the future specification process are to be set (step 2). Here, a list of critical success factors may help to decide how to proceed with the analysis. Besides the well described studies on analyzing lead time performances, recourse utilization etc., the less discussed issue of strongly varying pre and cost calculations is further investigated. To evaluate how successful completed projects were and to what percentage the manufacturing costs were affecting these results, the analysis of GMs and the distribution of manufacturing costs is suggested.

In case the KPIs fluctuate stronger than the company's business strategy allows, in the third step, traditionally one ore more TOBE specification processes are to be drawn. With the focus on cost calculations, here, we additionally propose TOBE calculation processes and a subsequent business case, where the most suitable scenario is chosen (step 4). Finally, in step 5, a plan of action is to be created ensuring the continuation of the project. Having briefly described the proposed framework, the following sections explain how the methods have been applied on a real case.

4   Case Description

4.1   Introducing the Company and the Business Environment

The studied company is a leading Danish producer of precast concrete elements for buildings, where customized products are offered for various building types, e.g. industrial buildings and warehouses or apartments and offices. Being successful on the market for many years, the company has gained a lot of expertise and working know-how. But because of the changing requirements in the construction business, the company is asked to respond to this dynamic situation efficiently. The manufacturer is intending to redesign its product portfolio and the way it is doing business. However, like in most companies, product development and development of business processes have been planned separately. This is especially common for the construction industry, which is regarded as being a project-based business sector [5]. Here, the product development is typically done in projects, where the individual products are being developed with more or less random reuse of previous solutions and knowledge.

Being aware of the present challenges, the research group is entering the development process to assist the domain experts and to apply and verify the newly developed hypothesis. Especially the combination of a dynamically changing business environment with low level of automation and IT experience promises many potential research achievements. The gained findings are summarized in the following sections.

4.2   Investigating the Current Specification Process

To improve the operational performance and thereby to reduce the complexity of the business processes, the precast manufacturer is considering the use of IT tools, such as configuration systems. In order to facilitate the success of the planned configuration project (see section 1), a clear defined scope has to be developed. Thus, following the procedure introduced in section 3, in the beginning, the most important specification processes have been studied.

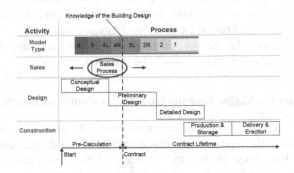

Fig. 3. Main activities in the precast industry

Figure 3 illustrates a high level representation of major procedures in the precast industry, where in addition to the actual design process, common management practices have been established to create "Models" of the same basic processes across the enterprises [3-4]. The contract between a contractor and the precast manufacturer is made on the basis of these models. They determine to what extend the manufacturer is involved in the design process of the building. In "Model 6" for instance, the manufacturer is supporting the design process from the very first beginning, making structural analysis for the entire building based on a given design intent from the architects. In contrast, in "Model 1", the foregoing design activities are done by the collaborating partners, while the precast manufacturer is focusing only on the detailed design for the concrete elements, including the reinforcement and installations.

4.3   Analyzing Deviations between Gross Margins and Pre- and Post-calculations

Regardless of the model type, the precast manufacturer and his client typically agree on a contract at a point of time where the preliminary or even conceptual design of a building is still made. The sales department is using its experience to pre-estimate the amount and type of concrete elements that are needed to construct the designed building. Based on their pre-estimations, the price for delivering the required precast elements is negotiated. Because of the complexity of construction projects [38], estimating the correct sales price is challenging. In case the price is set too high, the precast manufacturer will not be able to compete on the market. On the other hand, if the sales department is offering a too low sales price, the profit will be reduced or the company might even produce with loss. In sum, because at this stage no detailed design information is available, uncertainty and high risk for changes on the design hamper making accurate cost estimations.

Apparently, the sales process in the precast industry is rather complex, as each project requires different products and most of the decisions are made at a point of time, where only little knowledge about the final building design is available and uncertainties about upcoming changes are present. This leads to the obvious assumption that the pre-estimated prices are often not representing the actual costs. In accordance with the developed research question, the results from the qualitative analysis are verified through a quantitative data analysis, leading to the question: how could the company benefit from implementing a configuration system in support of the sales process? By analyzing the current performance quantitatively, the developed assumptions can either be proven or rebutted, so that the highest potential of implementing product configurators would be revealed.

4.4   Identifying the Major Benefits from Using Configuration Systems

To verify the evidence from the qualitative oriented analysis, a nearly complete sample of projects performed over the last 2 years is investigated. Since the objective was to identify how good or bad the accuracy of the cost estimation is, the two proposed indicators from section 3 (depended variables) were set in relation to possible cause (independent variables), e.g. the project size.

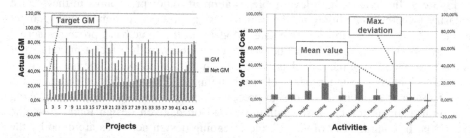

Fig. 4. Deviation between gross margin and pre- and cost calculation

As displayed in Figure 4 above, the projects' GMs and the relative allocation of the costs compared to the total cost were evaluated. To obtain a clear cost picture for each project, only direct and indirect variable costs were considered, leaving out fixed costs, e.g. for administration, and overheads. The graph to the left compares the total

GMs with those when the material costs are excluded (Net GM). The deviation shows that the labor cost do not behave proportional to the project size, as the GM and the one without the material costs do not change correspondingly.

The graph to the right illustrates the strong deviation of the relative activity cost. Here, the pre-work (project management, engineering, design) is not equally distributed across the sizes, but is significantly higher for smaller projects. Also the relative costs for production (casting, reinforcement, forms and material) highly vary with the project size and have the highest percentage for mid-range projects (not in the graph). In sum, the analysis shows a surprisingly high variation of actual GMs and relative cost deviations, where the labor cost is not proportional to the produced elements. Assuming well functioning manufacturing process, the result indicates that the current pre estimation of both sales prices and labor recourses is not being done sufficiently.

5   Contribution to Future Specification Process and Cost-Benefit Analysis

When designing the TOBE specification process, the pattern for the deviations has to be revealed. To identify the cause-effect relationship for the strongly varying indicators from the first analysis, the domain experts were asked to provide additional information to the projects and the way they can be compared. Hence, apart from the project sizes, it was decided to consider an estimated complexity factor (based on the produced elements), the model type (see section 4.2) and the project type, e.g. apartments or malls.

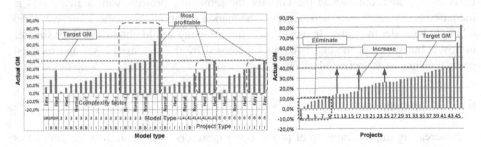

Fig. 5. Analytical results and cost-benefit investigation

The left graph in Figure 5 illustrates the deviation of the GM in relation to the most influencing factor, the model type. The analysis shows that the actual GM is much lower (24%) than the one the company is aiming for. Besides, it becomes clear that the company is most profitable for a certain combination of model type, complexity factor and project type. For these types of projects, the actual GMs were higher and the done pre calculations were more accurate.

Once all influencing factors have been detected, a more precise price calculation model that better reflects the actual cost picture can be designed and incorporated in a configuration system. The right graph in Figure 5 indicates the scope for decision-making, when deciding on a scenario for the right calculation method. Here, the company has to determine: 1. what would be the minimum GM, which would cover the fixed costs and overheads, and 2. how much could the company benefit from a more accurate price calculation. Indeed, we believe that a sufficient calculation model leads to a stronger negotiating position with the customer and thus helps to increase the GM towards the targeted one. Therefore, more precisely, depending on the company's strategy, we argue that the following possible scenarios shown in Table 2 below might be feasible for the introduced case.

Table 2. Cost-benefit analysis

	Scenario 1	Scenario 2	Scenario 3		Scenario 4		Scenario 5	
Description	Keep current status	Remove all <10% GM	Remove all <5% GM, increase 5-10% by 5%		Remove all <5% GM, increase 5-10% by 10%		Remove all <5% GM, increase 5-30% by 5%	
Turnover	280	247	275		277		285	
Gross Margin	24%	67	28%	69	26%	72	27%	75	27%	77
Fixed costs and overheads	42	37	41		42		43	
EBIT	28	32	30		33		34	

To ensure anonymity, the total numbers have been changed in the table, whereas their relative ratios have been kept accordingly. A for this industry common EBIT of 10% has been assumed in scenario 1 [36], which further serves as the comparison measurement. Then, in scenario 2 to 5, different combinations of rejected projects with a simultaneous increase of the remaining GMs are proposed. As expected, in the current case, "Scenario 5" appears to be most profitable for the company, where all projects with less than 5% GMs are rejected, but instead the GM for the remaining projects with 5-30% GMs is increased by 5%. However, depending on the market situation, "Scenario 2" or "Scenario 4" might be easier to realize. In these cases, the company would obtain a slightly smaller turnover, but which at the end is overcompensated by the increased GMs and reduced costs, resulting in an increased EBIT.

6   Conclusion

The increasing implementation of product configurators over the last two decades has proven a number of potential benefits for companies providing customized goods. However, in academia and practice only little analytical methods have been used to actually uncover these benefits and thus to utilize the maximum capability of the supportive configuration systems. The framework presented in this paper therefore reveals an evident opportunity for better scoping planned configuration projects and thereby to lowering the risk of abandoning projects. To this end, a less discussed investigation of deviations between gross margins and pre- and post-calculations have been applied on an industry case, an ETO manufacturer providing complex building products. The analyses confirmed how a well structured quantified approach, supported by a cost-benefit analysis, can determine the potential advantage of more accurate cost and price calculations and thus lead to improved sale processes.

References

1. Haug, A., Hvam, L., Mortensen, N.H.: A classification of strategies for the development of product configurators. In: 5th World Conference on Mass Customization & Personalization, Proceedings of MCPC, Helsinki (2009)
2. McCutcheon, D.M., Raturi, A.S., Meredith, J.R.: The customization-responsiveness squeeze. Sloan Management Review 35(2), 89–99 (1994)
3. Sacks, R., Eastman, C.M., Lee, G.: Process Model Perspectives on Management and Engineering Procedures in the Precast/Prestressed Concrete Industry. Journal of Construction Engineering and Management 130(2), 206–215 (2004)
4. Byggeri - informationsteknologi - produktivitet – samarbejde, http://www.bips.dk
5. Scherer, R., Sharmak, W.: Process Risk Management Using Configurable Process Models. In: Camarinha-Matos, L.M., Pereira-Klen, A., Afsarmanesh, H. (eds.) PRO-VE 2011. IFIP AICT, vol. 362, pp. 341–348. Springer, Heidelberg (2011)
6. Ardissono, L., Felfernig, A., Friedrich, G., Goy, A., Jannach, D., Petrone, G., Schäfer, R., Zanker, M.: A Framework for the Development of Personalized, Distributed Web-Based Configuration Systems. Artificial Intelligence Magazine 24, 93–108 (2003)
7. Forza, C., Salvador, F.: Application support to product variety management. International Journal of Production Research 46(3), 817–836 (2008)
8. Forza, C., Salvador, F.: Configuring products to address the customization-responsiveness squeeze: A survey of management issues and opportunities. International Journal of Production Economics 91, 273–291 (2004)
9. Forza, S., Salvador, F.: Managing for variety in the order acquisition and fulfillment process: The contribution of product configuration systems. International Journal of Production Economics 76, 87–98 (2002)
10. Forza, C., Salvador, F.: Product information management for mass customization: connecting customer, front-office and back-office for fast and efficient customization. Palgrave Macmillan, New York (2007)
11. Haug, A., Hvam, L., Mortensen, N.H.: Definition and evaluation of product configurator development strategies. Computers in Industry 63(5), 471–481 (2012)
12. Haug, A., Hvam, L., Mortensen, N.H.: The impact of product configurators on lead times in engineering-oriented companies. Artificial Intelligence for Engineering Design, Analysis and Manufacturing 25(2), 197–206 (2012)
13. Hvam, L., Pape, S., Nielsen, M.K.: Improving the quotation process with product configuration. Computers in Industry 57(7), 607–621 (2006)
14. Hvam, L., Bonev, M., Denkena, B., Schürmeyer, J., Dengler, B.: Optimizing the order processing of customized products using product configuration. Production Engineering 5(6), 595–604 (2011)
15. Song, Z., Kusiak, A.: Optimizing product configurations with a data-mining approach. International Journal of Production Research 47(7), 1733–1751 (2009)
16. Tenhiälä, A., Ketokivi, M.: Order Management in the Customization-Responsiveness Squeeze. Decision Sciences Journal 43(1), 173–206 (2012)
17. Trentin, A., Perin, E., Forza, C.: Overcoming the customization-responsiveness squeeze by using product configurators: Beyond anecdotal evidence. Computers in Industry 62(3), 260–268 (2011)
18. Helo, P.T.: Product configuration analysis with design structure matrix. Industrial Management Data Systems 106(7), 997–1011 (2006)
19. Yang, D., Miao, R., Wu, H., Zhou, Y.: Product configuration knowledge modeling using ontology web language. Expert Systems with Applications 36(3), 4399–4411 (2009)

20. Kropsu-Vehkampera, H., Haapasalo, H., Jaaskeleinen, O., Phusavat, K.: Product Configuraion Management in ICT Companies: The Practicioners' Perspective. Techology and Investment 2, 273–285 (2011)

21. Blecker, T., Abdelkafi, N., Kreuter, G., Friedrich, G.: Product Configuration Systems: State-of-the-Art, Conceptualization and Extensions. In: Génie Logiciel & Intelligence Artificielle. Eight Maghrebian Conference on Software Engineering and Artificial Intelligence (MCSEAI 2004), pp. 25–36 (2004)

22. Trentin, A., Perin, E., Forza, C.: Product configurator impact on product quality. International Journal of Production Economics 135(2), 850–859 (2012)

23. Hvam, L., Malis, M., Hansen, B., Riis, J.: Reengineering of the quotation process application of knowledge based systems. Business Process Management Journal 10(2), 200–213 (2004)

24. Tiihonen, J., Soininen, T., Männistö, T., Sulonen, R.: State-of-the-practice in product configuration - a survey of 10 cases in the Finnish industry. In: Mäntylä, M., Finger, S., Tomiyama, T. (eds.) Knowledge Intensive CAD, vol. 1, pp. 95–114. TJ Press, Padstrow (1996)

25. Pine, B.J., Victor, B., Boynton, A.C.: Making Mass Customization Work. Harvard Business Review 71(5), 108–119 (1993)

26. Hvam, L., Mortensen, N.H., Riis, J.: Product Customization. Springer, Berlin (2008)

27. Blecker, T., Abdelkafi, N., Friedrich, G., Kaluza, B., Kreutler, G.: Information and Management Systems for Product Customization. Springer, New York (2005)

28. Tseng, M., Jianxin, J.: Mass customization. In: Salvendy, G. (ed.) Industrial Engineering Handbook: Technology and Operations Management, 3rd edn., pp. 684–709. Wiley, New York (2001)

29. Heiskala, M., Tiihonen, J., Paloheimo, K.S., Soininen, T.: Mass customization with configurable products and configurators. In: Blecker, T., Friedrich, G. (eds.) Mass Customization Information Systems in Business, Hershey, PA, pp. 1–32 (2007)

30. Coughlan, P., Coghlan, D.: Action research for operations management. International Journal of Operations & Production Management 22(2), 220–240 (2002)

31. Hartman, T., van Meerveld, H., Vossebeld, N., Adriaanse, A.: Aligning building information model tools and construction management methods. Automation in Construction 22, 605–612 (2012)

32. Jacobs, M.A., Swink, M.: Product portfolio architectural complexity and operational performance: Incorporating the roles of learning and fixed assets. Journal of Operations Management 29(7-8), 677–691 (2011)

33. Crosby, P.B.: Quality Is Free. Mentor, New York (1980)

34. Mortensen, N.H., Hvam, L., Boelskifte, P., Lindschou, C., Frobenius, S., Haug, A.: Making product customization profitable. International Journal of Industrial Engineering: Theory Applications and Practice 17(1), 25–35 (2010)

35. Nudurupati, S., Arshad, T., Turner, T.: Performance measurement in the construction industry: An action case investigating manufacturing methodologies. Computers in Industry 58(7), 667–676 (2007)

36. Balatbat, M., Lin, C., Carmichael, D.: Comparative performance of publicly listed construction companies: Australian evidence. Construction Management and Economics 28(9), 919–932 (2010)

37. van der Aalst, W.M., Stoffele, M., Wamelink, J.W.: Case handling in construction. Automation in Construction 12(3), 303–320 (2003)

A Two-Step Hierarchical Product Configurator Design Methodology

Yue Wang and Mitchell M. Tseng

Advanced Manufacturing Institute, The Hong Kong University of Science and Technology,
Hong Kong
{yacewang,tseng}@ust.hk

Abstract. Product configurators are widely used to elicit custmer requirements in the form of tangible product specification in product design. However customers are heterogeneous in terms of preferences and needs. To meet the diversified customer needs, companies have provided a wider variety of product variants. In this situation, there may be too many choices for customers to specify during configuration process. The huge number of choices may lead to mass confusion among customers, and less customer satisfaction. To handle this issue, we propose a hierarchical product configuration design method which operates in a from-coarse-to-fine manner. K-means method is used to classify the whole product attribute sets into several interested clusters. The coarse configuration process will be used to identify customers' preferences cluster. The fine configuration stage will be conducted within each cluster to fine tune the product configuration. Thus the confusion caused by the large number of choices can be mitigated.

Keywords: Configurator, mass customization, k-means.

1   Introduction

Configurators have become generally accepted as powerful product development toolkit to capture customers' requirements. Configurators operate on a set of pre-defined attribtues. It takes customers' specifications as input and the output is the product variant they want while at the meantime satisfying a set of requirements and constraints. Configuration has long been researched in computer science community as an effort to achieve design efficiency and quality simultaneously. Artificial intelligence techniques are used extensively, e.g. rule-based reasoning, cased-based reasoning, model-based reasoning etc [1-3]. Studies of customers' decision making processes have also shown that customers have higher satisfaction with the configuring process than with traditional selection, not only with the results but also with the decision process [4].

Nowadays, global marketplace has become more and more competitive. The fulfillment of customer needs becomes a key factor for customer's purchase decision. This phenomenon is particularly true in today's markets where the power is gradually shifted from producers to consumers. To remain competitive, managers in many

L. Chen et al. (Eds.): ISMIS 2012, LNAI 7661, pp. 343–348, 2012.
© Springer-Verlag Berlin Heidelberg 2012

industries use product variety as a key strategic lever [5]. Huffman presents a case, Choice Seating Gallery, where large assortment eventually makes customers feel overwhelmed and dissatisfied [6]. Choice Seating Gallery is a customized sofa provider and allows the customers to make a choice among abundant choices: 500 styles, 3000 fabrics, and 350 leathers. Hence, there are 1,500,000 different fabric sofas and 175,000 different leather sofas available for the customers to choose. Facing such a huge number of selections, customers tend to have difficulty evaluating alternatives and identifying what they want. In this paper, we try to provide a new product configurator to prevent customer from the tedious and lengthy configuration process and handle the complexity of variety without too much burden.

The remainder of the paper is organized as follows. In section 2, we will introduce the methodology for the new two-step configuring approach, including the representative configuration selection. A numerical example is presented in section 3 to verify the proposed approach. Session 4 concludes the whole paper.

2   Methodology

2.1   Basic Idea

For some complicated product, customers may face a large number of possible choices for each attribute in product configuring process. However, there may be similarities in the choices. In this paper, we want to firstly eliminate some redundant choices by selecting representative alternatives and narrow down the choices to an acceptable level. Then the configuring task is performed on the much simplified representative choices set. The objective of this step is to identify the customer's general interest and preferences. In the next step, the configurator will only show the attribute alternatives in the potentially interested cluster rather than the whole choices set. In this way, customers may not need to configure from the huge number of alternatives.

Let's still use Choice Seating Gallery to illustrate the idea. In the initial configurator, customers need to choose from 500 styles, 3000 fabrics, and 350 leathers, a really challenging task. We divide the configuring task into two steps. Firstly we cluster the possible products into a certain number of product clusters and each cluster has a representative product configuration by k-means clustering approach. These representative products only cover a small portion of all the attribute alternatives, for example 20 styles, 30 fabrics, and 15 leathers. Then for a new customer, the configurator on this shrunk attribute space is much easy to carry out. After performing the coarse level selection, we can identify which product cluster will contain the customer's target product. Then the fine-tune level configuring process is only performed on that particular cluster. It is also a much simpler configurator than the original one. In summary, we use two simple configurators to replace one complicated configurator while at the meantime achieve the similar result without too much burden for customers. The idea is illustrated in Figure 1.

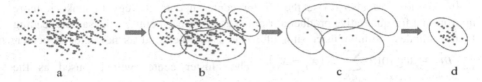

Fig. 1. Idea of the two-step configuring process. 1(a): the original configuration space; 1(b): the clustered space; 1(c): representatives for each cluster are identified and the first step configuring process is conducted on the representative space; 1(d): the second step configuring process is conducted only on the interested cluster.

2.2   k-Means Clustering

k-means is a data clustering method of to partition n data into k clusters in which each data sample belongs to the cluster with the nearest mean [7]. Given a set of data samples $(x_1, x_2, ..., x_n)$ and an initial set of cluster center/means $(m_1, m_2, ..., m_k)$, k-means method tries to minimize the within-cluster sum of squares. The optimal clusters set S should minimize $\sum_{i=1}^{k} \sum_{x_j \in S_i} d^2(x_j - m_j)$ where d is the distance function and m_j is the center of the jth cluster. Traditional k-means operated in two steps iteratively as follows;

Assignment step: For each data sample, assign it to the cluster with the nearest distance to the cluster center.

Update step: update the cluster center. The typical update criteria can be the mean of the cluster, i.e., $m_{t+1} = \dfrac{1}{|S_i^t|} \sum_{x_j \in S_i^t} x_j$ where m_{t+1} is the cluster center after t iterations and S_i^t is the corresponding clustering result.

These two steps are iterated until there is no change to the cluster center.

2.3   Application of k-Means to Representative Configuration Selection

In this paper, we will apply k-means to construct several clusters of the whole product. Within each cluster, one or two representative products will be selected and the corresponding attribute choices will be narrowed down to a small portion of the original choices set. Since attributes are usually qualitative, we apply Hamming distance to quantify the distance between attribute alternatives. Hamming distance between two equal-length strings is the number of different symbols in the corresponding positions [7]. For example, the Hamming distance of two four-tuple attributes vector (A1, B2, C2, D1) and (A2, B2, C1, D1) is 2 because there are two different values in the corresponding positions.

In addition, we also revise the cluster center update strategy to make it more appropriate for qualitative product attribute case. For each cluster, the sample leading to the smallest within-cluster sum of squares will be used as new cluster center, i.e., $m_i = \arg\min_{x_k} \sum_{x_j \in S_i} d^2(x_j - x_k)$. The cluster center will be used as the representative for the corresponding cluster.

If the cluster centers remain unchanged after rounds of iterations, we say the k-means converges. The k cluster centers are selected as the representatives for the whole data samples.

In summary the k-mean based representatives selection process is as follows;

Initialization: select k samples as the cluster center

Assignment step: For each data sample, assign it to the cluster with the nearest distance to the cluster center.

Update step: update the cluster center. For each cluster, the sample leading to the smallest within-cluster sum of squares will be used as new cluster center, i.e., .

Iterate the above two steps until the cluster center remain unchanged.

2.4   Two-Stage Configuring Process

In the coarse level of product configuring process, only the attributes alternatives appeared in representatives will be included for customer to specify. In this way, the attribute space is narrowed down to a much smaller scale. It is relatively easy for a customer to configure the product he/she wants. Since the configurator is conducted in a representative attribute space, we call it coarse level configuring process.

The result of the coarse level configuring process will be used as an indicator of the customer's preferences. The fine level configuring will be carried out in the cluster nearest to the coarse level configuration, instead of the whole attribute space. In this way, we use configuring steps to trade off the selection difficulty when customers face a large number of alternative choices.

3   Case Study

We use a simplified customized living area design as an example to illustrate this approach. The case is adapted from a customized apartment project sponsored by a real estate company in Hong Kong. There are six attributes to be specified in the configurator, namely Style (A), Floor Tile (B), Electric appliance (C), Wall paper (D), Cabinet (E), Layout (F). The corresponding numbers of choices for each attribute are 8 for style, 25 for floor tile, 16 for electric appliance, 20 for wall paper, 9 for cabinet and 5 for layout. Each product configuration is represented by a six-tuple, such as (A2, B13, C1, D8, E2, F4). We interviewed 494 respondents to conduct a survey format configurator.

The methodology introduced in previous session is adopted to construct the two-level configurator. In this paper, we set the number of clusters to be 6 due to the consideration of efficiency and effectiveness. After the k-means clustering, six representative product configurations are identified. These representatives covers 4 styles, 5 for floor tile, 3 for electric appliance, 5 for wall paper, 4 for cabinet and 2 for layout. The representative space is much smaller than the original choice space. The potential number of product variants decreases from 8*25*16*20*9*5=2,880,000 to 4*5*3*5*4*2=2,400. This is much easier for customer to configure the preferred attributes.

After the first round configuring process, the corresponding interested cluster is identified for each customer. Then the fine-tune stage configuring process is carried out. The scale of each cluster is still much smaller than the original attribute choice set.

In this paper, we use the total number of choices screened by the customer as a metric to measure the increase of efficiency. In the original setting, each customer needs to screen 8+25+16+20+9+5=83 attributes values in the whole configuring process. In the proposed approach, 4+5+3+5+4+2=23 attribute values need to be screened and compared in the first round of configuration process. The number of choices in the second rounds of configuration for each cluster differs. The average number of attribute alternatives in each cluster is 38.5. Therefore the total number of attributes to be screen is 23+38.5=61.5 which is smaller than the original one 83. The scale of the configurator in the case is still moderate. It can be anticipated that if the scale is much bigger like the Choice Seating Gallery case, the improvement would be much bigger.

4   Conclusion

In this paper, we propose a two-stage product configurator design methodology to avoid confusing customers with too many choices in each configuration step. The two-stage product configurator operates in a from-coarse-to-fine manner. In the coarse stage, the total product choices are pruned to a small set of representative product configurations. Customers' preliminary specifications lead to potential shrunk consideration set. The fine configuration stage will conducted in the consideration set and customers are no longer need to screen all the possible alternatives. It should be noted that this paper focuses on the product configuration process. To simplify the analysis, constraints in the product configuration are not considered in this paper. In future work, we will incorporate the proposed approach to the constraint satisfaction framework.

Acknowledgements. This research is supported by Hong Kong Research Grants Council (RGC CERG HKUST 620609).

References

1. Sabin, D., Weigel, R.: Product Configuration Frameworks - A Survey. IEEE Intelligent Systems, 42–49 (July/August 1998)
2. Wang, Y., Tseng, M.M.: Adaptive attribute selection for configurator design via Shapley value. Artificial Intelligence for Engineering Design, Analysis and Manufacturing 25(1), 185–195 (2011)

3. Wang, Y., Tseng, M.M.: Customized products recommendation based on Probabilistic relevance model. Journal of Intelligent Manufacturing (accepted, 2012), doi:10.1007/s10845-012-0644-7
4. Kurniawan, S.H., Tseng, M.M., So, R.H.Y.: Modeling consumer behavior in the customization process. In: Piller, F., Tseng, M. (eds.) The Customer Centric Enterprise, pp. 267–282. Springer, Heidelberg (2003)
5. Bayus, B.L., Putsis Jr., W.P.: Product proliferation: An empirical analysis of product line determinants and market outcomes. Marketing Science 18(2), 137–153 (1999)
6. Huffman, C., Kahn, B.E.: Variety for sale: Mass customization or mass confusion? Journal of Retailing 74(4), 491–513 (1998)
7. Duda, R.O., Hart, P.E., Stork, D.G.: Pattern Classification. John Wiley & Sons, New York (2001)

Anonymous Preference Elicitation
for Requirements Prioritization

Gerald Ninaus, Alexander Felfernig, and Florian Reinfrank

Institut of Software Technology,
Graz University of Technology,
Inffeldgasse 16b/II, 8010 Graz, Austria
{gerald.ninaus,alexander.felfernig,florian.reinfrank}@ist.tugraz.at
http://www.ist.tugraz.at

Abstract. Requirements Engineering is a very critical phase in the software development process. Requirements can be interpreted as basic decision alternatives which have to be negotiated by stakeholders. In this paper we present the results of an empirical study which focused on the analysis of key influence factors of successful requirements prioritization. This study has been conducted within the scope of software development projects at our university where development teams interacted with a requirements prioritization environment. The major result of our study is that anonymized preference elicitation can help to significantly improve the quality of requirements prioritization, for example, in terms of the degree of team consensus, prioritization diversity, and quality of the resulting software product.

Keywords: Requirements Prioritization, Group Decision Making.

1 Introduction

Requirements Engineering (RE) is the branch of software engineering concerned with the real-world goals for functions of and constraints on software systems [1]. RE is considered as one of the most critical phases in software projects, and poorly implemented RE is one major risk for project failure [2]. Requirements are the basis for all subsequent phases in the development process and high quality requirements are a major precondition for the success of the project [3].

Today's software projects still have a high probability to be canceled or at least to significantly exceed the available resources [4]. As stated by Firesmith [5], the phase of requirements engineering receives rarely more than 2-4% of the overall project efforts although this phase has a significant impact on project success rates. A recent Gartner report [6] states that *requirements defects are the third source of product defects (following coding and design), but are the first source of delivered defects. The cost of fixing defects ranges from a low of approximately $70 (cost to fix a defect at the requirements phase) to a high of $14.000 (cost to fix a defect in production). Improving the requirements gathering process can reduce the overall cost of software and dramatically improve time to market.*

L. Chen et al. (Eds.): ISMIS 2012, LNAI 7661, pp. 349–356, 2012.

Requirements can be regarded as a representation of decision alternatives or commitments that concern the functionality and qualities of the software or service [7]. Requirements Engineering (RE) is a complex task where stakeholders have to deal with various decisions [8]:

- *Quality decisions*, e.g., is the requirement non-redundant, concrete, and understandable?
- *Preference decisions*, e.g., which requirements should be considered for the next release?
- *Classification decisions*, e.g., which topics have a high importance and which requirements belong to these topics?
- *Property decisions*, e.g., is the effort estimation for this requirement realistic?

Stakeholders are often faced with a situation where the amount and complexity of requirements outstrips their capability to survey them and to reach a decision [9]. The amount of knowledge and number of stakeholders involved in RE processes tend to increase as well. This makes individual as well as group decisions much more difficult. The focus of this paper will be *preference decisions*, i.e., we want to support groups of stakeholders in the context of *prioritizing software requirements* for the next release. Typically, resource limitations in software projects are triggering the demand of a prioritization of defined requirements [2]. Prioritizations support software project managers in the systematic definition of software releases and to resolve existing preference conflicts among stakeholders.

Only a systematic prioritization can guarantee that the most essential functionalities of the software system are implemented in-time [10]. Typically, requirements prioritization is a collaborative task where stakeholders in a software project collaborate with the goal to achieve consensus regarding the prioritization of a given set of requirements. The earlier requirements are prioritized, the higher is the probability to avoid the implementation of irrelevant requirements and the higher is the amount of available resources to implement the most relevant requirements [10].

Establishing consensus between stakeholders regarding the prioritization of a given set of requirements is challenging. Prioritizations do not only have to take into account business process related criteria but as well criteria which are related to technical aspects of the software. Especially in larger projects, stakeholders need a tool-supported prioritization approach which can help to reduce influences related to psychological and political factors [10]. Requirements prioritization is a specific type of group work which becomes increasingly important in organizations [11].

Prioritization decisions are typically taken in groups but this task is still ineffective due to reasons such as social blocking, censorship, and hidden agendas [11]. The major contribution of this paper is to show how *anonymity* in group decision processes can help to improve the quality of requirements prioritizations. Furthermore, anonymous preference elicitation increases the probability of detecting hidden profiles [12], i.e., increases the probability of exchanging decision-relevant information [13].

The remainder of this paper is organized as follows. In Section 2 we provide an overview of the basic functionalities of the INTELLIREQ requirements engineering environment developed at our university. In Section 3 we introduce the basic hypotheses that have been investigated within the scope of our empirical study; in this context we also provide details about the study design. In Section 4 we report the major results of our empirical study. With Section 5 we conclude the paper.

2   IntelliReq Decision Support

INTELLIREQ is a group decision environment that supports computer science students at our university in deciding on which requirements should be implemented within the scope of their software projects. For this task 219 students enrolled in a course about *Object-Oriented Analyse and Design* at the Graz University of Technology had to form groups of 5–6 members. Unfortunately, it is not possible to evaluate the knowledge and experience of the students and the resulting groups but the course is typically attended by students in the third semester of an informatics programme or similar. We therefore distributed the resulting groups randomly on the different evaluation pools and assume that the knowledge and experience is equally distributed on each pool. Each group had to implement a software system with an average effort of about 8 man months.

In our study, 39 software development teams had to define a set of requirements which in the following had to be implemented. These requirements had to be prioritized and the resulting prioritization served as a major criteria for evaluating the delivered software product at the end of the project.

The requirements prioritization process consisted of three different phases (see Figure 1) denoted as *construction* (collection of individual stakeholder preferences), *consensus* (discussion of prioritization alternatives and adaptation of own preferences), and *decision* (group decision defined and explained by the project manager). This decision process structure results in about 15.000 stakeholder decisions and 798 corresponding group decisions (39 groups with approximately 20 final decisions per group) taken by the team leaders (project managers). On the basis of this scenario we conducted an empirical evaluation with the goal to analyze the effects of supporting anonymized requirements prioritization. The basic settings of this study will be presented in the following section.

3   Empirical Study

Within the scope of our empirical study we wanted to investigate the impact of *anonymous preference elicitation* on the decision support quality of the INTELLIREQ environment. The study was conducted during the course *Object Oriented Analysis and Design* at the Graz University of Technology and the stakeholder part of the customer was impersonated by four course assistants. These assistants were not aware of the study settings and had to review the software functionality

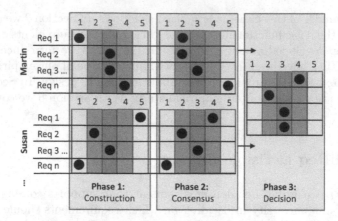

Fig. 1. INTELLIREQ Prioritization (Decision) Process. *Construction*: stakeholders define their initial preferences; *Consensus*: stakeholders adapt their preferences on the basis of the knowledge about preferences of other stakeholders. *Decision*: project managers take the final group decision. Preferences represent the wish of a stakeholder to implement a requirement (1: lowest, 5: highest)

developed by different teams. This evaluation did not include a code review. Instead, we wanted to evaluate the user experience of the final product and which important functionality was supported. This approach of evaluating the prioritization quality is more deductive as the selected requirements have a stronger impact on the functional range of the software product than on the quality of the written program code. For the evaluation the quality value was represented on a scale between 0..30 credits.

Consequently, each project team interacted with exactly one of two existing preference elicitation interface types. One interface (*type 1: non-anonymous preference elicitation*) provided an overview of the personal preferences of team members where each team member was represented by her/his name. In the second interface type (*type 2: anonymous preference elicitation*) the preferences of team members were shown in anonymized form where the name of each individual team member were substituted with the terms $person_1$, $person_2$, etc. The hypotheses (H1–H5) used to evaluate the decision process are summarized in Figure 2. These hypotheses was evaluated on the basis of the following observation variables.

Anonymous preference elicitation. This variable indicates with which type of prioritization interface the team members were confronted (either summarization of the preferences of the team members including the name of the team members or not including the name of the team members).

Consensus and Dissent. An indication to which extent the team members managed to achieve consensus – see the second phase of the group decision process in Figure 1 – is provided by the corresponding variables. We measured the *consensus of a group* on the basis of the standard deviation of requirement-specific team member preferences and decisions. Formula 1 can be used to determine the *dissent* of a group x which is defined in terms of the normalized sum of the standard deviations (sd) of the requirement-specific votings. The group *consensus* can then be interpreted as the counterpart of dissent (see Formula 2). As the consensus is the simple inversion of the dissent, we will only take into account the consensus in the remaining paper.

$$dissent(x) = \frac{\sum_{r \in Requirements} sd(r)}{|Requirements|} \tag{1}$$

$$consensus(x) = \frac{1}{dissent(x)} \tag{2}$$

Decision Diversity. The decision diversity of a group can be defined in terms of the average over the decision diversity of individual users in the consensus phase (see Figure 1). The latter is defined in terms of the standard deviation derived from the decision d_u of a user and the mean value $\bar{d}$ of her/his preferences – a decision consists of the individual requirements prioritizations of the user (see Formula 3).

$$sd(d_u) = \sqrt{\frac{\sum_{u \in UserDecisions}(d_u - \bar{d})^2}{|UserDecisions| - 1}} \tag{3}$$

$$diversity(x) = \frac{\sum_{u \in Users} sd(d_u)}{|Users|} \tag{4}$$

Output Quality. The output quality of the software projects conducted within the scope of our empirical study has been derived from the criteria such as degree of fulfilment of the specified requirements. We also weighted the requirements according to their defined priority in the prioritization task. E.g. not including a very high important requirement enormously decreases the output quality. On the other hand, low priority requirements will only have a small impact on the output quality. Consequently, a requirement, for which a high priority was defined but is of minor importance for the software product, has to be implemented anyway. Hence, the requirements prioritization has a direct impact on the quality value. The quality of the project output has been determined by teaching assistants who did not know to which type of preference elicitation interface (anonymous vs. non-anonymous) the group has been assigned to. These assignments were randomized over all teaching assistants, i.e., each teaching assistant had to evaluate (on a scale of 0..30 credits) groups who interacted with an anonymous and a non-anonymous interface.

Within the scope of our study we wanted to evaluate the following hypotheses.

H1: Anonymous Preference Elicitation increases Consensus. The idea behind this hypothesis is that anonymous preference elicitation helps to increase the consensus by decreasing the commitment [14] related to an individual decision taken in the preference construction phase (see Figure 1), i.e., changing his/her mind is easier with an anonymous preference elicitation interface.

H2: Consensus increases Decision Diversity. As a direct consequence of an increased exchange of decision-relevant information (see Hypothesis H1), deeper insights into major properties of the decision problem can be expected. As a consequence, the important differentiation between important, less important, and unimportant requirements with respect to the next release [9] can be achieved.

H3: Decision Diversity increases Output Quality. Group decision diversity is assumed to be a direct indicator for the quality of the group decision. With this hypothesis we want to analyze the direct interrelationship between prioritization diversity and the quality of the resulting software.

H4: Consensus increases Output Quality. From Hypothesis H3 we assume a positive correlation between the degree of consensus and the diversity of the group decision. The diversity is an indicator for a meaningful triage [9] between important, less important, and unimportant requirements.

H5: Anonymous Preference Elicitation increases Output Quality. Finally, we want to explicitly analyze whether there exists a relationship between the type of preference elicitation and the corresponding output quality.

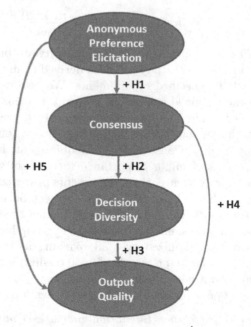

Fig. 2. Hypotheses defined to evaluate the IntelliReq Decision Support

4   Study Results

We analyzed the hypotheses (H1–H5) on the basis of the variables introduced in Section 3.

H1. The degree of group consensus in teams with anonymous preference elicitation is significantly higher compared to teams with non-anonymous preference elicitation (Mann-Whitney U-test, $p < 0.05$). An explanation model can be the reduction of commitment [14] and a higher probability of discovering hidden profile information which improves the overall knowledge level of the team.

H2. There is a positive correlation between the group consensus and the corresponding decision diversity (correlation 0.523, $p < 0.01$). This can be explained if we take a look on the aggregation process of initial preferences in the consensus phase. A study [15] about the prediction quality of group recommendation heuristics showed that the final decision is often similar to the average of the initial preferences. If we assume that there are a lot of requirements with a high dissent, the resulting compromises will not have a high diversity but will rather be arranged in the center of the possible options.

H3. In our analysis we could detect a positive correlation between group decision diversity (diversity of prioritization) and the corresponding output quality (correlation 0.311, $p < 0.01$). Decision diversity can be seen as an indicator of a reasonable triage process and reasonable prioritizations result in software products with a high quality.

H4. Consensus in group decision making increases the output quality (correlation 0.399, $p < 0.01$). An overlap in the personal stakeholder preferences can be interpreted as an indicator of a common understanding of the underlying set of requirements. This leads to a better prioritization and a higher quality of the resulting software components.

H5. Groups with anonymous preference elicitation defined significantly better requirements prioritization compared to groups with a non-anonymous preference elicitation (independent two-sample t-test, $p < 0.05$). This resulted in a better set of functions and in a better user experience measured by the teaching assistants.

5   Conclusions

Requirements prioritization is an important task in software development processes. In this paper we motivated the application of requirements prioritization and discussed issues related to the aspect of anonymizing group decision processes in requirements prioritization. The results of our empirical study clearly show the advantages of applying anonymized preference elicitation, for example in terms of higher-quality software components, and can be seen as a step towards a more in-depth integration of decision-oriented research in the requirements engineering process.

References

1. Zave, P.: Classification of research efforts in requirements engineering. ACM Computing Surveys 29(4), 315–321 (1997)
2. Hofmann, H., Lehner, F.: Requirements engineering as a success factor in software projects. IEEE Software 18(4), 58–66 (2001)
3. Felfernig, A., Zehentner, C., Ninaus, G., Grabner, H., Maalej, W., Pagano, D., Weninger, L., Reinfrank, F.: Group Decision Support for Requirements Negotiation. In: Ardissono, L., Kuflik, T. (eds.) UMAP Workshops 2011. LNCS, vol. 7138, pp. 105–116. Springer, Heidelberg (2012)
4. Yang, D., Wu, D., Koolmanojwong, S., Brown, A., Boehm, B.: Wikiwinwin: A wiki based system for collaborative requirements negotiation. In: HICCS 2008, Waikoloa, Big Island, Hawaii, p. 24 (2008)
5. Firesmith, D.: Prioritizing requirements. Journal of Object Technology 3(8), 35–47 (2004)
6. Group, G.: Hype cycle for application development: Requirements elicitation and simulation (2011)
7. Aurum, A., Wohlin, C.: The fundamental nature of requirements engineering activities as a decision-making process. Information and Software Technology 45(14), 945–954 (2003)
8. Regnell, B., Paech, B., Aurum, C., Wohlin, C., Dutoit, A., Ochdag, J.: Requirements means decision! In: 1st Swedish Conf. on Software Engineering and Practice (SERP 2001), Innsbruck, Austria, pp. 49–52 (2001)
9. Davis, A.: The art of requirements triage. IEEE Computer 36(3), 42–49 (2003)
10. Wiegers, K.: First things first: Prioritizing requirements. Software Development (1999)
11. Pinsonneault, A., Heppel, N.: Anonymity in group support systems research: A new conceptualization, measure, and contingency framework. Journal of Management Information Systems 14, 89–108 (1997)
12. Greitemeyer, T., Schulz-Hardt, S.: Preference-consistent evaluation of information in the hidden profile paradigm: Beyond group-level explanations for the dominance of shared information in group decisions. Journal of Personality & Social Psychology 84(2), 332–339 (2003)
13. Mojzisch, A., Schulz-Hardt, S.: Knowing other's preferences degrades the quality of group decisions. Journal of Personality & Social Psychology 98(5), 794–808 (2010)
14. Cialdini, R.: The science of persuasion. Scientific American 284, 76–81 (2001)
15. Felfernig, A., Ninaus, G.: Group recommendation algorithms for requirements prioritization. In: Workshop on Recommender Systems for Software Engineering (RSSE 2012), Zurich, Switzerland, pp. 1–4 (2012)

Clustering Users to Explain Recommender Systems' Performance Fluctuation

Charif Haydar, Azim Roussanaly, and Anne Boyer

Université de Lorraine, Laboratoire Loria, Bâtiment C,
Equipe KIWI 615, rue du jardin botanique
54600 Vandœuvres-lès-Nancy, France
{charif.alchiekhhaydar,azim.roussanaly,anne.boyer}@loria.fr
http://kiwi.loria.fr

Abstract. Recommender systems (RS) are designed to assist users by recommending them items they should appreciate. User based RS exploits users behavior to generate recommendations. Users act in accordance with different modes when using RS, so RS's performance fluctuates across users, depending on their act mode. Act here includes quantitative and qualitative features of user behavior. When RS is applied in an e-commerce dedicated social network, these features include but are not limited to: user's number of ratings, user's number of friends, the items he chooses to rate, the value of his ratings, and the reputation of his friends. This set of features can be considered as the user's profile.

In this work, we cluster users according to their acting profiles, then we compare the performance of three different recommenders on each cluster, to explain RS's performance fluctuation across different users' acting modes.

Keywords: Recommender system, collaborative filtering, trust-aware, trust, reputation, user profile, clustering, item popularity, abnormality.

1 Introduction

Recommender systems (RS) [4] are designed to assist users by recommending them items they should appreciate. User based RS exploit users behavior to generate recommendations. Different users act in accordance with different modes when using RS, so RS's performance fluctuates across users, depending on their act mode. Act here includes quantitative and qualitative features. when RS is applied in an e-commerce dedicated social network, these features include but are not limited to: the number of ratings a user does, the number of friends he has, the items he chooses to rate, the value of his ratings, and the reputation of his friends. This set of features can be considered as the user's profile.

We use the epinion.com[1] dataset. epinion.com is a consumers opinion website where users can rate items in a range of 1 to 5, and write reviews about them.

---

[1] http://www.epinion.com

L. Chen et al. (Eds.): ISMIS 2012, LNAI 7661, pp. 357–366, 2012.

Users can also express their trust towards reviewers whose reviews seem to be interesting to them.

Many recommender systems were tested on this corpus, such as collaborative filtering (CF) [7], trust-aware [6,8], and hybrid recommenders [18].

In this paper, we apply a clustering algorithm over users to characterize essential acting modes, then we compare the performance of three different recommenders (collaborative filtering, trust-aware, hybrid) on each cluster. We try to explain why on some clusters the performance of all recommenders gets better/worse, or why a recommender performance gets better on a given cluster while others' get worse on the same cluster. Globally, we try to give explanation to recommenders' performance fluctuation as a function of users' acting mode in the system.

The outline of the paper is organized as follows: in section 2, we discuss recommenders structures and users analysis. In section 3, we explain the details of the used dataset, the context of the experiments, and the analyses of the results. Finally, the last section is dedicated to conclusion and future works.

2   State of Art

Although, the choice of recommendation approach is to much related to the context, collaborative filtering (CF) [7] is one of the most used approaches, because of its efficiency and high performance in various contexts. The arise of social networks in the last several years opened the door to a new approach called trust-aware recommenders [6,8], which uses the information offered by these social networks to generate recommendations.

In some contexts, more than one recommenders can be appropriate. Several proposition were made to hybridize RSs, so make use of their qualities together. [1] proposed a taxonomy of hybridizing strategies.

The following sub sections, are limited to explain only the approaches used in this paper.

2.1   Collaborative Filtering Recommenders

CF is based on the similarity of users' preferences (usually expressed by rating items). CF used a $m \times n$ ratings matrix, where m is the number of users, and n is the number of items. Rating matrix is used to compute smiliarity between users' preferences. Similar users are called also neighbors.

[7] proposed equation 1 to predict the ratings that user u_a will give to item r depending on the ratings given to r by the neighbors of u_a.

$$p(u_a, r) = \overline{v_{u_a}} + \frac{\sum_{u_j \in U_r} f_{simil}(u_a, u_j) \times (v_{(u_j, i)} - \overline{v_{u_j}})}{card(U_r)} \qquad (1)$$

Where:

$f_{simil}(u_a, u_j)$: the similarity between u_a and u_j, we use Pearson similarity coefficient [7].
U_r: the set of users who have rated r.
$card(U_r)$: is the number of users in U_r.

Neighbors, in this approach, are computed automatically. By consequence, the approach is sensible to user's rating choices. Cold start [9] is one of the essential drawbacks of this approach. It consists in the difficulty to generate recommendations to users who did not rate enough items (called cold start users), because it is difficult to find neighbors to them.

RS performance can also fluctuate because of certain styles of ratings, such as rating rarely rated items, which make finding neighbors a complicated issue, or appreciating items that are globally unappreciated by the community, which complicates the prediction of ratings values.

2.2 Trust Aware Recommenders

Trust-aware recommenders (TAR) make use of the structure of the social network, so uses the trustee friends instead of neighbors in CF [8,17]. Neighbors (friends) are chosen by the user himself, this yields the system more controllable by the user, and more robust to malicious attacks.

Compared to CF, Trust-aware recommenders are less concerned by the cold start problem. Many studies show that they surpass the performance of CF [10,2,11,12,5,6].

Trust can be propagated. TAR considers not only the user's friends, but their friends and so on. Many models to propagate trust where proposed in the literature [6,13,15,14].

In our studied case, trust is simply a binary value. Thus we choose the model MoleTrust [6]. This model is adapted and tested to our dataset. In MoleTrust, each user has a domain of trust where he adds his trustee users. In this context, user can either fully trust other user or not trust him at all. The model considers that trust is transitive, and that its value decline according to the distance between the source user and the destination user. The only initializing parameter is the maximal propagation distance d.

If user A added user B to his domain, and B added C, then the trust of A in C is given by the equation:

$$Tr(A, C) = \begin{cases} \frac{(d-n+1)}{d} & \text{if } n \le d \\ 0 & \text{if } n > d \end{cases} \tag{2}$$

Where n is the distance between A and C ($n = 2$ as there two steps between them; first step from A to B, and the second from B to C).
d is the maximal propagation distance.
Consider $d = 4$ then: $Tr(A, C) = (4 - 2 + 1)/4 = 0.75$.

2.3   Hybridization

In [1], author identifies seven strategies to hybridize recommenders.

In [18], we applied five hybridization strategies on epinion dataset, and compared them to CF and TAR. Most of those strategies improved the prediction coverage, without a serious decrease in the accuracy. The best score was obtained by weighted hybridization strategy, shown in the equation 3, with ($\alpha = 0.3$).

$$score(u_a, u_j) = \alpha \times simil(u_a, u_j) + (1 - \alpha) \times trust(u_a, u_j) \qquad (3)$$

2.4   Users Behavior Analysis

The performance of RS fluctuate across users. This fluctuation was commonly attributed to quantitative features, such as the number of ratings and trust relations. Nevertheless, some users keep receiving poor recommendation despite their numerous ratings/trust relations. In [19], we considered other qualitative features, and showed how does the performance relates to each of them. We considered the popularity of items that the user rates, the difference between user's rating and the average rating of an item (abnormality), and the reputation of his trustee friend.

In this paper we aim to study these features as a set that defines a user profile.

3   Experiments and Performance Evaluation

3.1   DataSet

Epinion dataset contains 49,290 users who rated a total of 139,738 items. users can rate items in a range of 1 to 5, the total number of ratings is 664,824. Users can also express their trust towards others (binary value), the dataset contains 487,182 trust ratings. We eliminate users having no ratings because we can not evaluate their recommendations. We keep only 32424 users.

We divide the corpus to two parts randomly, 80% for training and 20% for evaluation (a classical ratios in the literature). We took into consideration that every user has 80% of his ratings in the training corpus and 20% in the evaluation corpus, this is important to us to analyse the recommendation accuracy by user.

3.2   User Profile

We present here the features of user profile:

– Number of ratings: is a quantitative feature. The number of ratings by user is an important feature to recommendation accuracy, especially for CF. It is generally considered as a unique feature to explain fluctuation. Even though, We think that it is not sufficient, and has to be accompanied by qualitative features.

– Rated items' popularity: We define item's popularity as the number of users
having rated this item. Users tend to rate popular items more than unpopular
ones [3], this behavior creates an important bias in items popularity. Hence,
RS tends to recommend popular items more than others. This can limit the
choices of users and reduce the serendipity in the RS.

This current feature concerns the influence of the popularity of rated items,
on the recommender's performance for the user.

We define user's ratings' popularity as the average of the popularity of
items rated by this user. Ratings' popularity is a qualitative feature.

$$urpop(u) = \frac{\Sigma_i pop(i)}{cord(i)}$$

pop(i): the popularity of the item i (the number of users having rated i).
– User abnormality: This feature detects users of particular tastes. It focuses
on the user's orientation versus the global orientation of the community.

Formally, we compute the average rate of the item.

$$avgr(i) = \frac{\Sigma_a v_{ai}}{cord(a)}$$

Where v_{ai} is the rate given by the user a to the item i.
then the difference between the rate supplied by the current user and this
average. The Abnormality coefficient of the user is the average of differences
between his ratings and the average rate of each item he rates.

$$Abn(u) = \frac{\Sigma_i |v_{ui} - avgr(i)|}{cord(i)}$$

– Number of trusted friends: We have shown in a later work [19] that the
relation between this feature and the recommendation accuracy is not linear,
and that having more friends is more beneficial for new users, than it for
users who have already much friends.
– Reputation of trusted friends: This feature measures the impact of trusting
reputed /not reputed people on the quality of recommendations.

We consider a primitive metrics of reputation; the reputation of a user is
the number of users who trust him.

$$Rep(u_i) = Nb.trusters_{u_i} \tag{4}$$

Where: $Nb.trusters_{u_i}$ is the number of people how trust u_i.

We think that even when a user trusts few people, this can be more beneficial
when they are well reputed persons. Therefore, our current feature $Trep(u_a)$
is the average of the reputations of users that the user u_a trusts.

$$Trep(u_a) = \frac{\Sigma_i^N Rep(u_i)}{N} \tag{5}$$

$u_i \in D(U_a)$ (the group of users who are trusted by u_a).

In [19], we showed the influence of each of the precedent features performance of RS separately, the main contribution of this paper is to study if those features can, together, define classes of users. these classes define a particular ratings and trusting strategy for their members, so we can compare the performance of the recommenders by class of users, to find which recommender is more adapted to each class.

3.3   Clustering

User vector is composed by the five precedent feature, We use Kmeans clustering [20]. As Kmeans does not compute automatically the optimal number of classes, we employ Davies-Bouldin evaluation metrics[21] to optimize the clusters number. We initialize Kmeans between 4 and 10 clusters, Davies-Bouldin is minimized with 8 clusters, which implies the optimal number of clusters. Table 1 illustrates the results of this clustering, with the size of each cluster (number of users), and the average value of each user profile feature. The second column illustrate the averages for the entire corpus.

Table 1. Kmeans clustering with 8 clusters

cluster	All	0	1	2	3	4	5	6	7
size	32424	3221	765	4458	5547	7909	1674	8324	526
percentage	100%	9.93%	2.36%	13.75%	17.11%	24.39%	5.16%	25.67%	1.62%
ratings	20.27	14.8	6.63	34.09	11.12	22.3	4.47	8.98	251.01
popularity	80.86	79.43	453.76	64.67	54.19	88.24	26.66	79.8	44.02
abnormality	0.8	1.26	0.84	0.72	0.42	0.8	2.18	0.81	0.76
trust	13.86	13.57	9.16	28.26	10.05	16.88	6.7	0.08	136.47
T-rep	143.65	109.76	90.13	320.4	82.97	74.02	119.47	1126.51	225.77

3.4   Performance Evaluation Metrics

Performance evaluation includes two aspects; coverage and accuracy. The coverage is percentage of users to whom RS could generate recommendations, whereas the accuracy is about how much the predicts values are close to real ones.

To measure accuracy, we make use of the mean absolute error metrics (MAE) [16]. MAE is a widely used predictive accuracy metrics. It measures the average absolute deviation between predicted and real values. We use a specific form of it, called User mean absolute error (UMAE) [17], which computes MAE by user:

$$UMAE(u) = \frac{\Sigma_{i=1}^{N}|p_u i - r_{ui}|}{N} \tag{6}$$

Where: p_i is the rating value predicted by the recommender to the item i. r_i is the real rating value supplied by the user to the item i.
N: The number of predicted items to the user u.

Then, to evaluate the accuracy of a recommender we compute global UMAE or GUMAE, which is the average of UMAE of all users. In order to aggregate both aspects in one metrics, we propose a F-metrics like measurement, which compromise between recall and precision. Recall is the percentage of the cases to which the system could reply, to the total number of cases, so it is simply the coverage in our case. Precision is the percentage of relevant replies, to the total number of replies, we represent it by the UMAE in the current case.

Recall and precision must be within the range [0,1]. UMAE varies in the range [4,0] so we need to normalize it, using the following equation:

$$precision = \frac{4 - UMAE}{4}$$

The coverage is already within the range [0,1] so the F measurement will be:

$$F = \frac{2 * precision * recall}{precision + recall}$$

3.5 Clusters' Analyzing

Table 2 shows the UMAE, coverage, and F values for the three recommenders by cluster:

Table 2. Kmeans clustering with 8 clusters

	cluster	All	0	1	2	3	4	5	6	7
UMAE	CF	**1.00**	1.24	1.07	0.94	0.88	0.97	1.46	1.03	0.88
	Trust	**0.86**	1.05	0.98	0.81	0.76	0.84	1.15	0.90	0.78
	Hybrid	**0.87**	1.05	098	0.81	0.76	0.83	0.15	0.92	0.78
Coverage	CF	**0.64**	0.69	0.66	0.85	0.55	0.82	0.12	0.49	0.99
	Trust	**0.69**	0.90	0.90	0.98	0.89	0.92	0.76	0.02	1
	Hybrid	**0.82**	0.93	0.92	0.98	0.91	0.96	0.77	0.47	1
F	CF	**0.69**	0.69	0.70	0.81	0.65	0.79	0.21	0.60	0.88
	Trust	**0.73**	0.81	0.82	0.88	0.85	0.85	0.74	0.06	0.89
	Hybrid	**0.80**	0.83	0.83	0.88	0.86	0.87	0.74	0.59	0.89

- Cluster 0 contains about 10% of the population, the particularity of this class is the high abnormality value with a score of 1.26. This abnormality score causes an augmentation in UMAE value, which should pull the F value down. Nevertheless, F values for this cluster are slightly higher. This results from the high coverage values of the cluster (compared to the entire corpus). The cluster is more dense than the corpus, users are closer to the center of the cluster, this improves the value of coverage, so the value of F.
- Cluster 1 contains 2.36% of users. It's users have a low number of ratings (6.6), but they rate popular items (popularity average is 453.76). In [19], we showed that rating popular items has a positive impact for cold start users, and a negative impact for users who rate alot of items. Users in this class are cold start users, which explains the slight performance augmentation for all recommenders.

- Cluster 2, has 13.75% of users. We note in this cluster a considerable augmentation in the number of ratings, number of trust relationships and friends' reputation. Whereas popularity and abnormality are slightly under their averages. All three recommenders improve their performance on this cluster, this is normal regarding the good quality values in almost all features.

- Cluster 3, with 17% of population. The ratings of this class users are closer to the orientation of the community (abnormality=0.42), their number of ratings is about 11 ratings by user, which is neither high nor very low. The items they rate, are generally not very popular. The performance of CF is slightly lower than normal. We explain this because of the low ratings popularity, and the limited number of ratings together. Hence, finding similar users is difficult. On the other hand, the performance of the trust-aware and hybrid recommenders improves because they profit of the low abnormality, and - at the same time - they are less concerned by the ratings and popularity issue, while the have an alternative way to find friends.

- Cluster 4, with about 24% of users. All features are slightly higher than their averages in the corpus, except the reputation of friends which is in it's lowest level (74.02). As we shown in [19], the influence of friends reputation feature tends to be stable when it is higher than 10. The performance of the three recommenders is better because they profit the slight augmentation in the other four features, whereas the decline of friends reputation does not has a strong impact.

- Cluster 5 contains 5.16% of users. These users have the lowest number of ratings and items' popularity and the highest abnormality average among clusters. One of the qualities of the trust-aware recommenders is reducing the impact of cold start, because one trust expression can be more informative than many items ratings. This is obvious in this cluster, where the average of trust relationships is 6.7 and that of ratings is 4.47 (both are relatively low). Hence. the loss in CF performance is farther than that in trust-aware (0.21 versus 0.74). It is true also that other features play a role in this bad performance of CF, when users rate a small number of relatively unpopular items, finding neighbors in CF becomes complicated (same problem as cluster 3). Even when RS finds the similar users, the predicting ratings is weak because of the abnormal behaviour of the users in this cluster.

- Cluster number 6 is the biggest cluster with 25% of users, it contains users who tend to rate items, more than trusting other users (about 9 ratings and 0.08 trust relationships by user), only 251 users over 8324 in this cluster have trusted others. Nevertheless, they tend to trust very reputed users. Hence, CF is the best recommender for these users, it depasses slightly the performance of the hybrid recommender (0.6 versus 0.59), whereas the trust recommender is far behind them, which can be explained by the lake of trust relations to the trust-aware recommender. The slight difference to the favour of CF over the hybrid system can be explained by the inutility of the information supplied to the hybrid system because trusting users who are trusted by everybody can be uninformative and disturbing to the hybrid system. The performance of the hybrid reommender on this cluster is the worst compared to other clusters.

- The last and the smallest cluster is cluster 7, with 1.62% of users. Users in this cluster have the highest average of ratings, they rate averagely popular items, thier abnormality average is close to the general average. They have a considerable number of trust relationships with a well reputed friends. All these values make this cluster the best quality over the 8 clusters, and the three recommenders achieve their best performance on this cluster.

4   Conclusion and Future Works

The main contribution of this paper is to detect different models of users' behavior in RS, and their impact on RS's performance. We have shown the relation between the performance fluctuation and the behavioral features of users. In some clusters, it was necessary to analyze more than one feature together to explain RS performance, for example clusters 3 and 5, these phenomena are not easy to detect and explain without the clustering phase.

We also referred to the usability of qualitative features in explaining the fluctuation. Rating unpopular items causes difficulties in finding neighbors, so recall problem to CF recommenders, trusting weakly reputed persons causes the same problem for TAR, being abnormal affects precision negatively in all recommenders. We think that regarding user behavior feature assist to build an adaptive recommender that agree with their mode of behavior within the social network.

References

1. Burke, R.: Hybrid Web Recommender Systems. In: Brusilovsky, P., Kobsa, A., Nejdl, W. (eds.) Adaptive Web 2007. LNCS, vol. 4321, pp. 377–408. Springer, Heidelberg (2007)
2. Burke, R., Mobasher, B., Zabicki, R., Bhaumik, R.: Identifying attack models for secure recommendation. In: Beyond Personalization: A Workshop on the Next Generation of Recommender Systems (2005)
3. Steck, H.: Item popularity and recommendation accuracy. In: Proceedings of the Fifth ACM Conference on Recommender Systems (RecSys 2011), pp. 125–132. ACM, New York (2011)
4. Resnick, P., Varian, H.R.: Recommender systems. Commun. ACM 40(3), 56–58 (1997)
5. Miller, B.N., Konstan, J.A., Riedl, J.: PocketLens: Toward a personal recommender system. ACM Trans. Inf. Syst. 22(3), 437–476 (2004)
6. Massa, P., Bhattacharjee, B.: Using Trust in Recommender Systems: An Experimental Analysis. In: Jensen, C., Poslad, S., Dimitrakos, T. (eds.) iTrust 2004. LNCS, vol. 2995, pp. 221–235. Springer, Heidelberg (2004)
7. Resnick, P., Iacovou, N., Suchak, M., Bergstrom, P., Riedl, J.: GroupLens: an open architecture for collaborative filtering of netnews. In: Proceedings of the 1994 ACM Conference on Computer Supported Cooperative Work (CSCW 1994), pp. 175–186. ACM, New York (1994)

8. Golbeck, J., Hendler, J.: FilmTrust: movie recommendations using trust in web-based social networks (2006)
9. Maltz, D., Ehrlich, K.: Pointing the way: active collaborative filtering. In: Katz, I.R., Mack, R., Marks, L., Rosson, M.B., Nielsen, J. (eds.) Proceedings of the SIGCHI Conference on Human Factors in Computing Systems (CHI 1995), pp. 202–209. ACM Press/Addison-Wesley Publishing Co., New York (1995)
10. Mobasher, B., Burke, R., Bhaumik, R., Williams, C.: Toward trustworthy recommender systems: An analysis of attack models and algorithm robustness. ACM Trans. Internet Technol. 7(4), Article 23 (October 2007)
11. Lam, S.K., Riedl, J.: Shilling recommender systems for fun and profit. In: Proceedings of the 13th International Conference on World Wide Web (WWW 2004), pp. 393–402. ACM, New York (2004)
12. O'Mahony, M., Hurley, N., Kushmerick, N., Silvestre, G.: Collaborative recommendation: A robustness analysis. ACM Trans. Internet Technol. 4(4), 344–377 (2004)
13. Golbeck, J.: Personalizing applications through integration of inferred trust values in semantic web-based social networks. In: Semantic Network Analysis Workshop at the 4th International Semantic Web Conference (November 2005)
14. Kuter, U., Golbeck, J.: Using probabilistic confidence models for trust inference in Web-based social networks. ACM Trans. Internet Technol. 10(2), Article 8, 23 pages (2010)
15. Ziegler, C.-N., Lausen, G.: Spreading Activation Models for Trust Propagation. In: Proceedings of the 2004 IEEE International Conference on e-Technology, e-Commerce and e-Service (EEE 2004), pp. 83–97. IEEE Computer Society, Washington, DC (2004)
16. Herlocker, J.L., Konstan, J.A., Terveen, L.G., Riedl, J.T.: Evaluating collaborative filtering recommender systems. ACM Trans. Inf. Syst. 22(1), 5–53 (2004)
17. Massa, P., Avesani, P.: Trust-Aware Collaborative Filtering for Recommender Systems. In: Meersman, R. (ed.) OTM 2004, Part I. LNCS, vol. 3290, pp. 492–508. Springer, Heidelberg (2004)
18. Haydar, C., Boyer, A., Roussanaly, A.: Hybridising collaborative filtering and trust-aware recommender systems. In: WEBIST 2012 - Proceedings of the 8th International Conference on Web Information Systems and Technologies, Porto, Portugal, April 18-21 (2012)
19. Haydar, C., Roussanaly, A., Boyer, A.: Analyzing Recommender System's Performance Fluctuations across Users. In: Quirchmayr, G., Basl, J., You, I., Xu, L., Weippl, E. (eds.) CD-ARES 2012. LNCS, vol. 7465, pp. 390–402. Springer, Heidelberg (2012)
20. Alsabti, K., Ranka, S., Singh, V.: An efficient k-means clustering algorithm. Electrical Engineering and Computer Science, Paper 43 (1997)
21. Davies, D.L., Bouldin, D.W.: A cluster separation measure. IEEE Trans. Pattern Anal. Machine Intell. 1, 224–227 (1979)

Network-Based Inference Algorithm on Hadoop

Zhen Tang, Qingxian Wang, and Shimin Cai

Web Sciences Center, School of Computer Science and Engineering, University of Electronic
Science and Technology of China, Chengdu 611731, P.R. China
{tangzhen8803,shimin.cai81}@gmail.com

Abstract. Network-based inference (NBI) algorithm is a new but effective
personalized recommendation algorithm based on bipartite networks, and it
performs better than global ranking method (GRM) and collaborative filtering
(CF).However, the complexity of NBI is high thus hinder NBI's use in large
scale system. In this paper, we implement NBI algorithm on a cloud computing
platform, namely Hadoop, to solve its scalability problem. We use MapReduce
model to distribute the NBI algorithm into serial parallel MapReduce jobs, and
implement them in parallel on Hadoop platform. Through performing extensive
experiments on the data sets of Netflix, the result shows that the NBI algorithm
can scale well and process large datasets on commodity hardware effectively.

Keywords: Network-based inference, recommender system, cloud computing,
Hadoop, MapReduce.

1 Introduce

With the rapid improvement of Internet, personalized recommendation has been more and
more popular in the field of e-commerce [1]. Network-based inference (NBI) algorithm is a
new but effective personalized recommendation algorithm based on bipartite networks [2].
Many works have been done to improve the performance. However, a big problem is its
scalability, i.e., when the volume of the dataset is very large, the computation cost of NBI
would be very high. In these years, cloud computing, like Amazon Web Service (AWS)
[3], Google App Engine (GAE) [4] etc, have become more and more popular, and it can
be used to overcome the problem of large scale computation task and solve the scalability
problem. Cloud computing is the provision of dynamically scalable and often virtualized
resources as a service over the Internet [5]. Users need not to have many knowledge of the
technology of infrastructure in the "cloud" that supports them, and users can start to build
the application quickly. Cloud computing services often provide common business
computing capability and storage services for users.

In order to solve scalability problem of the recommender system, we implemented
the Network-based inference (NBI) algorithm on a cloud-computing platform. There
are several cloud-computing platforms available, for example, the Dynamo [6] of
amazon.com, the Nettune [7] of Ask.com and the Dryad [8] of Microsoft, etc. In this
paper, we choose the Hadoop [9] platform as the base of our implementation. Because
the Hadoop platform is an open source cloud computing platform, which is widely
used by Yahoo, Twitter etc, it implements the MapReduce framework which was

L. Chen et al. (Eds.): ISMIS 2012, LNAI 7661, pp. 367–376, 2012.

proposed by Google Lab. The Hadoop platform implements a distributed file system, Hadoop Distributed file System (HDFS). HDFS is fault-tolerant, and designed to deploy on low-cost hardware, it provides high throughput access to application program's data, and applies to these application programs which have large data set. We can also make our program in parallel easily. In order to do that, we can use the MapReduce framework to break down big problems into small ones, and improve the speed of computing through the parallel computing. Because the NBI algorithm is a new personal recommendation algorithm which was proposed these years, there is almost no implementation of NBI algorithm on cloud computing platform [10-12].

In this paper, we study the implementation of NBI algorithm on cloud computing platform. The works we have done are summarized as follow. Firstly, we designed a NBI algorithm for the MapReduce program framework, and implemented the algorithm on the Hadoop platform. Secondly, we tested our implementation on Hadoop platform.

The rest of the paper is organized as follow: Section2 discusses the NBI process and the MapReduce programming model. Section 3 presents the implementation of the NBI algorithm on Hadoop platform. Section 4 shows our experiment results and our analysis. Section 5 concludes the paper with pointers to future work.

2   NBI and MapReduce

2.1   Network-Based Inference

Network-based inference algorithm is a new personal recommendation algorithm based on bipartite networks [2, 13], but it has been proved more effective than global ranking method (GRM) and collaborative filtering (CF) [14, 15]. Network-based inference algorithm is based on the following assumptions: we do not consider the content features of the users and items, we only treat them as abstract nodes. All information which is used in algorithm is hided in the choice relation between users and items. We can use the user-item bipartite networks to construct the relation between users and items and do the resource allocation to solve the problem of data sparsity, partly.

Consider a recommender system constructed by m users and n items (for example, books, movies, web pages etc). In the system, if user i have collected item j, we can link i and j with an edge $a_{ji} = 1(i=1, 2\ldots m; j=1, 2\ldots n)$, otherwise, $a_{ji} = 0$. So we can construct a bipartite networks with $m+n$ nodes to express. To any target user i, recommendation algorithm sorts all items uncollected by i with the degree of like, and recommends i with the items which lead in the rank list. We assume that, all items collected by i have the ability that they can recommend other items. The abstract ability can be considered some separable resource in related items, items which hold resource will give more resource to the items they prefer. If we use w_{ij} express the resource quotas which item j will give to item i. This can be written as

$$w_{ij} = \frac{1}{k_j} \sum_{l=1}^{m} \frac{a_{il}a_{jl}}{k_l},$$ (1)

Where k_j is the degree of item j, and k_l is the degree of user l.

To a given target user, we can set the initial resource of items he has collected with 1, others with 0.So we can have an n-dimensional 0/1 vector, which express initial resource allocation of the user. Obviously, the vector expresses personalized information, which is different to different users. We can denote the vector with f, and the final resource allocation of this process can rewritten as

$$f' = Wf, \tag{2}$$

We can sort all items which target user has not collected yet with the vector f' corresponding element, more value means the user like more. The items which rank front can be recommended to the target users.

From the above steps we can find that the calculation process of NBI algorithm would consume intensive computing time and computer resource. When the data set grows very large, the calculation would continue for several hours or even longer. Therefore, we propose a new method to implement the NBI algorithm on Hadoop platform to solve that problem.

2.2   MapReduce Overview

The MapReduce model is a distributed implementation model which is proposed by Google [16, 17]. Here, we introduce the MapReduce model and describe its working mechanism on Hadoop platform.

The MapReduce model is inspired by the Lisp programming language. The MapReduce model abstract the calculation process into two phase: map phase and reduce phase. In the Map phase, the Map function which is written by the user, takes a set of input key/value pairs, and produces a set of output key/value pairs. The MapReduce library groups together all intermediate values associated with the same intermediate key and passes them to the Reduce phase. In the Reduce phase, the function also written by the user, accepts an intermediate key I and a set of values for that key. It merges together these values to form a possibly smaller set of values. Typically, none output value or only one is produced per Reduce invocation. MapReduce allows for distributed processing of the map and reduction operations. Provided each mapping operation is independent of the others, all mappers can be performed in parallel – though in practice it is limited by the number of independent data sources and/or the number of CPUs near each source. Similarly, a set of reducers can perform the reduction phase - provided all outputs of the map operation that share the same key are presented to the same reducer at the same time. While this process can often appear inefficient compared to algorithms that are more sequential, MapReduce can be applied to significantly larger datasets than "commodity" servers can handle a large server cluster can use MapReduce framework to sort one PetaByte of data in only a few hours. Before partition, users can set combine class to process intermediate key, most like Reduce phase. Combine able to reduce the intermediate results <key, value> the number, thereby reducing network traffic.

We analyze the scheduling mechanism of the Hadoop. In the Hadoop platform, the HDFS block size is 64MB as default [18, 19], which can be reset by the Hadoop configuration, it also determines the maximum input size of one mapper. When the size of the input file is larger than 64MB, the platform would split it into a number of small files which size is less than 64MB automatically. For one input file, the Hadoop platform initializes a mapper to deal with it, the line number of the file as the key and the content of that line as the value. In the map phase, the function which is defined by users deals with the input key/value and passes the intermediate key/value to the reduce phase, and all the intermediate key/value would be sum up in the reduce phase. When the task blocks are complete, the Hadoop platform would kill the corresponding mapper. However, if the documents are not been done all, the platform would choose one new file to initialize a new mapper to deal with it. The Hadoop platform should be circulate the above process until the map task is completed.

3   MapReduce for NBI

In this section we present how to implement the NBI algorithm with MapReduce framework. According to the introduction of the second section, we know that the calculation process of NBI algorithm is not easy to use the MapReduce model decomposition directly. So we can divide the task into 4 Hadoop jobs, every Hadoop job must include one mapper and one reducer function, each job finishes one step of the task, and then the recommendation task will be done at last. The later job input may use the earlier job output, so we use the sequence file as the intermediate file text format. It provides a persistent data structure of binary key/value pair, and it can quickly serialize any data type to a file to be the input of the next mapper. And we also implement the combiner function in NBI algorithm on Hadoop. Available network bandwidth in clusters limits the number of MapReduce jobs, so we should avoid the network transfer between Map task and Reduce task. The advantage of the combiner functions is combining the output key/value pairs in Map phase to reduce the cost of network transfer in Reduce phase.

The first job calculates the item list for every user from the user-item log file to make preparation for further computing. In the Map phase, it will be a user log filter, the input is the raw log file with the consumer records of the users, the key is the offset in bytes of this record to the start point of the data file, and the value is a string of the content of this record. The dataset is split and globally broadcast to all mappers. And the output will be the <*user, item*> key/value pairs. Then sum all mapper file up to Reduce phase, and generate the item list *{user, {item1, item2... item N}}* for every user. Figure 1 shows us the MapReduce procedure of the first job.

The second job calculates the degree for each item, and stores all of them in a temporary directory for the third job. The item degree can be treated as the count of the edges linked with the item. In the Map phase, the input is still the raw log file with the consumer records of the users, the output will be the <*item, user*> key pairs, and then sum all mapper file up to Reduce phase, count all users in value which the key is the same item, and generate the item degree pair <*item, item degree*> for each item. Figure 2 shows us the MapReduce procedure.

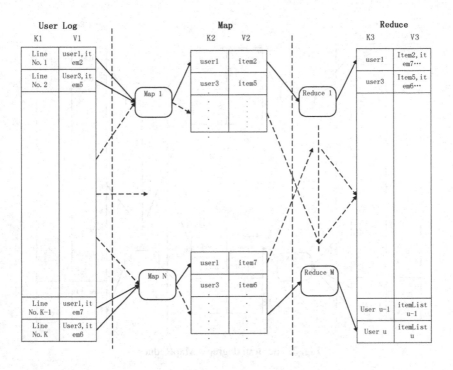

Fig. 1. The item list's MapReduce

The input of the third job is the output of the first job. In the mapper phase, we calculate the concurrence between item i and item j. The MapWritable, which is one of four collection classes in package org.apache.hadoop.io, will store the mapping relationship. In the reducer phase, the job will load the item degree list which has been computed by the second job in the distribute cache. Distribute cache can be used to distribute simple, read-only data/text files and/or more complex types such as archives, jars etc. Archives (zip, tar and tar.gz files) are un-archived at the slave nodes. Jars may be optionally added to the classpath of the tasks, a rudimentary software distribution mechanism. In the reduce phase we sum up the concurrence between every two items. In the end of the job, the output is the resource quotas matrix W_{ij} as we have mentioned in section 2.1. Figure 3 shows us the MapReduce procedure of the third job.

The last job we load the user-item source record which has been computed in the first job and stored in a temporary dictionary into the distribute cache. If the item has been collected by user, the item computation for the user will be skipped; otherwise, the item will multiply with the resource quotas matrix. At last, the output is the rating vector for every user. The front of the vector will be recommended to corresponding user. Figure 4 shows us the MapReduce procedure of the fourth job.

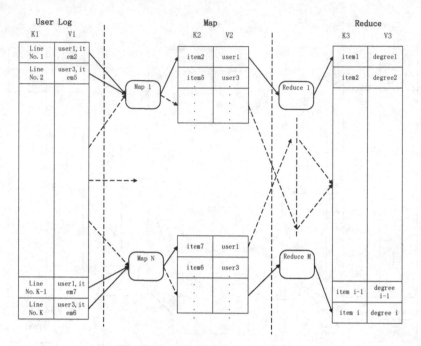

Fig. 2. The item degree's MapReduce

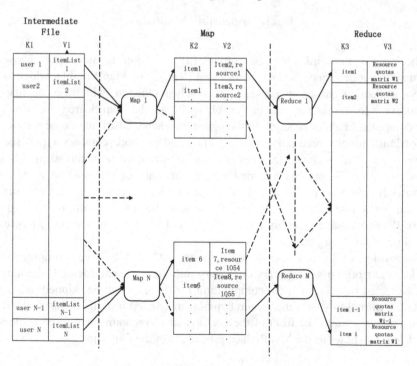

Fig. 3. The resource quotas matrix's MapReduce

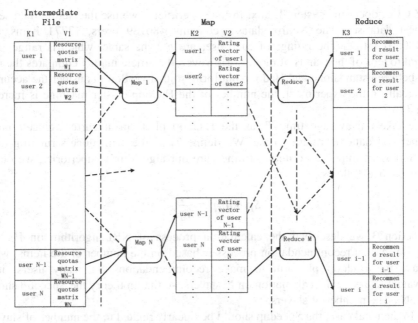

Fig. 4. The recommendation result's MapReduce

3.1   Algorithm Complexity Analysis

We assume that the count of items is n, that the count of users is u, that we have m training records, and that there are P cores. So we can do our running-time analysis, for single-core NBI algorithm, we have to compute the item list and item degree for $O(m)$ in the first and second job, compute the resource quotas matrix for $O(un^2)$ in the third job, and compute the recommendation results for $O(n^3)$ in the last job. So the time computation complexity of single-core NBI algorithm is $O\ (m+un^2+n^3)$. For paralleled NBI algorithm, we have assumed that the resource quotas matrix computing can be sped up by a factor of P' on P cores. (In practice, we expect $P'\approx P$.) The reduce phase can minimize communication by combining data as it's passed back; this accounts for the $log\ (P)$ factor. The reduce step incurs a resource quotas matrix communication cost of $O\ (n^2)$. So the time computation complexity of paralleled NBI algorithm is $O\ (\frac{m}{P} + \frac{un^2}{P'} + \frac{n^3}{P'} + n^2\ log(P))$. The space computation complexity is $O\ (n^2)$.

4   Experiment Results and Analysis

In this section we show our experiment results. We implement NBI algorithm on the Eclipse platform, we describe the NBI'S process at section 3. Our Hadoop compute-cluster was composed by 7 computers, one computer as a master controller and other 6 computers as slavers. Each computer's memory is 2 GB and INTEL Dual-core 2.6

GHZ CPU, operating system Linux. In the experiments we use the Netflix data set as the test data set. The Netflix dataset contains 480189 users, 17771 items, and 100480507 rating. The ratings of each user are not the same, which is range from several to tens of thousands of movies. However, the article mainly compares the runtime between standalone and Hadoop platform, so that we don't consider the accuracy and recall. In the experiment, the number of the Hadoop cluster machine is increase from 3 to 7.

We take the average time T_a as the Hadoop platform at current cluster nodes number and data set running time. We define T_s as the standalone's running time. Speedup as an important criterion to measure our algorithm's superiority, we define the speedup as follows:

$$Speedup = \frac{T_s}{T_a},\qquad(3)$$

At section 3 we described the calculation process of NBI algorithm on Hadoop platform. As the recommendation process is based on the number of the items, when we use the Hadoop platform to make recommendations for many users, it is equivalent to assign the calculation on N slavers, so that in theory the Speedup should be N, N is the number of slavers.

In the ideal case, the Speedup should be linearly related to the number of slavers. But we should take network cost and the file size into consideration. Every time, we select M (M=500, 1000, 2000, 3000, 5000) items (movies) randomly to test the performance. According to Heap's Law, with the increase in number of items which we select, the size of data set grows exponentially.

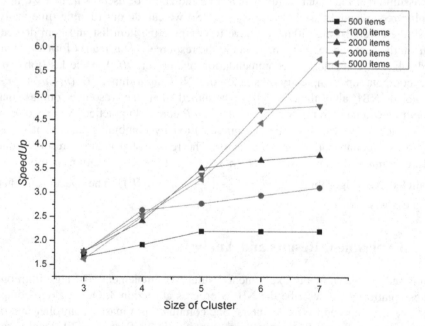

Fig. 5. Speedup of NBI on Hadoop

From Figure 5, that with the increase in the size of cluster, the Speedup increase linearly. With the increase in number of items, e.g. M=500, 1000, 2000, 3000, the size of data set increases, and the speedup increase linearly with the size of cluster. But when the size of cluster continues to increase to a threshold, the speedup will not grow any more, and stabilize at a certain speedup. The size of data set limits the number of tasks, when the size of cluster is greater than the number of tasks, the tasks cannot be fully parallelized to all nodes in the cluster, so the speedup will no longer increase. When the number of items continue to increase, e.g. M=5000, the speedup increase linearly with the size of the cluster. So the NBI algorithm can split the computing tasks effectively, and scale out linearly on Hadoop.

5   Conclusions

In this paper, we mainly use the MapReduce model to parallelize NBI algorithm to solve the scalability problem. As a stateless programming model, MapReduce cannot directly express NBI algorithm. To achieve our goal, we have divided the calculation into four jobs. Every job finishes each step of the computing work, so we can run NBI algorithm on Hadoop platform in parallel. The experiment results shows that our design algorithm in Hadoop platform to take the good performance.

Through the experiments we find that the main drawback of the MapReduce framework is that we should input the file of enough size to initialize the enough number of mapper file to process in parallel. If the size of input is too small to split enough numbers of files, MapReduce framework shows its strong ability of distributed computing.

In our future works, we are planning to improve the MapReduce process, to enable the NBI algorithm run on Hadoop platform more effectively.

Acknowledgements. This work is partially supported by the National Natural Science Foundation of China (Grant Nos. 60973069, 90924011, 61073099).

References

1. Adomavicius, G., Tuzhilin, A.: Toward the next generation of recommender systems: A survey of the state-of-the-art and possible extensions. IEEE Trans. on Knowledge and Data Engineering 17(6), 734–749 (2005)
2. Zhou, T., Ren, J., Medo, M., Zhang, Y.C.: Bipartite network projection and personal recommendation. Phys. Rev. E 22(5), 419–434 (2007)
3. Amazon Web Service, http://aws.amazon.com/
4. Google App Engine, https://appengine.google.com/
5. Katsaros, D., Mehra, P., Pallis, G., Vakali, A.: Cloud Computing: Distributed Internet Computing for IT and Scientific Research. IEEE Internet Computing 13(5), 10–13 (2009)
6. DeCandia, G., Hastorun, D., Jampani, M., Kakulapati, G., Lakshman, A., Pilchin, A., Sivasubramanian, S., Vosshall, P., Vogels, W.: Dynamo: Amazons highly available key-value store. In: Proc. of the 21st ACM Symposium on Operating Systems Principles, pp. 205–220. ACM Press, New York (2007)

7. Chu, L.K., Tang, H., Yang, T., Shen, K.: Optimizing data aggregation for cluster-based Internet services. In: Proc. of the ACM SIGPLAN Symposium on Principles and Practice of Parallel Programming, pp. 119–130. ACM Press, NewYork (2003)
8. Isard, M., Budiu, M., Yu, Y., Birrell, A., Fetterly, D.: Dryad: Distributed data-parallel programs from sequential building blocks. In: Proc. of the 2nd European Conf. on Computer Systems (EuroSys), pp. 59–72 (2007)
9. Apache Hadoop, http://hadoop.apache.org/
10. Apache Mahout, http://mahout.apache.org/
11. Chu, T.C., Kim, S.K., Lin, Y.A., Yu, Y.Y., Bradski, G., Ng, A., Kunle, O.: Map-reduce for machine learning on multicore. In: NIPS, pp. 281–288 (2006)
12. Zhao, Z.D., Shang, M.S.: User-based collaborative filtering recommendation algorithms on hadoop. In: Proc. of International Workshop on Knowledge Discovery and Data Mining, pp. 478–481 (2010)
13. Petter, H., Fredrik, L., Christofer, E., Beom, K.: Network bipartivity. Phys. Rev. E 68, 056107 (2003)
14. Su, X.Y., Taghi, K.: A survey of collaborative filtering techniques. Advances in Artificial Intelligence, 421–425 (2009)
15. Linden, G., Smith, B., York, J.: Amazon.com Recommendations. Item-to-item Collaborative Filtering. IEEE Internet Computing 7(1), 76–80 (2003)
16. Jeffrey, D., Sanjay, G.: MapReduce: Simplified data processing on large clusters. Communications of the ACM 51(1), 107–113 (2005)
17. Dean, J., Ghemawat, S.: Distributed programming with Mapreduce. In: Oram, A., Wilson, G. (eds.) Beautiful Code, pp. 371–384. O'Reilly Media, Inc., Sebastopol (2007)
18. Ranger, C., Raghuraman, R., Penmetsa, A., Bradski, G., Kozyrakis, C.: Evaluating MapReduce for Multi-core and Multiprocesor Systems. In: Proc. of the 10th IEEE International Symposium on High Performance Computer Architecture, pp. 13–24 (2007)
19. Sanjay, G., Howard, G., Shun-Tak, L.: The Google File System. In: Proc. of the 19th ACM Symposium on Operating Systems Principles, pp. 29–43. ACM Press, New York (2003)

A Hybrid Decision Approach to Detect Profile Injection Attacks in Collaborative Recommender Systems

Sheng Huang, Mingsheng Shang, and Shimin Cai

Web Sciences Center, School of Computer Science and Engineering, University of Electronic Science and Technology of China, Chengdu 611731, P.R. China
{huangshenguestc,shang.mingsheng,shimin.cai81}@gmail.com

Abstract. Collaborative filtering is a vitally central technology of personalized recommendation, yet its recommended result is so sensitive to users' preferences that the recommender system has significant vulnerabilities. To overcome the addressed issue, this paper proposes a hybrid decision approach to effectively and efficiently detect profile injection attacks in collaborative recommender systems. Through modifying the algorithms of RDMA (Rating Deviation from Mean Agreement) and WDMA (Weighted Deviation form Mean Agreement), the hybrid decision approach is integrated from these modified algorithms and the UnRAP (Unsupervised Retrieval of Attack Profiles) algorithm. The extensive experiments based on three common attack models show that the proposed detection algorithm is the best comparing with the modified RDMA and WDMA or origin ones, by which the detecting accuracy significantly increases almost 35%, 25%, and 8% than the RMDA, WMDA, and UnRAP algorithms, respectively. Furthermore, for the mixed attack model, we compare it with the UnRAP algorithm and improve the 10% accuracy.

Keywords: Collaborative Recommender system, profile injection attacks, hybrid decision, attack model.

1 Introduction

Collaborative recommender systems are vulnerable to profile injection attacks due to their natural openness [1]. In these attacks, some malicious users artificially inject a large number of attack profiles into the systems in order to bias the recommended results to their advantage. Thus, how to effectively and efficiently identify and resist the profile injection attacks has become an urgent need to resolve problem for the well development and extensive application of collaborative recommender systems.

Recent works in this area have focused on detecting and preventing profile injection attacks. Chirita et al. [2] proposed several metrics, e. g. RMDA (Rating Deviation from Mean Agreement) and WMDA (Weighted Deviation form Mean Agreement), for analyzing the rating patterns of malicious users and evaluate their potentials for detecting such attacks. Su et al. [3] developed a spreading similarity algorithm in order to detect groups of similar attackers. Williams et al. [4] trained three supervised classifiers to detect the attacks by extracting the features of user profiles. Although these classifiers

L. Chen et al. (Eds.): ISMIS 2012, LNAI 7661, pp. 377–386, 2012.
© Springer-Verlag Berlin Heidelberg 2012

can successfully detect the attacks of various attack sizes and filer sizes, they suffer from low precision and need a large number of training samples [5]. O'Mahony et al. [6] presented several techniques to defend against the attacks, including new strategies for neighborhood selection and similarity weight transformations. Hurley et al. [7] designed supervised and unsupervised Neyman-Pearson detectors based on statistical detection theory, respectively. These detectors can only detect the attack profiles that have the certain distributions. Mehta et al. [8] believed that the information contributed to a recommender system by attack profiles was less than that of genuine profiles and showed a PCA (Principal Component Analysis) based detection algorithm to filter the attack profiles. Bryan et al. [9] cast the detection problem as a problem of detecting anomalous structure in network analysis and proposed a detection algorithm, which is called UnRAP (Unsupervised Retrieval of Attack Profiles), to identify the attack profiles by introducing the variance adjusted mean square residue. The UnRAP algorithm is fairly accurate in detecting various standard attacks, however, the recall of this algorithm declines sharply when detecting a mixed attack. In Refs. [11-15], they compute the reputation or influence of user in recommender systems to find the attack profiles.

In the current work, the hybrid decision approach to detect the profile injection attacks includes: (1) we first modify the RMDA and WMDA algorithms, and name the new algorithms with RMDA_mod and WMDA_mod. (2) The two modified algorithms and the UnRAP algorithm is integrated together to construct the hybrid decision approach. (3) We conduct experiments on the MovieLens dataset and verify the proposed approach is effectively and efficiently to detect the random, average, bandwagon and mixed attacks.

The rest of the paper is organized as followings: An overview of attack models used in this paper is described in section 2. The strategies of attack detection are introduced in section 3. The experiments and comparing results of strategies are shown in section 4. Finally, we conclude our work in last section.

2   Attack Models

An attack against a collaborative recommender system consists of a set of attack profiles, biased profie data associated with fictitious user identities, and a target item. The attacker usually wishes that the target item is recommended more high by the system (i.e. a push attack), or hindered from the recommendation of system (a nuke attack). Herein, we concentrate on the push attacks based on a number of attack models, which are constructed from the knowledge of the recommender system, including the rating data, items and users, and will evaluate the strategy of attack detection in the context of the widely studied random, average and bandwagon attacks [10]. To keep our work self-contained, the following parts simply introduce these attack models.

2.1   Random Attack

In random attack model, the attacker firstly selects a target item to rate a reservation value. Then, the attacker randomly chooses a subset from a certain percentage of items of the system, and rates them with random values following a distribution with mean being the average of all user ratings across all items. The domain knowledge required to launch random attack is quite minimal, especially the average rating for all users in many systems are opened, which makes attackers easy get these information. However, the random attack is not particularly effective in real B2C environment.

2.2   Average Attack

The average attack is similar to the random attack that filler items are selected uniformly from the system item set. The only difference is the way rating the value of the selected items. In average attack, rating for each selected item is computed based on more specific domain knowledge of the individual mean for each selected item. So the average attack required the more domain knowledge. Average attack has been found to outperform than the random attack in Ref. [10].

2.3   Bandwagon Attack

The basic idea of Bandwagon attack is that in the systems a few items can attract most users. The bandwagon attack can be viewed as an extension to random attack, where it has to utilize some additional knowledge, which is used to identify a few of the most popular items in a particular domain. In real B2C environment, the attackers choose those popular or selling items as part of the attack profile, and rate these items with r_{max} (the maximal value of rating scale). Then, the attackers select a target item and randomly pick up a subset from a certain percentage of items of the system of which the rates randomly follow within the rating scale like that of random attack. The attack profiles of Bandwagon attack have a good probability of being similar to a large number of users. It has been demonstrated that the bandwagon attack significantly outperforms than both random and average attacks, and needs the low cost of domain knowledge.

3   Detection Algorithm Based Hybrid Decision Approach

3.1   Basic Definition

Before illustrating the detection algorithm based hybrid decision approach, we have to introduce the basic definitions, RMDA-score, WMDA-score, and UnRAP-score, according to the detection algorithms of RMDA, WMDA, and UnRAP [2, 9].

Definition 1 RMDA-score : Given a rating matrix D, the RMDA-score of a user u in D is computed by using the following equation:

$$RDMA-score(u) = \frac{\sum_{i=0}^{u} \frac{r_{ui} - \overline{r_{.i}}}{NR_i}}{N_u}, \tag{1}$$

where r_{ui} denotes the rating of user u on item i, N_u indicates the number of the items rated by user u, $\overline{r_{.i}}$ indicates the average rating of all users in D on item i, NR_i is the whole value of the item i given by the users.

Definition 2 WMDA-score : Given a rating matrix D, the WMDA-score of a user u in D computed by using the following equation:

$$WMDA - score(u) = \frac{\sum_{i=0}^{u} \frac{r_{ui} - \overline{r_{.i}}}{NR_i^2}}{N_u}, \tag{2}$$

where r_{ui} denotes the rating of user u on item i, N_u denotes the number of the items rated by user u , $\overline{r_{.i}}$ indicates the average rating of all users in D on item i, NR_i denotes the whole value of the item set i given by the users.

Definition 3 UnRAP-score : Given a rating matrix D, the UnRAP-score of a user u in D is computed by using the following equation:

$$UnRAP - score(u) = \frac{\sum_{i \in I} (r_{ui} - \overline{r_{.i}} - \overline{r_{u.}} + \overline{r_{UI}})^2}{\sum_{i \in I} (r_{ui} - \overline{r_{u.}})^2}, \tag{3}$$

where r_{ui} denotes the rating of user u on item i, $\overline{r_{.i}}$ indicates the average rating of all users in D on item i, $\overline{r_{u.}}$ and $\overline{r_{UI}}$ are the average of rating scores voted by user u on all item set I and rating scores in the whole D, respectively.

3.2   Modified RMDA and WMDA Algorithms

Based on the basic definitions, we herein modify the RMDA and WMDA algorithms, which are named by RMDA_mod and WMDA_mod, respectively. Because their modified procedures are similar, we only show one of them with following steps:

Step 1: User profile scoring using the RMDA-score (WMDA-score): We need compute the RMDA-score and WMDA-score to all users in the database according to Equations (1) and (2).

Step 2: Retrieval of target item: In the second step, the top-n highest scoring users from the sorted user list are examined to identify the target item. In the experiment, regardless of attack or filler size, we only examine the top-20 users (i.e., n=20). The target item i_t can generally be detected by retrieving the item which deviates most from the mean user rating. In general, the item with the largest consistent deviation over the top-20 users is selected as target item i_t.

Step 3: Retrieval of attack profiles: A sliding window (of 10 users in size) is passed along the sorted user list. At each step, the window is shifted by one user and the sum of the rating deviation for the target item over the current 10 users is computed. The iteration is terminated when the deviation of the target item reaches or exceeds zero. The final filtering procedure is applied to reduce the users who haven't rated the target item. After that, the remaining user set is returned as attack profiles for the target item i_t.

3.3   The Integrated Hybrid Decision Approach

The hybrid decision approach is based on the idea that a user profile may be an attack profile if it conforms to the distributing feature of attack profiles when detected by three (or more) unsupervised detection algorithms. In Fig. 1, we illustrate the hybrid decision approach (denote MIX) based on the modifying algorithms of RDMA and WMDA and UnRAP algorithm.

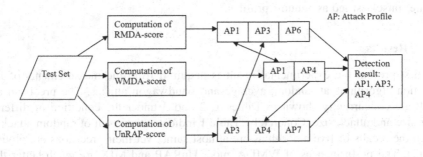

Fig. 1. Hybrid common decision approach

Its details of process are described as below:

(1) Given a rating database D, the RMDA-score, WMDA-score and UnRAP-score of each user profile in D are computed.

(2) Based on the RMDA-score, WMDA-score and UnRAP-score, we obtain three groups of the potential attack profiles.

(3) Based on the potential attack profiles groups acquired from step two, we decide the attack profile if it at least belongs to two groups.

4   Experiments and Evaluations

4.1   Design of Experiments

In our experiments, we use the publicly available Movie-Lens100K data set [16]. The data set totally consists of 100,000 ratings on 1682 movies by 943 users. Ratings are integer values between 1 (low) and 5 (high). The testing date set includes the users who have rated at least 20 movies.

In this study we generate the random, average and bandwagon and mixed attack strategies according to the attack models described in Section 2. The size of attack profiles injection into the system are 2%, 5% and 10% and filler sizes of each attack profile are 3%, 10% and 40% are generated. All the results are averaged over 100 randomly selected target items.

To measure the detection performance, the standard measurements of precision and recall are used, which are shown as follows:

$$\Pr ecision = \frac{TP}{TP + FP} \tag{5}$$

$$\mathrm{Re}call = \frac{TP}{TP + FN} \tag{6}$$

where TP is the number of attack profiles correctly detected, FP is the number of genuine profiles misclassified as attack profiles, and FN is the number of attack profiles misclassified as genuine profiles.

4.2   Results

We first present the experimental result is acquired from the measurements of five detection algorithms at random, average and bandwagon attacks. The precision and recall are concurrently shown in Tables 1, 2 and 3 under the condition of different filler size and attack size. Concretely, Table 1 indicates the result of random attack, in which the recalls of five algorithms are almost same, yet their precisions are obvious distinct. The performances of WMDA_mod, UnRAP and MIX are much better than the RMDA, WMDA, which suggests that the random attack is almost uncovered by the MIX. Following that, Table 2 describes the result of average attack, in which the RMDA, WMDA, and their modification are useless for the average attack according to the detecting precisions, and the MIX behaves the best for all indices. Finally, we present the result of bandwagon attack in Table 3. The RMDA_mod, WMDA_mod, and MIX show the better performance. For the smaller, the MIX behaves the best precision, while for the middle filler size the RMDA_mod overcome the MIX. The WMDA_mod and MIX perform the same the precision at the large attack size and filler size. By consider the whole performance of three attack strategies, we can conclude that the detecting accuracy of MIX significantly increases almost 35%, 25%, and 8% than the RMDA, WMDA, and UnRAP, respectively.

In the experiments, we also consider the mixed attack strategy. Table 4 and 5 describe the precisions and recalls of MIX comparing with the UnRAP at the random-average mixed attack and average -bandwagon mixed attack, respectively. As shown in Table 4, the precision and recall of the MIX are better than the UnRAP in all cased at the random-average mixed attack. For the average-bandwagon mixed attack, only with smaller attack sizes (e.g. attack size 2%), the UnRAP overcomes the MIX. Thus, the detecting results show that the MIX is also able to effectively and efficiently detect the mixed attack, and can improves 10% comparing with the UnRAP algorithm under the condition of mixed attack.

Table 1. Precision and Recall on test set with random attack of various attack sizes and filler sizes

	Filler Size	3%			10%			40%		
	Attack Size	2%	5%	15%	2%	5%	15%	2%	5%	15%
Precision	RMDA	0.07	0.23	0.66	0.07	0.2	0.56	0.07	0.16	0.45
	RMDA_mod	0.61	0.82	0.95	**0.99**	0.99	**1.0**	0.97	0.99	0.99
	WMDA	0.13	0.35	0.76	0.12	0.3	0.69	0.09	0.22	0.54
	WMDA_mod	**1.0**	**1.0**	**1.0**	**0.99**	0.99	**1.0**	**0.99**	**1.0**	**1.0**
	UnRAP	0.98	0.99	**1.0**	**0.99**	0.99	0.99	**0.99**	0.99	**1.0**
	MIX	0.99	0.99	**1.0**	**0.99**	**1.0**	**1.0**	**0.99**	**1.0**	**1.0**
Recall	RMDA	1.0	1.0	1.0	1.0	1.0	1.0	1.0	1.0	1.0
	RMDA_mod	1.0	1.0	1.0	1.0	1.0	1.0	1.0	1.0	1.0
	WMDA	1.0	1.0	1.0	1.0	1.0	1.0	1.0	1.0	1.0
	WMDA_mod	1.0	1.0	1.0	1.0	1.0	1.0	1.0	1.0	1.0
	UnRAP	0.99	0.99	1.0	0.99	1.0	1.0	0.99	1.0	1.0
	MIX	0.99	1.0	1.0	1.0	1.0	1.0	1.0	1.0	1.0

Table 2. Precision and Recall on test set with average attack of various attack sizes and filler sizes

	Filler Size	3%			10%			40%		
	Attack Size	2%	5%	15%	2%	5%	15%	2%	5%	15%
Preci-sion	RMDA	0.03	0.04	0.0	0.01	0.0	0.0	0.0	0.0	0.0
	RMDA_mod	0.0	0.0	0.0	0.0	0.0	0.0	0.0	0.0	0.0
	WMDA	0.06	0.18	0.55	0.09	0.22	0.54	0.09	0.22	0.54
	WMDA_mod	0.02	0.05	0.18	0.02	0.09	0.24	0.05	0.11	0.26
	UnRAP	0.96	0.98	0.98	0.84	0.98	0.98	**0.98**	**0.99**	0.99
	MIX	**0.97**	**0.99**	**0.99**	**0.98**	**1.0**	**1.0**	0.97	**0.99**	**1.0**
Recall	RMDA	0.49	0.26	0.0	0.04	0.0	0.0	0.0	0.0	0.0
	RMDA_mod	0.0	0.0	0.0	0.0	0.0	0.0	0.0	0.0	0.0
	WMDA	**1.0**	**1.0**	**1.0**	**1.0**	**1.0**	**1.0**	**1.0**	**1.0**	**1.0**
	WMDA_mod	0.86	0.92	0.92	0.89	0.92	0.96	0.92	0.99	**1.0**
	UnRAP	0.74	0.82	0.88	0.76	0.94	0.95	0.99	**1.0**	**1.0**
	MIX	**1.0**	**1.0**	**1.0**	**1.0**	**1.0**	**1.0**	**1.0**	**1.0**	**1.0**

Table 3. Precision and Recall on test set with bandwagon attack of various attack sizes and filler sizes

	Filler Size	3%			10%			40%		
	Attack Size	2%	5%	15%	2%	5%	15%	2%	5%	15%
Preci-sion	RMDA	0.06	0.18	0.55	0.02	0.06	0.22	0.0	0.0	0.04
	RMDA_mod	0.61	0.82	0.95	**0.99**	**0.99**	**1.0**	**0.97**	0.99	0.99
	WMDA	0.11	0.29	0.66	0.11	0.28	0.65	0.09	0.21	0.52
	WMDA_mod	0.74	**0.99**	1.0	0.81	**0.99**	1.0	0.84	**1.0**	**1.0**
	UnRAP	0.09	0.18	0.31	0.05	0.12	0.23	0.05	0.11	0.16
	MIX	**0.82**	**0.99**	**1.0**	0.8	0.9	0.99	0.77	**1.0**	**1.0**
Recall	RMDA	0.53	0.49	0.55	0.17	0.28	0.43	0.0	0.0	0.09
	RMDA_mod	0.51	0.57	0.76	0.99	0.99	0.98	0.98	0.99	0.94
	WMDA	**1.0**	**1.0**	**1.0**	**1.0**	**1.0**	**1.0**	**1.0**	**1.0**	**1.0**
	WMDA_mod	0.82	0.99	1.0	0.86	**1.0**	**1.0**	0.87	**1.0**	**1.0**
	UnRAP	0.84	0.97	0.99	0.44	0.63	0.76	0.43	0.52	0.63
	MIX	0.93	**1.0**	**1.0**	0.91	**1.0**	**1.0**	0.88	**1.0**	**1.0**

Table 4. Precision and Recall on test set with random and average mixed attack of various attack sizes and filler sizes

	Filler Size	3%			10%			40%		
	Attack Size	2%	5%	15%	2%	5%	15%	2%	5%	15%
Preci-sion	UnRap	0.68	0.98	0.99	0.91	0.97	0.99	0.98	0.99	0.99
	MIX	0.98	0.98	0.99	0.98	0.99	0.99	0.95	0.99	0.99
Recall	UnRap	0.3	0.45	0.48	0.41	0.48	0.49	0.45	0.48	0.49
	MIX	0.62	0.86	0.97	0.77	0.98	0.87	0.82	0.52	0.51

Table 5. Precision and Recall on test set with average-bandwagon mixed attack of various attack sizes and filler sizes

	Filler Size	3%			10%			40%		
	Attack Size	2%	5%	15%	2%	5%	15%	2%	5%	15%
Preci-sion	UnRap	0.71	0.96	0.96	0.74	0.98	0.99	0.73	0.98	0.99
	MIX	0.48	0.98	0.98	0.57	0.97	0.99	0.71	0.94	0.99
Recall	UnRap	0.28	0.39	0.54	0.39	0.45	0.47	0.56	0.66	0.77
	MIX	0.41	0.86	0.97	0.57	0.99	0.99	0.77	0.95	0.98

5   Conclusions

Collaborative recommender systems are vulnerable to profile injection attacks. In this paper, based on the RMDA and WMDA algorithms, we first make their modification, named by RMDA_mod and WMDA_mod. Furthermore, we integrate the RMDA_mod, WMDA_mod and UnRAP algorithms into a hybrid decision approach. The hybrid decision approach overcome the algorithm in three separate attacks, and can effectively and efficiently detect the standard attack and mixed attacks. Through extensive experiments, the results proved our algorithm is effective and efficient, which show that the detecting accuracy based on hybrid decision approach significantly increases almost 35%, 25%, and 8% than the RMDA, WMDA, and UnRAP algorithms, respectively, and improves 10% comparing with the UnRAP algorithm under the condition of mixed attack.

We think the reason for the hybrid decision approach improves the detecting accuracy is that the single detecting algorithms just consider one factor to detect the attack users, but the attack users may have more than one feature. The hybrid decision approach analysis the factors by synthesis, so the result of detecting can be better than the single detecting algorithms.

Although the hybrid decision approach proposed obviously improves the detecting accuracy, it cannot solve all of the attack issues in real B2C environment. Our work is still based on the User-Item bipartite network to find attack profiles. In the future work, we will take into account the established the User-User network and study the network structure in order to find the differences between attack profiles and genuine profiles.

Acknowledgements. This work is partially supported by the National Natural Science Foundation of China (Grant No. 60973069, 90924011, 61073099). SM-C appreciates the financial support by the National Natural Science Foundation of China (Grant No. 61004102).

References

1. Mobasher, B., Burke, R., Bhaumik, R., Williams, C.: Toward trustworthy recommender systems: An analysis of attack models and algorithm robustness. ACM Transactions on Internet Technology 7(4), 23 (2007)
2. Chirita, P., Nejdl, W., Zamr, C.: Preventing shilling attacks in online recommender systems. In: ACM WIDM, pp. 67–74 (2005)
3. Su, X., Zeng, H., Chen, Z.: Finding group shilling in recommendation system. In: WWW, pp. 960–961 (2005)
4. Williams, C., Mobasher, B., Burke, R.: Defending recommender systems: Detection of profile injection attacks. Service Oriented Computing and Applications 1(3), 57–170 (2007)
5. Mobasher, B., Burke, R., Bhaumik, R.: Effective attack models for shilling item-based collaborative filtering systems. In: ACM WebKDD (2005)

6. O'Mahony, M., Hurley, N., Silvestre, G.: Utility-based neighborhood formation for efficient and robust collaborative filtering. In: ACM EC, pp. 260–261 (2004)
7. Hurley, N., Cheng, Z., Zhang, M.: Statistical attack detection. In: ACM RecSys, pp. 149–156 (2009)
8. Mehta, B., Nejdl, W.: Unsupervised strategies for shilling detection and robust collaborative filtering. User Modeling and User-Adapted Interaction 19(1), 65–79 (2009)
9. Bryan, K., O'Mahony, M., Cunningham, P.: Unsupervised retrieval of attack proles in collaborative recommender systems. In: ACM RecSys, pp. 155–162 (2008)
10. Burke, R., Mobasher, B., Williams, C., Bhaumik, R.: Classification features for attack detection in collaborative recommender systems. In: ACM SIGKDD, pp. 542–547 (2006)
11. Jamali, M., Ester, M.: Trustwalker: a random walk model for combining trust-based and item-based recommendation. In: ACM SIGKDD, pp. 397–406 (2009)
12. Jamali, M., Ester, M.: A matrix factorization technique with trust propagation for recommendation in social networks. In: ACM RecSys, pp. 135–142 (2010)
13. Au Yeung, C., Iwata, T.: Strength of social influence in trust networks in product review sites. In: ACM WSDM, pp. 495–504 (2011)
14. Cha, M., Haddadi, H., Benevenuto, F., Gummad, K.P.: Measuring user influence on twitter: The million follower fallacy. In: ICWSM (2010)
15. Pu, P., Chen, L., Hu, R.: A user-centric evaluation framework for recommender systems. In: ACM RecSys, pp. 157–164 (2011)
16. Movielens, http://movielens.umn.edu

Recommendation of Leaders
in Online Social Systems

Hao Liu[1,2], Fei Yu[2], An Zeng[2,*], and Linyuan Lü[1,2]

[1] Institute of Information Economy, Hangzhou Normal University, 310036 Hangzhou,
People's Republic of China
[2] Department of Physics, University of Fribourg, Chemin du Musée 3,
CH-1700 Fribourg, Switzerland
an.zeng@unifr.ch

Abstract. The online social systems are now playing a more and more important role in our daily life. Information coming from such systems is more personalized and preferable than those from search engines and portals. Those systems are normally described by directed networks where the nodes represent users and the information spreads from leaders to followers. Therefore, the selection of suitable leaders determines the quality of the coming information. In this paper, we propose a leader recommendation method based on a local structure consisting of 4 nodes and 3 directed links. The simulation results on real networks show that our method can accurately recommend the potential leaders. Moreover, further investigation on recommendation diversity indicates that our recommendation method is very personalized. Finally, we remark that our method can be easily extended to improve the existing link prediction algorithms in directed networks.

Keywords: Leader recommendation, Online social systems, Subgraph.

1 Introduction

With the development of the online social websites such as Digg.com and Twitter.com where users select others as information sources and import stories or tweets from them, more and more people rely on these systems for information. A wise choice of leader will provide the user with very personalized and interesting information, while a random choice of leader will end up receiving irrelevant and boring information [1].

In the past few years, more and more attentions have been paid to investigate what is the efficient topology structure for such systems [2,3]. In principle, users are more likely to select the most similar users as their leaders since they often share common interests [4,5]. Accordingly, a new type of approach called *social recommendation* emerges where the system will recommend the potential leaders to the user instead of directly recommending news or products [6].

---

* Corresponding author.

L. Chen et al. (Eds.): ISMIS 2012, LNAI 7661, pp. 387–396, 2012.

Recently, a framework has been proposed to mimic the news spreading process in online social systems. An adaptive news recommendation model is designed to recommend users with the most suitable leader candidates by analyzing their historical reading records [7]. More specifically, when a user in this model reads a news, she can either approve or disapprove it. If approved, the news further spreads to followers of this user. Each user has an evolving set of leaders (i.e., information sources, according to the terminology of the original paper) and can become a leader of other users. Simultaneously with spreading of news, the leader-follower network evolves with time to best capture similarity of users. In [7] they applied agent-based modelling to simulate the system's behaviour and test model's performance. Based on this framework, the user reputation [8], implicit ratings [9], link reciprocity [10], scale-free leadership [11] and local rewiring algorithm [12] are detailedly investigated.

However, these adaptive models require the information of users' reading behaviors, namely approval or disapproval of the news. The deficiency of such information will definitely decrease the recommendation accuracy. Furthermore, this model cannot work in the systems without any reading information. In this case, how to well utilize the network topology to recommend leaders is a crucial problem as well as a challenge for very large-scale systems [13]. Motivated to design an algorithm of both effectiveness and efficiency, in this paper, based on a local structure consisting of 4 nodes and 3 directed links, we propose a leader recommendation method with preferential selection, named the Preferential Motif-based Recommendation (PMR) method. We examine our method on three typical real networks, Political blog, Slashdot and Wikipedia. The results suggest that our method can accurately recommend the potential leaders for users. In addition, through the investigation on recommendation diversity of the PMR algorithm, it shows that the recommendation from PRM is very personalized.

From the practical point of view, the PMR method evolving local information will significantly speed up the recommendation process compared to the traditional reading-record-based method. Thus it is applicable to large-scale system due to its low computational complexity. Additionally, since how to predict missing links in directed networks is still a not well-solved problem [14], we finally remark that our method can be easily extended to further improve the existing link prediction methods [15].

2   Method and Data

Previous studies showed that the "Bi-fan" structure, a subgraph consisting of 4 nodes and 3 links, plays very important role in directed networks [16]. The prediction based on this structure is much more accurate than some other subgraphs [15]. Inspired by these results, we here propose a "Bi-fan"-based method to recommend potential leaders to target user in online social system with the consideration of preferential attachment mechanism. We therefore call it Preferential Motif-based Recommendation method. In the following, our method will be referred as PMR.

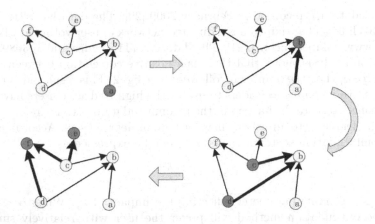

Fig. 1. A simple example to illustrate how to find the "Bi-fan" structure for a target user a

An online social system can be represented by a directed network where nodes are users and links represent follower-leader relationships. If user i selects user j as leader, there is a link from i to j, and the information will spread from j to i. In this directed network, the meaning of the "Bi-fan" structure can be interpreted in the following way. For a target user i, we first find the leaders that i has already chosen. Then the users who have at least one common leader with user i are found and defined as "relevant users of i". Finally, the leaders of the relevant users are candidates that will be recommended to user i. According to this process, there will be a path from the target user to the final candidates, namely "$i \rightarrow i$'s leader $\leftarrow$ relevant user of $i \rightarrow$ candidate j". Clearly, if a candidate j is involved in many of these paths starting from the target user i, j is more suitable to be i's leader since most of i's relevant users have chosen j as a leader. If user i accepts j as leader, a "Bi-fan" structure will be generated. Therefore, for each candidate j, what we need to do is to calculate c_{ij} as the number of generated "Bi-fan" subgraphs by adding the directed link from i to j, namely the number of above mentioned paths from i to j. Larger c_{ij} indicates higher probability to be recommended. Figure 1 gives an example of how to calculate the number of "Bi-fan" structure. Let node a be the target user. The first step is to find the leader of user a. In the second step we find nodes c and d as the relevant users of a. Then the leaders of the relevant users, namely nodes e and f are potential candidates. For node e there is only one path "a$\rightarrow$ b $\leftarrow$ c $\rightarrow$ e", while for node f there are two paths "a$\rightarrow$ b $\leftarrow$ c $\rightarrow$ f" and "a$\rightarrow$ b $\leftarrow$ d $\rightarrow$ f", indicating that node f is more likely to be recommended.

In this paper, we consider three real datasets: political blog, slashdot and wikipedia. The political blog is a directed network of hyperlinks between weblogs on US politics, recorded in 2005 [19]. The data has 1224 nodes and 19090 links. The slashdot is a directed network of users' friendship in a news sharing website (www.slashdot.org), recorded in February 2009 [20]. The data has 82168 nodes and 948464 links. The wikipedia is a directed network of user voting in wikipedia website (www.wikipedia.com) [21]. The data has 7115 nodes and 103689 links.

Empirical analysis shows that there is a positive correlation between c_{ij} and j's in-degree k_j (i.e., the number of followers), see fig. 2. This result indicates that if we use c_{ij} to recommend leaders to user i, the high in-degree users have much higher chance to be ranked ahead in the recommendation list. Therefore, we use the user in-degree k_j to adjust the bias for high in-degree users. Accordingly, the final recommendation score for each user j can be expressed as

$$s_{ij} = c_{ij}k_j^{\alpha}, \tag{1}$$

where α is a tunable parameter reflecting the impact of k_j. When $\alpha < 0$, the leader recommendation method will prefer the user with relatively small in-degree, and vice versa.

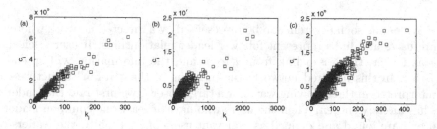

Fig. 2. The correlation between c_{ij} and user in-degree k_j in (a) Political blogs, (b) Slashdot and (c) Wiki dataset

To test the algorithm performance we randomly remove a certain rate (p) of the links as the probe set denoted as E^P [17]. We then apply the algorithm to the remainder (the training set denoted as E^T) to produce a recommendation list for each user. In order to test the robustness of our method, we calculate the different cases where $p = 10\%$, $p = 20\%$ and $p = 30\%$.

3   Results

In this section, we are going to discuss the performance of the PMR method from the aspect of recommendation accuracy and diversity. The exact value of each metric discussed below can be seen in table I.

3.1   Recommendation Accuracy

In order to measure the accuracy of the recommendation result, we make use of the *ranking score* index [18]. For a target user i, the recommender system will

Table 1. The performance of the PMR method in *Political blog*, *Slashdot* and *Wiki* data. p is the proportion of the probe set. For the accuracy metrics, R is the overall ranking score of the system and $R_{k\leq10}$ is the average ranking score of the users who have at most 10 followers (i.e., in-degree$\leq$ 10). For the diversity metrics, H is the hamming distance and N is the novelty. The recommendation list length is set as $L = 20$. The results under the optimal parameter is marked in black and the performance of the case $\alpha = 0$ is given for comparison.

Network	p	$R^{\alpha=0}$,	R^*,	$R_{k\leq10}^{\alpha=0}$,	$R_{k\leq10}^*$,	$H^{\alpha=0}$,	H^*,	$N^{\alpha=0}$,	N^*
	10%	0.0591	**0.0579**	0.2031	**0.1829**	10.05	**10.68**	131.24	**122.13**
Political Blogs	20%	0.0638	**0.0628**	0.2096	**0.1911**	10.07	**10.65**	117.59	**109.38**
	30%	0.0673	**0.0666**	0.2053	**0.1878**	10.08	**10.47**	100.79	**92.27**
	10%	0.0867	**0.0853**	0.2385	**0.2300**	16.97	**18.36**	499.66	**262.26**
Slashdot	20%	0.1101	**0.1100**	0.2600	**0.2532**	18.07	**18.75**	287.44	**155.01**
	30%	0.1261	**0.1261**	0.2791	**0.2738**	18.03	**18.74**	251.42	**135.54**
	10%	0.0347	**0.0298**	0.1760	**0.1281**	12.42	**13.92**	162.73	**82.52**
Wiki	20%	0.0360	**0.0321**	0.1656	**0.1190**	11.85	**13.27**	146.42	**72.44**
	30%	0.0382	**0.0359**	0.1612	**0.1175**	11.55	**12.67**	120.41	**60.10**

return a ranking list of all her unconnected users according to the recommendation scores. For each hidden leader-follower relation (i.e., the link in probe set), we measure the rank of the leader in the recommendation list of this user i. For example, if there are 1000 unconnected user for the target user i, and user j is at 10th place, we say the position of this user j is 10/1000, denoted by $R_{ij} = 0.01$. A successful recommendation result is expected to highly recommend the real leader in the probe set, and thus leading to small ranking score. Averaging over all the hidden leader-follower relations, we obtain the mean value of ranking score to evaluate the recommendation accuracy, namely

$$\langle R \rangle = \frac{1}{|E^P|} \sum_{ij\in E^P} R_{ij}, \qquad (2)$$

where ij denotes the probe link connecting user i and user j. Clearly, the smaller the ranking score, the higher the algorithm's accuracy, and vice versa.

The dependence of recommendation ranking score R on the parameter α is reported in fig. 3. As we can see, R in all three data sets are close to 0, which indicates that the recommendation accuracy is quite high. Moreover, we find that the lowest ranking score denoted by R^* can be obtained at the optimal parameter α^* which is negative in all three datasets. With a negative α, the recommendation scores of high in-degree candidates are depressed. In this way, some relevant yet not so popular users can have the opportunity to be recommended and the recommendation accuracy is enhanced accordingly. Similar idea has been applied to design a preferential diffusion method for recommendation in user-object bipartite networks. Readers are encouraged to see Ref. [22] for more details.

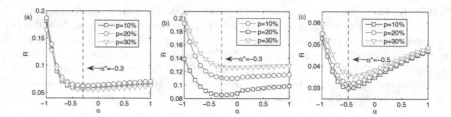

Fig. 3. The dependence of ranking score R on the parameter α in three datasets: (a) Political blog, (b) Slashdot and (c) Wiki

Moreover, we compare the PMR method with two very straight-forward but non-personalized leader recommendation methods: (i) Popularity-based method will recommend the users according to their in-degree (i.e., number of followers). Users with more followers are ranked ahead. (ii) PageRank-based method [25] will recommend the users according to the PageRank score. Users with higher PageRank score are ranked ahead. The comparison of the ranking score of these three methods are shown in table II. Obviously, the PMR performs best.

Table 2. The comparison of the recommendation accuracy (ranking score) of the popularity-based method, pagerank-based method and PMR method. The results for PMR method is under the optimal parameter α^*. The best performance is marked in black.

dataset	Political Blogs			Slashdot			Wiki		
p	10%	20%	30%	10%	20%	30%	10%	20%	30%
popularity-based	0.0863	0.0972	0.0976	0.2439	0.2324	0.2336	0.0746	0.0739	0.0730
pagerank-based	0.0871	0.0943	0.0945	0.2870	0.3089	0.3195	0.0776	0.0765	0.0752
PMR	**0.0579**	**0.0628**	**0.0666**	**0.0853**	**0.1100**	**0.1261**	**0.0298**	**0.0321**	**0.0359**

An universal challenge for recommendation systems is the cold-start problem. That is to say, the new users or small in-degree users (referred as cold users) are quite difficult to be correctly recommended to the right users since they have only limit historical information. To monitor how accurate the small in-degree users are recommended by our method, we further make use of the local ranking score $R_{k\leq 10}$ which is the average ranking score of the recommended users who have at most 10 followers (i.e., in-degree$\leq$ 10). The dependence of $R_{k\leq 10}$ on parameter α are reported in fig. 4 (a)-(c). As we can see, the ranking score increases with the increasing of α. With a negative α, the recommendation accuracy on cold users can be largely improved.

To show how the ranking score varies on users with different value of in-degrees, we additionally investigate the degree-dependent ranking score R_k. The R_k is defined as the average ranking score over users with the same value of in-degree k. In fig. 4 (d)-(f), the relation between R_k and the user's degree

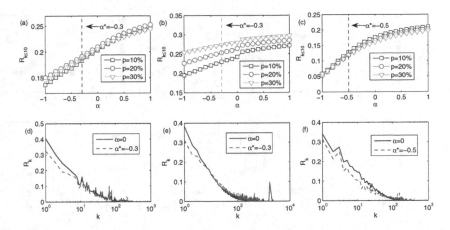

Fig. 4. The dependence of local ranking score $R_{k \leq 10}$ on the parameter α in three datasets: (a) Political blog, (b) Slashdot and (c) Wiki. The degree-dependent ranking score R_k is shown in (d) Political blog, (e) Slashdot and (f) Wiki.

k is displayed at the optimal parameter α^*. Besides, we also plot the results under the case where $\alpha = 0$ (pure recommendation based on "Bi-fan" structure) for comparison. Obviously, the ranking score of small in-degree users can be significantly decreased by introducing a negative parameter α. Although the improvement becomes more and more unnoticeable with the increasing of k, the recommendation accuracy of popular users can also be effectively preserved by the PMR method.

3.2   Recommendation Diversity

Another important aspect for recommendation is the diversity of the recommendation results. Here, we consider two kinds of diversity measurements. One is called *interdiversity*, which considers the uniqueness of different users' recommendation lists. Given two users u_i and u_j, the difference between their recommendation lists can be measured by the Hamming distance,

$$H_{ij}(L) = 1 - \frac{T_{ij}(L)}{L}, \qquad (3)$$

where $T_{ij}(L)$ is the number of common leaders in the top-L places of both lists. Clearly, if u_i and u_j have the same list, $H_{ij}(L) = 0$, while if their lists are completely different, $H_{ij}(L) = 1$. Averaging $H_{ij}(L)$ over all pairs of users, we obtain the mean distance $H(L)$, for which bigger or smaller values mean, respectively, greater or lesser personalization of users' recommendation lists.

Highly accurate recommendations itself might not be satisfied by users. A diverse recommender algorithm should be able to find the new or not so popular leaders who cannot be easily known by other ways yet match users' taste

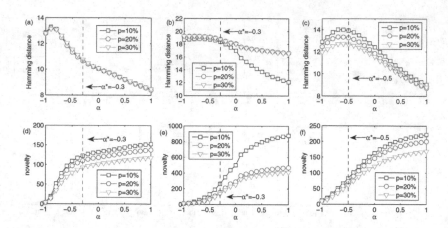

Fig. 5. The performance of our method in recommendation diversity measured by Hamming distance and Novelty in (a)(d) Political blog, (b)(e) Slashdot and (c)(f) Wiki datasets

and forward many relevant news or stories to the followers. The metric *novelty* quantifies the capacity of an algorithm to generate novel and unexpected results, that is to say, to recommend less popular leaders unlikely to be already known. The definition reads

$$N_i(L) = \frac{1}{L} \sum_{o_\alpha \in O_\alpha^i} k_{o_\alpha} \qquad (4)$$

where O_α^i is the recommendation list for user u_i .Clearly, lower popularity indicates higher novelty and surprisal. Averaging $N_i(L)$ over all users, we obtain the mean popularity $N(L)$ for the system.

In this paper, we set the recommendation list $L = 20$ and report the results of hamming distance and novelty in fig. 5. Clearly, with a negative α, our method significantly outperforms the case where only "Bi-fan" structure is considered (i.e. $\alpha = 0$). More specifically, the hamming distance generally decreases with the increasing of α while the novelty increases with α. For the sake of recommendation accuracy, the optimal α^* is set as a negative value where the hamming distance is larger than the case where $\alpha = 0$ and the novelty is also smaller than the case where $\alpha = 0$. Taken together, the recommendation diversity can be largely enhanced by introducing the in-degree preferential parameter α in the PMR method.

4   Conclusion

Leader recommendation is a very timely and important problem due to the rapid development of the online social networks for information sharing. Most of the previous works focused on using the historical reading records to calculate the similarity between users and make the recommendation accordingly. Apparently, the recommendation accuracy is heavily depend on the historical information. The absence of such information will definitely hamper the application of this kind of methods. Therefore how to well utilize the network topology to recommend leaders is a challenge problem especially for very large-scale systems. In this paper, we propose a preferential motif-based recommendation method, which has high performance in both recommendation accuracy and diversity.

From the practical point of view, our method can significantly improve the algorithm efficiency comparing with the traditional leader recommendation methods for its lower memory space (only the network topology information is required) and computational complexity. Finally, we remark that our method can be easily extended to solve the problem of link prediction and noisy link identification in directed networks [23,24].

Acknowledgement. We would like to thank Prof. Yi-Cheng Zhang for helpful discussions. This work is supported by the Swiss National Science Foundation (No. 200020-132253). HL and LL acknowledge the Pandeng Project of Hangzhou Normal University Grant PD12001003001-4.

References

1. Billsus, D., Pazzani, M.J.: Adaptive News Access. In: Brusilovsky, P., Kobsa, A., Nejdl, W. (eds.) The Adaptive Web. LNCS, vol. 4321, pp. 550–570. Springer, Heidelberg (2007)
2. Zeng, A., Yeung, C.H., Shang, M.-S., Zhang, Y.-C.: The reinforcing influence of recommendations on global diversification. Europhys. Lett. 97, 18005 (2012)
3. Zhang, C.-J., Zeng, A.: Behavior patterns of online users and the effect on information filtering. Physica A 391, 1822–1830 (2012)
4. Gualdi, S., Yeung, C.H., Zhang, Y.-C.: Tracing the evolution of physics on the backbone of citation networks. Phys. Rev. E 84, 046104 (2011)
5. Sinha, R., Swearingen, K.: Comparing recommendations made by online systems and friends. In: Proc. DELOS-NSF Workshop on Personalization Recommender Systems in Digital Libraries (2001)
6. Golbeck, J.: Weaving a Web of Trust. Science 321, 1640 (2008)
7. Medo, M., Zhang, Y.-C., Zhou, T.: Adaptive model for recommendation of news. Europhys. Lett. 88, 38005 (2009)
8. Cimini, G., Medo, M., Zhou, T., Wei, D., Zhang, Y.-C.: Heterogeneity, quality, and reputation in an adaptive recommendation model. Eur. Phys. J. B 80, 201 (2011)
9. Wei, D., Zhou, T., Cimini, G., Wu, P., Liu, W., Zhang, Y.-C.: Effective mechanism for social recommendation of news. Physica A 390, 2117 (2011)
10. Cimini, G., Chen, D.-B., Medo, M., Lü, L., Zhang, Y.-C., Zhou, T.: Enhancing topology adaptation in information-sharing social networks. Phys. Rev. E 85, 046108 (2012)
11. Zhou, T., Medo, M., Cimini, G., Zhang, Z.-K., Zhang, Y.-C.: Emergence of scale-free leadership strcuture in social recommender systems. PLoS One 6(7), e20648 (2011)
12. Chen, D.-B., Gao, H.: An improved adaptive model for information recommending and spreading. Chin. Phys. Lett. 29, 048901 (2012)

13. Lü, L., Medo, M., Yeung, C.H., Zhang, Y.-C., Zhang, Z.-K., Zhou, T.: Recommendation systems. Phys. Rep., doi:10.1016/j.physrep.2012.02.006
14. Guimera, R., Sales-Pardo, M.: Missing and spurious interactions and the reconstruction of complex networks. Proc. Natl. Acad. Sci. USA 106, 22073 (2009)
15. Zhang, Q.-M., Lü, L., Wang, W.-Q., Zhu, Y.-X., Zhou, T.: Potential Theory for Directed Networks. arXiv:1202.2709v1 (2012)
16. Milo, R., Shen-Orr, S., Itzkovitz, S., Kashtan, N., Chklovskii, D., Alon, U.: Network Motifs: Simple Building Blocks of Complex Networks. Science 298, 824 (2002)
17. Zhou, T., Kuscsik, Z., Liu, J.-G., Medo, M., Wakeling, J.R., Zhang, Y.-C.: Solving the apparent diversity-accuracy dilemma of recommender systems. Proc. Natl. Acad. Sci. 107, 4511–4515 (2010)
18. Herlocker, J.L., Konstan, J.A., Terveen, K., Riedl, J.T.: Evaluating collaborative filtering recommender systems. ACM Transactions on Information Systems 22, 5–53 (2004)
19. Adamic, L.A., Glance, N.: The political blogosphere and the 2004 U.S. election: divided they blog. In: Proceedings of the WWW-2005 Workshop on the Weblogging Ecosystem (2005)
20. Leskovec, J., Lang, K., Dasgupta, A., Mahoney, M.: Community Structure in Large Networks: Natural Cluster Sizes and the Absence of Large Well-Defined Clusters. Internet Mathematics 6(1), 29 (2009)
21. Leskovec, J., Huttenlocher, D., Kleinberg, J.: Predicting positive and negative links in online social networks. In: CHI (2010)
22. Lü, L., Liu, W.: Information filtering via preferential diffusion. Phys. Rev. E 83, 066119 (2011)
23. Lü, L., Zhou, T.: Link prediction in complex networks: a survey. Physica A 390, 1150–1170 (2011)
24. Zeng, A., Cimini, G.: Removing spurious interactions in complex networks. Phys. Rev. E 85, 036101 (2012)
25. Brin, S., Page, L.: The anatomy of a large-scale hypertextual Web search engine. Comput. Networks ISDN Systems 30, 107–117 (1998)

Application of Recommendation System: An Empirical Study of the Mobile Reading Platform

Chun-Xiao Jia, Chuang Liu, Run-Ran Liu, and Peng Wang*

Institute for Information Economy and Alibaba Business College,
HangZhou Normal University, Zhejiang 310036, China
wangpenge@gmail.com

Abstract. Mobile reading on 'smart' terminals (like smartphones and tablet computers)is an increasing popular subject, and the recommendations of e-books for users also begin to attract more attentions. In this paper, we mainly demonstrate the performance of the personalized recommendation on the mobile reading platform, based on the analysis of the reading records on mobile phones. The analysis results of the feedback of users for the recommendations show that the personalized recommendation based on the mass diffusion algorithm is much better than the algorithm of the mobile company used before. In particular, both the number of the motivated page views and the motivated users have a dramatically increase. All these results indicate that the mass diffusion algorithm has an outstanding performance on the mobile reading recommendation, which can help users quickly find the books they are interested in. Meanwhile, it help the company enlarge the customer volume and improve the customer experience.

Keywords: personalized recommendation, network-based algorithm, mass diffusion, mobile reading platform.

1 Introduction

Over the last decades, recommendation system as a hot area not only has received the attention of the researchers, but also as an important technology has been widely used by many e-commerce companies to deal with information overload and provide personalized services. These applications include recommending books by douban [1], movies by Netflix [2], and other products at Amazon.com [3] or taobao.com [4]. Moreover, with the rapid development of mobile technology, more and more people use their mobile phones for browsing, shopping and reading. These provide opportunities for retailers to drive sales through mobile websites and applications, and also provide a new platform for recommender applications.

Motivated by the significance in economy and society [5], the design of an efficient recommendation algorithm becomes a joint focus from engineering science to marketing practice. Various kinds of algorithms have been proposed,

---

* Corresponding author.

L. Chen et al. (Eds.): ISMIS 2012, LNAI 7661, pp. 397–404, 2012.

including collaborative filtering [6–8], content-based analysis [9], spectral analysis [10], iteratively self-consistent refinement [11], principle component analysis [12], and so on. Recently, some physical dynamics, including heat conduction process [13] and mass diffusion [14–19], have been introduced into personalized recommendation. These physical approaches exhibit both high efficiency and low computational complexity.

In this paper, we mainly adopt the mass diffusion algorithm [14] to achieve the personalized recommendations for users. In Section 2, we do a series analysis of the reading data of the mobile phone users, and achieve some valuable results. Then, using the mass diffusion algorithm we calculate the recommended list, and push the recommendations to users by short message. In Section 3, we analyse the effects of the feedback of the recommendation. In Section 4, the summary and discussion are given.

2   Data Analysis and Algorithm Description

2.1   Data Analysis

In order to optimize the performances of the recommendation algorithm, we made a series of analysis on the data of the e-book reading records of the mobile users in advance. Figure 1 gives the time series of the users' page view (pv for short). When a user views a page, it is recorded one pv. The records include 24003 users related with 41519 books, ranging from March 1, 2011 to July 19, 2011. From Fig. 1 we can find that there are obvious local peaks, especially for the charged-pv (it is charged for viewing pages). We call this weekend effect, which means the users tend to read e-books at weekend. Based on this finding, we infer that weekend maybe a good time for recommendation. It's worth noting that the weekend effect is not obvious for free-pv(it is free for viewing pages), because the reading willingness of the users for free pages not strong and their reading behaviors are easily affected by various factors. For example, there are two special periods, i.e., from April 28 to May 10 and July 1 to July 19, in which the weekend effect for free-pv is nearly disappear, that is considered mainly due to the holidays (Labor Day and summer holiday).

We also analyzed the age distribution of users. The results show that users aged 20-30 account for almost half, and aged 30-40 account for a quarter. In addition, the investigation of the matching of mobile phones platform shows that the conversion of the reading users(all the reading users) to the active users(their total page views from March 1,2011 to July 19,2011 are more than 20) is the most significant for the mid-range smartphones. This is because the high-end smartphones have too many entertainment functions that scattering off the energy of users, and the reading functions of the low-end phones are not good and unsuitable for reading. All these analysis can be seen as the market analysis, coupled together with a good recommendation algorithm, they can guarantee to achieve good marketing feedbacks.

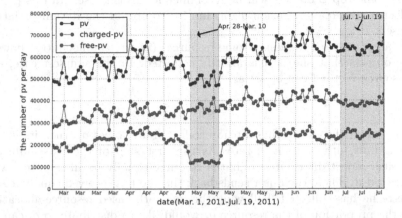

Fig. 1. The varies of the *pv* from March 1,2011 to July 19,2011. The folding line of Blue, Green and Red represent *pv*, charged-*pv* and free-*pv* respectively. These dash lines correspond to the *pv* on Sunday. The two shadow areas represent the days from April 28 to May 10 and July 1 to July 19, respectively.

2.2 Data for Recommendation

The data used in the mobile reading recommendation is collected from March 1,2011 to September 1,2011, and consists of $m = 258215$ users, $n = 79723$ e-books (objects), and the records of the users' *pv* . If a user reads only a few pages of a book, which can be seen that she/he has read this book and dose not like it. There are some screening processes for users, therefore, we apply a coarse-graining method [14, 16]: E-books are considered as a suitable taste for a user if and only if the given *pv* is at least 3 (i.e., the user read this e-book at least 3 pages). After the coarse gaining, the data contains 103800 users and 36727 e-books. Based on this data set, every user is recommended an e-book that she/he has never read. These personalized recommendations are sent to users by short messages.

2.3 Algorithm Description

Generally, a recommendation system consists of m users and n objects (e-books in the application context of this paper). The relationship between users and objects can be represented by a bipartite network [14]. The two sets of nodes in the bipartite networks are presented the users and objects respectively. If a user u_j has collected an object o_i, there is an edge between o_i and u_j, and the corresponding element a_{ij} in the adjacent matrix A is set as 1, otherwise it is zero.

The aim of recommendation algorithm is to guess a user's personalized tastes of those objects that she/he has not yet collected and recommend the user's favorites to her/him. In other words, for a target user, a recommendation algorithm ranks all the objects she/he has never collected before, according to her/his preference. The top L objects will be recommended to this user. In this paper, we implement the mass diffusion algorithm[14] to obtain the recommendation lists of users.

For a target user u_i, we assign some resources (i.e., recommendation power) on those objects that she/he has already collected. In the simplest case, the initial resource vector $\mathbf{f}$ can be set as

$$f_j = a_{ji}, \tag{1}$$

That is to say, if the user u_i has already collected an object o_j, then its initial resource is unit, otherwise it is zero. As the first step of the mass-diffusion process, each object distributes its initial resource equally to all the users who have collected it. Then, each user sends back that he has received equally to all the objects he has collected. Following this network-based resource-allocation process, the proportion of the resource o_j would like to distribute to o_i can be expressed as:

$$w_{ij} = \frac{1}{k(o_j)} \sum_{l=1}^{m} \frac{a_{il} a_{jl}}{k(u_l)}, \tag{2}$$

where $k(o_j)$ denotes the degree of the object o_j and $k(u_l)$ denotes the degree of the user u_l.

After the mass-diffusion process, all u_i's uncollected objects o_j ($1 \leq j \leq$ n, a_{ji} = 0) are sorted in the descending order of their final resources. In this way, we obtain the recommendation lists of users in this recommendation system [14].

3   Online Feedback and Results

Based on the reading records and the mass diffusion algorithm, we recommend each user an e-book that she/he has never collected and send this personalized recommendations to users by the short messages on Sep. 3 and 4, 2011, and most users was recommended on Sep. 4. In order to test the performance of the personalized recommendation, we divide the users into two groups. The e-books are recommended for the two groups respectively using the mass diffusion algorithm and the company's algorithm, which randomly choose an e-book among the top ten most popular e-books to the users. We analysis the feedback results of the motivated pv and the motivated users. The user who has produced at least one pv for the recommended e-books is considered as a motivated user. All the page views motivated by the recommended e-books are regarded the motivated pv.

From Fig. 2, we can see that the numbers of the motivated pv by personalized recommendation are much larger than that of the mobile company's method at Sep. 3 to Sep. 13. For the company's algorithm, there are only a few motivated pv and decaying very rapidly. The total number of the motivated pv by personalized recommendation is 35.9(17341/482) times as many as that of the company's algorithm. These results strongly suggested that the personalized recommendation of mass diffusion algorithm has a good performance on stimulating user's reading interests, and it is much better than the company's algorithm. It is worth mentioning that there are no longer recommendation for the users after Sep. 4, but there are still high motivated pv for the days after Sep. 4, and the decay is slow for the personalized recommendation. Here we use the next-day-conservation rate to describe the persistence of motivated pv, which is defined as the ratio of the pv of Sep. 5 to Sep. 3 and 4. Here, we can see the next-day-conservation rate of the mass diffusion algorithm is 58.5%(2346/4013) and that of the company's algorithm is 7.1%(28/393), which means that the next-day-conservation rate of the mass diffusion algorithm is 8 times as high as that of the algorithm used by the mobile company.

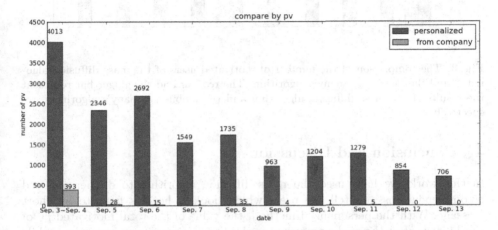

Fig. 2. The comparison of the number of motivated pv of the mass diffusion algorithm and the mobile company's algorithm. The red bar and the ginger bar represent the results of the mass diffusion algorithm and the mobile company's algorithm, respectively.

In addition, we also compare the number of motivated users in Fig. 3, where a user is defined as a motived user if her/he read the recommended book. We can see that the number of motivated users is up to 421 for the mass diffusion algorithm, but only 45 for mobile company's algorithm, which means that the number of motivated users of personalized recommendation is 9 times as large as that of the company. Meanwhile, a large proportion of motivated users can be preserved in next days, and the decay is slow. These results indicate that the users interest in some recommended e-books is stimulated and sustained over time. However, for the company's method, the number of motivated users is small, and the retention rate is very low. The next day conservation rate of motivated users of the personalized recommendation is 33%(139/421) and that of the company is 22.2%(10/45), which means that the next-day-conservation rate of the personalized recommendation is nearly 1.5 times as large as that of the company's algorithm.

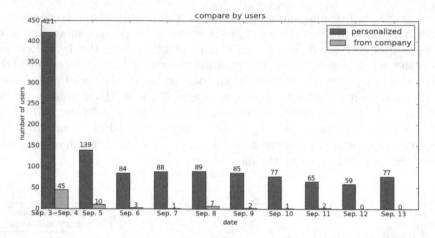

Fig. 3. The comparison of the number of motivated users of the mass diffusion algorithm and the mobile company's algorithm. The red bar and the ginger bar represent the results of the mass diffusion algorithm and the mobile company's algorithm, respectively.

4   Conclusion and Discussion

In this work, we have used the mass diffusion algorithm to do personalized recommendations and tell the users which book is her/his favorite by short message. With the personalized method, the value of the total motivated pv of the 11 days after the message pushed is close to 35 times larger than the mobile company's recommendation. The next-day conservation rate is about 7 times. In addition, using the personalized algorithm, the number of the motivated users in the first day and the next-day-conservation rate are 9 times and 1.5 times respectively as large as that of the company's recommendation method. It is also found that there is a slow decay of the number of motivated users, which indicates that the user's interesting aroused by the personalized recommendation can be preserved for a long time. The feedback results show that the personalized recommendation is very efficient. It can not only attract more and more new users but also improve the experience of the users.

The personalized recommendation is widely concerned in both academia and industry. Besides the mobile reading in this work, the personalized recommendation techniques have been used in many fields especially the e-commerce. How to design the recommendation algorithm is the most important problem in the recommendation system. In this work, we use mass diffusion method to obtain the recommendation list. Comparing with the traditional methods, such as the global rank and the collaborative filtering, the mass diffusion method is more efficiency and simplicity in implementation. Recently, many researches have focused on the hybrid of the mass diffusion [14] and heat conduction [13] techniques, which fulfills simultaneously high accuracy and large diversity [20]. Applying the numerous improvement methods [16–19, 21, 22] and the hybrid method in the mobile reading recommendation has been put forward in our group. The problem is that there is a parameter λ in the most of these algorithms, which determine the performance of the recommendation. The cost for searching the optimal λ and the profit for the accuracy improvement with the optimal parameter should be considered in the algorithm design. Moreover, we can also consider the social relation network of users for mobile reading recommendation and build the user-link-object tripartite graphs [23], which could improve the effect of the recommendation further. Of cause, there are many other good algorithms, such as probabilistic Latent Semantic Analysis (pLSA) [24] and Latent Dirichlet Allocation [25]. The applications of these algorithms need to be further developed.

In summary, it is very encouraging that the feedback results of the mass diffusion algorithm on the mobile reading platform. Though the number of the pv and the users have a great increase by using the mass diffusion algorithm, there are many problems in the research, such as the algorithm design, cold start, data sparsity and so on. The personalized recommendation is an important issue both in theory and applications. In this paper, we showed the significant effect of the mass diffusion algorithm on mobile reading platform by the company's assessment criteria, which may shed light on the application of the recommendation system research.

Acknowledgement. We would like to thank the Hangzhou branch company of China mobile and Shanghai inflow information technology limited company for providing us the data. Chun-Xiao Jia,Run-Ran Liu and Peng Wang are supported by the research start-up fund of Hangzhou Normal University under Grant Nos. 2011QDL29, 2011QDL31, and 2011QDL30. Run-Ran Liu is also supported by the Zhejiang Provincial Natural Science Foundation of China under Grant No. LY12A05003.

References

1. http://www.douban.com
2. http://www.netflix.com
3. http://www.amazon.com
4. http://www.taobao.com
5. Schafer, J.B., Konstan, J.A., Riedl, J.T.: E-commerce recommendation applications. Data Min. Knowl. Disc. 5, 115–152 (2001)
6. Herlocker, J.L., Konstan, J.A., Terveen, L.G., Riedl, J.T.: Evaluating collaborative filtering recommender systems. ACM Trans. Inform. Syst. 22, 5–53 (2004)
7. Liu, R.-R., Jia, C.-X., Zhou, T., Sun, D., Wang, B.-H.: Personal Recommendation via Modified Collaborative Filtering. Physica A 388, 462–468 (2009)
8. Sun, D., Zhou, T., Liu, J.-G., Liu, R.-R., Jia, C.-X., Wang, B.-H.: Information filtering based on transferring similarity. Phys. Rev. E 80, 017101 (2009)
9. Pazzani, M.J., Billsus, D.: Content-Based Recommendation Systems. In: Brusilovsky, P., Kobsa, A., Nejdl, W. (eds.) The Adaptive Web. LNCS, vol. 4321, pp. 325–341. Springer, Heidelberg (2007)

10. Maslov, S., Zhang, Y.-C.: Extracting Hidden Information from Knowledge Networks. Phys. Rev. Lett. 87, 248701 (2001)
11. Ren, J., Zhou, T., Zhang, Y.-C.: Information Filtering via Self-Consistent Refinement. Europhys. Lett. 82, 58007 (2008)
12. Ken, G., Theresa, R., Dhruv, G., Chris, P.: Eigentaste: A Constant Time Collaborative Filtering Algorithm 4, 133–151 (2001)
13. Zhang, Y.-C., Blattner, M.: Heat Conduction Process on Community Networks as a Recommendation Model. Phys. Rev. L 99, 154301 (2007)
14. Zhou, T., Ren, J., Medo, M., Zhang, Y.-C.: Bipartite network projection and personal recommendation. Phys. Rev. E 76, 046115 (2007)
15. Zhang, Y.-C., Medo, M., Ren, J., Zhou, T., Li, T., Yang, F.: Recommendation model based on opinion diffusion. Europhys. Lett. 80, 68003 (2007)
16. Zhou, T., Jiang, L.-L., Su, R.-Q., Zhang, Y.-C.: Effect of initial configuration on network-based recommendation. Europhys. Lett. 81, 58004 (2008)
17. Jia, C.-X., Liu, R.-R., Sun, D., Wang, B.-H.: A new weighting method in network-based recommendation. Physica A 387, 5887–5891 (2008)
18. Zhou, T., Su, R.-Q., Liu, R.-R., Jiang, L.-L., Wang, B.-H., Zhang, Y.-C.: Accurate and diverse recommendations via eliminating redundant correlations. New J. Phys. 11, 123008 (2009)
19. Liu, R.-R., Liu, J.-G., Jia, C.-X., Wang, B.-H.: Personal recommendation via unequal resource allocation on bipartite networks. Physica A 389 (2010) 0378-4371
20. Zhou, T., Kuscsik, Z., Liu, J.-G., Medo, M., Wakeling, J.R., Zhang, Y.-C.: Solving the apparent diversity-accuracy dilemma of recommender systems. Proc. Natl. Acad. Sci. USA 107(10), 4511–4515 (2010)
21. Liu, J.-G., Zhou, T., Guo, Q.: Information filtering via biased heat conduction. Phys. Rev. E 84, 037101 (2011)
22. Lü, L.-Y., Liu, W.-P.: Information filtering via preferential diffusion. Phys. Rev. E 83, 066119 (2011)
23. Zhang, Z.-K., Liu, C., Zhang, Y.-C., Zhou, T.: Solving the cold-start problem in recommender systems with social tags. Europhys. Lett. 92, 28002 (2010)
24. Hofmann, T.: Latent semantic models for collaborative filtering. ACM Transactions on Information Systems 22, 89–115 (2004)
25. Griffiths, T.L., Steyvers, M.: Finding scientific topics. Proceedings of the National Academy of Sciences of the United States of America 101, 5228–5235 (2004)

Measuring Quality, Reputation and Trust
in Online Communities

Hao Liao, Giulio Cimini, and Matúš Medo

Physics Department, University of Fribourg, CH-1700, Switzerland
{hao.liao,giulio.cimini,matus.medo}@unifr.ch

Abstract. In the Internet era the information overload and the challenge to detect quality content has raised the issue of how to rank both resources and users in online communities. In this paper we develop a general ranking method that can simultaneously evaluate users' reputation and objects' quality in an iterative procedure, and that exploits the trust relationships and social acquaintances of users as an additional source of information. We test our method on two real online communities, the EconoPhysics forum and the Last.fm music catalogue, and determine how different variants of the algorithm influence the resultant ranking. We show the benefits of considering trust relationships, and define the form of the algorithm better apt to common situations.

Keywords: complex networks, ranking algorithm, reputation, trust.

1 Introduction

Nowadays, ranking techniques and reputation systems are widely employed in e-commerce services, where buyers and sellers may give each other a score after a completed transaction—encouraging good behavior in the long term [1]. Other reputation systems are content-based, in the sense that users are evaluated by their contribution [2]. In the field of search engines, PageRank [3], the most successful algorithm for ranking web pages, is basically a random walk process on the directed graph of websites and hyperlinks. HITS (Hyperlink-Induced Topic Search [4]), a predecessor of PageRank, instead assigns to web pages two different scores: as hub and as authority. Thanks to this twofold nature of the score, HITS was later generalized [5] to bipartite graphs, an important class of systems where entities are divided in two disjoint sets such that interactions happen only between entities in different sets. Examples of systems modeled by bipartite graphs include reviewers and movies in rental websites, scientists and papers in citation networks, customers and products in e-commerce services, and so on. In these systems, each set is endowed with only one kind of score, and if the two sets consist of users and objects it is natural to associate these scores with reputation and quality, respectively. However, bipartite networks are often embedded in the social network of the participant users: for instance, in websites like Digg.com or Last.fm, users can select other users as their friends; also, citation networks are

L. Chen et al. (Eds.): ISMIS 2012, LNAI 7661, pp. 405–414, 2012.

naturally influenced by the professional relationships among scientists. The underlying social network represents an additional source of information a ranking algorithm may exploit, as social links can be associated with trust relationships among users. This is similar to recently proposed recommendation techniques that make use of social ties to obtain recommendations [6].

In this work, we propose a novel and generalized ranking algorithm for bipartite systems to assign quality values to objects and reputation values to users. Such method, which we name QTR (Quality, Trust and Reputation), also exploits the information coming from the users' social relationships. QTR is a generalized algorithm in the sense that it can be easily adapted to different situations (e.g. by giving more weight to certain kind of actions, or to a particular behavior of users). We test our method on two different datasets, the EconoPhysics forum online community and the Last.fm online radio and social network, which are particularly suited for our generalized algorithm—as will be explained later. The results of our study are twofold. We first confirm that ranking is a difficult task, and that an improper algorithm or a peculiarly-structured dataset can lead to extremely biased results. Hence we propose a form of the QTR which is efficient in avoiding such bias. In addition, we show that social relationships can play a valuable role in improving the quality of the ranking.

The rest of this paper is organized as follows. Section 2 presents the QTR ranking method, including its relation with HITS. Section 3 reports the description of each dataset used for testing, followed by the results of our analysis. We conclude with possible generalizations of the QTR algorithm in section 4.

2   Generalized QTR Algorithm

Before presenting our ranking method, we describe the underlying bipartite system and introduce some notations. A bipartite network consists of two disjoint sets of entities (nodes), which for convenience we name as users (labeled by latin letters, $i = 1, \ldots, N$) and objects (labeled by greek letters, $\alpha = 1, \ldots, M$). An interaction between user i and object α is represented by a link connecting the two. Using the formalism of the adjacency matrix, we say that $a_{i\alpha}$ equals 1 if an interaction has occurred, and 0 otherwise. More generally, such interaction can be represented by a weighted link $w_{i\alpha}$, where the strength of the link depends on which particular interaction has occurred or how important/demanding that interaction was. We can further define the degree of user i as the number of objects that user has interacted with: $k_i = \sum_\alpha a_{i\alpha}$, and the degree of object α as the number of users who interacted with it: $k_\alpha = \sum_i a_{i\alpha}$. The total weight of user i is instead defined as $k_i^W = \sum_\alpha w_{i\alpha}$, and of object α as $k_\alpha^W = \sum_i w_{i\alpha}$.

Aside from the bipartite network, users interact with each other in a monopartite social network, where we say that the adjacency matrix element b_{ij} equals 1 if user i is a friend of user j or trusts user j (note that in the first case the matrix is symmetric, whereas it is not in the second). As before, we can introduce a weighted link T_{ij} which represents the "amount of trust" user i puts in user j. The number of friends of/users who trust user j is $f_j = \sum_i b_{ij}$, whereas the total weight of user j is $f_j^W = \sum_i T_{ij}$.

2.1 Definition and Interrelation of Quality and Reputation

We shall now define the meaning of the ranking scores the algorithm assigns to objects and users: quality and reputation, respectively.

Quality it is not an inherent property of an object, rather it is constructed through interactions of the community with the object itself. *Reputation* represents the general opinion of the community towards a user, hence it is ascribed by others and assessed on the basis of the user's actions. We use these conceptual definitions to write down the equations for the QTR ranking method:

$$Q_\alpha = \frac{1}{k_\alpha^{\theta_Q}} \sum_{i=1}^{N} w_{i\alpha}[R_i - \rho_R \bar{R}] \tag{1}$$

$$R_i = \frac{1}{k_i^{\theta_R}} \sum_{\alpha=1}^{M} w_{i\alpha}[Q_\alpha - \rho_Q \bar{Q}] + \frac{1}{f_i^{\theta_T}} \sum_{j=1}^{M} [R_j - \rho_R \bar{R}][T_{ji} - \rho_T \bar{T}] \tag{2}$$

where $\bar{Q} = \sum_\alpha Q_\alpha / M$, $\bar{R} = \sum_i R_i / N$, $\bar{T} = \sum_{ij} T_{ij}/[N(N-1)]$ are the average values of quality, reputation and trust in the community, and θ_Q, θ_R, θ_T, ρ_Q, ρ_R, ρ_T are control parameters—all varying in the range $[0,1]$.

Since equations (1) and (2) are mutually interconnected, quality and reputation values can be determined iteratively. Starting with evenly distributed scores $Q_\alpha^{(0)} = 1/\sqrt{M} \; \forall \alpha$, $R_i^{(0)} = 1/\sqrt{N} \; \forall i$, values of quality and reputation at iteration step $n+1$ are computed from the values at the previous step n by:

$$Q_\alpha^{(n+1)} \leftarrow \frac{1}{k_\alpha^{\theta_Q}} \sum_{i=1}^{N} w_{i\alpha}[R_i^{(n)} - \rho_R \bar{R}^{(n)}]$$

$$R_i^{(n+1)} \leftarrow \frac{1}{k_i^{\theta_R}} \sum_{\alpha=1}^{M} w_{i\alpha}[Q_\alpha^{(n)} - \rho_Q \bar{Q}^{(n)}] + \frac{1}{f_i^{\theta_T}} \sum_{j=1}^{M} [R_j^{(n)} - \rho_R \bar{R}^{(n)}][T_{ji} - \rho_T \bar{T}]$$

To avoid divergence, normalization is applied at the end of each step so that:

$$\sum_{\alpha=1}^{M} [Q_\alpha^{(n+1)}]^2 = 1 \qquad\qquad \sum_{i=1}^{N} [R_i^{(n+1)}]^2 = 1$$

The iterative procedure stops when the algorithm converges to a stationary state:

$$\sum_{\alpha=1}^{M} |Q_\alpha^{(n+1)} - Q_\alpha^{(n)}| + \sum_{i=1}^{N} |R_i^{(n+1)} - R_i^{(n)}| < \delta$$

The QTR algorithm just introduced is a generalization of the notorious HITS algorithm for bipartite graphs [5], namely:

$$Q_\alpha = \sum_i w_{i\alpha} R_i \qquad\qquad R_i = \sum_\alpha w_{i\alpha} Q_\alpha \tag{3}$$

QTR reduces to standard HITS when all parameters θ_Q, θ_R, θ_T, ρ_Q, ρ_R, ρ_T go to zero and $T_{ij} = 0$ $\forall i, j$. However we shall see in what follows that, although making analytical treatment hard, these parameters are extremely valuable in controlling the outcome of the ranking, and that trust represents additional information which is worth to consider.

3   Experimental Results

In this section we test the QTR algorithm on two different datasets, the Econo-Physics forum community and the Last.fm online radio and social network. We present the rankings obtained for objects and users for different values of the model parameters. In order to better describe our results, we make use of Pear-son correlation coefficients between various pairs of quantities: Q_α—k_α (c_{Qk}), Q_α—k_α^W (c_{Qw}), R_i—k_i (c_{Rk}), R_i—k_i^W (c_{Rw}), and only for Last.fm R_i—f_i (c_{Rf}).

3.1   EconoPhysics Forum

The EconoPhysics Forum (http://unifr.ch/econophysics/) is an online platform for interdisciplinary collaboration between physicists and social scientists. Users of the forum can share different resources related to econophysics and complexity science. In what follows, we will consider as objects only the papers uploaded to the forum. As a consequence, a user action can be either uploading, downloading, or viewing a paper. To obtain the dataset of interactions, we analyzed the forum's weblogs dating from 6th July 2010 until 1st June 2012. We removed all entries corresponding to web bots (which cause approximately 75% of the traffic) and repeated access (a user viewing/downloading the same paper several times). We also removed all papers uploaded before 6th July 2010 (for which we do not know the uploader) and all actions associated with them. Finally, we removed the users who both did not upload any paper and have only one view or download action. Altogether, our refined data contains 3511 users, 597 papers and 19578 links.

Among the three types of users' access considered (uploading, downloading and viewing a paper), the first is obviously the more demanding, whereas the second reflects the user's interest in the paper much better than the latter. Hence we can associate to each action a different weight. In what follows, we set $w = 1.0$ for upload actions, $w = 0.1$ for download actions and $w = 0.05$ for view actions. Of course this is just a particular choice, which we consider as reasonable, and we are going to investigate different weighting system in future works. We are also currently running an online survey[1] to determine how these different actions are perceived by scientists—this will allow for a more justified choice of the weights. In any case, the freedom to chose the particular set of weights[2] is what makes

---

[1] Available at http://surveys.soh.surrey.ac.uk/limesurvey/
   index.php?sid=14327&lang=en

[2] We remark that one is nevertheless constrained to a region of the weight space, because if some action(s) become dominant then the other(s) lose their significance and the graph becomes much sparser (as there were only dominant actions).

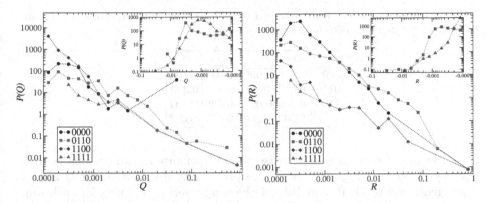

Fig. 1. Probability distributions of Q (left) and R (right) values in EconoPhysics forum data for different configurations of the QTR algorithm. Insets: prolongation to negative values.

Table 1. Top-2 papers (top) and users (bottom) obtained by different configurations of QTR for EconoPhysics forum data. Top papers are: 295 (A. Storkey, Machine Learning Markets, 2011), 102 (R. Tsekov, Brownian markets, 2010), 260 (T. Preis, Switching processes in financial markets, 2011), 263 (T. Preis, Econophysics - complex correlations and trend switchings in financial time series, 2011), 525 (M. Hisakado, Two kinds of Phase transitions in a Voting model, 2012) and 530 (A. Zeng, Enhancing network robustness for malicious attacks, 2012).

			0000		0110		1100		1111			
α	k	k^W	rank	Q	rank	Q	rank	Q	rank	Q	uploader	cited
295	384	29.6	1	4.52E-02	9	5.29E-02	522	3.54E-05	23	-3.74E-04	180	6
102	214	18.1	2	4.48E-02	31	6.05E-03	563	2.09E-05	30	-4.44E-04	180	0
260	172	15	536	1.13E-03	1	5.35E-01	557	2.42E-05	21	3.72E-04	1161	16
263	138	12.75	535	1.17E-03	2	4.97E-01	565	1.83E-05	28	-4.18E-04	1161	1
525	13	1.95	594	9.39E-05	597	-1.49E-02	1	1.00E+00	1	9.99E-01	3200	0
530	4	1.3	597	5.91E-05	98	-3.11E-05	2	2.39E-02	2	2.13E-02	2036	0

i	k	k^W	rank	R	rank	R	rank	R	rank	R		
180	533	527.4	1	9.95E-01	3505	-7.64E-03	33	1.42E-04	3508	-1.77E-03		
17	139	13.05	2	2.38E-02	3023	-8.38E-04	1149	7.43E-06	3119	-1.44E-04		
1161	5	4.05	1332	4.50E-04	1	7.22E-01	566	1.22E-05	3499	-5.69E-04		
1550	1	1	3437	2.18E-05	2	2.30E-01	902	8.81E-06	3504	-1.05E-03		
3200	1	1	3472	4.07E-06	3511	-3.14E-02	1	9.97E-01	1	9.97E-01		
3201	2	0.2	3511	6.63E-07	3499	-1.76E-03	2	5.11E-02	2	5.09E-02		

the EconoPhysics dataset an ideal candidate for testing QTR, despite it does not contain information about users' social or trust relationships.

We test the QTR algorithm on these data with different values of the parameters θ_Q, θ_R, θ_T, ρ_Q, ρ_R, ρ_T. Since social relationships are absent ($T_{ij} = 0$ $\forall i, j$), θ_T and ρ_T are meaningless. The particular form of the algorithm under consideration will be labeled by the parameter values used: for instance, 0000 means $\theta_Q = 0$, $\theta_R = 0$, $\rho_Q = 0$, $\rho_R = 0$ (which corresponds to standard HITS). Apart from HITS, we make use of other three configurations: 1100, 0110, 1111 (we do not use 0011 as it shows convergence problems). 0110 was chosen instead

Table 2. Correlation coefficients obtained by different configurations of the QTR algorithm on the EconoPhysics forum data

	c_{Rk}	c_{Rw}	c_{Qk}	c_{Qw}
0000	0.714	0.999	-0.044	-0.035
0110	-0.017	-0.005	0.320	0.340
1100	-0.006	0.001	-0.030	-0.031
1111	-0.007	-0.001	-0.017	-0.018

of 1001 as in our opinion is more reasonable to penalize high-degree users than high-degree objects. Figure 1 shows the probability distributions of Q and R values generated by QTR, and Table 1 the top-2 users and papers for each configuration. For R, we immediately notice that one extremely high value (very close to one) is present in all cases. In 0000, the top user is the system administrator, which is the uploader of many papers, and this is why all his uploads get the same (high) score—the algorithm is not able to distinguish between them. In both 1100 and 1111, top users have very low degree and this is also an undesirable feature: a single good action shouldn't be enough to obtain high reputation. At the same time, top papers here are very recent works that attracted the attention of a few highly-reputed users. In 0110 finally we obtain the best situation where the scores are distributed more evenly, top users have a non negligible number of contributions and top papers have on average more citations than in the other settings (although we do not consider citation count as a perfect benchmark for quality). Table 2 further shows that this is the only case in which c_{Qk} and c_{Qw} are positive, whereas c_{Rk} and c_{Rw} are close to 0.

3.2   Last.fm

Last.fm (http://www.last.fm/) is a music website which records details of the songs users listen to (form Internet radio stations, personal computers and portable music devices), and provide them with personalized recommendations. The site also offers a social networking feature, in which users can become friends with each other and join groups. The dataset we analyzed is available online[3] and was generated by the Information Retrieval Group at Universidad Autonoma de Madrid [7]. It contains 1892 users, 17632 artists, 92834 artist listening records and 12717 bi-directional friend relations. A peculiar feature of the data is that the users' degree is almost always equal to 50. This is because Last.fm service is free for users in UK, US and Germany, but users in other countries require a subscription to use the radio service and have to pay a fee after a 50 track free trial.

Since the artist listening records from users are labeled by the total listening counts, the weighting system for the bipartite network comes out automatically. Instead the social network only contains the friendship relation of the users. In order to have the two terms in the sum of equation (2) of the same magnitude,

---

[3] http://www.grouplens.org/node/462

we set $T_{ij} = \bar{w} (\bar{k}/\bar{f})$ whenever $a_{ij} = 1$, and $T_{ij} = 0$ otherwise (here $\bar{w}$ is the average of all weights in the bipartite network, $\bar{k}$ is the average users' degree in the bipartite network and $\bar{f}$ is the average users' degree in the monopartite social network). Within this framework, ρ_T loses its meaning while θ_T does not. To be consistent with the previous analysis, we set here $\rho_T = \theta_T = 0$ and use the same configurations as before. To better illustrate the role of trust, we consider both the cases in which $T_{ij} = 0 \; \forall i, j$ ("without trust") and $T_{ij} \neq 0$ ("with trust").

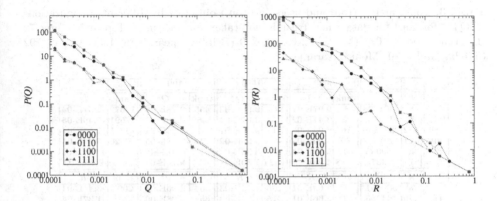

Fig. 2. Probability distributions of Q (left) and R (right) values in Last.fm data for different setting of the QTR algorithm, and when trust is not taken into account

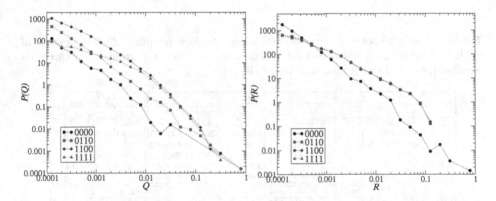

Fig. 3. Probability distributions of Q (left) and R (right) values in Last.fm data for different setting of the QTR algorithm, and when trust is taken into account

Figures 2 and 3 show the probability distributions of Q and R values, and Tables 3 and 4 the top-2 users and artists for each configuration. When trust is not taken into account, we notice again the presence of isolated and extremely high values, especially for Q (this effects is less evident for 0110 and also for 0000 here—because of the absence of an overwhelmingly active user/popular artist). However if trust is considered, scores become distributed more evenly for each

QTR configuration. Table 5 gives additional confirmations of the benefit brought by considering trust: without trust, c_{Qk} and c_{Qw} are positive only for 0000 and 0110 (the latter is better), and c_{Rf} is always 0—as expected; with trust, c_{Qk} and c_{Qw} grow slightly for 0000 and considerably for 0110, whereas c_{Rf} is now always close to 1 as it should be—users with many friends/trusted by many should be indeed highly reputed. We remark that the effect of trust can be tuned by adjusting the weights of the friend relationships.

Table 3. Top-2 artists (top) and users (bottom) obtained by different configurations of QTR for Last.fm data when trust is not taken into account. Top artists are: 72 (Depeche Mode), 1072 (Martin L. Gore), 289 (Britney Spears), 89 (Lady Gaga), 792 (Thalia) and 2390 (Monica Naranjo).

α	k	k^W	0000 rank	Q	0110 rank	Q	1100 rank	Q	1111 rank	Q
72	282	1301308	1	9.97E-01	24	9.98E-03	495	2.78E-06	18672	-2.57E-04
1072	42	39658	2	3.41E-02	352	3.07E-04	1569	2.40E-07	17691	-5.49E-05
289	522	2393140	16	6.04E-03	1	9.85E-01	81	1.24E-04	18431	-1.17E-04
89	611	1291387	39	3.20E-03	2	9.26E-02	132	5.18E-05	17626	-5.33E-05
792	26	350035	309	1.73E-04	65	3.33E-03	1	1.00E+00	1	1.00E+00
2390	7	2437	4278	4.96E-07	1624	1.30E-05	2	9.76E-03	2	9.75E-03

i	k	k^W	rank	R	rank	R	rank	R	rank	R
1642	50	388251	1	8.69E-01	166	3.53E-03	364	3.06E-06	2097	-3.41E-04
446	50	244556	2	2.66E-01	318	1.23E-03	351	3.38E-06	2085	-1.37E-04
542	7	133236	131	1.96E-03	1	9.42E-01	21	3.60E-04	2098	-4.67E-04
1307	1	34328	350	2.71E-04	2	1.61E-01	23	2.74E-04	2099	-5.07E-04
2071	50	338400	429	1.87E-04	253	1.89E-03	1	1.00E+00	1	1.00E+00
1057	50	19207	840	4.46E-05	290	1.51E-03	2	2.14E-02	2	2.14E-02

Table 4. Top-2 artists (top) and users (bottom) obtained by different configurations of QTR for Last.fm data when trust is taken into account. New top artists: 292 (Christina Aguilera), 6373 (Tyler Adam) and 18121 (Rytmus).

α	k	k^W	0000 rank	Q	0110 rank	Q	1100 rank	Q	1111 rank	Q
72	282	1301308	1	9.97E-01	21	4.72E-02	441	1.30E-02	18560	-1.41E-02
1072	42	39658	2	3.41E-02	430	1.40E-03	2028	2.54E-03	16270	-3.09E-03
289	522	2393140	12	9.36E-03	1	7.00E-01	17	9.48E-02	20	8.72E-02
292	407	1058405	47	2.65E-03	2	3.46E-01	35	6.90E-02	28	6.82E-02
6373	1	30614	618	7.21E-05	203	3.76E-03	1	3.60E-01	2	2.36E-01
18121	1	23462	773	5.00E-05	196	3.88E-03	2	3.41E-01	1	2.63E-01

i	k	k^W	rank	R	rank	R	rank	R	rank	R	f
1642	50	388251	1	8.61E-01	474	9.93E-03	512	8.40E-03	578	6.64E-03	33
446	50	244556	2	2.72E-01	584	6.35E-03	595	6.03E-03	614	5.30E-03	19
542	7	133236	129	3.18E-03	1	1.46E-01	132	5.08E-02	134	4.97E-02	24
1300	50	124115	194	1.58E-03	2	1.30E-01	1	1.29E-01	1	1.29E-01	89
1023	50	41123	236	1.16E-03	3	1.20E-01	2	1.20E-01	2	1.20E-01	91

4 Further Generalizations

In this section we discuss two further generalizations of the QTR algorithm, which will be studied and tested in future works.

Table 5. Correlation coefficients obtained by different configurations of the QTR algorithm on the Last.fm dataset

	without trust					with trust				
	c_{Rk}	c_{Rw}	c_{Rf}	c_{Qk}	c_{Qw}	c_{Rk}	c_{Rw}	c_{Rf}	c_{Qk}	c_{Qw}
0000	0.0085	0.2436	0.0387	0.1192	0.3044	0.0074	0.2439	0.0496	0.1225	0.3088
0110	-0.1849	0.1480	0.0877	0.2922	0.6311	-0.0154	0.2572	0.8664	0.6052	0.8667
1100	0.0038	0.1418	-0.0051	-0.0001	0.0769	0.0205	0.2410	0.8846	-0.0016	0.2064
1111	0.0042	0.1408	-0.0054	-0.0001	0.0759	0.0211	0.2367	0.8840	-0.0019	0.1259

4.1   Time Decay

Bipartite systems and their related social networks are not static but instead evolve in time. This means that new users can join the community, whereas other users who are already members may become inactive after a while. On the other hand, newly appeared objects can become hits in almost no time, whereas old objects usually end up losing their attractiveness. Because of these features, a ranking algorithm should be able to handle time effects, for instance by avoiding giving high score to objects which were very popular in the past but whose relevance is currently negligible, or by giving low scores to users who were reliable in the past but then started to behave badly. We can hence introduce in the equations a decaying function of time $D(\tau)$:

$$Q_\alpha(t) = \frac{1}{k_\alpha(t)^{\theta_Q}} \sum_{i=1}^{N} w_{i\alpha}[R_i(t) - \rho_R \bar{R}(t)]D(\tau_{i\alpha}) \qquad (4)$$

$$R_i(t) = \varepsilon + \frac{1}{k_i(t)^{\theta_R}} \sum_{\alpha=1}^{M} w_{i\alpha}[Q_\alpha(t) - \rho_Q \bar{Q}(t)]D(\tau_{i\alpha})$$

$$+ \frac{1}{f_i(t)^{\theta_T}} \sum_{j=1}^{M} [R_j(t) - \rho_R \bar{R}(t)][T_{ji}(t) - \rho_T \bar{T}(t)]D(\tau_{ij}) \qquad (5)$$

where t is the current time, $\tau_{i\alpha} = t - t_{i\alpha}$ is the age of the interaction of user i and object α, $\tau_{ij} = t - t_{ij}$ is the age of the trust relationship between users i and j, and ε is the small positive reputation assigned to new members of the community (who do not have any interaction yet). The decay function $D(t)$ can have non-zero tail even when t is large, and the strength of the decay can be tuned to focus on a particular time window. Some examples of decay function include $D(t) = [1 + (t/\tau_0)^\beta]^{-1}$ or $D(t) = d_0 + (1 - d_0)\exp[-t/\tau_0]$, where τ_0 is the characteristic time scale of decay.

4.2   Projected Trust

Trust is the subjective opinion of one user towards another. We argue that, when no explicit assessments from users are available, trust relationships can be inferred form the bipartite network by measuring the similarity of users' actions, which essentially means by projecting the bipartite user-object network into the monopartite user-user network:

$$\tilde{T}_{ij}(t) = \frac{[R_j(t) - \rho_R \bar{R}(t)]}{k_j(t)^{\theta_R}} \sum_{\alpha=1}^{M} w_{i\alpha} w_{j\alpha} \frac{[Q_\alpha(t) - \rho_Q \bar{Q}(t)]}{k_\alpha(t)^{\theta_Q}} D(\tau_{i\alpha}) D(\tau_{j\alpha}) \qquad (6)$$

We name this term as "projected" trust. Despite the fact that projected trust values are computed with the same source of information used for quality and reputation assessment, preliminary results (not reported here) show that using $\tilde{T}$ instead of T values in a slightly modified version of the algorithm can bring to some improvements with respect to simple HITS, especially when the bipartite network is sparse.

5   Conclusion

In this work we introduced a general ranking method for bipartite networks that can simultaneously evaluate users' reputation and objects' quality. This is by no means the first attempt in the literature [5, 8, 9], however our method differs from the others by exploiting the trust relationships and social acquaintances of users as an additional source of information. Testing of our method on real datasets revealed which form of the algorithm gives more reasonable results. In addition, we showed that considering trust relationships indeed brings improvements to the resultant ranking. The positive results we obtained are encouraging. However, the number of parameters used by the algorithm, and in general the difficulties in assessing the reliability of a ranking method pose additional issues on the effectiveness of our method, which will require further tests and future studies.

Acknowledgments. This work was partially supported by the Future and Emerging Technologies programme of the European Commission FP7-COSI-ICT (project QLectives, grant no. 231200) and by the Swiss National Science Foundation (grant no. 200020-121848).

References

[1] Josang, A., Ismail, R., Boyd, C.: Decision Support Systems 43(2) (2007)
[2] Adler, B.T., et al.: Technical report UCSC-CRL-07-09, School of Engineering. University of California, Santa Cruz (2007)
[3] Brin, S., Page, L.: Comput. Netw. ISDN Syst. 30, 107 (1998)
[4] Kleinberg, J.: J. ACM 46, 604 (1999)
[5] Deng, H., Lyu, M.R., King, I.: In: Proceedings of the 15th ACM SIGKDD International Conference on Knowledge Discovery and Data Mining, pp. 239–248. ACM, New York (2009)
[6] Golbeck, J.: Science 321, 5896 (2008)
[7] Cantador, I., Brusilovsky, P., Kuflik, T.: In: Proceedings of the 5th ACM Conference on Recommender Systems. ACM, New York (2011)
[8] Zhou, Y., Lei, T., Zhou, T.: EPL 94, 48002 (2011)
[9] Zhou, Y., Lü, L., Li, M.: New J. Phys. 14, 033033 (2012)

Providing Timely Results with an Elastic Parallel DW

João Pedro Costa[1], Pedro Martins[2], José Cecilio[2], and Pedro Furtado[2]

[1] Polytechnic Institute of Coimbra
[2] University of Coimbra
{jpcosta,pmon,jcecilio,pnf}@dei.uc.pt

Abstract. OLAP analysis is a fundamental tool for enterprises in competitive markets. While known (planned) queries can be tuned to provide fast answers, ad-hoc queries have to process huge volumes of the base DW data and thus resulting in slower response times. While parallel architectures can provide improved performance, by using a divide-and-conquer approach, their structure is rigid and suffers from scalability limitations imposed by the star schema model used in most deployments. Therefore usually they are over-dimensioned with computational resources in order to provide fast response times. However, for most business decisions, it is more important to have guarantees that queries will be answered in a timely fashion. The star schema model physical representation introduces severe limitations to scalability and in the ability to provide timely execution, due to the well-known parallel join issue and the need to use solutions such as on-the fly repartitioning of data or intermediate results, or massive replication of large data sets that still need to be joined locally. In this paper, we propose PH-ONE an architecture that overcomes the scalability limitations by combining an elastic set of inexpensive heterogeneous nodes with a denormalized DW storage model organization, which requires a minimal set of predictable processing tasks, using in a shared-nothing scheme to remove costly joins. PH-ONE delivers timely execution guarantees by adjusting the number of processing nodes and by rebalancing the data load according to the nodes characteristics. We used the TPC-H benchmark to evaluate PH-ONE ability to provide timely results.

Keywords: DW, timely query results, elastic parallel DW.

1 Introduction

Data Warehouses (DW) typically follow a star schema storage organization, composed of a central fact table with business metrics and a set of foreign keys referencing business perspectives stored as dimension tables. The simplicity and straightforwardness of the model minimizes the users' learning curve and provides a hassle free model for datamining and business analysis. However, usability is constrained by its inability to provide timely results, especially for ad-hoc queries. While some time guarantees can be accurately provided for known queries, using timing from past executions, pre-computed data (materialized views), and index structures, the timely execution of ad-hoc queries cannot be guaranteed. The data volume produced by data intensive industries in competitive markets, such as telecom

L. Chen et al. (Eds.): ISMIS 2012, LNAI 7661, pp. 415–424, 2012.

and smart-grids, stresses the limits of DW systems. DBAs of such systems have to maintain a constant supervision to the query load, query pattern and selectivity to ensure top performance. However, the ability to provide timely results is not just a performance issue (high throughout), but also a matter of providing query results when expected before the business decision-making process.

Parallel architectures improve query performance by dividing the data volume among processing nodes and parallelizing query execution. While shared-nothing deployments allow that each node can compute partial results locally, its scalability is constrained by the star schema model used in most DW deployments. This divide-and-conquer strategy offers limited scalability, since in order to allow the local computation of partial results, only the central fact table should be split among nodes, whereas dimensions are replicated. However, the data placement of the star schema among nodes and the in-network join processing costs (network costs for exchanging temporary results between nodes) limits the system scalability, resulting in sub-linear speedup. And therefore, they are unable to provide timely results.

In this paper, we propose PH-ONE a parallel shared-nothing architecture that manages a elastic set of heterogeneous nodes, and adapts the degree of parallelism according to the timely execution targets, and redistributes the data load using the predictability and minimum processing requirements provided by the ONE [1] storage model, to deliver timely execution results with minimum system disturbance.

The paper is divided as follows: section 2 presents some related works; section 3 motivates the need for timely results and identify the factors that constrain current deployments, in particular the storage organization; section 4 presents our architecture that uses a elastic set of parallel nodes to provide timely execution; section 5 presents evaluation results and finally section 6 concludes and presents some future work.

2   Related Works

In recent decades, both academia and industry have investigated different methods and algorithms for speeding-up the time required to process queries that join several relations. Research in join algorithms includes sort-merge, hash join, grace-hash join and hybrid-hash join [2][3] and also adaptive [4], as well as access methods such as the Btree or Bitmap indexes [5]. Materialized views [6] use extra storage space to physically store aggregates concerning well-known and planned queries. Sampling [7] trades off precision for performance by employing the power offered by statistical methods to reduce the data volume that needs to be processed to compute an acceptable result. Research on data partitioning and data allocation, include [8][9] which exploit horizontal fragmentation and hash partitioning of relations and intermediate results to process parallel, multi-way join queries and increase parallelism; [10] exploits a workload-based data placement and join processing in parallel DW. A performance evaluation of parallel joins is presented in [11]. Vertical partitioning and column-wise store engines [12] proved to be effective in reducing the disk IO and thus boosting query performance. However, these works focus on improving query performance and minimize the cost of joining relations, not on providing predictable and invariant execution time.

Schema denormalization showed to be effective in providing predictable execution time [1, 13]. [1] evaluated the impact of full denormalization of star schemas in both storage space and query performance on a single server. [13] provides consistent response times, by partitioning data by frequency to achieve good partitioning using fixed-length codes. Other works on denormalization [14–16] do not focus on the denormalization limitations of the star schema model, and they do not offer a clear insight into the query performance predictability and scalability over parallel shared-nothing architectures.

3   Scalability Issues That Constrain Timely Results

Most DWs use the star schema model, which offers a trade-off between storage size and performance in a single processing node. Large DWs require parallel architectures to handle huge amounts of data with acceptable response times. They are usually deployed in shared-nothing parallel architectures, which yield good performance and scalability capabilities, by distributing data among nodes to maximize the local computation of partial results. While fact tables are partitioned into smaller partitions and allocated to nodes, dimensions (regardless of their size) are replicated into each node, as illustrated in Figure 1.

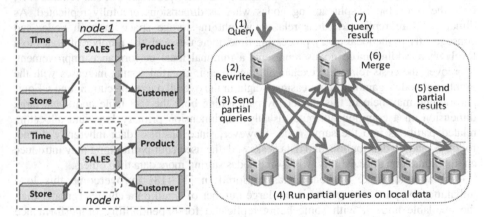

Fig. 1. Star-schema data placement   **Fig. 2.** Query processing in shared-nothing

Each node independently computes partial results locally, as shown in figure 2. A query (1) received by a submitter node is rewritten (2) and forwarded to the processing nodes (3). Each processing node executes the rewritten query against the local data (4) before sending the partial results (5) to the merger node. The merger node, which may be the submitter node, waits for the intermediate results, and merges them (6) to compute the final query result, before sending it (7) to the user.

Definition: Consider that for a query Q, t is the query execution time, t_{rw} the time to rewrite and build the partial queries (2), t_{tpq} the time to transfer the partial query to each node, t_n the local execution time on a node n; t_{tpr} the time for a node to transfer its partial results to the merger node; t_m the time required by the merger node to *receive and merge the partial results and to compute the final result and t_s the time required to send the final results. The execution time t for a query Q is computed as*

$$t(Q) = t_{rw}(Q) + \sum_{j=1}^{\eta} t_{tpq_j}(Q_p) + max\left(\left\{t_{n_j}(Q_p) + t_{tpr_j}(Q_{pR})\right\}_{j=1}^{\eta}\right)$$

$$+ t_{m_j}\left(\sum_{j=1}^{\eta} Q_{pR}\right) + t_s(Q_R) \tag{1}$$

Assuming that t_{rw}, t_{tpq} and t_s are negligible then t can be estimated as

$$t(Q) = max\left(\left\{t_{n_j}(Q_p) + t_{tpr_j}(Q_{pR})\right\}_{j=1}^{\eta}\right) + t_{m_j}\left(\sum_{j=1}^{\eta} Q_{pR}\right) \tag{2}$$

The overall query execution time t is mostly influenced by the local query execution of the slowest node, determined as $max\left(\left\{t_{n_j}(Q_p)\right\}_{j=1}^{\eta}\right)$, the number of nodes, the partial results' size and the cost of sending them to the merger node.

Local query execution time t_n can be improved by increasing the number of processing nodes of the parallel infrastructure, and thus reducing the amount of data that each node has to process. However, to allow independent local query processing of star schema DW, only the fact table is split among nodes whereas dimensions are fully replicated. As dimensions are replicated, their relative weight in storage space in each node, and consequently in the local query processing time, increases with the number of nodes to a level where adding more nodes represents a minimal local performance improvement. Moreover, the costs related to exchanging and merging partial results increases with the number of nodes, gaining an increasing weight in the overall query execution time t. Equi-partitioning may help in this matter, by partitioning both the fact table and some large dimension on a common attribute, usually the dimension primary key, and allocating related partitions into the same node. However, since business data inherently do not follow a random distribution, and data is skewed, the used equi-partitioning may introduce another limitation to scalability, with some nodes storing more data than others.

A map-reduce-like approach was explored in [17][18] to overcome this load unbalance, by partitioning data into a large number of small data chunks (greater than the available nodes), with some being replicated for dependability or performance reasons. Chunks are processed as nodes become available to process new data. However, it does not solve the increasing weight of dimensions and results in higher network costs since more partial results (one for each chunk) have to be sent.

Frequently large non-equi-partitioned dimensions are also partitioned and distributed among nodes, without being co-located according the fact table data. While this overcomes the storage scalability limitations, it introduces additional query complexity related to processing parallel network joins, and is sensitive to network costs related to exchanging intermediate results. A performance evaluation of parallel joins is presented in [11]. The scalability of parallel architectures is thus constrained by those correlated limitations: local data processing, node data unbalance, and in-network parallel joins, number of nodes, and number of partitions per node.

Two main paradigms are used to attain that goal: parallel DBMS and Map-Reduce (MR) frameworks. Pavlo et al. [19] present a comparison of approaches to large-scale data analytics, identifying their limitations both in performance and usability. While MR offers elastic scalability it lacks the expressiveness and usability of SQL. Both show scalability constraints in processing in-network joins and consequently they are unable to provide timely executions. As a consequence, the traditional approach of adding more nodes to parallel node infrastructures is insufficient to enforce timely guarantees, because of their inability to provide scale-up execution guarantees. Although, all the effort to devise improved parallel joins algorithms, this network scalability limitation introduced by parallel joins is not solved.

In this paper, we advocate that scalability can only be attained by eliminating parallel joins, and changing the data organization in order to provide a true shared-nothing infrastructure where timely results can be provide with a elastic set of heterogeneous nodes.

4   An Elastic Parallel DW

Deployment of large star-schema DW on complex and over-dimensioned clusters of parallel dedicated nodes may provide fast query results using an expensive brute-force approach. It follows the assumption that a large amount of top-edge hardware will be sufficient to have the job done. The over-dimensioning is lead by future performance and data load expectations, and by the fact that future upgrades are costly and may require a full architecture substitution, if similar (homogeneous) nodes are became obsolete (unavailable). And a good balance of the amount of data allocated to heterogeneous nodes in a parallel shared-nothing architecture is hard to accomplished, since query time is constrained by the performance of the slowest node (equation 2).

Our architecture (PH-ONE) manages an elastic set of heterogeneous nodes, added or removed according to the current timely execution targets, and re-balances data load between processing nodes until the variance of local query processing is below an acceptable threshold. Data in each node is stored using ONE [1] storage model which stores all the DW related data into a single denormalized relation, without all the key related overheads from the relational model (keys and related indexes). ONE, since no joins are required, provides predictable and almost invariant query execution times, with less demanding DBMS systems, and deployed on COTS hardware with minimal memory and processing requirements. [1, 13] demonstrated that schema denormalization provides predictable query execution times. Since ONE provides a high degree of predictability, and query processing is highly IO dependent, the denormalized relation can be freely partitioned according to each node's capacity.

Query processing of a denormalized relation in a shared-nothing fashion, illustrated in Fig. 4 is similar to Fig. 2 for the active nodes. The main difference in query processing resides on step 2, which rewrites a query into a set of partial queries to be executed by the local nodes. Besides rewriting, it also as to remove all join conditions and perform the necessary attribute mapping, translating all the existing references of the star schema model to the corresponding attributes in the denormalized relation. To ensure that dimension aggregates results do not include double-counting, some additional predicates have to be included.

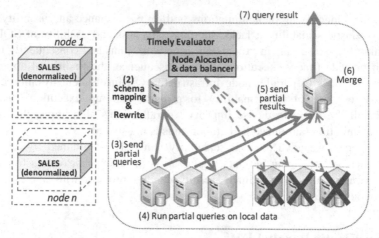

Fig. 3. PH-ONE partitioning and placement   **Fig. 4.** PH-ONE Query processing

In a parallel deployment with homogenous nodes, the amount of data is evenly divided between nodes to deliver the $t(Q)$ expected execution time. The number of nodes η is determined as a function of the desired timely execution, so that

$$t_n(Q_p) = t(Q) - t_{tpr}(Q_{pR}) - t_{m_j}\left(\sum_{j=1}^{\eta} Q_{pR}\right) \tag{3}$$

However, in parallel architecture with heterogeneous nodes, and since the ONE is highly IO dependent, data is allocated to each node in order to minimize the variance of the inter-node local query execution time.

$$max\left(\left\{t_{n_j}(Q_p)\right\}_{j=1}^{\eta}\right) = t(Q) - t_{tpr}(Q_{pR}) - t_{m_j}\left(\sum_{j=1}^{\eta} Q_{pR}\right) \tag{4}$$

The maximum local query execution local time (equation 4), is adjusted when the number of processing nodes increases, to account for the additional intermediate results that have to be exchanged and merged before computing the final result.

PH-ONE maintains a registry of a set of heterogeneous processing nodes, which are activated when additional processing power is required to deliver stricter timely query results. A Timely Evaluator module is constantly monitoring and assessing if the current PH-ONE deployment can deliver the user-specified temporal targets. When it cannot be guaranteed, the Node Allocation & Data Balancer module determines the maximum amount of data that each existing node can processed within the temporal targets, and redistributes the remaining data to additional new nodes. To reduce the amount intermediate that have to be send to the merger node, it starts by activating the fastest nodes (algorithm allocation_by_fastest).

Data is partitioned and allocated to each node as a set of chunks. However, to allow fine-tuning the amount of data allocated according to the nodes' characteristics, it can also be performed at row-level. While this process is performed, the Timely Evaluator continuous to monitor the current deployment to assess that the expected timely results is provided, repeating the process until the average inter-node variance falls below a given threshold. Fresh new data is inserted in round-robin fashion, or inserted into the faster nodes, and rebalanced afterwards if needed.

Algorithm. allocation_by_fastest

Require: n, $maxt_n$ #n->number of nodes, $maxt_n$ -> maximum Local Execution time

```
 1:    for j in 1 to noffline_nodes        # sorted descendent by tn_per_size(i); faster first
 2:    for i in 1 to n
 3:       dlredistribute (i) = tn(i) / maxtn(i) * dlsize(i)    # amount of data to re-distribute
 4:       If dlredistribute (i) >0 then
 5:          for j in 1 to noffline_nodes      # sorted descendent by tn_per_size(i); faster first
 6:             If tn(j) < maxtn then
 7:                dltoallocate = max(dlredistribute (i) - dlsize(j), dlmaxsize (j))
 8:                allocate_to_node (i,j, dltoallocate);
 9:                dlredistribute (i) = dlredistribute (i)- dltoallocate
10:             end if
11:             If dlredistribute (i) =0 then        break
12:          end if
13:          end for
14:       end if
15:    end for
```

---

5   Evaluation

In this section, we evaluate the effectiveness of PH-ONE in overcoming the scalability limitations and in providing timely results. The experimental evaluation was conducted using TPC-H [20] deployed on a parallel architecture, using two distinct RDBMS: PostgresSQL 9.0 and DbmsX, a well-known commercial system. The parallel infrastructure consisted of 30 Linux Server nodes, interconnected with a full duplex gigabit switch. An additional node was used as the submitter, controller and merger node.

We created two different types of schema: the base TPC-H schema (P-STAR) and PH-ONE composed by a single relation as proposed in [1]. The former was populated with DBGEN available at [20] and the latter with a modified version that generates the denormalized data as single flat file, with a scale factor of 100 (SF=100).

We created: PH-ONE with 3, 10, 20 and 30 nodes, containing respectively 1/3rd, 1/10th, 1/20th and 1/30th of the full denormalized data; and P-STAR with 3, 10, 20 and 30 nodes, containing respectively 1/3rd, 1/10th, 1/20th and 1/30th of tables ORDERS and LINEITEM equi-partitioned on O_ORDERKEY and fully replicating the remaining relations. Fig. 5b) shows, the storage space allocated to each node for each schema with varying number of nodes. It also depicts P-STAR(O) where only LINEITEM is partitioned, and P-STAR (LO+PS) which is similar to P-STAR but also with PARTSUPP equi-partitioned with PART.

While the overall storage requirements of PH-ONE remain constant with 627GB, regardless of the number of processing nodes and without storage overheads, we observe that *P-STAR* increased to about 900GB in a 30 node setup, and continues to increase with the number of nodes.

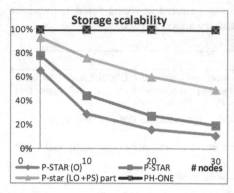

# Nodes	1	3	10	20	30
P-STAR	179	76,3	40,2	32,5	29,9
P-STAR (LO+PS)	179	64,0	23,5	14,9	12,0
P-STAR (O)	179	91,0	61,1	54,7	52,6
PH-ONE	627	209	62,7	31,4	20,9

Fig. 5. Storage scalability (a) data size in each node (in GB) (b)

PH-ONE presents higher loading time (roughly a 3x ratio). However, when taking into consideration the time required by TPCH schema to create all the key related indexes, the loading time is almost is similar. PH-ONE delivers better storage scalability.

Node Partial Execution Time

We evaluated the partial execution time in each node for queries 1..10. Fig 6 depicts the results for a varying number of nodes, using postgreSQL (a) and dbmsX (b).

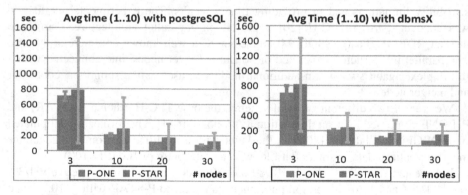

Fig. 6. Partial execution time in each node using a) postgreSQL and b) dbmsX

The results show that local execution time of PH-ONE decreases almost linearly, as the data volume diminishes. Increasing the number of nodes from 3 to 10, the local average execution time decreased from 709s to 208s, while P-STAR only decreased from 783s to 291s with postgreSQL. DbmsX, presents a similar behavior. Since PH-ONE uses a simpler storage model, both RDBMS engines deliver similar query execution times (with variations smaller than 0.2%) whereas the same does not happens with P-STAR. This characteristic allows that more complex shared-nothing architectures can be built using a set of heterogeneous database engines.

System Ability to Provide Timely Results

We changed the timely execution targets to evaluate the system ability to adjust the processing and data load to provide timely results. In a 3 node-setup we defined a timely execution target of 250s. The Timely Evaluator determined (from equation 3) that at least 10 nodes are required to provide such timely execution target (also shown in Fig. 6). The Node Allocator took around 10 minutes to add 7 additional nodes and to rebalance the data between the 10 nodes, according to each node's characteristics, before starting to deliver timely results within the specified target. On average, each additional node had to receive about 62Gb of data. Fig. 7 depicts the query execution for the different network deployments. The red line indicates the timely execution target. In each setup, the inter-node variance remained below 0.05%.

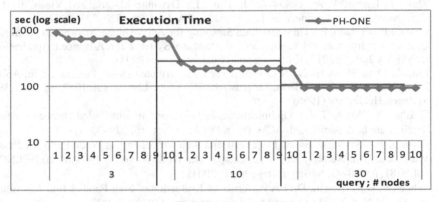

Fig. 7. PH-ONE execution time

Afterwards, we defined a stricter timely target of 100s. It determined that it needed at least 30 nodes to provide such timely execution targets. It took about 4 minutes to make all the necessary readjustments before starting to deliver timely results.

6   Conclusions and Future Work

In this paper we discussed the scalability limitations of the star schema model when deploying large DWs in parallel, shared-nothing architectures, and discussed the increasing demand to have a parallel architecture that can, without minimal costs and disturbance, to adapt in order to provide timely results. We propose an architecture that uses an elastic set of nodes, which may be heterogeneous, in conjunction with a denormalized storage model, in order to adapt the strength of the processing infrastructure according to the timely results requirements. We show that the architecture is able to overcome the scalability limitations and provide timely results.

PH-ONE delivers improved speedup without increasing the global data storage space, and can adjust the processing infrastructure in order to provide timely results. Since the approach provides almost linear speedup and predictable execution time, determined as a function of the data volume, it allows fine tuning of the amount of data allocated to each heterogeneous node.

References

1. Costa, J.P., Cecílio, J., Martins, P., Furtado, P.: ONE: A Predictable and Scalable DW Model. In: Cuzzocrea, A., Dayal, U. (eds.) DaWaK 2011. LNCS, vol. 6862, pp. 1–13. Springer, Heidelberg (2011)
2. Patel, J.M., Carey, M.J., Vernon, M.K.: Accurate modeling of the hybrid hash join algorithm. ACM SIGMETRICS Performance Evaluation Review (1994)
3. Harris, E.P., Ramamohanarao, K.: Join algorithm costs revisited. The VLDB Journal 5, 064–084 (1996)
4. Bornea, M.A., Vassalos, V., Kotidis, Y., Deligiannakis, A.: Double Index NEsted-Loop Reactive Join for Result Rate Optimization. In: Proceedings of the 2009 IEEE International Conference on Data Engineering, Washington, DC, USA, pp. 481–492 (2009)
5. Johnson, T.: Performance Measurements of Compressed Bitmap Indices. In: Proceedings of the 25th International Conference on Very Large Data Bases, pp. 278–289 (1999)
6. Zhou, J., Larson, P.-A., Goldstein, J., Ding, L.: Dynamic Materialized Views. In: Int. Conference on Data Engineering, Los Alamitos, CA, USA, pp. 526–535 (2007)
7. Costa, J.P., Furtado, P.: Time-Stratified Sampling for Approximate Answers to Aggregate Queries. In: International Conference on Database Systems for Advanced Applications (DASFAA 2003), p. 215. IEEE Computer Society, Kyoto (2003)
8. Liu, C., Chen, H.: A Hash Partition Strategy for Distributed Query Processing. In: Apers, P., Bouzeghoub, M., Gardarin, G. (eds.) EDBT 1996. LNCS, vol. 1057, pp. 371–387. Springer, Heidelberg (1996)
9. Shasha, D., Wang, T.-L.: Optimizing equijoin queries in distributed databases where relations are hash partitioned. ACM Trans. Database Syst. 16, 279–308 (1991)
10. Furtado, P.: Workload-Based Placement and Join Processing in Node-Partitioned Data Warehouses. In: Kambayashi, Y., Mohania, M., Wöß, W. (eds.) DaWaK 2004. LNCS, vol. 3181, pp. 38–47. Springer, Heidelberg (2004)
11. Schneider, D.A., Dewitt, D.J.: A Performance Evaluation of Four Parallel Join Algorithms in a Shared-Nothing Multiprocessor Environment, pp. 110–121 (1989)
12. Stonebraker, M., Abadi, D.J., Batkin, A., Chen, X., Cherniack, M., Ferreira, M., Lau, E., Lin, A., Madden, S., O'Neil, E., O'Neil, P., Rasin, A., Tran, N., Zdonik, S.: C-store: a column-oriented DBMS. In: Proceedings of the 31st International Conference on Very Large Data Bases, pp. 553–564 (2005)
13. Raman, V., Swart, G., Qiao, L., Reiss, F., Dialani, V., Kossmann, D., Narang, I., Sidle, R.: Constant-Time Query Processing. In: Proceedings of the 2008 IEEE 24th International Conference on Data Engineering, pp. 60–69 (2008)
14. Yma, P.: A Framework for Systematic Database Denormalization. Global Journal of Computer Science and Technology 9 (2009)
15. Sanders, G.L., Shin, S.: Denormalization Effects on Performance of RDBMS. In: Proceedings of the 34th Annual Hawaii International Conference on System Sciences, p. 3013. IEEE Computer Society, Washington, DC (2001)
16. Zaker, M., Phon-Amnuaisuk, S., Haw, S.-C.: Optimizing the data warehouse design by hierarchical denormalizing. In: Proc. 8th Conference on Applied Computer Science (2008)
17. Furtado, P.: Efficient, Chunk-Replicated Node Partitioned Data Warehouses. In: 2008 IEEE International Symposium on Parallel and Distributed Processing with Applications, Sydney, Australia, pp. 578–583 (2008)
18. Yang, C., Yen, C., Tan, C., Madden, S.: Osprey: Implementing MapReduce-Style Fault Tolerance in a Shared-Nothing Distributed Database. In: Proc. ICDE (2010)
19. Pavlo, A., Paulson, E., Rasin, A., Abadi, D.J., DeWitt, D.J., Madden, S., Stonebraker, M.: A comparison of approaches to large-scale data analysis. In: Proc. of the 35th SIGMOD International Conference on Management of Data, pp. 165–178 (2009)
20. TPC-H: TPC-H Benchmark, http://www.tpc.org/tpch/

Discovering Dynamic Classification Hierarchies in OLAP Dimensions*

Nafees Ur Rehman, Svetlana Mansmann, Andreas Weiler, and Marc H. Scholl

Department of Computer & Information Science
University of Konstanz, Germany
{nafees.rehman,svetlana.mansmann,andreas.weiler,
marc.scholl}@uni-konstanz.de
http://dbis.uni-konstanz.de/

Abstract. The standard approach to OLAP requires measures and dimensions of a cube to be known at the design stage. Besides, dimensions are required to be non-volatile, balanced and normalized. These constraints appear too rigid for many data sets, especially semi-structured ones, such as user-generated content in social networks and other web applications. We enrich the multidimensional analysis of such data via content-driven discovery of dimensions and classification hierarchies. Discovered elements are dynamic by nature and evolve along with the underlying data set.

We demonstrate the benefits of our approach by building a data warehouse for the public stream of the popular social network and microblogging service Twitter. Our approach allows to classify users by their activity, popularity, behavior as well as to organize messages by topic, impact, origin, method of generation, etc. Such capturing of the dynamic characteristic of the data adds more intelligence to the analysis and extends the limits of OLAP.

Keywords: Data Warehousing, OLAP, Data Mining, multidimensional data model, OLAP cube, OLAP dimensions.

1 Introduction and Motivation

The necessity to integrate OLAP and data mining was postulated in the late 90-es [5]. Today, a powerful data mining toolkit is offered as an integrated component of any mature data warehouse system, such as Microsoft SQL Server, IBM DB2 Data Warehouse Edition, Oracle, and others. Data mining tools require the input data to be consolidated, consistent and clean. OLAP cubes – where the extracted data undergoes exactly this kind of transformation – appear to be perfect candidates to harbor data mining algorithms. In a standard data warehouse system architecture, data mining functionality resides at the upper layer

* This work was partially supported by DFG Research Training Group GK-1042 "Explorative Analysis and Visualization of Large Information Spaces", University of Konstanz.

L. Chen et al. (Eds.): ISMIS 2012, LNAI 7661, pp. 425–434, 2012.

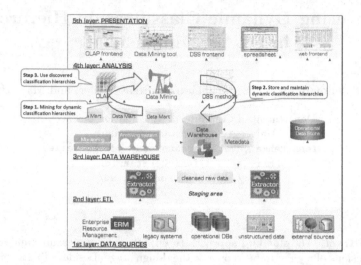

Fig. 1. Integrating a Data Mining Feedback Loop into OLAP Cubes

over the existing cubes and marts of the data warehouse layer as shown in Figure 1. Our proposed contribution is depicted as a 3-step feedback loop between the application and the data warehouse layers in the same figure. In the first step, data mining classification algorithms are applied to cluster dimensional data based on some dynamic characteristics (e.g., to group users by popularity, activity or interest). In the second step, the acquired classification is added as a new aggregation path to the respective dimension, leading to the third step of enabling this new aggregation path in OLAP queries. Introduction of discovered classifications to dimensional hierarchies raises a number of research challenges, such as their maintenance, evolution, temporal validity and aggregation constraints. These issues will be handled later on in this work.

Mining data cubes for dynamic classifications is a popular technique in OLAP applications dealing with customer trending, risk or popularity assessment, etc. However, traditional data mining applications return such classifications as the outcome of the analysis, whereas our approach is to feed this outcome back to the data warehouse as elements of the data model in their own right.

1.1   Tweet Analysis as Motivating Example

Twitter[1] is a popular social network with microblogging service for real-time information exchange. Twitter offers a set of APIs for retrieving the data about its users and their communication. Extreme popularity of the Twitter and the availability of its public stream have resulted in the multiplication of Twitter-related research initiatives as overviewed in the Related Work.

Twitter employs a rather simple data model that encompasses users, their messages (*tweets*), and the relationships between and within those two classes.

---

[1] http://twitter.com/

Users can be friends or followers of other users, be referenced (i.e., tagged) in tweets, be authors of tweets or retweet other users' messages. The third component is the timeline, which describes the evolution, or the ordering, of user and tweet objects. The structure of the original stream explicitly contains a rather small number of attributes usable as measures and dimensions, whereas a wealth of additional parameters, categories and hierarchies can be obtained using different computation methods, from simple derivations to complex techniques of knowledge discovery. Many of the characteristics (e.g., status, activity, interests, popularity, etc.) are dynamic and, therefore, cannot be captured as OLAP dimensions. However, from the analyst's perspective, such characteristics may represent valuable dimensions for the analysis.

The dataset delivered by the Twitter Streaming API is semi-structured using the JavaScript Object Notation (JSON). Each tweet is streamed as an object containing 67 data fields with high degree of heterogeneity. A tweet record encompasses the message itself along with detailed metadata on the user's profile and geographic location. A straightforward mapping of this set of attributes to a multidimensional perspective results in the identification of cubes *Tweet* and *TweetCounters* for storing the contents and the metadata of the messages and for storing the statistical measurements provided with each record, respectively.

1.2 Related Work

The work related to our contribution can be subdivided into two major sections: 1) research on integrating data warehousing and mining and 2) knowledge discovery from Twitter data.

A pioneering work on integrating OLAP with data mining was carried out by Han [5] who proposed a theoretical framework for implementing OLAP mining functions. His *mining then cubing* function is a predecessor of our approach. The idea is to enable application of OLAP operators on the mining results. An example of implementing such functionality can be found in the Microsoft SQL Server and is known as *data mining dimensions* [10]. The latter contain classifications obtained by applying clustering or other algorithms on the original cube and can be materialized and used (with some limitations) just like ordinary dimensions for OLAP. Usman et al. review the research literature on coupling OLAP and data mining in [17] and propose a conceptual model for combining enhanced OLAP with data mining systems. The urge to enhance the analysis by integrating OLAP and data mining was expressed in multiple publications in the past. Significant works in this area include [6], [18], [4], and [3]. It was Han et al. [6] who introduced the concept of integrating OLAP and data mining called Online Analytical Mining (OLAM).

Research contributions related to the Twitter analysis mostly focus on improving the search and navigation in a huge flow of messages as well as on discovering valuable information about the content and the users. We are more interested in the latter types of works. In 2007 Java et al. [8] presented their observations of the microblogging phenomena by studying the topological and geographical properties of Twitter's social network. They came up with a few categories for

Twitter usage, such as daily chatter, information and url sharing or news reporting. Mathioudakis and Koudas [14] proposed a tool called Twitter Monitor for detecting trends from Twitter streams in real-time by identifying emerging topics and bursty keywords. Recommendation systems for Twitter messages are presented by Chen et al. [2] and Phelan et al. [16]. Chen et al. studied content recommendation on Twitter to better direct user attention. Phelan et al. also considered RSS feeds as another source for information extraction to discover Twitter messages best matching the user's needs. Michelson and Macskassy [15] discover main topics of interest of Twitter users from the entities mentioned in their tweets. Hecht et al. [7] analyze unstructured information in the user profile's location field for location-based user categorization. While most Twitter-related contributions focus on mining or enhancing the contents of tweets, improving the frontend or generating meaningful recommendations, we exploit the advantages of coupling the OLAP technology with data mining to enable aggregation-centric analysis of the Twitter data.

2   Conceptual Modeling of Dynamic Elements

Data in a data warehouse is structured according to the aggregation-centric multidimensional data model, which uses numeric measures as its analysis objects [1]. A fact consists of one or multiple measures along with their descriptive properties referred to as *dimensions*. Values in a dimension can be structured into a *hierarchy* of granularity levels to enable drill-down and rollup operations.

The terms *fact* and *measure* are often used as synonyms in the data warehouse context. We distinguish between those terms to account for facts without measures. According to Kimball [9], a fact is given by a many-to-many relationship between a set of attributes. There exist many-to-many mappings in which no attribute qualifies as a measure. A classical example is an event record, where an event is given by a combination of simultaneously occurring dimensional characteristics. We use the notion *non-measurable fact type* introduced in [12] for facts with no measures. Back to the Twitter scenario, a non-measurable fact type could be used to capture the tweeting events with user, message and time/date as its dimensions.

A dimension is a *one-to-many* characteristic of a fact and can be of arbitrary complexity, from a single data field to a collection of related attributes, from uniform grain to a hierarchical structure with multiple alternative and parallel hierarchies [11,13]. OLAP does not support definition of dynamic, non-strict, or fuzzy dimension hierarchies. However, the extended Dimensional Fact Model (x-DFM) [12] makes provisions for modeling such hierarchy types at the conceptual level. We adopt the x-DFM notation for the concepts introduced in this section.

Figure 2 shows an example of modeling a cube for storing user activity statistics in x-DFM. A fact type is represented as a graph centered at the fact type node (*TweetCount*), which includes the measures (*#friends, #followers, #status, #favorited* and *#listed*) and a degenerated (i.e., consisting of a single data field) dimension (*FactID*). Dimensions are modeled as outgoing paths of the fact

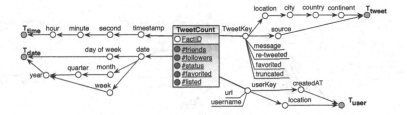

Fig. 2. Fragment of the tweet record in x-DFM

type node with edges as "rolls-up-to" relationships between hierarchy levels. Multiple aggregation paths are possible within a dimension, all converging in an abstract ⊤ node, which corresponds to the aggregated value *all*. A level node in a dimension consists of at least one key attribute, but may include further attributes shown as underlined terminal nodes.

In general, a datacube can be extended by adding new elements of type a) *measure*, b) *dimension*, or c) *hierarchy level*. Adding a new element can be rather trivial if its value is derived from the values of other elements within the same fact entry. We are interested in discovering non-trivial and hidden relationships in the dataset such as those that cannot be expressed by a derivation formula. Our approach is to apply data mining algorithms for discovering clusters or rules useful for defining new elements in the cube. For this purpose, the input set has to be transformed into a representation more generic than the one offered by the multidimensional model. The goal is to treat all elements symmetrically as potential input fields for discovering new categories. To achieve this, we "homogenize" the graph model of the cube and get rid of different types of nodes and edges based on the observation that all edges are of type "many-to-one" or "one-to-one" and all nodes are of type attribute.

Figure 3 (a) shows the transformed graph from Figure 2, describing the cube in terms of attributes and hierarchical relationships between them. The new graph is centered at the fact identifier attribute *FactID*, which uniquely identifies each fact entry (this may be an artificially generated attribute). The obtained representation of a data cube is suitable for specifying the input set for data mining algorithms by selecting a relevant subgraph and extracting the data behind it.

Consider an example of adding a dynamic category *re-tweet activity* to the *user* dimension defined as the frequency of re-tweeting relative to the period ellapsed since the creation of the user's account. This category should assign each user into one of four clusters: *mature-active*, *new-active*, *mature-passive*, and *new-passive* for users who registered long ago or recently and who re-tweet more or less frequently, respectively. Neither the time elapsed since the user registration nor the frequency of re-tweeting is explicit in the data set, but both are derivable from other data fields.

Figures 3 (b), (c), and (d) demonstrate the steps of obtaining the new aggregation path. Figure 3 (b) shows the subgraph relevant for discovering the desired category. Figure 3 (c) shows the derivation of the required fields *time elapsed*

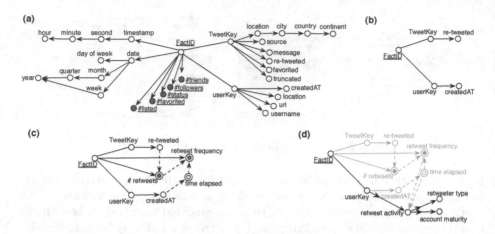

Fig. 3. Stages of acquiring new hierarchy levels

as the difference between the current and the account creation date, cumulative measure *# retweets* as the number of messages with the *re-tweeted* value set to true, and, finally, *retweet frequency* as *# retweets* divided by *time elapsed*. Figure 3 (d) shows the result of adding *re-tweet activity* as a hierarchy level to the *user* dimension. In the conceptual model, a discovered or derived category can be treated just as an ordinary one. For instance, we added parallel hierarchy levels *retweeter type* with member values *active* and *passive* and *account maturity* with member values *new* and *mature* on top of *re-tweet activity*. So far we have considered the presentation of discovering new structural elements at the the conceptual level in order to provide an abstract, generic and implementation-independent view on the data. However, there are significant differences in the behavior of static and dynamic elements in terms of their maintenance and usage in OLAP queries, as elaborated in the next section.

3   Maintenance Strategies for Dynamic Categories

Classically, dimensions in a data cube correspond to non-volatile characteristics of the data. This property ensures consistency and validity of pre-aggregation. In reality, however, the instance or even the structure of a dimension may evolve in time. The problem of *Slowly Changing Dimensions* (SCD)[9] is well elaborated in the literature, with various strategies proposed for maintaining the up-to-date or the historical view, or even both. More sophisticated strategies employ some kind of multiversioning to preserve various states of the aggregates. Dynamic dimensions proposed in our work may be considered a special case of SCD, in which the changes occur in a predictable fashion: discovered categories reflect a particular state of the cube and as such, have to re-computed on a regular or ad-hoc basis to stay consistent with the evolution of the underlying data set. Preservation of all previous states of a dynamic dimension appears crucial for correct aggregation. With this scheduled update behavior, the SCD methodology

Type 4 [9] appears an appropriate implementation option. This method offers unlimited history preservation by creating multiple records for a given natural key and storing the temporal validity bounds for each entry in history tables.

Another challenge is the recomputation of the dynamic category itself. Frequent and complete recomputation may impose an unaffordable burden on the system. A performance gain can be achieved by re-using the outcome rules of the data mining routines used for discovering the category. In our example, we could use the previously established threshold values for *account maturity* and *retweet frequency* for refreshing the assignment if *user* entries to *re-tweet activity*. This way, the data does not need to be mined repeatedly and the maintenance is reduced to simple computations and adjustments within the existing clusters.

A problem specific only to discovered categories is how to assign new member values in a dimension to the parent values of such a category. Depending on the definition of the discovered relationship, either a default assignment should be provided (for example, newly registered users are most likely to fit into *new and passive* cluster of re-tweet activity), or, if the rules of the dynamic assignment are available, these can be applied for assigning the new values.

Finally, there is a problem of quering the data along dynamic categories in the presence of its multiple versions of a dimension hierarchy. In our scenario, it is important to ensure correct analysis by matching the timeframes of the queried facts and those of the applied dimension hierarchies. For example, if we analyse user activity patterns in 2010 by applying the re-tweet activity hierarchy computed in 2012, we will obviously end up with historically incorrect aggregate values. A consistent result can be achieved by the matching the temporal characteristic of each fact entry with the matching version of the dynamic dimension hierarchy. The SCD implementation of Type 2 offers exactly this type of matching for ensuring historically correct aggregation.

4 Demonstration

We implemented the data warehouse for Twitter analysis using the Microsoft SQL Server system with its powerful set of analysis services including OLAP and data mining. We see a big gain in the ability to employ the existing DW technology and tools for enabling discovered dimensions. The dataset for the experiments was obtained via the Twitter Streaming API, which provides 10% of the total public stream of Twitter. We proceed by presenting two cases of discovering new categories in the process of analyzing events on Twitter.

Case 1 - Spatio-temporal analysis of tweeting during the Super Bowl 2012[2]. 2012's Super Bowl XLVI has been of much interest to many, not only sports fans but also to the social network analysts, as it was the top tweeting event to date, with its record value of 12,233 tweets per second. Tweets relevant to this event and with time-bounds of the game were extracted. One task was to find the top (i.e., with the highest number of tweets sent) tweeting cities in the

---

[2] The Super Bowl is the annual championship game of the National Football League (NFL), the highest level of professional American football in the United States.

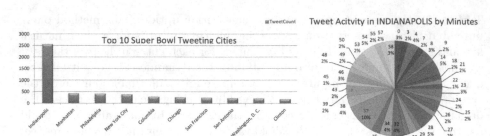

Fig. 4. Twitter Activity during the Superbowl 2012

US during Super Bowl championship. For this purpose, the *FactCount* cube was extended by the measure *TweetCount* and a hierarchical dimension *geolocation*. The input facts were filtered to the tweets originating from the *USA*.

The tweet activity of top 10 cities is plotted in the chart on the left hand side of Figure 4. *Indianapolis*, the city that hosted Super Bowl 2012 championship, remained the most active city during this game with 2559 tweets. The game venue has capacity for 70,000 spectators, which is a contributing factor to make *Indianapolis* the top tweeting city. One other task was to see peak activity along the timeline for the city hosting the championship. The chart on the right hand-side in Figure 4 plots Twitter activity by minutes only for Indianapolis where most tweets were sent in the 37th minute.

Case 2 - Types of Twitter users by geographic regions. Users on Twitter engage in many activities including 1) posting tweets 2) (un-)marking tweets as favorite 3) making other users friend, and 4) (un-)following other users. Our task was to explore geographical regions based on such activities. Figure 5 depicts the outcome of this analysis.

The first pie-chart shows the distribution of Activity (tweeting / status updates) by continent. South America with 37% share is the top active continent followed by North America with 26%. Please note that users on Twitter can exclude location specific data from the tweet. BLANK represents such tweets in the chart. The second pie-chart plots regions by favoriting activity. North Americans lead the way with 32% share and are followed by Asians with 26%. The third pie-chart shows regions by friendship. North American have most friends with 34% share. The last chart shows regions by number of followers with South America having 40% of the total and North America having about 26% of the total number followers, respectively. The mining structure consisted of fields *UserID, User-Created-At, Language* from *UserDIM* dimension and all the measures in the fact table. The mining model, however, contains *User-Created-At*(a Date field) and *StatusCount*. Microsoft Clustering Algorithm was configured to use scalable K-Means method and to have 4 clusters as to correspond to *Active & New, Passive & New, Active & Mature* and *Passive & Mature* categories.

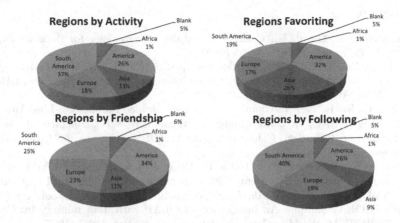

Fig. 5. Exploration of geographical regions by user activity

The presented cases demonstrate the advantages of coupling OLAP with data mining for discovering and analyzing dynamic data characteristics. Re-using the mining results as aggregation paths in OLAP queries enables new insights, which could not be obtained without the feedback loop at the level of data modeling.

5   Conclusions and Future Work

In this work we proposed to extend the classical approach to modeling OLAP dimensions by the inclusion of dynamic categories and hierarchies discovered from the data through the application of data mining algorithms and other computations. The discovered classifications reflect "hidden" relationships in the data set and thus represent new axes for exploring the cube's measures. We handled the process of adding discovered categories at the conceptual modeling level by transforming the cube schema into a homogeneous graph consisting of attribute nodes and hierarchical relationships between them. This representation allowed us to treat measures and dimensions symmetrically for the purpose of discovering interesting relationships and grouping options.

We tested our approach on the dataset of the Twitter's public stream focusing the analysis on the metadata represented by over 60 data fields about the message and its author. The presented application scenarios demonstrate how the original data can be enriched by discovered knowledge about the dynamic characteristics of the data set, such as activity and popularity of the users, Twitter usage patterns by geographical distribution, emergence and dissemination of events, etc. In contrast to the standard application of data mining tools where the outcome is used as the final result, we provide a feedback loop to integrate the obtained groupings into the data cube as additional aggregation paths. We expect our approach to enhancing multidimensional cubes with dynamic hierarchy paths to be a valuable contribution for numerous OLAP applications.

References

1. Chaudhuri, S., Dayal, U., Ganti, V.: Database technology for decision support systems. Computer 34(12), 48–55 (2001)
2. Chen, J., Nairn, R., Nelson, L., Bernstein, M.S., Chi, E.H.: Short and tweet: experiments on recommending content from information streams. In: Proc. CHI, pp. 1185–1194. ACM (2010)
3. Dehne, F., Eavis, T., Rau-Chaplin, A.: Coarse Grained Parallel On-Line Analytical Processing (OLAP) for Data Mining. In: Alexandrov, V.N., Dongarra, J., Juliano, B.A., Renner, R.S., Tan, C.J.K. (eds.) ICCS 2001. LNCS, vol. 2074, pp. 589–598. Springer, Heidelberg (2001)
4. Dzeroski, S., Hristovski, D., Peterlin, B.: Using data mining and OLAP to discover patterns in a database of patients with y-chromosome deletions. In: Proceedings of the AMIA Symposium, p. 215. American Medical Informatics Association (2000)
5. Han, J.: OLAP mining: An integration of OLAP with data mining. In: Proc. of the 7th IFIP 2.6 Working Conf. on Database Semantics, DS-7 (1997)
6. Han, J., Chee, S., Chiang, J.: Issues for on-line analytical mining of data warehouses. In: Proc. of the Workshop on Research Issues on Data Mining and Knowledge Discovery, Seattle, Washington, pp. 2:1–2:5 (1998)
7. Hecht, B., Hong, L., Suh, B., Chi, E.H.: Tweets from justin bieber's heart: the dynamics of the location field in user profiles. In: Proc. CHI, pp. 237–246 (2011)
8. Java, A., Song, X., Finin, T., Tseng, B.: Why we twitter: understanding microblogging usage and communities. In: Proceedings of the 9th WebKDD and 1st SNA-KDD 2007 Workshop on Web Mining and Social Network Analysis, pp. 56–65. ACM (2007)
9. Kimball, R.: The Data Warehouse Toolkit: Practical Techniques for Building Dimensional Data Warehouses. John Wiley & Sons, Inc., New York (1996)
10. MacLennan, J., Tang, Z., Crivat, B.: Mining OLAP Cubes, ch. 13, pp. 429–431. Wiley Publishing (2008)
11. Malinowski, E., Zimányi, E.: Hierarchies in a multidimensional model: From conceptual modeling to logical representation. Data & Knowledge Engineering 59(2), 348–377 (2006)
12. Mansmann, S.: Extending the OLAP Technology to Handle Non-Conventional and Complex Data. PhD thesis, Konstanz, Germany (2008)
13. Mansmann, S., Scholl, M.H.: Empowering the OLAP Technology to Support Complex Dimension Hierarchies. International Journal of Data Warehousing and Mining 3(4), 31–50 (2007) (Invited Paper)
14. Mathioudakis, M., Koudas, N.: Twittermonitor: trend detection over the twitter stream. In: Proceedings of the 2010 International Conference on Management of Data, pp. 1155–1158. ACM (2010)
15. Michelson, M., Macskassy, S.A.: Discovering users' topics of interest on twitter: a first look. In: Proceedings of the Fourth Workshop on Analytics for Noisy Unstructured Text Data, AND 2010 (in Conjunction with CIKM 2010), Toronto, Ontario, Canada. ACM (October 26, 2010)
16. Phelan, O., McCarthy, K., Smyth, B.: Using twitter to recommend real-time topical news. In: Proceedings of the Third ACM Conference on Recommender Systems, pp. 385–388. ACM (2009)
17. Usman, M., Asghar, S., Fong, S.: A conceptual model for combining enhanced OLAP and data mining systems. In: Fifth International Joint Conference on INC, IMS and IDC, NCM 2009, pp. 1958–1963. IEEE (2009)
18. Zhu, H.: On-line analytical mining of association rules. PhD thesis, Simon Fraser University (1998)

BIAccelerator – A Template-Based Approach for Rapid ETL Development

Reinhard Stumptner[1], Bernhard Freudenthaler[2], and Markus Krenn[1]

[1] BIA Business Intelligence Accelerator GmbH, Softwarepark 26,
4232 Hagenberg, Austria
{r.stumptner,m.krenn}@BIAccelerator.com
[2] Software Competence Center Hagenberg GmbH, Softwarepark 21,
4232 Hagenberg, Austria
bernhard.freudenthaler@scch.at

Abstract. Business Intelligence projects include a big variety of challenges for all involved parties. Due to individual and changeable customer requirements or lacking standardization, there is much legwork which BI developers or consultants have to spend time on. BIAccelerator's template-based ETL development approach tries to tackle this issue by providing developers with a flexibly extendable environment which helps to quickly set up prototypes and which allows simplified change management. The tool allows importing metadata from source or destination data storages, adding additional knowledge (e.g. source - destination mappings or definition of Slowly Changing Dimension attributes) and generating executable ETL packages based on custom templates and collected metadata. This way, any changes (e.g. modification of source/destination entities or changes in the ETL processes) can easily be deployed.

Keywords: Template-based ETL, Business Intelligence, Data Warehousing, Data Integration.

1   Introduction and Motivation

Business Intelligence (BI) consultants are confronted with a lot of challenges during their daily work, e.g. customer requirements are very individual, therefore (conventional) ETL templates cannot be used in a reasonable way and standardized procedures rarely exist. Furthermore, networking between BI consultants from different companies is not often done.

Also, during the ETL development there exist some challenges, e.g. transformation rules are not known in the analysis phase, bad data quality, high complexity, large number of tables as well as transparency, maintainability and documentation [3], [6].

SQL Server Integration Services (SSIS), a tool from Microsoft's BI bundle, is a secure and reliable ETL platform with high performance and scalability. High usability can be achieved with a comprehensive development environment, source control, visual debugging and a comprehensive library for transformation etc. Furthermore, SSIS is

L. Chen et al. (Eds.): ISMIS 2012, LNAI 7661, pp. 435–444, 2012.

extensible e.g. with user defined tasks, source/transformation/destination components and so on (e.g. access to SAP data). But, there also are a lot of challenges when developing with Microsoft SSIS:

- *No satisfactory storage for ETL-relevant information*, e.g. for mappings, transformation rules etc. An integrated change management support or a metadata repository in SSIS does not exist.
- *For development of SSIS packages*, a lot of legwork is necessary, e.g. data type conversions, mappings. And concepts for dynamisation could be more powerful.
- *Individuality of BI projects* and therefore *high costs*.
- *Templates for standard ETL processes* do not exist. Standard process models can rarely be used.

To tackle these challenges a new software and framework called BIAccelerator was developed. BIAccelerator supports the user with standardized procedures and helps to realize BI projects faster. The vision of the approach is an application which lets the developer model BI solutions on a high level and generates data bases and corresponding load processes automatically. Thereby, all relevant information about the involved systems should be recorded, all definitions like names of attributes, transformation rules or dynamic statements (the dynamic statement concept will be explained later in the text) are entered e.g. through different kinds of graphs or diagrams and finally, the data base(s) and load packages of the BI system should be generated with the generator module of BIAccelerator.

The current version of BIAccelerator (1.0) does not support the automated generation of data schemes, but the SSIS package generator is fully featured. Objects can be generated and re-generated very quickly at any time taking into account possible changes of data schemes or changes to the ETL process. Furthermore, BI developers have an optimally documented system and a significantly higher flexibility by using the automated generation from the beginning of a BI project.

BIAccelerator's aim is to drastically reduce development time for ETL packages of BI systems. The time DT which can be saved depends on the relation between the number of automatically generated packages GP, the number of manually created packages MP and the number of templates T. If a template implements a certain ETL process, then in case of any changes only the template has to be modified and the according packages are re-generated. Thus, the expected time saving rate DT can be estimated as follows:

$$DT = 1 - \frac{T + MP}{GP + MP} \quad (\text{T} \leq \text{GP})[1] \tag{1}$$

For instance, if the number of manually created packages is 20 (MP), the number of automatically generated packages is 50 (GP) and the number of used templates is 10 (T), then a reduction of ETL development time for about 57% can be expected.

---

[1] (GP + MP): Number of built packages in case of manual development.

(T + MP): Number of built packages in case of template-based development.

This contribution is structured as follows. Section 2 gives an overview of related work or related tools. Section 3 introduces the template-based ETL modeling approach and in section 4 the technical description of the BIAccelerator architecture is presented. Finally, some results are shown in section 5 and conclusion and intended future work are discussed in section 6.

2   Related Research

There exist a certain number of extract, transform, and load (ETL) [6] package generators, but most of these generators create packages from provided metadata.

Vulcan[2], a small Business Intelligence Framework available on the Open Source Project Community Codeplex[3], generates SSIS packages from provided definitions, which makes it possible to create BI template solutions in some sense. Extending this approach, the current version of BIDS Helper[4] offers the functionality to create packages from a metadata file (in BIML – Business Intelligence Markup Language – format[5]). Hereby metadata refers to data which describes (and only describes) the designated SSIS package by means of defining all SSIS objects (tasks, components of data and control flow etc.) and relations between them (dependencies, data flow). This approach allows the user to edit the text-only representation of packages, having the operations of word processing tools available. One may think that this is not a big advantage, but even the simple "replace" functionality can be a big advantage and spare a lot of legwork. However, the main idea of BIAccelerator is different. It basically relies on Microsoft's metadata-driven ETL approach [1], which shall be introduced in the following.

Microsoft SQL Server 2008 Integration Services (SSIS) is an environment which allows to deploy and to manage high-performance data integration solutions like data warehouses or business intelligence applications in general. The ETL process has the goal of bringing heterogeneous and asynchronous data sources into a homogeneous environment [4]. SSIS offers a very powerful environment for developing ETL applications, but still there are some issues which SSIS developers face, especially when building large data warehouses. Metadata-driven ETL tries to tackle these issues.

The metadata-driven ETL process is described as follows and shown in figure 1 (according to [1]):

1. Definition of source and destination of the ETL process (databases, tables, etc.).
2. Automatic or manual collection of data schemes (columns, index, constraints).
3. The mapping between source and destination entities has to be defined.
4. Choose full or delta load.
5. For each entity the steps for performing the extraction have to be defined.

---

[2] http://vulcan.codeplex.com/
[3] http://www.codeplex.com/
[4] http://bidshelper.codeplex.com/
[5] http://www.varigence.com/documentation/biml

6. An execution plan for the ETL packages is provided.
7. The system creates executable SSIS packages.
8. The system deploys the packages to the execution environment.
9. The system executes the ETL packages and logs their status.

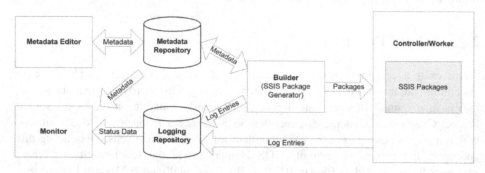

Fig. 1. Metadata-driven ETL Platform Architecture [1]

The design goals of a metadata-driven ETL platform aim at "improving the productivity of developers, enforcing ETL standards, supporting a cost-effective way to deploy large data warehouses on commodity hardware, and providing a centralize metadata repository for lineage tracking" [1].

The key components of the platform are [1]:

- *Metadata repository*: Store for ETL definitions, which describe data sources and destinations, source-destination mappings, data transformations, and orchestration processes.
- *Metadata editor*: This component is the user interface to edit data from the metadata repository.
- *Builder*: The builder is a SSIS package generator, which creates packages based on the definitions from the metadata repository.
- *Controller/worker runtime environment*: Distributed execution of SSIS packages and unified logging management.
- *Logging repository*: Store for package execution logging data.
- *Monitor*: Monitoring and reporting of the ETL execution status.

Basically BIAccelerator follows the main concepts of metadata-driven ETL approach, but still it offers more flexibility especially in steps 4 and 5 of the metadata-driven ETL process. Hereby the tool lets the ETL developer design her/his individual ETL processes, which can be defined as templates for the certain extraction processes. This concept is introduced in the following section.

3   Template-Based ETL Modeling Approach

The starting point of the template-based ETL process is a "normal" SSIS package. The tasks and data flows of the package work correctly (necessary tests executed) for a certain source-destination relationship. Let us assume that there exist a number of source-destination relationships with similar requirements (commonly given in the staging layer of a data warehouse). What one could do is to dynamise certain properties of the ETL-Package in order to make it work for all these relationships. Which properties have to be dynamic? Properties of the source component for instance. In case of a source database this would be a table (or view, procedure, etc.) name or a SQL select statement or a file name in case of flat file sources.

Statement Type	SSIS Object	Dynamic Statement
Substitution of SSIS-Object-Properties	Executables.Update Statistics.SQLStatementSource	Update Statistics of destination table
Substitution of SSIS-Object-Properties	Executables.Update DestinationTable.SQLStatementSource	Update changed records
Substitution of SSIS-Object-Properties	OLE DB Source.OLE DB Source.SqlCommand	Select new records
Substitution of SSIS-Object-Properties	OLE DB Destination.OLE DB Destination.OpenRowset	Destination table
Update of Mappings in Destination-Task	Executables.Data Flow Task.Components.OLE DB Destination	Destination Mapping

Fig. 2. Dynamisation of SSIS Objects

SSIS templates consist of a set of static and dynamic objects. Consequently all instances have static and individual ETL process components. BIAccelerator supports dynamisation of properties of frequently used SSIS objects. Thereby the user has to specify, which properties of which SSIS objects have to be dynamised in which way. Figure 2 illustrates this approach.

The properties are dynamised by following a special syntax (dynamic statements), written in T-SQL and supporting special dynamic tags, e.g. "{SourceTable}". The following example shall illustrate the idea of dynamic SQL statements.

Let us assume we have an SSIS package which contains a task (e.g. an "Execute SQL Task") which should update a table with data from a certain source. In our metadata repository we defined mappings for a "Customer" entity as follows:

Table 1. Metadata and Definition

Source (Database-name: L1)			Destination(Database-name: L2)		
dbo.Customer			**dbo.Customer**		
PK	CUST_ID	varchar(12)	PK	CustomerID	varchar(12)
	CUST_NAME	nvarchar(80)		CustomerName	nvarchar(80)
	ADD_CITY	nvarchar(50)		AdressCity	nvarchar(80)
	ADD_STR	nvarchar(80)		AdressStreet	nvarchar(80)
Mapping					
		CUST_ID	CustomerID		
		CUST_NAME	CustomerName		
		ADD_CITY	AdressCity		
		ADD_STR	AdressStreet		

The following SQL statement performs an update as described before:

```
UPDATE d
SET d.{DestinationNonPKColumn} = s.{SourceNonPKColumn}
FROM {SourceTable} s
INNER JOIN {DestinationDatabase}.{DestinationTable} d
  ON s.{SourcePKColumn} = d.{DestinationPKColumn}
```

Now, having assigned this statement to the property of a SSIS object ("SQL Statement" of the "Execute SQL Task"), the concrete SQL statement can be deduced from dynamic statement and metadata:

```
UPDATE d
SET d.CustomerName = s.CUST_NAME, d.AdressCity =
s.ADD_CITY, d.AdressStreet = s.ADD_STR
FROM L1.dbo.Customer s
INNER JOIN L2.dbo.Customer d ON s.CUST_ID = d.CustomerID
```

For the according source-destination mapping the concrete SQL statement could be created. Now, one can imagine that this generation of statements can be applied to a set of n source-destination mappings. What could be done is to create n SSIS packages containing "Execute SQL Task" objects with the according SQL statements to update the particular tables defined by the mappings. This is the basic principle of the template-based ETL modeling approach. Every time, when a dataflow can be applied to a number of n source-destination relations, the template-based modeling approach is a useful procedure which helps to save time and costs. Because the developer can design a template SSIS package for one of the relations, verify its correctness and reproduce it for the rest of the relations. This reduces not only the development time, but also testing efforts, because generated packages in principle imitate the functionality of their template package, which generally can be assumed correct. Besides reducing development time, the template-based ETL modeling approach also aims at simplified change management. In case of changed source or destination entities (e.g. additional or removed attributes, changed data types, etc.), the SSIS developer performs these changes in one place only – in the metadata. After having adapted all affected metadata entries (all source destination mappings in particular), the user simply re-generates and deploys all concerned ETL packages. Furthermore, the application of changes in the ETL process is drastically simplified, because such changes are applied to the template and can be propagated to its derived packages automatically. However, relying on the BIAccelerator approach one can efficiently and flexibly generate 1:1 copies of a dataflow for multiple source-destination relations. Of course, in practice 1:1 copies of a package rarely are directly applicable - without having to do any adaptions. Therefore, an extension of the template-based ETL approach will be introduced in section 6, which should allow metadata-driven assembling of ETL packages based on predefined template SSIS snippets. Nevertheless, due to the homogeneous nature of ETL processes in the staging layer of a data warehouse, the generation of 1:1 copies of ETL processes is well applicable.

4   Technical Description

BIAccelerator is fully integrated into Visual Studio, Microsoft's well-known software development environment, or rather into the Microsoft Business Intelligence Development Studio (BIDS). It shall be mentioned that the template-based ETL modeling approach is a general one. However, BIAccelerator is an implementation for the Microsoft BI environment. Implementation and testing of the SSIS packages is done in Visual Studio – as usual. Furthermore the user can apply dynamisation concepts to her/his tasks and components using the same environment. The following process model describes the fundamentals of the approach on the development of a BI system with BIAccelerator:

1. *Requirements analysis*: Systematic collection of all requirements and storing them in the metadata repository.
2. *Import of metadata from data sources and analysis of data quality*: All required metadata from relevant data sources are imported using the according wizards. A metadata provider component is used for accessing different systems and their data (metadata particularly).
3. *Definition of transformation rules*: All relevant metadata on source and destination systems imported via respective wizards. Afterwards mappings and transformation rules are specified.
4. *Generation of SSIS packages*: A main part of the needed SSIS packages can be generated automatically with the SSIS generator by using the defined transformation rules and mappings. A set of predefined SSIS templates (frequently used ETL processes – best practice) are delivered together with the tool.
5. *Test*: The BI system is tested. Testing is simplified because packages which were derived from a correct template package normally should work correctly.
6. *Documentation*: With the use of integrated reports the BI system basically is well-documented automatically. Furthermore, technical descriptions and general information can be added.

On the basis of this process model, figure 3 describes the BIAccelerator architecture with its modules:

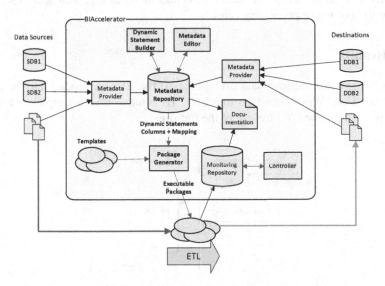

Fig. 3. BIAccelerator Architecture

- *Metadata repository*: The BIAccelerator framework manages information about data sources and metadata in an internal data structure. This metadata repository is used to store e.g. technical and business metadata, destination definition, data movement and pattern definition, and orchestration process definition.
- *Dynamic statement builder:* The dynamic statement builder is a wizard which helps the user to define and (context-dependently) test dynamic statements (see section 3 for details on the dynamic statement concept).
- *Metadata editor*: The metadata editor offers the possibility to a user to edit imported metadata and to add her/his custom definitions in the central metadata repository. Furthermore, one has the possibility to propagate changes in source systems (e.g. new/deleted tables/attributes, changed data types, etc.) to the metadata repository and to get a mark on affected templates or packages.
- *Package generator*: The builder is designed to automatically generate SSIS packages and instances based on metadata definitions. All possible data sources and destinations are represented by a manageable number of classes to support all data objects from SSIS. The package generator generates the data models, the BIA data bases, the ETL processes and the client modules.
- *Metadata provider*: The metadata provider is used to access and extract metadata from the certain target systems (SQL Server, Oracle, SAP, flat files, etc.).
- *Controller*: The controller module provides a user with an interface to monitor and report the warehouse's status.
- *Templates*: A set of pre-defined and user-defined template SSIS packages.
- *Monitoring repository*: SSIS packages write their execution status information to uniform data storages (monitoring repository). Reports can be generated to show the current and historical statistics of a warehouse in easily consumable formats.
- *Documentation*: Generally, documentation automatically is up-to-date (because it is derived from metadata and definitions) and includes e.g. technical descriptions of the BI solution or processing rules.

5   Results

In the following test scenario data from about 20 flat files are loaded into the staging layer of our Data Warehouse (see figure 4). To achieve this, a template package is created performing the ETL for one of the flat files.

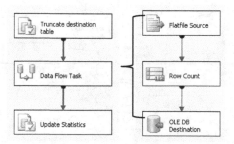

Fig. 4. Template Package

In a second step some of the task's and component's properties are dynamised as follows:

Statement Type	SSIS Object	Dynamic Statement
Definition of Attributes in Flatfile-Source	Executables.Data Flow Task.Components.Flatfilequelle	Source Attributes
Substitution of SSIS-Object-Properties	Executables.Truncate destination table.SQLStatementSource	Truncate destination table
Substitution of SSIS-Object-Properties	Executables.Update Statistics.SQLStatementSource	Update Statistics of destination table
Substitution of SSIS-Object-Properties	Global Variables.txtFile_Name.Value	Flatfile name
Substitution of SSIS-Object-Properties	OLE DB Destination.OLE DB Destination.OpenRowset	Destination table
Update of Mappings in Destination-Task	Executables.Data Flow Task.Components.OLE DB Destination	Destination Mapping

Fig. 5. Dynamisation

According to figure 5, source attributes should be set in the flat file source component, the truncate table SQL statement (see figure 6) is assigned to the according Execute SQL task and the same for update statistics, and so on. Finally, the source destination mapping is set in the OleDb Destination component.

Statement for [dbo].[AllocationKeys]

TRUNCATE TABLE [dbo].[AllocationKeys]

Close

Fig. 6. Truncate Destination Table

Having defined this and having specified source destination mappings (see figure 7) for all files and target tables, packages for the rest of the flat files (one separate package for each) can be generated automatically.

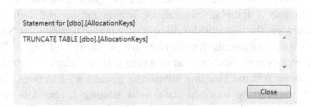

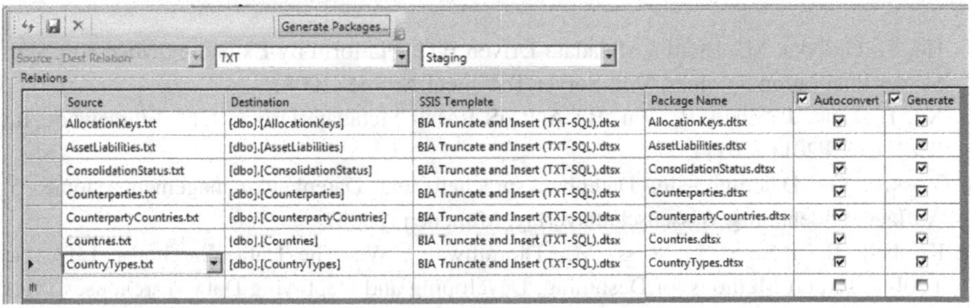

Fig. 7. Source Destination Mapping

Executing these packages, the certain destination tables are filled with the according data without having to apply any changes to the generated packages.

6   Conclusion and Future Work

Template-based ETL modeling promises reduced development and maintenance efforts in all stages of Data Warehousing projects. Due to the reason that main testing efforts can be transferred from the individual packages to their templates, implementation and change management consumes much less time. Packages which were deduced from correctly working templates generally can be assumed correct. And of course the overall development time for SSIS packages is reduced, because each type of dataflow (ETL process) is modeled one single time and accordingly duplicated. Especially in the staging layer dataflows frequently are simple 1:1 data-copy-processes, which can be supported very well by the template-based ETL modeling approach. Besides this, based on reports from the metadata and logging repository, the Data Warehouse's documentation always is up-to-date and available for all involved persons or parties (developer, administrator, management, end-user, etc.). The dynamic statements concept guarantees according flexibility and broad applicability. This quite powerful concept makes it possible, that each and any SSIS package can be defined to be a template by the developer. As soon as a dataflow is applicable to more than one dataflow (source-destination relation), it is more efficient to use a template instead of implementing the individual packages manually. Furthermore, the concept should be easily understandable for SSIS developers, because the dynamic statements themselves are written in T-SQL. BIAccelerator's current further development efforts aim at supporting BI developers with integrated semi-automatic warehouse scheme building (database generator) functions. In many cases, e.g., in case of transferring external data to a staging layer, data schemes in the warehouse are easily deducible from given schemes. In such cases BIAccelerator should provide automatic schema generator functionalities and generate the needed ETL packages. Besides this, also more complex scheme generation should be supported which can be applied to data transfer processes between staging and data warehouse layers.

References

1. He, T., Gudyka, M.: Build a Metadata-Driven ETL Platform by Extending Microsoft SQL Server Integration Services. SQL Server Technical Article (2008)
2. Xu, L., Liao, J., Zhao, R., Wu, B.: A PaaS Based Metadata-Driven ETL Framework. In: IEEE CCIS 2011 (2011)
3. Vossen, G.: Datenmodelle, Datenbanksprachen und Datenbankmanagementsysteme. 5. Auflage. Oldenbourg Wissenschaftsverlag, München (2008)
4. Kimball, R., Reeves, L., Ross, M., Thornthwaite, W.: The Data Warehouse Lifecycle Toolkit: Expert Methods for Designing, Developing and Deploying Data Warehouses. John Wiley & Sons, New York (1998)
5. Elmasri, R., Navathe, S.B.: Fundamentals of Database Systems, 5th edn. Addison-Wesley, Boston (2007)
6. Kimball, R., Ross, M.: The Data Warehouse Toolkit: The Complete Guide to Dimensional Modeling, 2nd edn. John Wiley & Sons, New York (2002)

BPMN Patterns for ETL Conceptual Modelling and Validation

Bruno Oliveira and Orlando Belo

ALGORITMI R&D Centre, University of Minho, Braga, Portugal
id4103@alunos.uminho.pt
obelo@di.uminho.pt

Abstract. ETL systems continue to suffer from a lack of a simple and rigorous approach for modelling and validation of populating processes for data warehouses. In spite of the efforts that researchers have been done, there is not yet a convinced and simply approach for modelling (conceptual and logical views), validating and testing ETL processes before conduct them to implementation and roll out. In this paper we explored the use of BPMN for ETL conceptual modelling and validation. Basically, we intended to provide a set of BPMN meta-models (patterns) especially designed to map standard data warehousing ETL processes and testing them before constructing the final system. We think this is a practical approach to reduce significantly the inadequacy of an ETL system in its real world scenario. A surrogate key pipelining process was selected as a case study to demonstrate the use and utility of the ETL modelling approach presented here.

Keywords: Data Warehouses, ETL Conceptual Modelling, Verification, Validation, Analysis and Testing of ETL Systems and BPMN.

1 Introduction

Software systems modelling [10, 13] is a delicate activity, often complex, requiring qualified personnel with proven experience and knowledge. The models produced are valuable tools whose utility can be demonstrated in almost all phases of a software system development process. The level of abstraction that conceptual models provide us is very useful in any kind of conversation we may have with future users of the system under development. They are an excellent groundwork for validation of software systems' requirements and functioning models. As such, it is not surprising the great importance they have in the daily life of every professional that develops its activity designing and implementing software systems. As we know, ETL *(Extract, Transform, and Load)* systems [7] are very particular systems of software. Their life cycle includes the steps of most typical phases of any software process development. However, their specific features, transform them into a very special piece of software, frequently complex and very difficult to implement. All this because, not only its operational requirements are different from other types of software – usually, ETL systems are running in batch, requiring strong mechanisms for treatment and error recovery, they are high concurrent processes and potentially parallel, generally act as

L. Chen et al. (Eds.): ISMIS 2012, LNAI 7661, pp. 445–454, 2012.

opportunistic and autonomous systems, etc. – but also because in most cases its operation and control is performed only by a single user with very high credentials: the data warehousing system administrator; normal users do not act directly on this system. Every ETL system is a particular case. We could even say that there aren't two separate data warehouses with the same ETL system. Each warehouse has its own user community, people having different ways of acting and thinking, and linked, in some way, to a specific set of decision-making processes. These characteristics make these users specific data clients. Different ways of decision making, usually require different dimensional models, and therefore different data sets and decision models [15]. Thus, even if we adopt a same standard solution for the implementation of a data warehousing system (which happens often in areas such as telecommunications, banking, or retail, for instance), we just have a different user community, and so we need to adjust the original data models already built in the solution accordingly the requisites of those decision makers. Changing dimensional models implies to change ETL systems. This does not help anything in an ETL system building process. It is known that long that the success of a data warehousing system depends heavily on the adequacy of its ETL system, which impose an extreme care and concern over its architects and software engineers in its planning, architecture, design, and implementation.

With this work we tried to study a method that could help to attenuate the negative effects of a less suitable planning for the development of an ETL system. Accordingly, we developed a set of meta models using BPMN (*Business Process Model Notation*) [5, 11] specifications for some ETL processes that we recognize as standard, which means that their can be found commonly in almost ETL systems for real world applications. The choice of BPMN 2.0 was mainly due to the clarity of its simplicity of notation for process representation, coupled with his power of expressiveness, implementation capacity, and control tasks included in a model. Thus, we can specify any ETL process conceptually, in a very concrete way, and then validate it by running the model that we defined in BPMN. Then, we can assess, not only how to analyse their results, but also we can evaluate *a priori* their performance. In this article we expose and explain our approach to conceptual modelling of ETL systems, demonstrating its feasibility and usefulness through a case study involving one of the most common process that usually occurs in any ETL system: a surrogate key pipelining process. So, after a brief exposure of some related work with this specific issue (Section 2), we present how we can build a meta model for an ETL system with its notation, and then introduce and explain the conceptual model built for the case study referred as well as analyzing its implementation and execution (Section 3). Finally, we end the article with some pertinent findings and indicating some directions to follow in terms of future work (Section 4).

2   Related Work

The use of BPMN in the specification of ETL systems conceptual is not new. As far as we know, Akkaoui and Zimány presented a paper [1] in 2009 exploring this approach, proposing a conceptual model for a specific ETL system developed using the BPMN notation, and showing how such model could be implemented through the use of BPEL (*Business Process Execution Language*) [4]. Their approach was very interesting, showing it was possible to construct properly a conceptual model in BPMN, easy to read and to understand, for an ETL system. In fact, using a very practical way, they showed us how the BPMN notation, originally designed for modelling business processes, could be successfully adopted in modelling systems for populating data warehouses. Furthermore, they also implemented successfully, the major drawback that usually exists between modelling conceptual specifications and their implementation, demonstrating how we can execute them using BPEL. In turn, the authors of [6] introduced a layered methodology to support the design of ETL processes for operational business intelligence systems using a unified formalism. Basically, following their methodology's steps, we start modelling business processes and their services' requirements and objectives, proceeding forward passing by model logical design to its physical implementation. Later, in [16] was exposed a very practical way to guide BPMN specifications for the definition of conceptual models of ETL systems. In that paper, the authors presented a set of BPMN components specifically oriented for ETL tasks specification. However, the materialization of the specifications produced still some way off from its practical implementation. The gap between system modelling and system implementation continued to exist, and crossing it only be possible developing some (significant) translation efforts to produce a source code that could be executed in some platform. Calling and using Web services through BPEL remained a constant on these cases, which was confirmed in [2]. They referred once more the existence of a lack of harmonization and the inexistence of an ETL implementation process that obeys to an integrated development strategy, covering all the essential parts of a design for a ETL system project, ranging from its conceptual modelling to its physical implementation. In this sense, the authors of [2] proposed in 2011 a new framework to support the development of ETL systems based on models – a model-driven approach. The framework proposed covers the entire ETL process development, including the automatic generation of source code for specific computer platforms. Many other important initiatives have been taken in this field. It should be noted, for example, the works reported in [17, 18, 12, 14, 9], which raised many interesting aspects that we must take into account when developing conceptual and logical models for ETL systems. But despite their many contributions, we still note a lack of a complete proposal that will permit the conceptual modelling of an ETL system as an initial work and discussion of its main features, allowing posteriorly the generation of the corresponding logical model and the generation of a physical model with the possibility of being executed.

3   ETL Based on BPMN Patterns

3.1   Meta Model Design and Implementation

Modelling ETL processes need to reflect the different flows of control and data between the various tasks embedded in them. The use of BPMN in this type of modelling, before its version 2.0, allowed modelling the flow of control between the various activities of an ETL process without great problems, as well as the characterization and description of the activities to implement. However, the description of their data flows was dependent on other tools. As referred in [16], the real challenge was to do the combination of these two types of flows in one single representation model and, therefore, in a single tool. Today, BPMN 2.0 already includes specific features for the representation and description of data flows, and also allow for aggregation of flows.

```
rows[] <- FactTable.getRows();
for each Rows in row
   for each n_key in row.getOperationalKeys(){
      mapping_table <-  n_key.identifyMappingTable();
      //Invalid operational keys
      if (check_integrity(n_key) = false)
         mapping_table.generateQuarantineKey(n_key);
      else
        //valid natural keys
        if (mapping_table.getSurrogateKey()
           n_key.replaceKey(mapping_table.getSurrogateKey()));
          else
            mapping_table.generateSurrogateKey(n_key);
          end if
      end if
   end for each
```

Fig. 1. An algorithm for a Surrogate Key Pipelining

With the primary purpose of providing a set of formal specifications for standard ETL processes, and taking into consideration the ideas provided in [1, 2], we present an approach for conceptual mapping and execution validation of a standard ETL process: a *Surrogate Key Pipeline* (SKP). We used the BPMN notation to construct a first conceptual model, revealing the necessary bridges to translate it into the correspondent physical model, already prepared for execution through a specific set of BPMN primitives. This will provide the necessary means to validate the model presented for a SKP algorithm implementation like the one presented in [7].

Due to many reasons, the feeding process of a fact table requires frequently to convert operational keys that come from information sources into specific surrogate keys. This imposes the maintenance of a map keeping all the valid correspondences between operational and surrogate keys. Otherwise we would lose important information about the origin of the facts. Therefore, it is quite common to specify mapping tables for each dimension that we have presented in the data warehouse, keeping in it data about its own operational-to-surrogate keys mappings. Usually, a SKP process begins working the data records of a fact table, analyzing what kind of substitution must be applied (or not) to the natural keys of each dimension presented in its schema. For each attribute in a fact table having a natural key, a specific query is launched over its corresponding mapping table (a typical lookup operation) in order to get the corresponding surrogate key. If a surrogate key does not exist for the current operational key, then the mapping table will be updated reflecting the generation of a new surrogate key - executing, for instance, an auto numbering operation. Once the mapping table is updated, the SKP process replaces all the facts that match the operational key in the attribute by its correspondent surrogate key value. Complementarily, we must also identify natural keys that are inconsistent across operating systems. For example, a same operational key can refer two or more different separate registers, imposing its verification in order to maintain data consistency. Kimball and Caserta [7] referred that such keys must be marked as "unknown". In Fig. 1 we present the pseudo code of an algorithm, illustrating the most important steps of a conventional SKP process. The algorithm presented is simple, making an approach to the problem typically sequential. For this kind of ETL task, this approach is not the best one. The nature of the operations involved, especially the ones involving data manipulation, as well as the large number of records involved, makes this solution a clear bottleneck in terms of system performance. Thus, in the case of a real implementation, other solutions should take into account, using more sophisticated algorithms that provide parallel execution of multiple threads and process, simultaneously, involving several groups of records (or chunks) at same time. In this manner we could run at the same time so many threads as the number of operational keys, performing each one of them as an autonomous process. This can reduce significantly the I/O cost per record, influencing positively the global performance of the ETL system. However, given the complexity of the BPMN modelling process, the SKP algorithm presented here is quite enough to satisfy our demonstrations purposes.

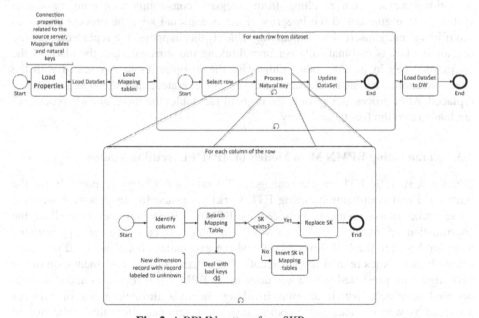

Fig. 2. A BPMN pattern for a SKP process

3.2   A BPMN Case Study

In order to represent all the implementation steps of the SKP algorithm presented, we used the BPMN notation for defining a meta-model (a pattern) for general ETL use, which can be instantiated and executed independently of the logical and physical structure of the data involved with. In Fig. 2 we show the basic operations that were defined for the execution of the referred standard process. This pattern can be used in every ETL system requiring the implementation of one or more SKP processes. The formal specification and standardization of high-level models can increase the efficiency of a process, reducing its complexity and susceptibility to errors in the definition of standard ETL processes. The SKP process begins loading the data required for its regular execution, which is constituted essentially by some configuration and functioning parameters such as the data staging area server connection, the fact table metadata, and all the necessary mappings for operational keys substitution. Next, all the records of the fact table will be read in order to identify the target operational keys, perform lookup operations on the mapping tables involved, and retrieve the correspondent surrogate keys. The BPMN model presented in Fig. 2 has two expanded sub-processes, representing the surrogate key manipulation process. Iteratively, the first sub-process handle all the records contained in the fact table and, one by one, invokes another sub-process that is in charge to select the attributes that contain the operational keys. For each operational key identified it accesses its mapping table to get the corresponding surrogate key. During the identification of the operational keys' attributes, there may occur a compensation event, which will be fired whenever they are some inconsistent operational keys, often resulting from integrity constraints violations in source systems. Once guaranteed the integrity of the operational key, the process checks to see if it is already inserted in its mapping table. If this happens, the replacement of the operational key is automatically performed taking the surrogate key the place of the operational key in the fact table. Otherwise the mapping table is updated with the new operational instance, the surrogate key is generated, and the operational key replaced. After processing entirely the original fact table, the facts, already processed, are loaded into the fact table destiny.

3.3   Translating BPMN Meta Models to BPMN Execution Models

Usually, a specific ETL engine manages ETL processes, being responsible for the interaction and coordination among ETL workflow tasks. In our approach we used some orchestration elements provided by BPMN 2.0 for services that allow the coordination of multiple tasks. However, in order to carry out an appropriated mapping between the high-level model and the executable model we need to ensure some characteristics related to the execution of activities. The implementation of the SKP algorithm presented before was done using BPMN specifications and a service-oriented approach, which allowed invoking methods through a set of services provided by web services. Once coordinated, these services allow the application of the necessary transformations for the implementation of the SKP algorithm through the use of an intermediate layer of stored procedures that ensures services interaction with all the storage structures that the algorithm needs. Beyond the specification of all atomic activities represented in the conceptual model, the definition of the executable model also requires an additional specification of activities, simply because it will has the right characteristics to support its own implementation. For example, activities that communicate with a Web service must be defined as service tasks. This type of activity is performed by the system without human intervention, and it is used regularly to maintain communication with external interfaces. In Fig. 3 we present a proposal to convert the sub-process 'Key Natural Process' included in the pattern shown previously (Fig. 2) for a BPMN 2.0 model that can be executed using the BizAgi tool (http://www.bizagi.com/). The implementation of the model is based on a data storage structure especially conceived for this process, with a layer SOA (*Service Oriented Architecture*) acting as a bridge between the model and the BPMN storage structure. All the activities of a model have a Web Service method associated with them. These methods are responsible to invoke the most appropriated *Stored Procedure* in order to perform its related activity over the target fact table. All the stored procedures were created using *Dynamic SQL*.

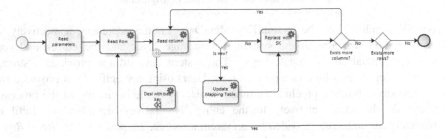

Fig. 3. The SKP BPMN model

The process begins to read the configuration parameters of its implementation, which involves usually access and data structuring information (server name, access port, and access credentials), the name of the fact table to process, the list of operational keys to replace, and the mapping table with the surrogate keys. All this metadata can be filled in the BizAgi configuration environment or retrieved in a table containing the identifier of the column and the name of the associated mapping table. The processes 'Load dataset' and 'Load Mapping Tables' are actually unnecessary in the specification of the execution model, since the records of the table of facts, and their own mapping tables, do not need to be copied to the BPMN runtime environment. The interaction with the mapping tables and the records of the fact table is coordinated by the BPMN process that runs using the methods implemented in the Web services selected. The processes extended (Fig. 3), which represent a repeating structure for a given sub-process, is identified in the execution model with the use of structured cycle available in the BizAgi tool [3]. These cycles (looping structures) are modelled using gateways with transitions, which allow repeating a set of predefined activities accordingly to some particular condition. In the model we designed, the process checks all the fact tables records, one by one, and for each of them, it follows replacing all the operational keys accordingly the mappings expressed in the configuration parameters when the global process had begin. BizAgi also provides us with a set of structures especially oriented to support the execution of BPMN models. After the model definition it is possible to construct a special oriented data structure to receive the information used in all the activities included in the model, as simple input data or as support business rules. In the structure of a model is possible to define entities and relationships as we usually do in a relational model. The data structures for the SKP algorithm implementation can be represented through a very simple diagram (Fig. 4) - in fact, all data structures can be represented in a very simple way.

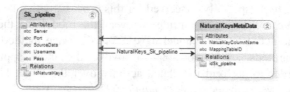

Fig. 4. Data structures for model configuration

The 'SK_pipeline' and the 'NatualKeysMetaData' entities represent, respectively, a link with a data structure in the system data storage layer and the correspondence between operational and surrogate keys. The system data storage provides a strong structure to keep data as long as a process runs. Users using a specific form provided by BizAgi introduce the 'SK_pipeline' instances manually at the beginning of the process. They can do the same relatively to the entity 'NaturalKeysMetaData' or fulfil it automatically through data provided by a system's source. To perform the *'Read Row'* *activity* it is called a *Web Service* method that gets a fact table record using the predefined data access. Next, the *'Read column' activity* use another *Web service* to check the existence of the operational-surrogate key mapping, using the operational key value and the identification of the source that generated the fact record. As mentioned earlier, it is possible that a given operational key is identified as inconsistent and it will be marked with 'unknown'. Otherwise, a specific gateway is used for comparison applying a specific business rule. The flow rules represented in the model provide the path to follow based on a decision made on a given gateway. Thus, if an operational key does not exist in the mapping system, the process invokes a service that will update the mapping table involved, and then invoke a new service that will replace the operational key for the correspondent surrogate key. This process is repeated for each dimension key in the fact table. The looping structure was implemented by means of two gateways, which operate as two conventional structured cycles.

3.4   Results Analysis and Evaluation

Modelling and execution of business processes is still a subject under strong discussion, especially in terms of the definition of boundaries between their conceptual representation and the corresponding execution model. In [8] Frank Leymann addressed several issues that have been discussed by several authors, with very contradictory views, regarding the implementation of business processes using the BPEL language or the execution of processes through BPMN 2.0. In our opinion, independently from the adopted approach, there is a clear distinction between the definition of the conceptual model of a business process specification and its corresponding implementation. Conceptual modelling should not consider any kind of implementation infrastructure, nor any criteria associated with its execution. For example, when a business process manager defines a specific conceptual model, usually does not take into account if the activity he wants to be executed is set to a Web service or to a script. The infrastructures that support the implementation of conceptual models are related to a specific class of users involving therefore the application of specific constructors. Taking into account our previous experience converting BPMN models into BPEL execution models, we find that BPEL fits more to modelling an executable process, mainly due to the nature of its constructors (e.g. looping structures). In turn, the BPMN notation gives us a greater freedom in building business processes, but affect negatively the application of execution semantics in the later stages of the work, mainly due to the number of possible combinations in the implementation of those constructors.

4   Conclusions and Future Work

In 2009 Akkaoui and Zimány proposed the use of the BPMN notation and BPEL to develop conceptual models for ETL systems [1]. Later, they reinforce their ideas proposing to model ETL generic processes independently from the platform [2]. At the time of reading these two works, we found that their ideas very interesting, having a huge potential particularly in what is concerned with the mapping of conceptual models in a set of execution primitives. After some time of research in the field, we thought we could extend a little bit more the ideas of those two authors, creating specific conceptual models, generally applicable, for each of the standard ETL processes we knew, including tasks such as change data capture, slowly changing dimensions, surrogate key pipelining, or data quality coverage. We designed these meta models as BPMN patterns for ETL systems. Following this line of researching, we developed this work using the same methodology that Akkaoui and Zimány used, demonstrating the effectiveness of our BPMN patterns proposal for ETL systems, explaining and illustrating them with one of the BPMN patterns we designed and implemented: the SKP (Section 4).

In our opinion, the specification of a conceptual model for a SKP is quite appropriate, once it's one of the most common ETL processes used in a data warehouse populating process. Our contribution allows the creation of models that can be defined as containers of operations in which we specified an input of data for tasks in there, producing an output as a result of a set of pre-established activities. Our goal is to have in the short term, an extended family of BPMN patterns that allow us to build a complete ETL system, covering all its working areas. Furthermore, we believe that the use of BPMN patterns in the specification of ETL conceptual models will facilitate its implementation, and will contribute significantly to increase its quality of construction, reduce functional and operational errors, and decrease significantly its costs. The use of BPMN 2.0, in particular the use of the BizAgi tool, allows the specification of very detailed models in which we point out the component for data modelling. With this component you can put in memory a set of structured data based on rules for ER models definition, allowing to load data records into memory, as well as the mapping tables for key substitution, and enabling the incorporation of optimizations in the process SKP itself.

References

[1] Akkaoui, Z., Zimanyi, E.: Defining ETL worfklows using BPMN and BPEL. In: Proceedings of the ACM Twelfth International Workshop on Data Warehousing and OLAP, Hong Kong, China (2009)

[2] Akkaoui, Z., Zimànyi, E., Mazón, J., Trujillo, J.: A model-driven framework for ETL process development. In: Proceedings of the ACM 14th International Workshop on Data Warehousing and OLAP, pp. 45–52. ACM, Glasgow (2011)

[3] Bizagi: Bizagi BPM Suite workflow patterns, Bizagi (2012), http://www.ibm.com/developerworks/library/specification/ws-bpel/ (accessed on June 17, 2012)

[4] BPEL, Business Process Execution Language for Web Services (2012), http://www.bizagi.com/docs/Workflow%20Patterns%20using%20BizAgi%20Process%20Modeler.pdf (accessed on June 15, 2012)

[5] BPMN, Object Management Group Business Process Model and Notation (2012), http://www.bpmn.org/ (accessed on June 15, 2012)

[6] Dayal, U., Wilkinson, K., Simitsis, A., Castellanos, M.: Business Processes Meet Operational Business Intelligence (2010)

[7] Kimball, R., Caserta, J.: The Data Warehouse ETL Toolkit - Pratical Techniques for Extracting, Cleaning, Conforming, and Delivering Data. Wiley Publishing, Inc. (2004)

[8] Leymann, F.: BPEL vs. BPMN 2.0: Should You Care? In: Mendling, J., Weidlich, M., Weske, M. (eds.) BPMN 2010. LNBIP, vol. 67, pp. 8–13. Springer, Heidelberg (2010)

[9] Muñoz, L., Mazón, J.-N., Pardillo, J., Trujillo, J.: Modelling ETL Processes of Data Warehouses with UML Activity Diagrams. In: Meersman, R., Tari, Z., Herrero, P. (eds.) OTM 2008 Workshops. LNCS, vol. 5333, pp. 44–53. Springer, Heidelberg (2008)

[10] Scacchi, W.: Process Models in Software Engineering. In: Marciniak, J.J. (ed.) Encyclopedia of Software Engineering (2001)

[11] Silver, B.: BPMN Method and Style: A levels-based methodology for BPM process modeling and improvement using BPMN 2.0. Cody-Cassidy Press (2009)

[12] Simitsis, A., Vassiliadis, P.: A Methodology for the Conceptual Modeling of ETL Processes. In: Eder, J., Missikoff, M. (eds.) CAiSE 2003. LNCS, vol. 2681, pp. 305–316. Springer, Heidelberg (2003)

[13] Sommerville, I.: Software Engineering, 8th edn. Pearson Education, Addison Wesley (2006)

[14] Trujillo, J., Luján-Mora, S.: A UML Based Approach for Modeling ETL Processes in Data Warehouses. In: Song, I.-Y., Liddle, S.W., Ling, T.-W., Scheuermann, P. (eds.) ER 2003. LNCS, vol. 2813, pp. 307–320. Springer, Heidelberg (2003)

[15] Weske, M., van der Aalst, W., Verbeek, H.: Advances in Business Process Management. Data & Knowledge Engineering 50(1) (2004)

[16] Wilkinson, K., Simitsis, A., Castellanos, M., Dayal, U.: Leveraging Business Process Models for ETL Design. In: Parsons, J., Saeki, M., Shoval, P., Woo, C., Wand, Y. (eds.) ER 2010. LNCS, vol. 6412, pp. 15–30. Springer, Heidelberg (2010)

[17] Vassiliadis, P., Simitsis, A., Skiadopoulos, S.: On the Logical Modeling of ETL Processes. In: Pidduck, A.B., Mylopoulos, J., Woo, C.C., Ozsu, M.T. (eds.) CAiSE 2002. LNCS, vol. 2348, pp. 782–786. Springer, Heidelberg (2002)

[18] Vassiliadis, P., Simitsis, A., Skiadopoulos, S.: Modeling ETL activities as graphs. In: Proceedings of the 4th Intl. Workshop on Design and Management of Data Warehouses 2002, DMDW 2002, Toronto, Canada, pp. 52–61 (2002)

Efficiently Compressing OLAP Data Cubes via *R*-Tree Based Recursive Partitions

Alfredo Cuzzocrea[1] and Carson K. Leung[2]

[1] ICAR-CNR and University of Calabria, Italy
cuzzocrea@si.deis.unical.it
[2] University of Manitoba, Canada
kleung@cs.umanitoba.ca

Abstract. This paper extends a quad-tree based multi-resolution approach for two-dimensional summary data by providing a *novel OLAP data cube compression approach* that keeps the critical novelty of relying on *R-tree based partitions* instead of more constrained classical kinds of partition. This important novelty introduces the nice amenity of (i) allowing end-users to exploit the *semantics of data* and (ii) obtaining compressed representations of data cubes where more space can be invested to describe those ranges of multidimensional data for which they retain a *higher degree of interest*. Hence, this paper can be considered as an important advancement over the state-of-the-art approaches. Experimental results confirm the benefits of our proposed approach.

1 Introduction

OLAP data cube compression (e.g., [6,7,8,11,13]) is a solid and well-understood research area in the context of database and data warehousing research [15,16]. Basically, the goal of compression is to reduce the size of the target data cube in order to gain efficiency during query evaluation. There are a wide range of approaches dealing with data cube compression issues. Recently, a renewed interest from the research communities is due to novel computational paradigms adhering to the MapReduce framework [19] (e.g., [1]), or analytics over multidimensional data (e.g., [12]). For instance, Buccafurri et al. [5] proposed a quad-tree based data cube compression approach. It first recursively partitions the input data cube into a hierarchical quad-tree based partition that minimizes the *sum of squared errors* (*SSE*) of the final buckets; it then generates a *quad-tree summary* (*QTS*) in a way similar to the data cube compression (e.g., [2,4,14]) obtained by using classical histograms. Furthermore, this compressed representation is further enhanced by equipping the terminal buckets of the QTS with indices [5] that provide a compact description of data distributions within these buckets and thus improve the accuracy of retrieved *approximate answers*. Consequently, the *indexed quad-tree summary* (*IQTS*) data structure is generated [5].

L. Chen et al. (Eds.): ISMIS 2012, LNAI 7661, pp. 455–465, 2012.

Key contributions of our current paper include our extension of the quad-tree based multi-resolution approach for two-dimensional summary data [5]. We do so by (i) first pointing out the limitations of the quad-tree based partitioning scheme and (ii) then proposing *a novel R-tree based approach for compressing OLAP data cubes* that introduces the critical advantage of *relying on unconstrained (i.e., arbitrary) data cube partitions*. This novelty leads to several benefits. For instance, when using *R*-tree based partitions, more space can be saved to describe those ranges of multidimensional data in which end-users have a *higher degree of interest*. We also conducted experiments on real-life data cubes to compare our proposed approach with state-of-the-art data cube compression approaches. Experimental results confirm the benefits of our proposal. Although we present our approach for two-dimensional OLAP data cubes, it can be easily extended to multidimensional OLAP data cubes.

The remainder of this paper is organized as follows. Section 2 points out the limitations of the existing quad-tree based partitioning scheme. We propose an *R*-tree based partition approach and the Square Chunk Partition in Sections 3 and 4, respectively. We further propose an Indexed R-Tree Summary (IRTS) in Section 5. Finally, experimental results and conclusions are presented in Sections 6 and 7, respectively.

2   Limitations of the Existing Quad-Tree Based OLAP Data Cube Partitioning Model

A weakness of the quad-tree based partitioning scheme [5] is that the generated partitions do *not* take into account the *end-user goals and interests* (as partitions are formed by minimizing SSE (see Section 1)). Consequently, the compression process may generate buckets that are not useful to the end-users (as the partitioning criterion only investigates *properties of data* and does not pay attention to the end-users' interests). Another weakness is that this scheme does *not* take into account the *semantics of data*. It focuses on investigating the *nature* of the data distributions of the target data cube.

Very frequently in OLAP scenarios, end-users retain a great interest in some data domains than others. Typically, in decision support systems (DSS)—"natural" application context for OLAP analysis, tasks adhere to a departmental subject-oriented analysis, and thus, implicitly define a *partitioned organization* of the whole data domain of interest. Another weakness of the quad-tree partitioning scheme [5] is that, although it is a very useful scheme for many OLAP applications (e.g., mainly for those who want to have a "bird's eye view" on *all* the domain of interest), it does *not* target for end-users who are especially interested in browsing a particular business subject. Note that OLAP aggregation is also a very popular way of analyzing multidimensional data.

As highlighted in this section, the quad-tree based partitioning scheme [5] suffers from some limitations that make it unsuitable to modern OLAP applications.

3   *R*-Tree Based Partitioning of OLAP Data Cubes

In this paper, we propose a scheme for building partitioned representations of two-dimensional OLAP views based on the *R*-tree as the core data structure instead of the quad-tree. With this scheme, the partition of the target data cube D is generated by the end-user based on his interest in the different portions (thus, different decision support-oriented subjects of analysis) of the whole data domain of D. After the two-dimensional view from D has been selected by editing a suitable OLAP query, the end-user defines the partitions for both dimensions of the view. Consequently, the partitioned representation is generated. In general, as the generating process is totally random, *rectangular* (i.e., arbitrary) buckets (than *square* buckets as in the quad-tree partitioning scheme [5]) are obtained.

Due to its generating process, we name it the *user-driven partitioning* scheme, meaning that the end-user drives the generation of the partition according to his interests. From a data-engineering point of view, an *R*-tree (instead of a *quad-tree*) is used as a core data structure. Given a two-dimensional data cube D, the compressed representation of the *R*-tree partition is called *R-Tree Summary* ($RTS(D)$), which can be reasonably considered as a general extension of the IQTS data structure [5].

To build the data structure $RTS(D)$ for a given data cube D, the end-user must define a *hierarchical partition*—denoted by $HP(d_0)$ and $HP(d_1)$—for each of the two dimensions of D (namely, d_0 and d_1). Particularly, $HP(d_1)$ is generated following a *bottom-up approach* (i.e., starting from the members of the target dimension d_i, and iteratively grouping them by generating new levels on d_i). Note that, according to the above described process, each new level (of members) contains other (sub-) levels (of members), thus originating a hierarchical organization. Moreover, $HP(d_i)$ can be represented as an *R*-tree, called *Partitioning Tree* denoted by $PT(d_i)$. In more detail, we obtain *generic trees* (i.e., trees such that each node n has at most m child nodes). For quad-trees, $m=4$ (i.e., quad-tree partitions are a special instance of *R*-tree partitions). Furthermore, at each step of the partitioning process, the current bucket is split into rectangular child buckets (as opposed to the quad-tree based approach where each bucket is square). Hence, the resulting representative tree is an *R*-tree instead of a quad-tree.

Note that the process is very similar to that for editing data cubes against multidimensional databases. The basic difference is that, according to the *R*-tree partitioning scheme, the end-user defines *customized* hierarchies on the dimensions of interest on the basis of his analytical goals, without being constrained to the hierarchies exposed from the OLAP server. In other words, this scenario captures the situation in which the end-user is interested in a *view* that is different from that defined on the OLAP server, thus leading to the definition of a "new" (two-dimensional) *virtual data cube*, which was not defined on the OLAP server. On the contrary, the quad-tree partitioning scheme works on "pre-defined" data cubes (i.e., without building customized *views*).

Computational requirements could represent a possible limitation for the *R*-tree partitioning scheme. In possible reference OLAP architecture, partitioning algorithms run on the server-side (thus, on hosts with high computational performances), whereas the lightweight presentation layer runs on the client-side (where computational performances are usually low).

In more detail, our approach builds $RTS(D)$ for a given two-dimensional data cube D as follows. The partitioning trees $PT(d_0)$ and $PT(d_1)$—which define the (final) partition $PT(D)$ of D—is built by combining nodes that belong to the corresponding levels of the two trees. The R-tree partitioned representation of D is obtained by aggregating data items in D according to the scheme defined by $PT(D)$, and $RTS(D)$ is the compressed data structure representing such a partition.

As an example, consider Figs. 1(b) and 1(c), which show the $HP(d_0)$ and $HP(d_1)$ partitions on the dimensions d_0 and d_1 of the two-dimensional data cube D presented in Fig. 1(a). The resulting $PT(D)$ is shown in Fig. 1(d), whereas three OLAP aggregation levels of the corresponding R-tree partition are depicted in Figs. 1(e)-1(g).

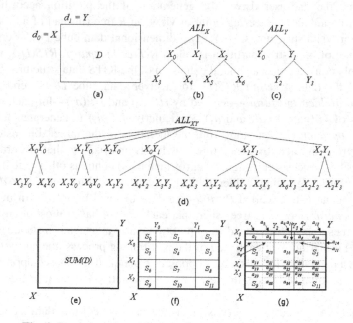

Fig. 1. R-tree based partitioning of a two-dimensional data cube

Given a two-dimensional data cube D, representing $RTS(D)$ is more complex than representing the analogous $IQTS(D)$ [5]. This is because, for each bucket b of the R-tree partition, we need to store (i) the coordinates of the upper-left vertex of b, (ii) the size of the two boundaries of b, and (iii) the sum of b. Obviously, we can generate a compressed representation of $RTS(D)$ by avoiding to store the sums that are equal to zero, and the sums of the last child node of each level because they can be derived from the sum of the parent node and the sums of the sibling nodes. Given a bucket b, we cannot avoid storing information about the geometry of b, even if the sum of b is zero, because we only need to know such information at query time (as buckets are rectangular).

As regards the in-memory-representation of the structure of $RTS(D)$, we follow the same approach as for $IQTS(D)$ [5]. In addition, we also store (i) the coordinates of the upper-left vertex and (ii) the sizes of the two boundaries for each bucket. As regards (i) the storage space occupancy and (ii) the number of splits during the compression process, analogous considerations to those provided for the quad-tree partitioning scheme (see Section 1) can be done for the *R*-tree partitioning scheme. Finally, range query on compressed data cubes generated by the *R*-tree partitioning scheme are evaluated in the same way as in the quad-tree partitioning scheme [5].

To summarize, while the quad-tree based scheme guarantees a better utilization of the available storage space (i.e., it allows us to approximate the *whole* data cube by mediating between (i) the need of creating a large number of buckets in the partition and (ii) the need of representing terminal buckets with a finer grain), the *R*-tree based scheme is more suitable for typical OLAP subject-oriented analysis. As a result, the *R*-tree partitioning scheme is more suitable to address end-users' goals, even at the price of inefficient allocation of the available storage space. However, both schemes handle different (and equally useful) scenarios for modern OLAP applications in emerging environments such as *mobile, grid, and cloud computing*.

4   Building Summarized Representations of Rectangular Terminal Buckets

An important problem arising from the *R*-tree based partitioning scheme described in Section 3 is the building of indices [5] (see Section 1) for rectangular terminal buckets. In the original quad-tree based approach [5], indices are suitable for square buckets only. On the other hand, obtaining a detailed representation of terminal buckets is a strict requirement in the context of summarized representations of data cubes for query processing purposes [6].

However, the user-driven partitioning scheme described in Section 3 can also generate square terminal buckets, for which indices are "native". As the model generally generates rectangular buckets, we must deal with this issue.

The technique we proposed for processing rectangular (terminal) buckets b consists of generating their *index-based summarized representation*, denoted by $ISR(b)$, based on simple geometrical issues. The key idea consists of recognizing that, from any rectangular bucket b, we can extract *one or more* square (sub-)buckets plus *one or more* rectangular (sub-)buckets called *chunk-buckets*, thus originating a partitioned representation of b, which we named *Square Chunk Partition SCP(b)*.

Devising a sub-optimal partitioning criterion of rectangular buckets into square buckets and chunk-buckets is a critical issue because there are many possibilities for a given rectangular bucket b (i.e., the number of $SCP(b)$ partitions obtained from b is exponential to the size of the bucket boundaries). To lower such a number, we propose to extract square buckets with boundaries of size equal to the *smaller size* of the input bucket boundaries. Under the conditions of our proposed criterion, each of the resulting rectangular buckets is partitioned into *one or more* square buckets but *only one* chunk-bucket, thus bounding a large possible number of partitioning buckets.

The ratio of this criterion is obtained from a summarized representation of the *whole* set of terminal buckets belonging to the R-tree partition, thus achieving a good level of approximation in summarizing the input data cube D. In fact, generating square and chunk-buckets based on the "extreme" fragmentation of *each* of the terminal bucket would seriously undermine the accuracy of the proposed approach in summarizing D.

Even with such a criterion, we may still result in more than one possible partition $SCP(b)$ for a given bucket b. Consider Fig. 2, where two $\Delta X \times \Delta Y$ rectangular buckets b such that $\Delta X < \Delta Y$ are depicted (note that the same could be discussed considering the dual case, i.e., $\Delta X \geq \Delta Y$. In Fig. 2, the possible partitions $SCP(b)$ are also depicted for both target buckets. In particular, square (sub-) buckets (denoted by b_i) are filled with dark gray color, whereas chunk-buckets are filled with light gray color. The more the number of square buckets that can be extracted from an input bucket b, the more is the number of possible partitions $SCP(b)$ that can be defined on b.

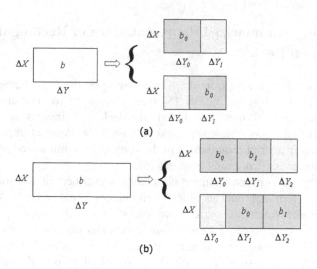

Fig. 2. Example of $SCP(b)$ partitions

Due to the combinatorial nature, we propose a *cost-based approach* for finding the sub-optimal partition among all the possible ones. Given a bucket b and the set of all the possible partitions $SCP(b)$, we consider the partition that minimizes the sum of the square errors as the "best" one—denoted by $SCP^*(b)$—due to the evaluation of a (large) set of queries—denoted by $QS(b)$—on the compressed representation of b (i.e., $SCP(b)$) with respect to computing the same queries on b. More formally, given a bucket b and a partition $SCP(b)$, we define the described metrics as follows:

$$\varepsilon_{SCP(b)}(b) = \sum_{i=0}^{|QS(b)|-1}\left(sum(q_i) - sum_{SCP(b)}(q_i)\right)^2 \qquad (1)$$

where (i) $QS(b)$ is the set of (test) queries defined on b, (ii) $sum(q_i)$ is the (exact) answer to $q_i \in QS(b)$ evaluated on b, and (iii) $sum_{SCP(b)}(q_i)$ is the (approximate) answer to $q_i \in QS(b)$ evaluated on $SCP(b)$.

To build the set $QS(b)$, in a way recalling the criterion for selecting the "best" *index* or a given bucket [5], we generate from the geometry of the input $\Delta X \times \Delta Y$ bucket b for *all* the *possible* $\frac{\Delta X}{N} \times \frac{\Delta Y}{N}$ rectangular queries, where N is an input (tunable) parameter, having (i, j)-coordinates in the two-dimensional space of b, such that $0 \leq i \leq \Delta X$ and $0 \leq j \leq \Delta Y$, as upper-left vertices. We highlight that the $QS(b)$ generation policy is appealing as OLAP queries are typically of *historical* kind, i.e., on the two-dimensional space, they are characterized by being "large" so that considering the N-th fraction of the dimension size, with N tunable as required, can be reasonably considered as a good way to get an estimation about the capability of the summarized representation in providing approximate answers to "typical" OLAP queries. Note that queries or part of queries involving the chunk-bucket of the current partition $SCP(b)$ to be evaluated are answered by applying the well-known *Continuous Value Assumption* (CVA) [8].

When all the *possible* partitions $SCP(b)$ are evaluated, the one having the minimum $\varepsilon_{SCP(b)}(b)$ value (i.e., $SCP^*(b)$) is selected, and the corresponding index-based summarized representation $ISR^*(b)$ is generated by storing (i) an index for each square bucket b_i, and (ii) the sum of the chunk-bucket, along with information concerning the bucket geometry (i.e., upper-left vertex plus sizes of boundaries). For square buckets, we need to store one boundary size only, thus keeping a certain (additional) compression.

5   The *IRTS* Data Structure

Analogously to the IQTS data structure mentioned in Section 1, combining *R*-tree partitioning schemes and summarized representations of rectangular terminal buckets based on indices [5] as described in Section 4, allows us to generate—given a data cube D—the *Indexed R-Tree Summary IRTS(D)*, which extends the capabilities of $IQTS(D)$ by also supporting user-driven partitioning schemes defined on D beyond the native quad-tree partitioning schemes [5] that is still supported by our proposed data structure.

The $IRTS(D)$ data structure represents a way of compressing a given OLAP data cube D, while still keeping the flexibility of representing user-driven partitions. The in-memory-representation of $IRTS(D)$ is thus composed by (i) the space needed for representing non-zero sums, (ii) the space needed for representing indices, and (iii) the space needed for representing the structure of the *R*-tree partition.

Given a two-dimensional data cube D and an amount of available storage space, $IQTS(D)$ [5] allows us to better summarize D *on a global fashion* rather than $IRTS(D)$ because $IQTS(D)$ requires (i) much more space for keeping information about the structure of the *R*-tree partition(which is more complex than any quad-tree partition because of bucket geometry must be stored) and (ii) much more space for storing the compressed representation of rectangular terminal buckets. On the other hand,

IRTS(D) allows us to (i) represent portions of data domains in *D* with a higher level of detail than others and (ii) keep a more prominent "heterogeneity" in the representation than *IQTS(D)*, thus better fitting end-users' goals into analyzing with a higher level of interest some domains than others.

6 Experimental Evaluation

In order to assess the effectiveness and the quality of our data cube compression technique, we conducted an experimental campaign on two-dimensional data cubes extracted from real-life data sets. We adopted the experimental framework of our previous work on OLAP data cubes compression [9]. In particular, we considered the real-life data sets *USCensus1990* and *ForestCoverType* from the UCI KDD Archive [17]. In more details, we generated 2,000 × 2,000 two-dimensional data cubes for both cases.

As regards the generation of user-driven partitions, we considered a *totally random process* that is capable of generating partitions of the input data cubes according to a *Zipf distribution* [20]. The well-understood asymmetries of Zipf distributions are well suited to model arbitrary partitions of several classes of end-users one can find in real-life OLAP applications, hence conveying reliability to our experimental assessment. As regards the input, we considered random populations of range-SUM queries Q_s, with a sufficient number of instances as to get "large-enough" observations.

As regards the metrics, we considered the *Average Relative Error* (*ARE*). For a query Q_k in Q_s, the ARE—denoted as $\bar{E}_{rel}(Q_s)$—is defined as follows:

$$\bar{E}_{rel}(Q_s) = \frac{|A(Q_k) - \tilde{A}(Q_k)|}{A(Q_k)} \tag{2}$$

where (i) $A(Q_k)$ is the exact answer to Q_k and (ii) $\tilde{A}(Q_k)$ is the approximate answer to Q_k. For the entire population of range-SUM queries Q_s, Equation (2) can be expressed as follows:

$$\bar{E}_{rel}(Q_s) = \frac{1}{|Q_s|} \cdot \sum_{k=0}^{|Q_s|-1} E_{rel}(Q_k) \tag{3}$$

Equation (3) is used as the ARE metric in our experiments. We considered (i) the *query selectivity* $\|\cdot\|$ [8], (which is a useful parameter for modeling how many queries are "difficult" to evaluate) and (ii) the *compression ratio* (which models the percentage occupancy of the space budget with respect to the overall occupancy of *D*).

As comparison techniques, we considered the following well-known techniques for compressing data cubes: MinSkew [2], Wavelets [18], and STHoles [4]. In particular, we exclude IQTS [5] from our comparison because IQTS is not suitable to capture arbitrary user-driven partitions originated by conventional OLAP tasks, hence it cannot fit the goals of our experimental assessment and analysis. In more detail, having fixed the space budget (i.e., the storage space available for housing the compressed representation of the input OLAP view), we derived—for each comparison technique—the configuration of the input parameters that respective authors consider the best in their papers. This ensures a *fair* experimental analysis (i.e., an analysis such that each comparison technique provides its *best* performance).

Figs. 3 and 4 show the experimental results on the data cubes derived from *USCensus1990* and *ForestCoverType*, respectively. As shown in Figs. 3 and 4, IRTS exposes a good performance under the ARE metric for both the cases of ranging the query selectivity $\|\cdot\|$ and the compression ratio. In addition, IRTS is capable of capturing user-driven partitions, hence introducing a capability that was not present in classical OLAP data cube compression techniques. This gives further merits to our research.

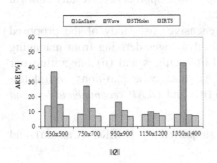 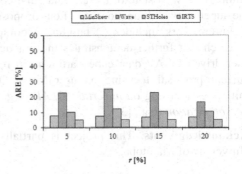

Fig. 3. ARE vs. query selectivity (left) and compression ratio (right) for the *USCensus1990* data cube ($N = 16$)

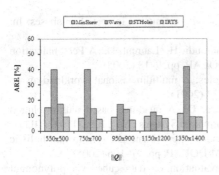 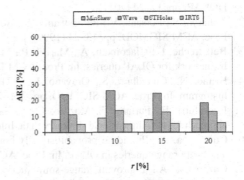

Fig. 4. ARE vs. query selectivity (left) and compression ratio (right) for the *ForestCoverType* data cube ($N - 16$)

To summarize, the retrieved results clearly confirm the superiority of our proposed approach over the other techniques we compared. Moreover, our proposed approach also generates flexible OLAP data cube partitions.

7   Conclusions and Future Work

In this paper, we proposed a further enhancement to Buccafurri's work [5] by proposing a novel R-tree based approach to compress OLAP data cubes. Our proposal generates user-driven partitions (captured by R-trees) that are suitable to support real-life OLAP analysis tasks over multidimensional domains where end-users expose a higher degree of interest for some ranges rather than others. This leads to the definition of unconstrained OLAP data cube partitions which can be completely captured by our IRTS data structure.

We also provided an experimental evaluation of IRTS on real-life data cubes in comparison with state-of-the-art data cube compression approaches. Results confirm the superiority and functionality of our proposal.

Future work includes (i) building a comprehensive case study of the proposed research that further demonstrates the practical advantages deriving from managing user-driven OLAP data cube partitions in real-life settings and (ii) integrating our current proposal focusing on user-driven OLAP data cube partitions with latest initiatives focusing on *uncertain OLAP* (e.g., [10]) and *OLAP recommendation and personalization* (e.g., [3]).

Acknowledgments. This project is partially supported by NSERC (Canada) and University of Manitoba.

References

1. Abelló, A., Ferrarons, J., Romero, O.: Building cubes with MapReduce. In: Proc. ACM DOLAP, pp. 17–24 (2011)
2. Acharya, S., Poosala, V., Ramaswamy, S.: Selectivity estimation in spatial databases. In: Proc. ACM SIGMOD, pp. 13–24 (1999)
3. Bellatreche, L., Giacometti, A., Marcel, P., Mouloudi, H., Laurent, D.: A Personalization framework for OLAP queries. In: Proc. ACM DOLAP, pp. 9–18 (2005)
4. Bruno, N., Chaudhuri, S., Gravano, L.: STHoles: a multidimensional workload-aware histogram. In: Proc. ACM SIGMOD, pp. 211–222 (2001)
5. Buccafurri, F., Furfaro, F., Mazzeo, G.M., Saccà, D.: A quad-tree based multiresolution approach for two-dimensional summary data. Information Systems 36(7), 1082–1103 (2011)
6. Cuzzocrea, A.: Providing probabilistically-bounded approximate answers to non-holistic aggregate range queries in OLAP. In: Proc. ACM DOLAP, pp. 97–106 (2005)
7. Cuzzocrea, A.: Improving range-sum query evaluation on data cubes via polynomial approximation. Data & Knowledge Engineering 56(2), 85–121 (2006)
8. Cuzzocrea, A.: Accuracy control in compressed multidimensional data cubes for quality of answer-based OLAP tools. In: Proc. of SSDBM, pp. 301–310 (2006)
9. Cuzzocrea, A.: Multiple-Objective Compression of Data Cubes in Cooperative OLAP Environments. In: Atzeni, P., Caplinskas, A., Jaakkola, H. (eds.) ADBIS 2008. LNCS, vol. 5207, pp. 62–80. Springer, Heidelberg (2008)
10. Cuzzocrea, A., Gunopulos, D.: Efficiently Computing and Querying Multidimensional OLAP Data Cubes over Probabilistic Relational Data. In: Catania, B., Ivanović, M., Thalheim, B. (eds.) ADBIS 2010. LNCS, vol. 6295, pp. 132–148. Springer, Heidelberg (2010)

11. Cuzzocrea, A., Serafino, P.: LCS-Hist: taming massive high-dimensional data cube compression. In: Proc. EDBT, pp. 768–779 (2009)
12. Cuzzocrea, A., Song, I.-Y., Davis, K.C.: Analytics over large-scale multidimensional data: the big data revolution! In: Proc. ACM DOLAP, pp. 101–104 (2011)
13. Cuzzocrea, A., Saccà, D., Serafino, P.: A Hierarchy-Driven Compression Technique for Advanced OLAP Visualization of Multidimensional Data Cubes. In: Tjoa, A.M., Trujillo, J. (eds.) DaWaK 2006. LNCS, vol. 4081, pp. 106–119. Springer, Heidelberg (2006)
14. Gunopulos, D., Kollios, G., Tsotras, V.J., Domeniconi, C.: Approximating multi-dimensional aggregate range queries over real attributes. In: Proc. ACM SIGMOD, pp. 463–474 (2000)
15. Leung, C.K.-S., Lee, W.: Efficient Update of Data Warehouse Views with Generalised Referential Integrity Differential Files. In: Bell, D.A., Hong, J. (eds.) BNCOD 2006. LNCS, vol. 4042, pp. 199–211. Springer, Heidelberg (2006)
16. Leung, C.K.-S., Lee, W.: Exploitation of Referential Integrity Constraints for Efficient Update of Data Warehouse Views. In: Jackson, M., Nelson, D., Stirk, S. (eds.) BNCOD 2005. LNCS, vol. 3567, pp. 98–110. Springer, Heidelberg (2005)
17. University of California, Irvine, Knowledge Discovery in Databases Archive, http://kdd.ics.uci.edu/
18. Vitter, J.S., Wang, M., Iyer, M.R.: Data cube approximation and histograms via wavelets. In: Proc. ACM CIKM, pp. 96–104 (1998)
19. Yang, H.-C., Dasdan, A., Hsiao, R.-L., Parker, D.S.: Map-Reduce-Merge: simplified relational data processing on large clusters. In: Proc. ACM SIGMOD, pp. 1029–1040 (2007)
20. Zipf, G.K.: Human Behavior and the Principle of Least Effort. Addison-Wesley Press (1949)

Author Index

Abelha, António   264, 274
Aldanondo, Michel   301
Ammar, Asma   81
Andruszkiewicz, Piotr   99

Belo, Orlando   445
Bembenik, Robert   162
Benton, Ryan   51, 61
Betliński, Paweł   21
Bonev, Martin   331
Boyer, Anne   357
Bratko, Ivan   41
Brocki, Łukasz   143
Burdziej, Jan   172

Cai, Shimin   367, 377
Cecilio, José   415
Chen, Jianhua   149
Choi, Jung-Whoan   293
Choi, Sung-Pil   284
Chun, Hong-Woo   284
Ciecierski, Konrad   234
Cimini, Giulio   405
Clark, Patrick G.   93
Colucci, Simona   192
Costa, João Pedro   415
Coudert, Thierry   301
Cuzzocrea, Alfredo   455

Donini, Francesco M.   192

Elahi, Mehdi   254
Elouedi, Zied   81

Feinerer, Ingo   321
Felfernig, Alexander   311, 349
Fournier-Viger, Philippe   31
Freudenthaler, Bernhard   435
Furtado, Pedro   415

Gao, Chao   125
Gawrysiak, Piotr   172
Georgiev, Dejan   41
Ghosh, Sujata   105
Giannini, Silvia   192

Giordana, Attilio   1
Gonçalves, Pedro   274
Grekow, Jacek   228
Groznik, Vida   41
Grzymala-Busse, Jerzy W.   93
Guid, Matej   41

Haydar, Charif   357
Heuchler, Sebastian   136
Hirano, Shoji   71
Homenda, Wladyslaw   218
Huang, Sheng   377
Hvam, Lars   331

Jastreboff, Pawel J.   244
Jia, Chun-Xiao   397
Jung, Hanmin   284, 293

Kim, Pyung   293
Kohli, Deepali   244
Kohout, Jan   87
Koržinek, Danijel   143
Krenn, Markus   435
Kuehnhausen, Martin   93
Kursa, Miron B.   208
Kwok, Lam-for   115
Kyaw, Thiri Haymar   105

Lavergne, Jennifer   51, 61
Lee, Sang Hwan   293
Lee, Seungwoo   293
Leung, Carson K.   455
Li, Wenjuan   115
Liao, Hao   405
Lingras, Pawan   81
Liu, Chuang   397
Liu, Hao   387
Liu, Jiming   125
Liu, Run-Ran   397
Lü, Linyuan   387

Machado, José   264, 274
Mansmann, Svetlana   425
Marasek, Krzysztof   143
Martins, Pedro   415

Medo, Matúš   405
Mendola, Dino   1
Meng, Yuxin   115
Metter, Joachim   136
Miranda, Miguel   274
Moio, Andrea   1
Možina, Martin   41

Neruda, Roman   87
Neves, José   274
Niederbrucker, Gerhard   321
Ninaus, Gerald   311, 349

Oliveira, Bruno   445

Peixoto, Hugo   264
Pestana, Gabriel   136
Pikuła, Mariusz   182
Pirtošek, Zvezdan   41
Pitiot, Paul   301
Podsiadły-Marczykowska, Teresa   162
Portela, Carlos   274
Protaziuk, Grzegorz   162
Przybyszewski, Andrzej W.   234
Punuru, Janardhana   149

Raghavan, Vijay V.   51, 61
Raś, Zbigniew W.   234, 244
Rauch, Jan   11
Rehman, Nafees Ur   425
Reinfrank, Florian   311, 349
Reis, Pedro   136
Ricci, Francesco   254
Rodrigues, Rui   274
Roussanaly, Azim   357
Ruan, Chun   202
Rubens, Neil   254
Rybiński, Henryk   162
Rybnik, Mariusz   218

Sadikov, Aleksander   41
Salzer, Gernot   321
Santos, Manuel   264, 274
Scholl, Marc H.   425
Sciascio, Eugenio Di   192
Seo, Dongmin   293
Shang, Mingsheng   377
Sisel, Tanja   321
Sitarek, Tomasz   218
Ślęzak, Dominik   21
Sobczak, Grzegorz   182
Song, Sa-Kwang   284, 293
Stumptner, Reinhard   435
Sydow, Marcin   155, 182

Tang, Zhen   367
Thompson, Pamela L.   244
Tinelli, Eufemia   192
Tseng, Mitchell M.   343
Tseng, Vincent S.   31
Tsumoto, Shusaku   71

Vareilles, Elise   301
Verbrugge, Rineke   105

Wang, Peng   397
Wang, Qingxian   367
Wang, Yue   343
Weiler, Andreas   425
Wieczorkowska, Alicja A.   208, 244
Wróblewska, Alina   155
Wróblewska, Anna   162

Yu, Fei   387

Zeng, An   387
Zhang, Linda   301